Group Theory in Quantum Mechanics

An Introduction to Its Present Usage

Volker Heine
University of Cambridge

Dover Publications, Inc.
Mineola, New York

Bibliographical Note

This Dover edition, first published in 1993 and reprinted in 2007, is an unabridged republication of the edition published by Pergamon Press, New York, 1960, in the *International Series of Monographs in Pure and Applied Mathematics.*

Library of Congress Cataloging-in-Publication Data

Heine, Volker.
Group theory in quantum mechanics : an introduction to its present usage / Volker Heine.— Dover ed.
p. cm.
Originally published: New York : Pergamon Press, 1960, in series: International series of monographs in pure and applied mathematics ; v. 9.
Includes bibliographical references and index.
ISBN 0-486-45878-4 (pbk.)
1. Group theory. 2. Quantum theory. I. Title.

QC174.5.H4 2007
530.1201'5122—dc22

2006053465

Manufactured in the United States of America
Dover Publications, Inc., 31 East 2nd Street, Mineola, N.Y. 11501

CONTENTS

PAGE

PREFACE vii

NOTATION ix

I. SYMMETRY TRANSFORMATIONS

1. The uses of symmetry properties 1
2. Expressing symmetry operations mathematically 3
3. Symmetry transformations of the Hamiltonian 6
4. Groups of symmetry transformations 12
5. Group representations 24
6. Applications to quantum mechanics 41

II. THE QUANTUM THEORY OF A FREE ATOM

7. Some simple groups and representations 48
8. The irreducible representations of the full rotation group 52
9. Reduction of the product representation $D^{(j)} \times D^{(j')}$ 67
10. Quantum mechanics of a free atom; orbital degeneracy 73
11. Quantum mechanics of a free atom including spin 78
12. The effect of the exclusion principle 89
13. Calculating matrix elements and selection rules 99

III. THE REPRESENTATIONS OF FINITE GROUPS

14. Group characters 113
15. Product groups 125
16. Point-groups 128
17. The relationship between group theory and the Dirac method 143

IV. FURTHER ASPECTS OF THE THEORY OF FREE ATOMS AND IONS

18. Paramagnetic ions in crystalline fields 148
19. Time-reversal and Kramers' theorem 164
20. Wigner and Racah coefficients 176
21. Hyperfine structure 189

V. THE STRUCTURE AND VIBRATIONS OF MOLECULES

22. Valence bond orbitals and molecular orbitals 206
23. Molecular vibrations 229
24. Infra-red and Raman spectra 245

VI. SOLID STATE PHYSICS

25. Brillouin zone theory of simple structures 265
26. Further aspects of Brillouin zone theory 284
27. Tensor properties of crystals 304

VII. NUCLEAR PHYSICS

PAGE

28. The isotopic spin formalism 313
29. Nuclear forces 321
30. Reactions 334

VIII. RELATIVISTIC QUANTUM MECHANICS

31. The representations of the Lorentz group 351
32. The Dirac equation 363
33. Beta decay 384
34. Positronium 397

APPENDICES

A. Matrix algebra 404
B. Homomorphism and isomorphism 410
C. Theorems on vector spaces and group representations 412
D. Schur's lemma 418
E. Irreducible representations of Abelian groups 420
F. Momenta and infinitesimal transformations 422
G. The simple harmonic oscillator 424
H. The irreducible representations of the complete Lorentz group 428
I. Table of Wigner coefficients $(jj'\ mm'|JM)$ 432
J. Notation for the thirty-two crystal point-groups 446
K. Character tables for the crystal point-groups 448
L. Character tables for the axial rotation group and derived groups 455

LIST OF GENERAL REFERENCES, WITH REVIEWS 457

BIBLIOGRAPHY 459

SUBJECT INDEX 464

PREFACE

The object of this book is to introduce the three main uses of group theory in quantum mechanics, which are: firstly, to label energy levels and the corresponding eigenstates; secondly, to discuss qualitatively the splitting of energy levels as one starts from an approximate Hamiltonian and adds correction terms; and thirdly, to aid in the evaluation of matrix elements of all kinds, and in particular to provide general selection rules for the non-zero ones. The theme is to show how all this is achieved by considering the symmetry properties of the Hamiltonian and the way in which these symmetries are reflected in the wave functions. In Chapter I the necessary mathematical concepts are introduced in as elementary and illustrative a manner as possible, with the proofs of some of the fundamental theorems being relegated to an appendix. The three uses of group theory above are illustrated in detail in Chapter II by a fairly quick run through the theory of atomic energy levels and transitions. This topic is particularly suitable for illustrative purposes, because most of the results are familiar from the usual vector model of the atom but are derived here in a rigorous and precise way. Also most of it, e.g. the introduction of spin functions and the exclusion principle, is fundamental to all the later more advanced topics. Chapter III is a repository for the theory of group characters, the crystallographic point-groups and minor points required in some of the later applications. Thus, after selected readings from chapter III according to his field of interest, the reader is ready to jump immediately to any of the applications of the theory covered in later chapters, namely: further topics in the theory of atomic energy levels (Chapter IV), the electronic structure and vibrations of molecules (Chapter V), solid state physics (Chapter VI), nuclear physics (Chapter VII), and relativistic quantum mechanics (Chapter VIII).

The level of the text is that of a course for research students in physics and chemistry, such as is now offered in many Universities. A previous course in quantum theory, based on a text such as Schiff *Quantum Mechanics*, is assumed, but the matrix algebra required is included as an appendix. In selecting the material for the applications in various branches of physics and chemistry in Chapters IV to VIII, I have restricted myself as far as possible to topics satisfying three criteria: (i) the topics should be simple applications that illustrate basic principles, rather than complicated examples designed to overawe the reader with the power of group theory; (ii) the material should be intrinsically interesting and of the sort that is suitable for inclusion in a general course of advanced quantum mechanics; and (iii) topics must not involve too much specialized background knowledge of particular branches of physics. The view adopted throughout is that group theory is *not* just a specialized tool for solving a few of the more difficult and intricate problems in quantum theory. In advanced quantum mechanics practically all general statements that can be made about a complicated system depend on its symmetry properties, and the use of group representations is just a systematic, unified way of thinking about and exploiting these symmetries. For this reason I have not hesitated to include simple results for which one could easily produce *ad hoc* proofs from first principles: indeed, it must always remain true that the use of group theory could be circumvented by detailed algebraic considerations on almost all occasions. However, the author is convinced that the essential ideas of group theory are sufficiently simple to make the time spent on acquiring this way of thinking well worth while.

A series of examples is appended to each section. Some of these are simple drill in the concepts introduced in the section; others, particularly in later chapters, indicate extensions of the theory and further applications. Those marked with an asterisk are more difficult or require additional reading, and are often suitable as topics for review essays (alias term papers).

With the three criteria for selection mentioned above, it has of course been quite impossible to do real justice to any of the applications to various branches of physics and chemistry that are touched on in Chapters IV to VIII. This appears to me unavoidable because of the amount of background knowledge required for many applications. It merely highlights the fact that in each of these specialized subjects there is a need for a monograph which uses group theory from the beginning as naturally and as freely as the Schrödinger equation itself. In this field the chemists have already led the way,† and the author hopes that the present book may hasten the day when the same applies in physics by providing a convenient basic reference text.

It is a pleasure to acknowledge my indebtedness to Professor B. L. Van der Waerden whose elegant book first inspired my interest in this subject. Also I am very grateful to Dr. S. F. Boys,

† See Eyring, Walter and Kimball (1944) *Quantum Chemistry*; and Wilson, Decius and Cross (1955) *Molecular Vibrations*.

Dr. G. Chew, Dr. R. Karplus, Dr. M. A. Ruderman, Dr. M. Tinkham and Mr. D. Twose, who have either patiently helped me to understand aspects of their special subject, or have read parts of the manuscript and made helpful comments. I am indebted to Mrs. M. Rogers and Mrs. M. Miller for undertaking the typing of the manuscript, and to Mr. J. G. Collins and Mr. D. A. Goodings who have generously helped with the correction of proofs. Dr. E. R. Cohen has kindly allowed the reproduction of his tables of Wigner coefficients, and D. Van Nostrand Co. similarly a figure.

Cambridge, England. V. HEINE

NOTATION

Note: e is taken as the charge on the *proton*: all angular momentum operators such as $\mathbf{L} = (L_x, L_y, L_z)$ have the dimensions of angular momentum and thus contain a factor $\hbar$ (except in § 18), whereas the quantum numbers L, M_L, etc., are of course pure numbers.

Chapter I

SYMMETRY TRANSFORMATIONS IN QUANTUM MECHANICS

1. The Uses of Symmetry Properties

Although this book has been titled "Introduction to the Present Use of Group Theory in Quantum Mechanics" in accordance with customary usage, a rather more descriptive title would have been "The Consequences of Symmetry in Quantum Mechanics". The fact that these symmetry properties form what mathematicians have termed "groups" is really incidental from a physicist's point of view, though it is vital to the mathematical form of the theory. It is in fact the *symmetries* of quantum mechanical systems that we shall be interested in.

The following three simple examples illustrate in a preliminary way what is meant by symmetry properties and what their main consequences are.

(i) It can be shown that the wave functions $\psi(\mathbf{r}_1, \mathbf{r}_2)$ (without spin) of a helium atom are of two types, symmetric and anti-symmetric, according to whether

$$\psi(\mathbf{r}_1, \mathbf{r}_2) = \psi(\mathbf{r}_2, \mathbf{r}_1) \qquad \text{or} \qquad \psi(\mathbf{r}_1, \mathbf{r}_2) = -\psi(\mathbf{r}_2, \mathbf{r}_1),$$

where $\mathbf{r}_1$ and $\mathbf{r}_2$ are the position vectors of the two electrons (Schiff 1955, p. 234). The corresponding states of the atom are also referred to as symmetric and anti-symmetric. Thus the eigenfunctions turn out to have well defined symmetry properties which can therefore be used in classifying and distinguishing all the different eigenstates.

(ii) There are three $2p$ wave functions for a hydrogen atom,

$$\psi(2p_x) = xf(r), \qquad \psi(2p_y) = yf(r), \qquad \psi(2p_z) = zf(r), \qquad (1.1)$$

where $f(r)$ is a particular function of $r = |\mathbf{r}|$ only (Schiff 1955, p. 85). Now in a free atom there are no special directions and we can choose and label the x-, y- and z-axes as we please, so that the three functions (1.1) must all correspond to the same energy level. If, however, we apply a magnetic field in some particular direction, the argument no longer holds, so that we may expect the energy level to be split into several different levels, up to three in number.

In this kind of way the symmetry properties of the eigenfunctions can determine the degeneracy of an energy level, and how such a degenerate level may split as a result of some additional perturbation.

(iii) The probability that the outer electron of a sodium atom jumps from the state ψ_1 to the state ψ_2 with the emission of radiation polarized in the x-direction is proportional to the square of

$$M = \int_{-\infty}^{\infty} \int_{-\infty}^{\infty} \int_{-\infty}^{\infty} \psi_1^* \, x \, \psi_2 \, dx \, dy \, dz \tag{1.2}$$

(Schiff 1955, p. 253). If the two states are the $4s$ and $3s$ ones, ψ_1 and ψ_2 are functions of r only. To calculate M in this case, we make the change of variable $x' = -x$ in (1.2) and obtain $M = -M$, i.e. $M(4s, 3s) = 0$. This transition probability is therefore determined purely by symmetry. The situation is rather different when the transition probability is not zero. Suppose ψ_1 and ψ_2 are the $4p_x$ and $3s$ wave functions $xf_1(r)$ and $f_2(r)$. Then (1.2) becomes

$$M(4p_x, 3s) = \int_{-\infty}^{\infty} \int_{-\infty}^{\infty} \int_{-\infty}^{\infty} f_1^*(r) \, x^2 \, f_2(r) \, dx \, dy \, dz. \tag{1.3}$$

By making the change of variable $x' = y$, $y' = x$, the x^2 in (1.3) can be replaced by y^2 or similarly by z^2. Thus by addition

$$M(4p_x, 3s) = \tfrac{1}{3} \int_{-\infty}^{\infty} \int_{-\infty}^{\infty} \int_{-\infty}^{\infty} f_1^*(r) \, r^2 f_2(r) \, dx \, dy \, dz. \tag{1.4}$$

Similarly the probabilities for all possible transitions from any $4p$ state to the $3s$ state or vice versa, with the emission or absorption of radiation, polarized circularly or linearly in any direction, can be reduced to the integral occurring in (1.4), the simple numerical factor in front being determined purely by the particular direction and $4p$ state chosen. Symmetry properties thus establish the *relative* magnitudes of several matrix elements of the form (1.2), their absolute values being then determined by the value of one integral. This type of argument explains why the intensities of the various components of a composite spectral line are often observed to bear simple ratios to one another.

These examples serve to illustrate what is generally true, namely that symmetry properties allow us to *classify and label the eigenstates* of a quantum mechanical system. They enable us to discuss qualitatively what *splittings* we may expect in a degenerate energy level under some perturbation. They help in calculating transition probabilities and other *matrix elements*, and, in particular, in setting up *selection rules* stating when these quantities are zero. In the following sections we shall develop these kinds of symmetry argument in a systematic fashion, and shall see how they can be used for the above three purposes in situations that are less elementary than the examples given above.

The real importance of symmetry arguments in such situations lies in the fact that for systems of interest the Schrödinger equation is usually too complicated to be solved analytically or even numerically without making gross approximations. For instance, for an atom with n electrons the equation contains $4n$ variables (including spin) which are not separable. However, the symmetry properties of the equation may be relatively simple, so that symmetry arguments can easily be applied to the problem. Another important point about symmetry arguments is that they are based on the symmetry of the Schrödinger equation itself, so that they do not involve approximations, in particular those used to obtain approximate eigenfunctions of the equation. In fact the beauty of the method lies in the fact that, for instance, an n electron problem can often be treated as simply and as rigorously as a one electron problem. At the present time the most spectacular illustrations of these two aspects of symmetry arguments occur in nuclear and fundamental particle physics. The shell-model theory of the energy levels of nuclei has been developed, with selection rules for various transitions, etc., all without an exact knowledge of the interaction between two nucleons. Similarly it is possible to discuss tentatively the relationships between the various fundamental particles and give selection rules for transitions between them, which are based purely on symmetry ideas, such as spin, charge conjugation, isotopic spin and parity, without the slightest understanding of the field equations describing the interactions of all these particles.

2. Expressing Symmetry Operations Mathematically

Many of the symmetry properties that we shall be concerned with involve rotations, so that we shall start by considering how a physical operation such as rotating a system is expressed mathematically.

Consider a body with a point P on it which has co-ordinates (x, y, z). If we rotate the body clockwise by an angle α (Fig. 1), i.e. we rotate by $-\alpha$ about the z-axis in the conventional sense, the point P moves to the position $P'(X, Y, Z)$, where

$$\begin{aligned} OA &= OC = OB\cos\alpha - BP'\sin\alpha \\ AP &= CP' = OB\sin\alpha + BP'\cos\alpha, \end{aligned} \tag{2.1}$$

i.e.

$$\boxed{\begin{aligned} x &= X\cos\alpha - Y\sin\alpha,\\ y &= X\sin\alpha + Y\cos\alpha,\\ z &= Z.\end{aligned}} \tag{2.2}$$

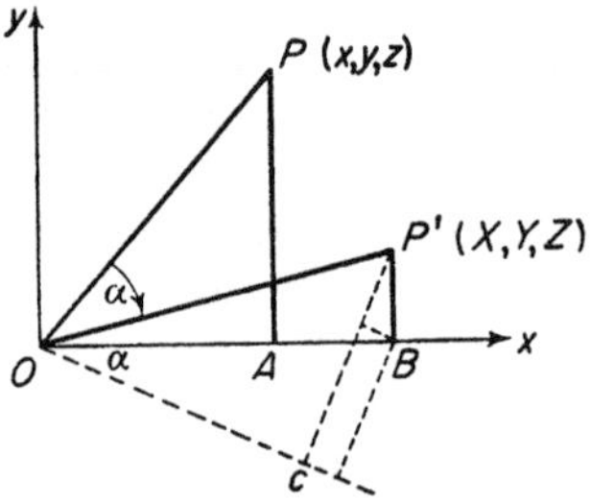

FIG. 1. Rotation of a point P to P'.

Here and elsewhere the x-, y- and z-axes are chosen to form a right-handed set. However, instead of rotating P, we can also consider the body and P as fixed, and refer all co-ordinates to a new pair of axes OX and OY which make an angle $+\alpha$ with Ox and Oy (Fig. 2).

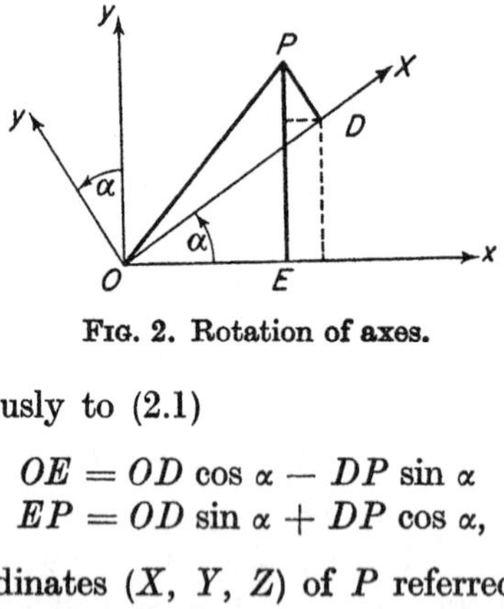

FIG. 2. Rotation of axes.

We have analogously to (2.1)

$$\begin{aligned} OE &= OD\cos\alpha - DP\sin\alpha\\ EP &= OD\sin\alpha + DP\cos\alpha,\end{aligned}$$

so that the co-ordinates (X, Y, Z) of P referred to the new axes are related to (x, y, z) again by (2.2).

Thus the single transformation (2.2) can represent either the change in the co-ordinates of a point when we *rotate a body* by an angle $-\alpha$, or the change in the co-ordinates of a fixed point when we *rotate the co-ordinate axes* by an angle $+\alpha$. The close relationship between these two operations is directly evident from the similarity between Figs. 1 and 2. The two different points of view also arise when considering the symmetry properties of a physical system.

Consider for instance a perfectly round plate without any markings on it: we say it is symmetrical about a vertical axis through its centre, say the z-axis. We can express this more precisely by saying that if we rotate the plate about its axis, we cannot tell that we have rotated it because it is completely round with no markings on it. On the other hand we could also say that for a fixed position of the plate, the various physical properties such as moments of inertia associated with the x- and y-axes must be the same, no matter in what directions these axes are chosen. In this example the first approach is perhaps more natural, but when discussing the symmetry of the Schrödinger equation for a physical system we shall adopt the second point of view. Anticipating a little, we shall be considering a given equation and the forms it takes when expressed in terms of different variables like x, y, z and X, Y, Z which correspond to using different co-ordinate axes. There are two reasons for this choice. Firstly, the Schrödinger equation is a mathematical relation and not like a plate so that we cannot rotate it in quite the same sense, though we could, of course, write down the equation for the rotated physical system. Expressing an equation in terms of different sets of co-ordinates is a more familiar concept. Secondly, we shall be considering some transformations of co-ordinates that have no simple physical analogue. For instance, we can carry out a rotational transformation of spin co-ordinates without altering the position vectors $\mathbf{r}_i$ of the electrons in an atom, but what does it mean physically to rotate an atom in spin space while holding it fixed in ordinary space? Nevertheless the transformations of co-ordinates which we shall apply to the Schrödinger equation will usually be suggested by and linked with the physical symmetry of the system in an obvious way.

When discussing linear transformations of co-ordinates, it is convenient to refer to them by a single symbol such as T. For instance, we shall call the transformation (2.2) *the transformation* R, or because it corresponds to a rotation, *the rotation* R. If it is necessary to be specific about the angle of rotation, we shall call (2.2) the rotation $R(\alpha, z)$ of $+\alpha$ about the z-axis because this sign corresponds to the change-of-axes point of view which we are adopting. We have already discussed in connection with Fig. 2 the effect of applying a transformation such as R on the co-ordinates of a point, and we shall now make the following preliminary definition of what it means to apply $R(\alpha, z)$ to a function of x, y, z. In § 5, reasons will appear for replacing this definition by a slightly enlarged concept. *Applying the transformation* $R(\alpha, z)$ (2.2) *to a function* $f(x, y, z)$ *means to substitute the expressions* (2.2) *for* x, y, z *in the function* *and thus express f in terms of X, Y, Z. This results in a function of X, Y, Z which in general displays a different functional form from* $f(x, y, z)$. For instance applying $R(\alpha, z)$ to the function $(x - y)^2$, we obtain

$$\begin{aligned}(x - y)^2 &= [(X \cos \alpha - Y \sin \alpha) - (X \sin \alpha + Y \cos \alpha)]^2 \\ &= [X(\cos \alpha - \sin \alpha) - Y(\cos \alpha + \sin \alpha)]^2, \qquad (2.3)\end{aligned}$$

which is a different function of X, Y, Z. Similarly we can apply a transformation to each side of an equation. For instance the equation

$$\frac{\partial}{\partial x}(x - y)^2 = 2(x - y) \qquad (2.4)$$

becomes†

$$\begin{aligned}&\left(\cos \alpha \frac{\partial}{\partial X} - \sin \alpha \frac{\partial}{\partial Y}\right)[X(\cos \alpha - \sin \alpha) - Y(\cos \alpha + \sin \alpha)]^2 \\ &\qquad = 2[X(\cos \alpha - \sin \alpha) - Y(\cos \alpha + \sin \alpha)], \qquad (2.5)\end{aligned}$$

which is still a correct equation as can easily be verified.

Problems

2.1 Apply the transformation $R(\alpha, z)$ (equation (2.2)) to each of the following functions: (a) $\exp x$; (b) $(x + iy)^2$; (c) $x^2 + y^2 + z^2$; (d) $xf(r)$, $yf(r)$, $zf(r)$.

2.2 Write down the linear transformation that corresponds to a rotation of α about the y-axis, and apply it to each of the functions of problem 2.1.

2.3 The Schrödinger equation for a simple harmonic oscillator of frequency ω is

$$\left(-\frac{\hbar^2}{2m}\frac{d^2}{dx^2} + \tfrac{1}{2}m\omega^2x^2\right)\psi(x) = E\psi(x),$$

where $\psi(x)$ is an eigenfunction belonging to the energy value E. By operating on the equation with the transformation $x = -X$, show that $\psi(-x)$ is also an eigenfunction belonging to the same energy level and so are $\psi(x) + \psi(-x)$ and $\psi(x) - \psi(-x)$.

3. Symmetry Transformations of the Hamiltonian

We shall now apply linear transformations like $R(\alpha, z)$ (2.2) to the time-independent Schrödinger equation

$$\mathscr{H}\psi = E\psi, \qquad (3.1)$$

† How to transform $\partial/\partial x$ is shown in any elementary calculus text.

where $\mathscr{H}$ is the Hamiltonian operator and E the energy value belonging to the eigenfunction ψ. It is convenient to consider first the effect of a transformation on the Hamiltonian $\mathscr{H}$.

The Hamiltonian for an atom with n electrons, considering the nucleus as fixed and omitting spin dependent terms, is (Schiff 1955, p. 284)

$$\boxed{\mathscr{H} = -\frac{\hbar^2}{2m}\sum_i^n \nabla_i^2 - \sum_i^n \frac{ne^2}{r_i} + \sum_{i<j}^n\sum^n \frac{e^2}{r_{ij}},} \tag{3.2}$$

where m is the mass of an electron, e the charge on a proton, and

$$\nabla_i^2 = \frac{\partial^2}{\partial x_i^2} + \frac{\partial^2}{\partial y_i^2} + \frac{\partial^2}{\partial z_i^2}, \tag{3.3}$$

$$r_{ij}^2 = (x_i - x_j)^2 + (y_i - y_j)^2 + (z_i - z_j)^2. \tag{3.4a}$$

If we apply the transformation $R(\alpha, z)$ (2.2) to the co-ordinates (x_i, y_i, z_i) of each of the n electrons, we have

$$\begin{aligned} r_i^2 &= x_i^2 + y_i^2 + z_i^2 \\ &= (X_i \cos\alpha - Y_i \sin\alpha)^2 + (X_i \sin\alpha + Y_i \cos\alpha)^2 + Z_i^2 \\ &= X_i^2 + Y_i^2 + Z_i^2. \end{aligned}$$

Similarly

$$r_{ij}^2 = (X_i - X_j)^2 + (Y_i - Y_j)^2 + (Z_i - Z_j)^2, \tag{3.4b}$$

and it can easily be shown that†

$$\frac{\partial^2}{\partial x_i^2} + \frac{\partial^2}{\partial y_i^2} + \frac{\partial^2}{\partial z_i^2} = \frac{\partial^2}{\partial X_i^2} + \frac{\partial^2}{\partial Y_i^2} + \frac{\partial^2}{\partial Z_i^2}. \tag{3.5}$$

Thus substituting these relations into (3.2), we see that the Hamiltonian has precisely the same form when expressed in terms of the (X_i, Y_i, Z_i) co-ordinates as in terms of the (x_i, y_i, z_i) co-ordinates, i.e.

$$\boxed{\mathscr{H}(x_i, y_i, z_i) = \mathscr{H}(X_i, Y_i, Z_i).} \tag{3.6}$$

This is expressed by saying that the transformation $R(\alpha, z)$ leaves $\mathscr{H}$(3.2) unchanged, or $R(\alpha, z)$ *leaves* $\mathscr{H}$ *invariant*, or $\mathscr{H}$ *is invariant under* $R(\alpha, z)$, or $R(\alpha, z)$ is a symmetry transformation of $\mathscr{H}$. A *symmetry transformation of a Hamiltonian is defined as a linear*

† The transformation of differential operators is discussed in any elementary calculus text.

transformation of co-ordinates which leaves that Hamiltonian invariant in the sense of equation (3.6).

The reason for applying linear transformations like $R(\alpha, z)$ (2.2) to a Hamiltonian now becomes a little clearer. We have seen that $R(\alpha, z)$ leaves the Hamiltonian (3.2) invariant. However, $R(\alpha, z)$ applied to the eigenfunctions of the Hamiltonian does *not* in general leave them invariant. Consider for instance the $2p$ wave functions for a hydrogen atom (example (ii) of §1). $R(\alpha, z)$ applied to $xf(r)$ gives $(X \cos \alpha - Y \sin \alpha) f(R)$ which has a different functional form. In particular for $\alpha = 90°$ we obtain $-Yf(R)$ so that $R(\alpha, z)$ has changed one eigenfunction into another. More generally consider a Schrödinger equation

$$\mathscr{H}(x_i, y_i, z_i)\, \psi_1(x_i, y_i, z_i) = E\psi_1(x_i, y_i, z_i). \tag{3.7}$$

Applying any symmetry transformation T we obtain

$$\mathscr{H}(X_i, Y_i, Z_i)\, \psi_2(X_i, Y_i, Z_i) = E\psi_2(X_i, Y_i, Z_i), \tag{3.8}$$

where ψ_2 in general has a different functional form from ψ_1. Thus $\psi_2(X_i, Y_i, Z_i)$ is an eigenfunction of $\mathscr{H}(X_i, Y_i, Z_i)$, but since $\mathscr{H}(X_i, Y_i, Z_i)$ and $\mathscr{H}(x_i, y_i, z_i)$ have the same form, we can also say from (3.8) that *$\psi_2(x_i, y_i, z_i)$ is an eigenfunction of $\mathscr{H}(x_i, y_i, z_i)$ and belongs to the same eigenvalue E as ψ_1*. An alternative method of wording this argument is to say that since (3.8) is a differential equation in terms of the variables X_i, Y_i, Z_i, we can replace X_i, Y_i, Z_i by x_i, y_i, z_i or any other set of symbols throughout without upsetting the validity of the equation. Thus (3.8) becomes

$$\mathscr{H}(x_i, y_i, z_i)\, \psi_2\, (x_i, y_i, z_i) = E\psi_2(x_i, y_i, z_i), \tag{3.9}$$

which is just our previous conclusion expressed in symbols. Thus we see that *the symmetry transformations of a Hamiltonian can be used to relate the different eigenfunctions of one energy level to one another* and hence to label them and to discuss the degree of degeneracy of the energy level. Before we can pursue this further (§ 6), we must discuss in greater detail the symmetry transformations of Hamiltonians (§§ 3 and 4) and their effect on wave functions (§ 5).

The Hamiltonian (3.2) has two other types of symmetry transformation besides the rotation R. The transformation

$$\boxed{\begin{aligned} (x_1, y_1, z_1) &= (X_2, Y_2, Z_2) \\ (x_2, y_2, z_2) &= (X_1, Y_1, Z_1) \\ (x_i, y_i, z_i) &= (X_i, Y_i, Z_i), \qquad i = 3, 4 \ldots n, \end{aligned}} \tag{3.10}$$

is called the interchange or permutation of the co-ordinates 1 and 2, and is a symmetry transformation of (3.2) as is obvious by inspection. Similarly any permutation of the co-ordinates x_i, y_i, z_i, $i = 1$ to n, is a symmetry transformation. The other symmetry transformation is the *inversion* transformation Π

$$\boxed{\Pi\colon\quad x_i = -X_i,\quad y_i = -Y_i,\quad z_i = -Z_i\quad \text{for all } i.} \tag{3.11}$$

This can be combined with the rotations. An ordinary rotation such as (2.2) is called a *proper rotation*, and the combination of a proper rotation with the inversion Π is called an *improper rotation*. As a particular example of an improper rotation, we have $\Pi R(180°, x)$ which is just the reflection m_x in the mirror plane $x = 0$, i.e.

$$\boxed{m_x\colon\quad (x_i, y_i, z_i) = (-X_i, Y_i, Z_i)\quad \text{for all } i.} \tag{3.12}$$

It can easily be verified that all improper rotations, as well as proper ones, leave the Hamiltonian (3.2) invariant. However, there are many simple and important transformations that are not symmetry transformations of (3.2), for instance the transformation to cylindrical polar co-ordinates

$$x_i = R_i \cos \Theta_i, \qquad y_i = R_i \sin \Theta_i, \qquad z_i = Z_i. \tag{3.13}$$

This transformation is in any case not a linear one because it involves products of R_i with trigonometric functions of Θ_i. Also ∇_i^2 becomes

$$\frac{1}{R_i}\frac{\partial}{\partial R_i}\left(R_i\frac{\partial}{\partial R_i}\right) + \frac{1}{R_i^2}\frac{\partial^2}{\partial \Theta_i^2} + \frac{\partial^2}{\partial Z_i^2}, \tag{3.14}$$

which is not identical in form with (3.3), so that (3.13) *is not a symmetry transformation*. Of course we may wish to express the Hamiltonian (3.2) in terms of cylindrical polar co-ordinates for some problem, but in the future we shall refer to such a transformation as a *change* to polar co-ordinates, so as to avoid confusion with symmetry transformations which we will be considering so much that it will be convenient to refer to the latter simply as transformations.

We must now indicate briefly what the symmetry transformations are for the Hamiltonians of physical systems besides free atoms and ions which we have been considering so far. An atom has complete spherical symmetry, i.e. it is invariant to any rotation about any axis (cf. problem 3.7), so that it has a higher degree of symmetry than molecules and crystal lattices which are usually only invariant to certain rotations about certain axes (cf. problems 3.4 and 3.5). Thus the latter have some of the symmetry transformations of the atom, but not any radically new ones except for the translational symmetry of a crystal lattice. We have therefore already mentioned in connection with (3.2) almost all the types of symmetry transformation which we shall discuss.

To sum up, the form of a Hamiltonian remains unchanged by certain linear transformations which are called symmetry transformations of the Hamiltonian. Symmetry transformations in general change the eigenfunctions of one energy level into one another.

PROBLEMS

3.1 Show that the following co-ordinate changes are *not* symmetry transformations of the Hamiltonian (3.2).

(a) $(x_i, y_i, z_i) = (2X_i, 2Y_i, 2Z_i), \quad i = 1 \text{ to } n.$

(b) $(x_1, y_1, z_1) = (-X_1, -Y_1, -Z_1),$
$(x_i, y_i, z_i) = (X_i, Y_i, Z_i), \quad i = 2 \text{ to } n.$

(c) $x_i = \exp X_i, \quad y_i = \exp Y_i, \quad z_i = \exp Z_i, \quad i = 1 \text{ to } n.$

(d) x_1, y_1, z_1 given in terms of X_1, Y_1, Z_1 by equation (2.2),
$(x_i, y_i, z_i) = (X_i, Y_i, Z_i), \quad i = 2 \text{ to } n.$

(e) $x_i = R_i \sin \Theta_i \cos \Phi_i, \quad y_i = R_i \sin \Theta_i \sin \Phi_i,$
$z_i = R_i \cos \Theta_i, \quad i = 1 \text{ to } n.$

3.2 Express the Hamiltonian (3.2) in terms of spherical polar co-ordinates r, θ, ϕ, where

$$x = r \sin \theta \cos \phi, \qquad y = r \sin \theta \sin \phi, \qquad z = r \cos \theta,$$

(Schiff 1955, p. 69).
Show that the inversion transformation Π takes the form

$$r_i = R_i, \qquad \theta_i = \pi - \Theta_i, \qquad \phi_i = \pi + \Phi_i, \qquad i = 1 \text{ to } n,$$

and express also the rotation (2.2) and the other symmetry transformations mentioned in § 3 in polar co-ordinates. Hence, verify that they again leave the Hamiltonian (3.2) invariant.

3.3 Write down the Hamiltonian without spin dependent terms for an ion of nuclear charge Z with n (not equal to Z) electrons, and show that it has the same symmetry transformations as the Hamiltonian (3.2). Do the same for the one-electron Hartree equation (Schiff 1955, p. 284) for the single valence electron of a sodium atom.

3.4 Write down the Hamiltonian without spin dependent terms for the two electrons in a hydrogen molecule, considering the two protons as fixed at the points $\pm$ (0, 0, a). Show that it is (a) invariant under any rotation about the z-axis, but only to 180° rotations about the x- or y-axes, (b) invariant under reflections in the plane through the origin perpendicular to the z-axis and in any plane containing the z-axis, (c) invariant under the inversion Π and the reflection in the z-axis

$$(x_i, y_i, z_i) = (-X_i, -Y_i, Z_i) \qquad i = 1, 2,$$

and (d) invariant under the interchange of co-ordinates 1 and 2.

3.5 In problem 3.4, assume that one of the protons has been replaced by a deuteron, and suppose that the deuteron has a slightly different charge from that of the proton. What effect does this have on the symmetry properties of the Hamiltonian? Although in reality the deuteron and proton have the same charge, they do have different masses and magnetic moments, and this would affect the symmetry of the problem in a similar way to the fictitious difference in charge if the interaction with the nuclear moments were included in the Hamiltonian.

3.6 Repeat the discussion of problem 3.4 in terms of spherical polar co-ordinates and in terms of cylindrical polar co-ordinates (equation (3.13)). Which set of co-ordinates do you think is most convenient for this problem?

3.7 A rotation about the origin can be defined mathematically as a linear transformation of co-ordinates that leaves invariant the distance of an arbitrary point (x, y, z) from the origin. Using this definition, show that the Hamiltonian (3.2) is invariant under any rotation about any axis. Show that the definition includes the improper as well as the proper rotations (Margenau and Murphy 1943, p. 310).

3.8 Show that an improper rotation of 180° about any axis is the same as a reflection in the plane through the origin perpendicular to that axis.

3.9 Write down the Hamiltonian for a hydrogen atom in small uniform electric and magnetic fields parallel to the z-axis (Schiff 1955, pp. 158, 292) omitting spin dependent terms and considering the nucleus as a fixed charge. Also assume that the $2p$ eigenfunctions

$$\psi_1 = \frac{x+iy}{\sqrt{2}} f(r), \qquad \psi_2 = \frac{x-iy}{\sqrt{2}} f(r), \qquad \psi_3 = zf(r),$$

where $f(r)$ is given by Schiff (1955, p. 85), are still eigenfunctions to a first approximation in the presence of the fields. Prove (a) the $2p$ level is three-fold degenerate in the absence of the external fields, (b) in the presence of the electric field only, ψ_1 and ψ_2 are degenerate with one another but need not be degenerate with ψ_3, (c) in the presence of the magnetic field only, symmetry arguments do not require any of the functions ψ_1, ψ_2 and ψ_3 to have the same energy so that the $2p$ level may be split into three levels. Hint: in each of the cases (a), (b) and (c) test whether the reflection in the plane $y = 0$ and the rotation of 90° about the y-axis are symmetry transformations. If they are, use them to apply the argument of equations (3.7), (3.8), (3.9) to each of the functions ψ_1, ψ_2, ψ_3. Try also the inversion Π, rotations about other axes and other reflections to ensure as far as possible that no degeneracy required by symmetry has been missed.

4. Groups of Symmetry Transformations

In this section we shall illustrate and define what is meant by a group in the mathematical sense of the word, and shall show what relevance this concept has to the symmetry transformations of Hamiltonians.

Example of a group

Let us first consider the symmetry properties of a particular physical object, namely an equilateral triangle cut out of a piece of cardboard having the same finish on both sides and lying on the table with its vertices at the points 1, 2 and 3 and its centre at the origin of co-ordinates (Fig. 3). *Ok*, *Ol*, *Om* are three other axes

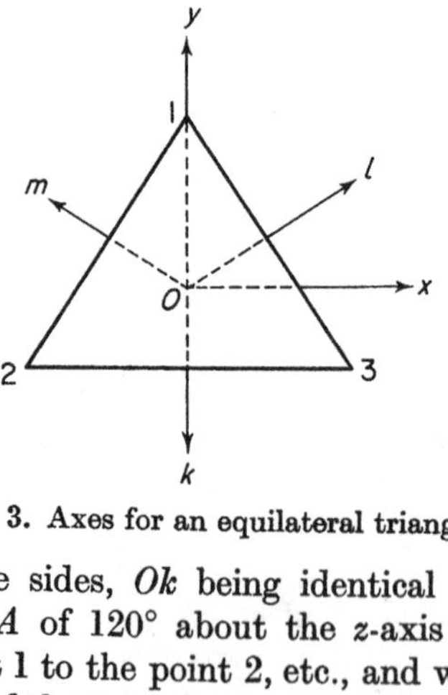

FIG. 3. Axes for an equilateral triangle.

perpendicular to the sides, *Ok* being identical with the negative y-axis. A rotation A of 120° about the z-axis moves the vertex that was at the point 1 to the point 2, etc., and we shall call this an equivalent position of the triangle since it is indistinguishable from the original position. It can easily be seen that the following rotations all leave the triangle in equivalent positions and that there are no other proper rotations that do this.

A: 120° about z-axis
B: 240° or −120° about z-axis
K: 180° about Ok axis
L: 180° about Ol axis
M: 180° about Om axis
E: 0° or 360° about any axis, i.e. no rotation. (4.1)

If we apply two rotations successively, for instance first A and then K, this moves the top vertex from position 1 first to position 2 and then to 3, the vertex at the position 2 to 3 and then to 2, the vertex at 3 to 1 and then to 1. Thus the combined operation A followed by K is identical with the single rotation L. Similarly K followed by A is the same as M, and it can easily be verified that combining any pair of the rotations (4.1) in either order gives another rotation which is also one of the ones listed in (4.1). If the rotation F applied First followed by S applied Second is equivalent to the single Combined rotation C, we write

$$SF = C, \tag{4.2}$$

where it is customary to write the S before the F in analogy with differential operators. For instance

$$x^2 \frac{d}{dx} f(x) \tag{4.3}$$

means *first* differentiating $f(x)$ and then multiplying the result by x^2. This is clearly not the same as

$$\frac{d}{dx} x^2 f(x), \tag{4.4}$$

and similarly when combining rotations it is important to follow the convention of (4.2). We have already seen that

$$KA = L, \qquad AK = M, \tag{4.5}$$

and similarly it is possible to write down a whole multiplication table (Table 1) where the rotation in the top row is applied first and the rotation in the left column second. There is an important feature of Table 1, namely that for every rotation P, there is also a rotation P^{-1}, say, which undoes the effect of P, and that P also undoes the effect of P^{-1}, i.e.

$$PP^{-1} = P^{-1}P = E. \tag{4.6}$$

In fact in every case P and P^{-1} are just two rotations by the same angle about the same axis but in opposite directions. When the angle is 180° this of course makes P and P^{-1} identical. It can also be verified from the multiplication table that the triple products

TABLE 1

Multiplication Table for the Group 32

	E	A	B	K	L	M applied first
E	E	A	B	K	L	M
A	A	B	E	M	K	L
B	B	E	A	L	M	K
K	K	L	M	E	A	B
L	L	M	K	B	E	A
M	M	K	L	A	B	E
applied second						

$P(QR)$ and $(PQ)R$ are always the same, so that they can be written unambiguously as PQR. Alternatively this follows directly from the physical nature of rotations as can easily be shown. These properties suffice to establish that the rotations E, A, B, K, L, M (4.1) are the *elements* of a *group*.

Definition of a group. *A group* $\mathfrak{G}$ *is a collection of elements* A, B, C, D, ... *which have the properties* (a) *to* (e) *below.* The elements in the simplest cases may be numbers. They may also be any other quantities such as matrices, physical operations like rotations, or mathematical operations such as making a linear transformation of co-ordinates.

(a) *It must be possible to combine any pair of elements* F *and* S *in a definite way to form a combination* C which we shall write

$$\boxed{C = SF} \tag{4.7}$$

where as before F is the first element, S the second element and C the combination, if the order of F and S is important. In our example with the elements (4.1), the law of combination was "first apply rotation F and then S". With other groups the law of combination may be matrix multiplication or like addition. If for two elements $PQ = QP$ then P and Q are said to *commute*, and if this is so for every pair of elements then the law of combination is *commutative* and the group is *Abelian.*

(b) *The combination* $C = SF$ *of any pair of elements* F *and* S *must also be an element of the group.* Thus a multiplication table among the group elements can always be set up like Table 1.

(c) *One of the group elements, E say, must have the properties of a unit element, namely*

$$EP = PE = P \tag{4.8}$$

for every element P. For instance omitting all reference to E would make it impossible to set up a complete multiplication table for the other rotations of (4.1) (cf. Table 1). This is related to the next property.

(d) *Every element P of the group must have an inverse P^{-1} which also belongs to $\mathfrak{G}$ with the property*

$$PP^{-1} = P^{-1}P = E. \tag{4.9}$$

(e) *The triple product PQR must be uniquely defined, i.e.*

$$P(QR) = (PQ)\,R = PQR. \tag{4.10}$$

This is true for all the kinds of elements and laws of combination that we shall wish to deal with, but there are examples where it does not hold, e.g. $24 \div (6 \div 2) \neq (24 \div 6) \div 2$!

Two simple examples of groups are all positive rational fractions excluding zero with the law of combination being multiplication, and all positive and negative integers including zero with the law of combination being addition. In the latter case it is interesting that zero plays the role of the unit element E. The permutations of n objects, i.e. the operations of rearranging them and not their different arrangements in a row, say, form the *permutation group* $\mathfrak{P}_n$ also known as the symmetric group S_n. The proper rotations by all possible angles about a fixed axis form the *axial rotation group.* This is clearly Abelian. The *full rotation group* (Chapter II) consists of all proper rotations about all axes through a point, and this becomes the *full rotation and reflection group* when all improper rotations are included. There are thirty-two groups of particular interest formed from a finite number of particular rotations about a point and are known as *point-groups* (§ 16). These clearly do not include all possible finite groups of rotations because, for instance, the rotations by $360\,r/n$ degrees about a fixed axis where $r = 1$ to n, always form a group of n elements. An example of a point-group is the group (4.1) which is called 32 (pronounced "three two", not "thirty-two") in the international notation, to denote that it includes some two-fold axes (rotations by 180°) perpendicular to a three-fold axis (120°, 240°). All the proper and improper rotations that move a cube to an equivalent position form the *full cubic group* m3m. In the older Schoenfliess notation these two point-groups are called D_3 and O_h. All square matrices of a given order and with non-zero determinant form a group, the law of combination being matrix multiplication. So do all unitary matrices (appendix A) of given order, and likewise all unitary matrices of given order and with determinant $+1$, as can easily be verified. Finally linear transformations of co-ordinates can form groups as we shall now see.

The group of symmetry transformations of a Hamiltonian. Consider three protons fixed at the points

$$\mathbf{r}_1 = [0, 2\sqrt{3}a, 0], \qquad \mathbf{r}_2 = [-3a, -\sqrt{3}a, 0], \qquad \mathbf{r}_3 = [3a, -\sqrt{3}a, 0] \tag{4.11}$$

forming an equilateral triangle about the origin (Fig. 3). The Hamiltonian for one electron moving in the field of the three protons is

$$\mathscr{H} = -\frac{\hbar^2}{2m}\nabla^2 + \frac{e^2}{|\mathbf{r}-\mathbf{r}_1|} + \frac{e^2}{|\mathbf{r}-\mathbf{r}_2|} + \frac{e^2}{|\mathbf{r}-\mathbf{r}_3|}. \tag{4.12}$$

This system is not one of physical importance but its symmetry is closely related to that of an ozone molecule† or that of an ion situated between three water molecules in the hydrated crystal of a salt, to which the following discussion can easily be extended (cf. problems 4.5 and 4.6). The physical system of three protons has the same rotational symmetry as the equilateral triangle already discussed, which suggests that the linear transformations E', A', B', K', L', M' corresponding to the rotations E, A, B, K, L, M (4.1) may be symmetry transformations of the Hamiltonian (4.12). These transformations can easily be found from (2.2) and simple extensions of the argument of §2. For instance A' is obtained from (2.2) by putting $\alpha = -120°$ in accordance with § 2 because A (4.1) is a *physical* rotation of $+120°$. We obtain:

$$\begin{array}{ll}
E'\colon (x, y, z) = (X, Y, Z) & \\
A'\colon x = -\tfrac{1}{2}X + \tfrac{1}{2}\sqrt{3}Y & B'\colon x = -\tfrac{1}{2}X - \tfrac{1}{2}\sqrt{3}Y \\
\quad\ \ y = -\tfrac{1}{2}\sqrt{3}X - \tfrac{1}{2}Y & \quad\ \ y = \tfrac{1}{2}\sqrt{3}X - \tfrac{1}{2}Y \\
\quad\ \ z = Z & \quad\ \ z = Z \\
K'\colon (x, y, z) = (-X, Y, -Z) & \\
L'\colon x = \tfrac{1}{2}X + \tfrac{1}{2}\sqrt{3}Y & M'\colon x = \tfrac{1}{2}X - \tfrac{1}{2}\sqrt{3}Y \\
\quad\ \ y = \tfrac{1}{2}\sqrt{3}X - \tfrac{1}{2}Y & \quad\ \ y = -\tfrac{1}{2}\sqrt{3}X - \tfrac{1}{2}Y \\
\quad\ \ z = -Z & \quad\ \ z = -Z
\end{array} \tag{4.13}$$

† Throughout this book we shall assume for illustrative purpose that the three oxygen atoms in ozone form an equilateral triangle, though this is not so in actual fact (see end of § 24).

It can easily be verified that all these transformations are indeed symmetry transformations of (4.12). The ∇^2 remains invariant under each as in § 3, and applying for instance A' to the other terms we obtain

$$\begin{aligned} |\mathbf{r} - \mathbf{r}_1|^2 &= x^2 + [y - 2\sqrt{3}a]^2 + z^2 \\ &= (-\tfrac{1}{2}X + \tfrac{1}{2}\sqrt{3}Y)^2 + [-\tfrac{1}{2}\sqrt{3}X - \tfrac{1}{2}Y - 2\sqrt{3}a]^2 + Z^2 \\ &= (X + 3a)^2 + [Y + \sqrt{3}a]^2 + Z^2 \\ &= |\mathbf{R} - \mathbf{r}_2|^2, \end{aligned}$$

$$|\mathbf{r} - \mathbf{r}_2|^2 = |\mathbf{R} - \mathbf{r}_3|^2 \quad \text{and} \quad |\mathbf{r} - \mathbf{r}_3|^2 = |\mathbf{R} - \mathbf{r}_1|^2,$$

whence (4.12) is invariant under A'.

The result that the transformation A' of (4.13) is a symmetry transformation of (4.12) is actually no accident and can be proved as follows using the corresponding rotation A without ever writing down the form of A' or substituting in (4.12). Let the potential in (4.12) due to the protons be

$$V(\mathbf{r}) = \sum_{1,2,3} \frac{e^2}{|\mathbf{r} - \mathbf{r}_i|},$$

and let $P(x, y, z)$ be any point. Consider the physical operation A of rotating the point P and the three protons but not the co-ordinate axes. The system of protons is moved into an equivalent position, one proton from $\mathbf{r}_1$ to $\mathbf{r}_2$, one from $\mathbf{r}_3$ to $\mathbf{r}_1$ and one from $\mathbf{r}_2$ to $\mathbf{r}_3$, and P is moved to the position (X, Y, Z). During this rotation the potential at P has remained constant because it depends only on the distances of P from the three protons, and these distances have not changed because P and the protons have been rotated as a rigid whole. Thus $V(x, y, z)$ due to the *initial* charge distribution is equal to the potential at (X, Y, Z) due to the *final* charge distribution. Since, however, A has moved the system of protons into an equivalent position, the initial and final charge distributions and potentials are identical, so that the potential at (X, Y, Z) due to the final charge distribution is $V(X, Y, Z)$, i.e. we have

$$\boxed{V(x, y, z) = V(X, Y, Z),} \tag{4.14}$$

where according to Fig. 1 of § 2, x, y, z and X, Y, Z are related by equations (2.2) with $\alpha = -120°$. It only remains to view (4.14) and (2.2) in terms of a change A' of co-ordinates, rather than in terms of physical rotations. This only involves a change in the interpretation of (4.14) and (2.2) and does not destroy their validity as correct mathematical relations. We therefore obtain that $V(\mathbf{r})$ is invariant under the co-ordinate transformation A'. This argument applies similarly to all the transformations (4.13), and indeed to any similar situation (cf. problem 4.8).

We can also verify from (4.13) or prove by the above type of argument that the transformation A' followed by K' is the same as the single transformation L'. For in detail this simply means that first expressing a function $f(x, y, z)$ in terms of X, Y, Z using A' (4.13) and then in terms of ξ, η, ζ using K'

$$(X, Y, Z) = (-\xi, \eta, -\zeta),$$

gives the same result as expressing it directly in terms of ξ, η, ζ using the combined transformation

$$\begin{aligned} x &= \tfrac{1}{2}\xi + \eta\sqrt{3}/2 \\ y &= \tfrac{1}{2}\sqrt{3}\xi - \tfrac{1}{2}\eta \\ z &= -\zeta, \end{aligned}$$

namely the transformation L'. In symbols $K'A' = L'$. Similarly, any of the transformations (4.13) can be combined, and their multiplication table is exactly the same as Table 1 for the point-group 32 of rotations as can be verified most easily by matrix multiplication. It is also easy to show that the transformations have all the other properties (a) to (e) above required for them to form a group, which we shall call the point-group 32 of transformations.

Theorem. We shall now generalize this result and prove that *the symmetry transformations of a Hamiltonian always form a group.* Suppose a Hamiltonian $\mathscr{H}$ is invariant under each of two symmetry transformations F and S. We shall first show that the combined transformation SF (first F, then S second) is also a symmetry transformation. Let the co-ordinates $x_1, y_1, z_1, x_2, y_2 \ldots z_n$ of the Hamiltonian be written for convenience $q_1, q_2, \ldots q_{3n}$, and let F be the transformation

$$q_i = F_{ij}Q_j \tag{4.15}$$

when written in terms of the summation convention (appendix A), and S be the transformation

$$q_i = S_{ij}Q_j. \tag{4.16}$$

Now the transformation SF means to substitute first for the q_i

in terms of the Q_i using (4.15) and then to substitute for the Q_i further in terms of some new variables ν_i where

$$Q_i = S_{ij}\nu_j. \tag{4.17}$$

Since F and S are both symmetry transformations,

$$\mathscr{H}(q_i) = \mathscr{H}(Q_i) = \mathscr{H}(\nu_i), \tag{4.18}$$

so that the composite transformation SF from the q_i direct to the ν_i is also a symmetry transformation. Thus the symmetry transformations satisfy the group requirements (a) and (b) above. We can indeed write down the transformation SF explicitly by eliminating the Q_i from (4.15) and (4.17), i.e. SF is

$$q_i = (F_{ik}S_{kj})\nu_j. \tag{4.19}$$

Further we always have the *identity transformation*

$$q_i = Q_i, \qquad i = 1 \text{ to } 3n, \tag{4.20}$$

having the property (4.8) of the unit element E, which verifies (c). As regards (d), if we substitute for the q_i in the initial Hamiltonian in terms of the Q_i using F (4.15) and obtain $\mathscr{H}(Q_i)$, then we can get back to $\mathscr{H}(q_i)$ by solving (4.15) for the Q_i and substituting into $\mathscr{H}(Q_i)$. But this is just applying the transformation F^{-1}

$$Q_i = (F^{-1})_{ij}q_j \tag{4.21}$$

which undoes the effect of F, and this is therefore also a symmetry transformation. In (4.21) the Q_i are now the initial variables and the q_j the new ones, and F^{-1} is the inverse of the matrix F_{ij}. To make the argument quite rigorous, we note that all the transformations in which we are interested are unitary (cf. appendix A, problem A.9), whence $|F| \neq 0$ and (4.15) can actually be inverted to give (4.21). This verifies (d), and (e) can easily be verified by writing out the transformation TSF

$$q_i = F_{ik}S_{kl}T_{lj}Q_j$$

in full without using the summation convention and noting that it does not matter where the brackets of (4.10) are inserted. This proves the theorem. It is now possible to give a precise meaning to the expression "the symmetry properties of a Hamiltonian" which has been used in a descriptive way up till now. *The symmetry properties of a Hamiltonian consist of the group of all symmetry transformations of the Hamiltonian.*

We shall now investigate the group of symmetry transformations of the Hamiltonian (4.12) in greater detail. Out of the six elements

(4.13), the elements E', A' and B' form a group in themselves as can easily be seen from the first three rows and columns of Table 1. These elements chosen from the bigger group (4.13) are said to form a *subgroup* of the larger group. Another subgroup of (4.13) is the group (E', K'), another one (E', L') etc. Similarly, (4.13) does not include all the possible symmetry transformations of the Hamiltonian, but is a subgroup of the group of all its symmetry transformations. For instance, a symmetry transformation not included in (4.13) is the reflection

$$(x, y, z) = (-X, Y, Z)$$

in the plane $x = 0$, i.e. in the plane kOz (Fig. 3). Further symmetry transformations are the reflections in the planes lOz, mOz and $z = 0$, and combinations of these with the rotations (4.13). These all together form the group of 12 elements called the point-group $\bar{6}m2$ of transformations, and they would appear to be all the symmetry transformations that the Hamiltonian (4.12) has. Incidentally, there is no certain method of ensuring that one has found all the symmetry transformations of a given Hamiltonian instead of just a subgroup: one can only try all the transformations one can think of. Some of the symmetry transformations may not be at all obvious, one case being the symmetry property of the pure Coulomb $1/r$ field in a hydrogen atom which gives rise to the degeneracy of all levels with the same principal quantum number n irrespective of their angular momentum quantum number l, unlike the more general situation in an alkali atom (Schiff 1955, p. 86; Fock 1935). Thus some of the more subtle symmetry transformations of certain systems have only been discovered relatively recently (Jauch and Rohrlich 1955, p. 143; Baker 1956; problem 24.11). If we now consider n electrons in the field of the three protons or of three identical charges of any magnitude similarly arranged, the Hamiltonian for this system would have the transformations $\bar{6}m2$ applied to each set (x_i, y_i, z_i), $i = 1$ to n, as symmetry transformations. It also has the $n!$ permutation transformations of the n variables x_i etc., and all combinations between the permutations and the point-group $\bar{6}m2$ transformations. Thus the group of all symmetry transformations would have a large number $(12n!)$ of elements, but it would be a simple combination of the groups $\bar{6}$m2 and $\mathfrak{P}_n$.

Isomorphism

It was shown above that the elements of the group (4.1), say the group $\mathfrak{G}$, and those of the group (4.13), $\mathfrak{G}'$ say, both multiply in the same way according to Table 1. This relationship between $\mathfrak{G}$ and $\mathfrak{G}'$ is called an *isomorphism* and can be described by saying that the elements $E, A, B, \ldots$ of $\mathfrak{G}$ can be paired off with the elements $E', A', B', \ldots$ of $\mathfrak{G}'$ such that the relationships between $E, A, B, \ldots$ as regards multiplication are in every way the same as the relationships between $E', A', B', \ldots$. Actually it requires more care to define isomorphism precisely and to distinguish it from the related concept of homomorphism (cf. appendix B), but the above description is sufficient for the present considerations. That $\mathfrak{G}$ and $\mathfrak{G}'$ have the same multiplication table is not accidental, but follows quite generally from the relationship between a physical rotation and the corresponding co-ordinate transformation. For if we apply to a system a rotation F which shifts $P(x = q_1, y = q_2, z = q_3)$ to $P'(Q_1, Q_2, Q_3)$ related by (4.15), and then a rotation S moving $P'(Q_1, Q_2, Q_3)$ to $P''(\nu_1, \nu_2, \nu_3)$ related by (4.17), then the combined rotation SF shifts $P(q_1, q_2, q_3)$ to $P''(\nu_1, \nu_2, \nu_3)$ related by (4.19). But we had previously seen that equations (4.15), (4.17) and (4.19) represents the combination of linear transformations of co-ordinates, which therefore combine in exactly the same way as the physical rotations, this argument being quite general. In connection with our study of the Schrödinger equation, it is *always the group of symmetry transformations of the Hamiltonian that we shall be interested in* (§§ 2, 3). However, the group of physical operations which move a system into equivalent positions can be used as an aid to the imagination and to suggest what the symmetry transformations of the Hamiltonian are. Thus in future we shall not distinguish the transformations (4.13) by primes from the rotations (4.1) and shall simply refer to either group as the point-group 32.

Another example of isomorphism is afforded by the *permutation group* $\mathfrak{P}_3$ *of order* 3, which consists of all the transformations (ijk)

$$\begin{aligned} (x_1, y_1, z_1) &= (X_i, Y_i, Z_i), \\ (x_2, y_2, z_2) &= (X_j, Y_j, Z_j), \\ (x_3, y_3, z_3) &= (X_k, Y_k, Z_k), \end{aligned} \qquad (4.22)$$

where ijk is some permutation of the numbers 1, 2 and 3. These transformations can again be combined according to equations (4.15), (4.16), (4.17) and (4.19), where F_{ij}, etc., are now matrices of order 9×9. The transformation (132) (231) means as usual the permutation transformation (231) followed by the permutation (132), and it is equal to the permutation (321) as can easily be verified. Similarly the whole multiplication table can be set up (Table 2). On examination, this table is seen to have exactly the

TABLE 2

Multiplication Table for the Group $\mathfrak{P}_3$

	(123)	(231)	(312)	(132)	(321)	(213) applied first
(123)	(123)	(231)	(312)	(132)	(321)	(213)
(231)	(231)	(312)	(123)	(213)	(132)	(321)
(312)	(312)	(123)	(231)	(321)	(213)	(132)
(132)	(132)	(321)	(213)	(123)	(231)	(312)
(321)	(321)	(213)	(132)	(312)	(123)	(231)
(213)	(213)	(132)	(321)	(231)	(312)	(123)
applied second						

same structure as Table 1. Indeed if we write in Table 2

E instead of (123), A instead of (231),
B instead of (312), K instead of (132),
L instead of (321), M instead of (213),

then Table 2 becomes identical with Table 1, and the group $\mathfrak{P}_3$ is thus isomorphic with the point-group 32. This situation often arises when one is considering groups with a small number of elements, but it must not be assumed that every group of six elements is isomorphic with the point-group 32. For instance, the group of numbers

$$1, \exp(i\pi/3), \exp(i2\pi/3), -1, \exp(i4\pi/3), \exp(i5\pi/3) \qquad (4.23)$$

is not, the law of combination being multiplication (cf. problem 4.2). However, it can be shown that any group of six distinct elements is isomorphic either with the point-group 32 or with the group (4.23).

References

All the matrix algebra required for this book is given in appendix A. Accounts of groups and their properties, more detailed than here but still introductory, may be found in Lederman (1953), Birkhoff and MacLane (1941) and other texts (see general references at the end of the book).

Summary

We have defined what is meant mathematically by a group. We have proved that all the symmetry transformations of a Schrödinger Hamiltonian form a group. In the examples studied so far, this group of symmetry transformations is directly related to the physical symmetry of the system to which the Hamiltonian applies.

PROBLEMS

4.1 Verify the multiplication table (Table 1) for the group of transformations (4.13).

4.2 Construct the multiplication table for the group of elements (4.23) and show that it cannot be made to have the same form as Table 1 by pairing off each of the elements (4.23) with one of the elements (4.1) in some way, i.e. that the groups (4.23) and (4.1) are not isomorphic. Note that this result also follows directly from the fact that the elements (4.23) all commute whereas the elements (4.1) do not.

4.3 Prove that all unitary matrices of order $n \times n$ form a group. Prove that all those with determinant $+1$ form a subgroup of this group (cf. problem A.8, appendix A).

4.4 What are two other simple arithmetic operations besides division for which the associative law (4.10) does not hold?

4.5 Write down the Hamiltonian for all the electrons in an ozone molecule omitting spin dependent terms and considering the nuclei as fixed in the form of an equilateral triangle. Show that the Hamiltonian is invariant under the point-group $\bar{6}$m2 of transformations applied simultaneously to the co-ordinates of each electron.

4.6 In the hydrated crystal of an inorganic salt, three molecules of water of crystallization are arranged in an equilateral triangle at a distance a from the centre with their permanent electric dipole moments pointing in towards the centre. Calculate the potential $V(\mathbf{r})$ near the centre in the form of a power series in x, y, z up to the cubic terms with the centre as origin, expressing it in the form

$$A_1 + A_2(x^2 + y^2 + z^2) + A_3z^2 + A_4i[(x + iy)^3 - (x - iy)^3].$$

Hence verify that $V(\mathbf{r})$ is invariant under the point-group $\bar{6}$m2 to this order of approximation, and show that this is also true for the exact expression for $V(\mathbf{r})$. Write down the Hamiltonian for an atom at the centre of the triangle and show that it is also invariant under $\bar{6}$m2.

4.7 In the benzene molecule, the carbon nuclei lie in a plane in the form of a regular hexagon. Discuss the symmetry of the Hamiltonian for the electrons in a benzene molecule.

4.8 Write out in clear English a proof of the following theorem. If a given distribution of charges remains unchanged by a group of physical rotations, then the potential $V(\mathbf{r})$ due to the charges remains invariant under the transformations corresponding to the rotations in the sense of § 2, and these transformations also form a group.

4.9 Which of the point-groups 6, $\bar{6}$, $\bar{3}$, 3m (§ 16) are isomorphic with the group (4.1) and which with the group (4.23)?

4.10 The elements A, B, C, . . . of a group $\mathfrak{G}$ are all distinct. If one element P is combined with each element of $\mathfrak{G}$, show that the set of combinations PA, PB, PC, . . . is identical with the group $\mathfrak{G}$. Hence show that each row and each column of the group multiplication table contains all the elements of the group once. Hint: every combination is equal to some element of $\mathfrak{G}$ by the group property (b); it is only necessary to show that all the combinations are distinct, for then each element of $\mathfrak{G}$ must occur exactly once among them. (Ledermann 1953, pp. 10, 32).

4.11 Are the point-groups 32 and $\bar{6}$m2 subgroups of the full rotation group? Is $\mathfrak{D}_n$ a subgroup of $\mathfrak{D}_m$ if $n < m$?

4.12 Prove that all groups consisting of two distinct elements are isomorphic. Do the same for all groups of three distinct elements. Hint: show that the multiplication table can only have one form by using result of problem 4.10.

5. Group Representations

Transformation of functions

In this section we shall study the effect of symmetry transformations on wave functions or other functions. They do not behave in a simple way like the Hamiltonian does, and it will be necessary to introduce several new concepts.

In § 2 we defined applying a transformation R like (2.2) to a function $\psi_1(x, y, z)$ to mean substituting the expressions (2.2) for x, y, z into $\psi_1(x, y, z)$, and obtaining in general a new function $\psi_2(X, Y, Z)$. Further in § 3 we saw that if $\psi_1(x, y, z)$ is an eigenfunction of a Hamiltonian $\mathscr{H}(x, y, z)$ (cf. equation (3.7)) and R is a symmetry transformation of this Hamiltonian, then $\psi_2(X, Y, Z)$ is an eigenfunction of $\mathscr{H}(X, Y, Z)$ (cf. equation (3.8)) with the same eigenvalue. Now to carry on using capital letters would be inconvenient and we might just as well say that $\psi_2(x, y, z)$ is an eigenfunction of $\mathscr{H}(x, y, z)$ (cf. equation (3.9)). This reversion to lower case letters is particularly useful in discussing the relationship between the functions ψ_1 and ψ_2. Accordingly we shall in the future short circuit the use of capital letters and say that applying the transformation R to $\psi_1(x, y, z)$ changes it into $\psi_2(x, y, z)$. This is written

$$R\psi_1(x, y, z) = \psi_2(x, y, z), \tag{5.1}$$

ψ_2 is called the *transformed function*, and we say that R *operating on* ψ_1 *gives* ψ_2, or R *transforms* ψ_1 *into* ψ_2. In more general terms we have the new definition: *applying a transformation* T *given by*

$$q_i = T_{ij}Q_j \tag{5.2a}$$

to a function $f(q_i)$ *means first substituting* (5.2a) *for the* q_i *and thus obtaining a new function* $F(Q_i) = f(T_{ij}Q_j)$ *of the* Q_i, *and secondly replacing each* Q_i *by* q_i *which gives* $F(q_i) = f(T_{ij}q_j)$. *In symbols*

$$\boxed{Tf(q_i) = f(T_{ij}q_j).} \tag{5.3}$$

In this notation the example of equation (2.3) becomes

$$R(\alpha, z)\,(x - y)^2 = [x(\cos\alpha - \sin\alpha) - y(\cos\alpha + \sin\alpha)]^2,$$

and (5.2a) itself can be written in the form

$$Tq_i = T_{ij}q_j. \tag{5.2b}$$

The transformation (5.2) can also be applied to differential operators as in §§ 2 and 3, so that in the new notation T *is defined to be a symmetry transformation of a Hamiltonian* $\mathscr{H}$ *if*†

$$\boxed{T\mathscr{H}(q_i) = \mathscr{H}(q_i).} \tag{5.4}$$

Representations of a group

Consider the transformations (4.13) operating on the functions

$$x\exp(-r),\ y\exp(-r) \qquad \text{where } r^2 = x^2 + y^2 + z^2.$$

For instance in terms of the notation (5.3) and dropping the prime on A in accordance with § 4, we have

$$\begin{aligned} A\,xe^{-r} &= -\frac{1}{2}\,xe^{-r} + \frac{\sqrt{3}}{2}\,ye^{-r}, \\ A\,ye^{-r} &= -\frac{\sqrt{3}}{2}\,xe^{-r} - \frac{1}{2}\,ye^{-r}. \end{aligned} \tag{5.5}$$

† Sometimes it is convenient to adopt a different notation from (5.4) for a transformed operator P. We write formally

$$T(P\psi) = (TPT^{-1})(T\psi), \tag{5.4†}$$

and define TPT^{-1} to be the transformed operator, which operating on the transformed function $T\psi$, gives the transformed product $T(P\psi)$. This notation will only be used on a few occasions in later sections.

Thus using the transformation A, we have generated the matrix†

$$D_{ij}(A) = \begin{bmatrix} -\frac{1}{2} & -\frac{\sqrt{3}}{2} \\ \frac{\sqrt{3}}{2} & -\frac{1}{2} \end{bmatrix} \tag{5.6}$$

where we have transposed the matrix of the coefficients in accordance with a convention to be explained below. In this way we can generate all the matrices shown in the first row of Table 3. If these matrices are multiplied together, it is seen that they have the property of multiplying according to Table 1. For instance

$$D_{ik}(K)\ D_{kj}(A) = D_{ij}(L),$$

corresponds to the relation

$$KA = L.$$

Because the matrices multiply in this way, they are said to form a *representation* of the group 32. This particular representation we shall call Γ (Table 3). More generally, *the square matrices* $D_{ij}(A)$, $D_{ij}(B)$, . . . *form a representation of a group* $\mathfrak{G}(A, B, \ldots)$ *if the matrix* $D_{ij}(C)$ *associated with any combination* $C = SF$ *is always equal to* $D_{ik}(S)\ D_{kj}(F)$. In terms of the nomenclature of appendix B, we can simply say that matrices of a representation form a group which is homomorphic to $\mathfrak{G}$. If the matrices are of order $n \times n$, the representation is of *dimension* n.

We can also operate with the transformations (4.13) on the function $z \exp(-r)$ and generate the representation $\mathscr{A}$ of 1×1 matrices (Table 3). The fact that the matrices are not all different does not prevent them from forming a representation, for it is easy to verify that they have the essential property, namely, if $SF = C$ then $D(S)\ D(F) = D(C)$ in every case. For instance $KA = L$ and $1 \times (-1) = -1$. The *identity representation* $\mathscr{I}$ (Table 3) is also one-dimensional and consists entirely of ones, and could for instance be generated by operating on the function $(x^2 + y^2) \exp(-r)$ (cf. problem 5.2). Clearly every group has such an identity representation. If the matrices of a representation (such as Γ) are all different, the representation is called faithful: if not all different (such as $\mathscr{A}$ or $\mathscr{I}$), it is unfaithful because we can find spurious relations such as $D(A)\ D(K) = D(L)$ which do not correspond to any relation in Table 1.

† All the matrix algebra and terminology which is assumed in the text is contained in appendix A.

TABLE 3

Irreducible Representations of the Group 32

Representation	Element					
	E	A	B	K	L	M
Γ	$\begin{bmatrix} 1 & \cdot \\ \cdot & 1 \end{bmatrix}$	$\begin{bmatrix} -\frac{1}{2} & -\frac{\sqrt{3}}{2} \\ \frac{\sqrt{3}}{2} & -\frac{1}{2} \end{bmatrix}$	$\begin{bmatrix} -\frac{1}{2} & \frac{\sqrt{3}}{2} \\ -\frac{\sqrt{3}}{2} & -\frac{1}{2} \end{bmatrix}$	$\begin{bmatrix} -1 & \cdot \\ \cdot & 1 \end{bmatrix}$	$\begin{bmatrix} \frac{1}{2} & \frac{\sqrt{3}}{2} \\ \frac{\sqrt{3}}{2} & -\frac{1}{2} \end{bmatrix}$	$\begin{bmatrix} \frac{1}{2} & -\frac{\sqrt{3}}{2} \\ -\frac{\sqrt{3}}{2} & -\frac{1}{2} \end{bmatrix}$
$\mathscr{A}$	1	1	1	−1	−1	−1
$\mathscr{I}$	1	1	1	1	1	1

Although we have introduced representations by transforming some functions, a representation is just a group of matrices with the right multiplication properties. Indeed the representations of Table 3 might have been found by inspection of the multiplication table of the group 32 (Table 1) without considering the group elements as linear transformations and without operating on any functions. Thus in the definition of a representation there is no mention of the functions that might have been used to obtain the representation. Nevertheless, in this book we are exclusively concerned with groups of linear transformations and with their effects on wave functions and other functions, so that in practice representations always will arise in the way they have been introduced, namely, through the transformation properties of some functions. To see this in a general way, consider a linearly independent set of functions $\phi_1, \phi_2, \ldots \phi_n$ and a group $\mathfrak{G}$ of linear transformations of which a typical one is T. If the functions and transformations are such that $T\phi_i$ can in every case be expressed as a linear combination of the set $\phi_1, \phi_2, \ldots \phi_n$, i.e.

$$\boxed{T\phi_j = D_{ij}(T)\phi_i,} \tag{5.7}$$

then we have immediately the set of matrices $D_{ij}(T)$. Furthermore these matrices form a representation of $\mathfrak{G}$, for if $C = SF$, then

$$C\phi_j = SF\phi_j = SD_{kj}(F)\phi_k = D_{kj}(F)S\phi_k = D_{ik}(S)D_{kj}(F)\phi_i$$

so that by (5.7)†

$$\boxed{D_{ij}(C) = D_{ik}(S)D_{kj}(F).} \tag{5.8}$$

The functions $\phi_1, \phi_2, \ldots \phi_n$ *are said to form a basis for the representation* $D_{ij}(T)$. We have said "*a*" *basis* for there are usually other sets of functions that give rise to the same matrices (cf. problems (5.2), (5.4)). However, we have used the words "*the*" *representation* because the matrices are uniquely determined. An alternative wording is to say that *the functions* $\phi_1\ \phi_2, \ldots \phi_n$ *transform according to the representation* $D_{ij}(T)$ *of* $\mathfrak{G}$. Thus in our first example (equations

† Although it might at first sight appear more natural to define the matrices of a representation by $T\phi_i = D_{ij}(T)\phi_j$ instead of (5.7), this would lead to the relation $D_{ij}(C) = D_{ik}(F)D_{kj}(S)$ which is the opposite of our convention (4.2) for the multiplication of group elements. This explains the necessity for using the transposed matrix of coefficients in (5.5), (5.6) and in (5.7). It may be helpful to write (5.7) in the form $T\phi_j = \phi_i D_{ij}(T)$, the set of functions ϕ_i behaving as a row matrix.

(5.5), (5.6)) $x \exp(-r)$ and $y \exp(-r)$ form a basis for the representation Γ (Table 3) of the group 32, or in other words they transform according to the representation Γ of the group 32. It is of course only special sets of functions that have the property (5.7) and hence can form a basis for a group representation. For instance the functions $\cos x$, $\sin x$, $\cos y$, $\sin y$ do not form a basis for a representation of the group 32, nor do $x^2 \exp(-r)$, $y^2 \exp(-r)$. The letters D (German, Darstellung) and Γ are most commonly used for representations.

Problems 5.1 to 5.7 are based on the preceding concepts.

Vector spaces

Suppose now that $x \exp(-r)$ and $y \exp(-r)$ are eigenfunctions of a Hamiltonian belonging to the same eigenvalue. If there are no further linearly independent eigenfunctions, the level is said to be two-fold degenerate. Nevertheless, there are an infinite number of different eigenfunctions belonging to this level because any linear combination

$$c_1 x e^{-r} + c_2 y e^{-r} \tag{5.9}$$

where c_1, c_2 are complex numbers is also an eigenfunction. All such functions together are said to form the *vector space* $[x \exp(-r), y \exp(-r)]$, and it is often more convenient to consider the whole vector space rather than a particular pair of functions as being associated with the energy level. More generally, *the vector space*† $\mathfrak{R}\ (\phi_1, \phi_2, \ldots \phi_n)$ *consists of all functions*

$$\boxed{c_1\phi_1 + c_2\phi_2 + \ldots + c_n\phi_n} \tag{5.10}$$

which are linear combinations of the $\phi_1, \phi_2, \ldots \phi_n$ *with complex coefficients.* The functions $\phi_1, \phi_2, \ldots \phi_n$ are the *base vectors* in the vector space, and they are also said to *span* the space $\mathfrak{R}$ to denote that any function belonging to $\mathfrak{R}$ can be expressed as a linear combination of the ϕ_j. A function (5.10) belonging to $\mathfrak{R}$ is a *vector* in or of the space $\mathfrak{R}$. Now any particular vector space can be referred to *different sets of base vectors*. For consider all the functions of the form (5.9). They could equally well be expressed in terms of different base vectors $(x \pm iy) \exp(-r)$ in the form

$$c'_1(x + iy)e^{-r} + c'_2(x - iy)e^{-r} \tag{5.11}$$

where $c'_1 = \frac{1}{2}(c_1 - ic_2)$, $c'_2 = \frac{1}{2}(c_1 + ic_2)$. Thus the two vector

† German, Raum.

spaces (5.9), (5.11) are equal and we can write

$$(xe^{-r}, ye^{-r}) = [(x + iy)e^{-r}, (x - iy)e^{-r}].$$

Similarly by choosing other linear combinations, the base vectors for this vector space can be chosen in an unlimited number of ways, and the same applies to the general vector space (5.10).

There is an important property of a vector space which is independent of the way the base vectors are chosen, and this is its dimension. Consider again the set of functions $\phi_1, \phi_2, \ldots \phi_n$. They are said to be *linearly independent* if it is impossible to find complex numbers $\alpha_1, \alpha_2, \ldots \alpha_n$ such that

$$\alpha_1\phi_1 + \alpha_2\phi_2 + \ldots \alpha_n\phi_n = 0, \tag{5.12}$$

i.e. if it is impossible to express one of the functions as a linear combination of the others. Thus $x \exp(-r)$, $y \exp(-r)$ are linearly independent functions, but $x \exp(-r)$, $y \exp(-r)$, $(x + iy) \exp(-r)$ are not because we have the relation

$$-xe^{-r} - iye^{-r} + (x + iy)e^{-r} = 0.$$

In fact it can easily be shown (problem 5.11) that it is always necessary to have precisely two linearly independent base vectors to span the whole space $[x \exp(-r), y \exp(-r)]$. More generally it can be proved (appendix C, Theorem 1) that the number n of linearly independent base vectors required to span the whole of a given vector space is the same no matter how the base vectors are chosen, so that n is a well-defined property of the vector space, and is called its dimension. Thus we have the definition, *the dimension of a vector space is the number of linearly independent base vectors required to span the space.* In the future we shall assume that the base vectors of a space have been chosen linearly independently unless the contrary is indicated. We shall also assume for convenience that they are orthogonal to one another and normalized. This involves no loss of generality. For suppose we are given a set of non-orthogonal base vectors $\phi_1, \phi_2, \ldots \phi_n$. We can construct an orthogonal set $\phi'_1, \phi'_2, \ldots \phi'_n$ as follows. Put $\phi'_1 = \phi_1$, then $\phi'_2 = \phi_2 + a\phi'_1$ and determine a such that ϕ'_2 is orthogonal to ϕ'_1. Then put $\phi'_3 = \phi_3 + b_1\phi'_1 + b_2\phi'_2$ with b_1 and b_2 such that ϕ'_3 is orthogonal to ϕ'_1 and ϕ'_2, etc. The functions $\phi'_1, \phi'_2, \ldots \phi'_n$ can then also be normalized.

Problems 5.8 to 5.13 are based on the above concepts.

Transformations in a vector space

We shall be interested almost exclusively in vector spaces $\mathfrak{R}$ $(\phi_1, \phi_2, \ldots \phi_n)$ in which the base vectors ϕ_i have the property that for every transformation T of a group $\mathfrak{G}$, $T\phi_j$ is expressible as a linear combination of the ϕ_i, i.e.

$$\boxed{T\phi_j = D_{ij}(T)\phi_i.} \tag{5.7}$$

More concisely, $T\phi_j$ always belongs to $\mathfrak{R}$. Then this property holds not only for the base vectors ϕ_i but for any vector $\phi = c_j\phi_j$, for

$$T\phi = c_jT\phi_j = c_jD_{ij}(T)\phi_i$$

and thus $T\phi$ belongs to $\mathfrak{R}$. Thus any transformation of $\mathfrak{G}$ turns every vector of $\mathfrak{R}$ into another vector also belonging to $\mathfrak{R}$, and we say that *the space $\mathfrak{R}$ is invariant under the group $\mathfrak{G}$ of transformations.* If $\mathfrak{R}$ is a one dimensional space (ϕ_1), the vector ϕ_1 is said to be an invariant vector. This implies that $T\phi_1 = c(T)\phi_1$ but the constant $c(T)$ is not necessarily unity.

There is another way of looking at transformations of functions in the vector space $\mathfrak{R}$ which we shall mention briefly but have no occasion to use except in §§ 20 and 32. Consider any function $\phi = c_j\phi_j$, and let $T\phi_j = \phi'_j$ given by (5.7). Now the ϕ'_j can be used as new base vectors in the space, and the untransformed function ϕ can be expressed as a linear combination of them with coefficients c'_j, i.e.

$$\phi = c_j\phi_j = c'_j\phi'_j. \tag{5.13}$$

The question is, how are the coefficients c'_j related to the $c_j\phi$? From (5.13)

$$c'_j\phi'_j = c'_jD_{ij}(T)\phi_i = c_i\phi_i,$$

whence

$$c_i = D_{ij}(T)c'_j$$

or

$$c'_i = [D(T)]_{ij}^{-1}c_j, \tag{5.14}$$

where $[D(T)]^{-1}$ is the reciprocal of the matrix $D(T)$. The difference between (5.7) and (5.14) is expressed by saying that the ϕ_i transform as base vectors and the c_i as coefficients according to the representation $D_{ij}(T)$.

Equivalence

We have seen that the functions $x\exp(-r)$, $y\exp(-r)$ form a basis for the representation Γ (Table 3) of the group 32. If now in the vector space $[x\exp(-r), y\exp(-r)]$ we choose different base vectors, say $(x \pm iy)\exp(-r)$, then we can use the matrices of Γ to determine the representation according to which the new base vectors transform. For instance from (5.5)

$$A(x + iy)e^{-r} = Axe^{-r} + iAye^{-r}$$
$$= \left(-\tfrac{1}{2} - i\frac{\sqrt{3}}{2}\right)(x + iy)e^{-r},$$

and in this way the matrices of the new representation can be written down. More generally, consider a vector space $\mathfrak{R}$ which is invariant under a group $\mathfrak{G}$ of transformations, so that the base vectors $\phi_1, \phi_2, \ldots \phi_n$ of $\mathfrak{R}$ form a basis for the representation $D_{ij}(T)$ according to (5.7). The question now arises, if we choose new base vectors ϕ'_i, say, what representation do they transform according to? The new base vectors can be expressed in terms of the ϕ_i,

$$\phi'_j = P_{ij}\phi_i,$$

since they belong to $\mathfrak{R}$. Also

$$\phi_j = P_{ij}^{-1}\phi'_i.$$

Then

$$T\phi'_j = TP_{lj}\phi_l = P_{lj}D_{kl}(T)\phi_k = P_{ik}^{-1}D_{kl}(T)P_{lj}\phi'_i.$$

Thus the ϕ'_i transform according to the representation $D'_{ij}(T)$ where in matrix form

$$\boxed{D'(T) = P^{-1}D(T)P,} \tag{5.15}$$

and this representation is said to be *equivalent* to the representation $D_{ij}(T)$. Definition: *two representations* $D_{ij}(T)$ *and* $D'_{ij}(T)$ *of a group are equivalent if a matrix* P *exists that relates the matrices of the two representations in the manner of* (5.15). We shall refer to (5.15) as the equivalence relation or as an equivalence transformation.

The equivalent representations $D_{ij}(T)$ and $D'_{ij}(T)$ are in fact so closely related that we shall usually not wish to distinguish between them. It is then convenient to use the expression *the representation* D to mean any or all of the representations that are equivalent to $D_{ij}(T)$. In fact if two representations $D(T)$ and $D'(T)$ are *equivalent*, we write

$$D = D'.$$

Thus we can say that the vector space $\mathfrak{R}$ transforms according to the representation D of $\mathfrak{R}$ without having to specify a particular set of base vectors or the actual representation. On the other hand when we want to discuss a particular set of base vectors $\phi_1, \phi_2, \ldots \phi_n$, we shall say that they transform according to *the particular representation* $D_{ij}(T)$. If the matrices of two particular representations $D(T)$ and $D'(T)$ are equal, we shall denote this by writing

$$D_{ij}(T) = D'_{ij}(T),$$

and shall describe the two representations as *identical*. Thus although we have so far used Γ to denote the particular representation given in Table 3, we may in the future also use it to refer to any representation equivalent to this one. It will also be convenient to use "different representations" to mean different non-equivalent representations, unless it is clearly specified that different particular equivalent representations are meant.

Reducibility of a representation

Consider the vector space $\mathfrak{R}_0$ $[x \exp(-r), y \exp(-r), z \exp(-r)]$ which is invariant under the group of transformations 32, so that the base vectors $x \exp(-r)$, $y \exp(-r)$, $z \exp(-r)$ form the basis of a representation Δ, say. Here Δ is a combination of Γ and $\mathscr{A}$ (Table 3), the matrix corresponding to a transformation T being

$$\Delta(T) = \left[\begin{array}{cc|c} \multicolumn{2}{c|}{\multirow{2}{*}{$\Gamma_{ij}(T)$}} & 0 \\ & & 0 \\ \hline 0 & 0 & \mathscr{A}(T) \end{array}\right]. \tag{5.16}$$

Such a representation is said to be reducible, in particular Δ is reducible into Γ and $\mathscr{A}$, which is written

$$\Delta = \Gamma + \mathscr{A}.$$

If in the vector space $\mathfrak{R}_0$ we had used the base vectors $(x + y) \exp(-r)$, $(y + z) \exp(-r)$, $(z + x) \exp(-r)$, the matrices of the corresponding representation Δ', say, would not have the simple reduced form (5.16), but we still say that Δ' is reducible into $\mathscr{A}$ and Γ and we write $\Delta' = \Gamma + \mathscr{A}$, because the form (5.16) can easily be achieved by choosing different base vectors, i.e. by applying a transformation of the type (5.15) to the representation. In general, *a representation* $D_{ij}(T)$ *of a group* $\mathfrak{G}$ *is reducible into the representations* $D^{(1)}, D^{(2)}, \ldots D^{(s)}$ *if a transformation of the type* (5.15) *exists which brings every matrix* $D_{ij}(T)$ *of the representation into the form*

$$\begin{bmatrix} D_{ij}^{(1)}(T) & 0 & 0 & 0 & \ldots & 0 \\ 0 & D_{ij}^{(2)}(T) & 0 & 0 & \ldots & 0 \\ 0 & 0 & D_{ij}^{(3)}(T) & 0 & \ldots & 0 \\ 0 & 0 & 0 & D_{ij}^{(4)}(T) & & 0 \\ \vdots & \vdots & \vdots & & & \\ 0 & 0 & 0 & 0 & & \end{bmatrix} \quad (5.17$$

with smaller matrices $D^{(i)}(T)$ *along the diagonal and zeros elsewhere.* This is written

$$\boxed{D = D^{(1)} + D^{(2)} + D^{(3)} + \ldots + D^{(s)},} \qquad (5.18)$$

and D is said to *contain* the representations $D^{(i)}$ or the $D^{(i)}$ *occur in* D. In (5.18) it is quite possible for some of the components to be equivalent or equal to one another, i.e. to occur in D more than once. For instance if we apply the group 32 of transformations (4.13) to the two sets of co-ordinates x_1, y_1, z_1 and x_2, y_2, z_2 simultaneously, then the functions

$$x_1 \exp(-r_1),\ x_2 \exp(-r_2),\ y_1 \exp(-r_1),\ y_2 \exp(-r_2)$$

transform according to a representation which is clearly equivalent to $\Gamma + \Gamma$.

A representation that cannot be reduced is termed irreducible. For instance any one-dimensional representation is automatically irreducible, and we shall prove at the end of this section that the representation Γ (Table 3) is irreducible. In fact the representations of Table 3 (and those equivalent to them) are the only possible irreducible representations of the point group 32 (cf. § 14). Thus every representation of the group 32 can be reduced into a sum of the representations Γ, $\mathscr{A}$ and $\mathscr{I}$, an irreducible representation being regarded as already in reduced form. It can also be shown quite generally (appendix C, Theorem 3) that if a representation D is reduced into irreducible components according to (5.18), the reduction is unique apart from equivalence; i.e. which $D^{(i)}$ occur in D and how many times each one occurs is unique, but the order in which the matrices $D_{ij}{}^{(i)}(T)$ are arranged in (5.17) and the particular elements $D_{ij}{}^{(i)}(T)$ are not unique.

Reducibility of a vector space

So far we have discussed what the reducibility of a representation means purely in terms of the matrices of the representation. Consider a vector space $\mathfrak{R}$ $(\phi_1, \phi_2, \ldots \phi_n)$ of which the base vectors ϕ_i transform according to the reducible representation $D_{ij}(T)$ of (5.18). *A vector space transforming according to a reducible* (or *irreducible*) *representation is called a reducible* (or *irreducible*) *vector space*, and we shall now discuss what the reducibility of the space $\mathfrak{R}$ implies. Since the representation $D_{ij}(T)$ is reducible, there exists by definition some transformation of the type (5.15) which brings the matrices of D into the reduced form (5.17). Such a transformation is equivalent to choosing new base vectors ϕ'_i in $\mathfrak{R}$, and these ϕ'_i therefore transform according to the matrices (5.17). Let $\mathfrak{R}^{(1)}$ be the space spanned by the first n_1 vectors ϕ_i where n_1 is the dimension of the representation $D^{(1)}$. From the form of (5.17), $\mathfrak{R}^{(1)}$ transforms according to the representation $D^{(1)}$, and similarly successive sets of vectors span spaces $\mathfrak{R}^{(i)}$ transforming according to the representations $D^{(i)}$. Thus it follows directly from the reducibility of the representation $D_{ij}(T)$ that the vector space $\mathfrak{R}$ can be split up or reduced into a series of invariant subspaces $\mathfrak{R}^{(i)}$ transforming according to the representations $D^{(i)}$. Expressed more descriptively, the transformations of $\mathfrak{G}$ transform the functions of each subspace $\mathfrak{R}^{(i)}$ into one another without mixing together the functions belonging to different subspaces. The importance of this arises very briefly as follows. In § 3 we saw that symmetry transformations change the eigenfunctions belonging to the same eigenvalue into one another. If now we consider all the eigenfunctions of a Hamiltonian as one big vector space $\mathfrak{R}$, then the functions belonging to different invariant subspaces $\mathfrak{R}^{(i)}$ are in general associated with different energy levels, as we shall discuss in detail in the next section.

Since the ϕ'_i span the whole space $\mathfrak{R}$, *any function in the space $\mathfrak{R}$ can be written as the sum of functions, one from each subspace $\mathfrak{R}^{(i)}$. This is written in symbols*

$$\boxed{\mathfrak{R} = \mathfrak{R}^{(1)} + \mathfrak{R}^{(2)} + \ldots + \mathfrak{R}^{(s)}.} \tag{5.19}$$

It can also be shown (appendix C, Theorem 2) that the reduction of $\mathfrak{R}$ has the further important property that the subspaces $\mathfrak{R}^{(i)}$ can always be made orthogonal to one another, i.e. any function from one $\mathfrak{R}^{(i)}$ is orthogonal to any function from any other $\mathfrak{R}^{(i)}$. Physically this corresponds to the fact that eigenfunctions belonging to different energy levels of a Hamiltonian are always orthogonal. It is also possible to establish the criterion (appendix C, Theorem 2) that if a space $\mathfrak{R}$ contains a subspace, then $\mathfrak{R}$ must be properly reducible.

The various points above are illustrated by the following example of a reducible vector space. Consider the space $\mathfrak{R}_0(x^2, y^2, z^2, yz, zx, xy)$ where to preserve the symmetry the base vectors have been neither orthogonalized nor normalized. This space is invariant under the group 32 of transformations (4.13) since $\mathfrak{R}_0$ consists of all functions of the second degree, so that the base vectors x^2, y^2, . . . xy transform according to some representation $D_{ij}(T)$, say. For instance the matrix $D_{ij}(A)$ is

$$\begin{bmatrix} \frac{1}{4} & \frac{3}{4} & \cdot & \cdot & \cdot & \frac{1}{4}\sqrt{3} \\ \frac{3}{4} & \frac{1}{4} & \cdot & \cdot & \cdot & -\frac{1}{4}\sqrt{3} \\ \cdot & \cdot & 1 & \cdot & \cdot & \cdot \\ \cdot & \cdot & \cdot & -\frac{1}{2} & \frac{1}{2}\sqrt{3} & \cdot \\ \cdot & \cdot & & -\frac{1}{2}\sqrt{3} & -\frac{1}{2} & \cdot \\ -\frac{1}{2}\sqrt{3} & \frac{1}{2}\sqrt{3} & \cdot & \cdot & \cdot & -\frac{1}{2} \end{bmatrix}.$$

This does not have the reduced form (5.17) and neither do the other matrices. However, the space $\mathfrak{R}_0$ must be reducible because by inspection it contains an invariant subspace, namely the vector $x^2 + y^2 + z^2$ transforming according to the identity representations $\mathscr{I}$ (Table 3). In fact it can easily be verified that the space is reduced by choosing the new base vectors

$$\begin{aligned} &\phi'_1 = 2xy \qquad\qquad \phi'_2 = x^2 - y^2 \\ &\phi'_3 = 2yz \qquad\qquad \phi'_4 = -2zx \\ &\phi'_5 = (2/15^{1/2})\,(x^2 + y^2 + z^2) \\ &\phi'_6 = (1/3^{1/2})\,[3z^2 - (x^2 + y^2 + z^2], \end{aligned} \tag{5.20}$$

which are all orthogonalized and normalized to $16\pi/15$ as regards the angular variables. For instance any transformation of the group 32 operating on a linear combination

$$c_1\phi'_1 + c_2\phi'_2$$

gives another linear combination of the same form without introducing any components involving the other base vectors. Thus

ϕ'_1, ϕ'_2 span an invariant subspace, and it may easily be verified that they transform according to the irreducible representation Γ (Table 3). Similarly ϕ'_3, ϕ'_4 transform according to Γ, ϕ'_5 according to $\mathscr{I}$, and ϕ'_6 according to $\mathscr{I}$. The whole set $\phi'_1, \ldots \phi'_6$ therefore transforms according to a representation with matrices $D'_{ij}(T)$ of the form

$$\begin{bmatrix} \multicolumn{2}{c}{\Gamma_{ij}(T)} & \cdot & \cdot & \cdot & \cdot \\ & & \cdot & \cdot & \cdot & \cdot \\ \cdot & \cdot & \multicolumn{2}{c}{\Gamma_{ij}(T)} & \cdot & \cdot \\ \cdot & \cdot & & & \cdot & \cdot \\ \cdot & \cdot & \cdot & \cdot & \mathscr{I}(T) & \cdot \\ \cdot & \cdot & \cdot & \cdot & \cdot & \mathscr{I}(T) \end{bmatrix}.$$

Thus we have

$$D = \Gamma + \Gamma + \mathscr{I} + \mathscr{I}, \tag{5.21}$$

and $$\mathfrak{R} = (\phi'_1, \phi'_2) + (\phi'_3, \phi'_4) + (\phi'_5) + (\phi'_6).$$

Irreducibility of Γ

We have so far discussed some of the properties of reducible and irreducible representations and vector spaces, without indicating how in practice one might reduce a given representation or prove it to be irreducible as the case may be. We shall see in § 14 that there are very elegant general methods for doing this by setting up all the different (i.e. non-equivalent) irreducible representations of a group, and testing whether each of them is contained in the given representation. However, as an illustration of the concepts of this section, we shall give an *ad hoc* proof from first principles that the representation Γ of Table 3 is irreducible.

Let ϕ_1, ϕ_2 be two vectors transforming according to the particular representation $\Gamma_{ij}(T)$ of Table 3. If Γ were reducible, it would have to be reducible into two one-dimensional representations, so that it would be possible to find two vectors in the space (ϕ_1, ϕ_2) each of which is invariant under the group 32. It is therefore sufficient to prove that there is no vector

$$\phi = c_1\phi_1 + c_2\phi_2 \tag{5.22}$$

which is invariant under all transformations of the group.

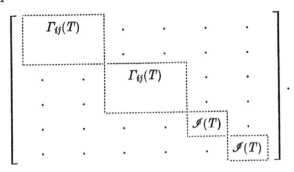

Since ϕ_1, ϕ_2 transform according to the representation $\Gamma_{ij}(T)$, we have from Table 3

$$K\phi_1 = -\phi_1, \qquad K\phi_2 = \phi_2. \tag{5.23}$$

If now we have a relation of the form

$$c_1\phi_1 + c_2\phi_2 = c'_1\phi_1 + c'_2\phi_2,$$

then this implies $c_1 = c'_1$, $c_2 = c'_2$; as otherwise we would have

$$\phi_1 = \frac{c_2 - c'_2}{c'_1 - c_1}\phi_2,$$

which substituted into (5.23) clearly leads to a contradiction.

Let us now suppose that there exists a vector ϕ (5.22) which is invariant under all transformations of the group 32. This means that

$$K\phi = \gamma\phi$$

where γ is some constant, i.e. from (5.23)

$$K(c_1\phi_1 + c_2\phi_2) = \gamma(c_1\phi_1 + c_2\phi_2) = -c_1\phi_1 + c_2\phi_2.$$

Therefore $\qquad -c_1 = \gamma c_1 \qquad$ and $\qquad c_2 = \gamma c_2,$

whence either $\qquad \gamma = -1, \qquad c_1 \neq 0, \qquad c_2 = 0,$

or $\qquad \gamma = 1, \qquad c_1 = 0, \qquad c_2 \neq 0:$

i.e. the invariant vector has to be either ϕ_1 or ϕ_2. However, from the matrix $\Gamma_{ij}(A)$ we see that neither ϕ_1 nor ϕ_2 is invariant under the transformation A, which gives a contradiction. Hence no invariant vector exists and the representation Γ is irreducible.

References

In appendix C, the theorems about representations and vector spaces which have been quoted in this section are proved rigorously. However, in keeping the proofs as elementary as possible, much of the beauty of the theory has been lost, and the reader is referred to Van der Waerden (1932) for an elegant treatment, as well as the other books mentioned in the list of general references immediately preceding the bibliography at the end of the book. Birkhoff and MacLane (1941) give an elementary discussion of vector spaces which covers some aspects of the present section in more detail.

Summary

We have defined what is meant by a representation of a group and by a vector space. We have shown how a vector space which is invariant under a group of transformations forms a basis for a representation of the group. Using different sets of base vectors in the space gives equivalent representations. Vector spaces and the corresponding representations may be reducible or irreducible. A reducible vector space can be split up into a sum of irreducible, mutually orthogonal subspaces.

This practically completes the number of new mathematical concepts required in this book.

Problems

Note: Throughout these problems the group 32 is the group of transformations (4.13), $\mathfrak{P}_3$ is the group of permutation transformations (4.22) of order 3, and $\mathfrak{G}$ is considered to be a general group of linear transformations.

5.1 Write the answers to problem 2.1 in terms of the new definition of transforming a function and the notation of equation (5.3).

5.2 Verify in detail that the functions $x \exp(-r)$, $y \exp(-r)$ transform according to the particular representation Γ of Table 3 under the group 32, and that $z \exp(-r)$ and $(x^2 + y^2) \exp(-r)$ transform respectively according to the representations $\mathscr{A}$ and $\mathscr{I}$. Write down some other functions transforming according to $\mathscr{A}$ and $\mathscr{I}$. Verify that the matrices of each of these representations multiply according to the group multiplication table (Table 1).

5.3 Show that the functions $x + y$, $x - y$ transform according to a representation of the group 32. Write down explicitly some of the matrices of this representation, and verify that they multiply according to the group multiplication table.

5.4 Show that $2xy$ and $x^2 - y^2$ transform according to the representation Γ (Table 3) of the group 32. Show that yz and $-zx$ do likewise. ϕ_1 and ϕ_2 transform according to Γ: if $\phi_2 = xyz$, what is ϕ_1?

5.5 Show that the functions x^2, y^2, z^2 do not form a basis for a representation of the group 32.

5.6 Show that the matrices of each of the representations of Table 3 form a representation of the group $\mathfrak{P}_3$, and identify the group element which each matrix represents. In this case the representations $\mathscr{I}$ and $\mathscr{A}$ are called respectively the symmetric and antisymmetric representations. More generally if the groups $\mathfrak{G}_1$ and $\mathfrak{G}_2$ are isomorphic with one another, show that a representation of the one group automatically forms a representation of the other.

5.7 Show that the functions $x_1x_2y_3 - x_1y_2x_3$ and $x_1x_2y_3 - y_1x_2x_3$ transform according to a representation of the group $\mathfrak{P}_3$ and write down explicitly some of the matrices of the representation.

5.8 Write down five different pairs of functions that span the same space [$x \exp(-r)$, $y \exp(-r)$], the two functions of each pair being orthogonal and normalized to the same value.

5.9 Show that the two functions of problem 5.7 are not orthogonal, and write down two functions spanning the same space that are orthogonal.

5.10 Show that the vector space ($x^2 - y^2$, $y^2 - z^2$, $z^2 - x^2$) has only the dimension two and not three.

5.11 Consider the vector space [$x \exp(-r)$, $y \exp(-r)$] and show from first principles

(i) it is impossible to span the whole vector space with one base vector;

(ii) any two vectors of the space always span the whole space provided they are not linearly dependent, i.e in this case they are not simple multiples of one another;

(iii) three vectors in the space are always linearly dependent.

5.12 Prove that the maximum number of linearly independent vectors that can be found in a vector space, is equal to the minimum number of base vectors required to span the whole space. (Both are equal to the dimension of the space).

5.13* Show that all functions of the form (5.10) form a group, the law of combination being addition. What is the unit element? Is the group Abelian? Show that viewing a vector space as a group in this way is a very powerful concept in setting up the mathematical theory of group representations. (Van der Waerden 1932, Chapter II.)

5.14 Using some of the functions of the preceding problems, write down some vector spaces that are invariant under the group 32 and that have dimensions from one up to five.

5.15 The pairs of functions $x \exp(-r)$, $y \exp(-r)$ and $(x \pm iy) \exp(-r)$ transform respectively according to the particular representations $\Gamma_{ij}(T)$ and $\Gamma'_{ij}(T)$ under the group 32. What is the matrix P which in the manner of equation (5.15) relates the matrices $\Gamma_{ij}(T)$ and $\Gamma'_{ij}(T)$? Verify equation (5.15) for some of the matrices $\Gamma_{ij}(T)$ and $\Gamma'_{ij}(T)$.

5.16 $D_{ij}(T)$ and $D'_{ij}(T)$ are the matrices of two equivalent representations of a group, defined by equation (5.15). Show that $D_{ii}(T) = D'_{jj}(T)$ (where the double suffix is summed on each side), and verify this relation using the two representations of problem 5.15. Verify also that if $D(S)\ D(F) = D(C)$, then $D'(S)\ D'(F) = D'(C)$, so that $D'(T)$ is really a representation of the group.

5.17 The vector space ($y_1x_2x_3$, $x_1y_2x_3$, $x_1x_2y_3$) transforms under the group $\mathfrak{P}_3$ according to the representation D. Express in your own words what is meant by the equation $D = \Gamma + \mathscr{I}$. Hint: the functions $x_1x_2y_3 - x_1y_2x_3$ and $-\sqrt{(1/3)}(x_1x_2y_3 + x_1y_2x_3 - 2y_1x_2x_3)$ transform according to Γ. See also problem 5.6.

5.18 Show that the vector space $(x^4, x^3y, x^2y^2, xy^3, y^4)$ is reducible under the group 32 by finding a one-dimensional subspace in it.

5.19 $\mathfrak{R}$ is the vector space of equation (5.19). Show that it is *not* true to say that any function of $\mathfrak{R}$ is equal to some function of $\mathfrak{R}^{(1)}$ *or* some function of $\mathfrak{R}^{(2)}$ *or* . . . etc.

5.20 Show that the permutation group $\mathfrak{P}_2$ has two one-dimensional representations, the symmetric one in which both elements are represented by $+1$, and the antisymmetric in which the elements (12) and (21) are represented by $+1$ and -1 respectively. Write down some functions that transform according to these representations, and prove that no other irreducible representations of $\mathfrak{P}_2$ exist.

5.21 Show that the identity transformation E and the inversion (3.11) form a group which is isomorphic with $\mathfrak{P}_2$ and hence that the symmetric and antisymmetric representations of problem 5.20 are the only irreducible ones. In the present case the irreducible representations are called the even and odd ones, and corresponding functions are said to have even or odd parity.

5.22 By considering functions of the form $(x + iy)^n \exp(-r)$, deduce six different one-dimensional representations of the point-group 6 (§ 16). This group consists of rotations by $2\pi r/6$ radians, $r = 0, 1, 2, 3, 4, 5$.

5.23 Show that the vector space $[x \exp(-r),\ y \exp(-r),\ z \exp(-r)]$ is invariant under the full rotation group, and that it transforms according to an irreducible representation of the group. Hint: if the space were reducible, it would contain an invariant vector. Show that this is impossible by considering rotations of 90° and 180° about the x-, y-, z-axes.

6. Application to Quantum Mechanics

In the last section, we introduced the concepts of representations and vector spaces, and we shall now establish what their relevance is to quantum mechanics. We have already seen in §§3 and 5 (e.g. equations (3.7) to (3.9)) that the eigenfunctions of a Hamiltonian belonging to one energy level are transformed into one another by symmetry transformations of the Hamiltonian, and this suggests that different invariant vector spaces are associated with different energy levels. We shall now use the concepts and results of the last section to embody this type of argument in three precise theorems.

THEOREM 1. *If a Hamiltonian is invariant under a group $\mathfrak{G}$ of symmetry transformations, then the eigenfunctions belonging to one energy level form a basis for a representation of $\mathfrak{G}$.*

Let q_i, $i = 1, 2, \ldots$ be the co-ordinates in which the Hamiltonian $\mathscr{H}$ is expressed, and let $\psi(q_i)$ be an eigenfunction belonging to the energy level E.

$$\mathscr{H}\psi(q_i) = E\psi(q_i). \tag{6.1}$$

If T is any symmetry transformation of $\mathfrak{G}$, then

$$T\mathscr{H} = \mathscr{H}. \tag{5.4}$$

Also let

$$T\psi(q_i) = \psi'(q_i).$$

Then operating with T on both sides of equation (6.1) we obtain

$$\boxed{\mathscr{H}\psi'(q_i) = E\psi'(q_i),}$$

so that $\psi'(q_i)$ is an eigenfunction belonging to the same energy level E as $\psi(q_i)$. Similarly, any eigenfunction belonging to E is transformed into another eigenfunction by the transformations of $\mathfrak{G}$. Also if ψ_1 and ψ_2 are eigenfunctions belonging to E, then so is $c_1\psi_1 + c_2\psi_2$. Hence all eigenfunctions belonging to one energy level form a vector space which is invariant under $\mathfrak{G}$ and which therefore forms a basis for a representation of $\mathfrak{G}$. This proves the theorem.

The importance of this theorem lies in the fact that we can label and describe an energy level and its eigenfunctions simply by naming the representation associated with it. This clearly does not tell us everything we may wish to know about the eigenfunctions such as detailed numerical tabulations, but it does indicate their symmetry properties which is frequently all that is of interest in establishing selection rules for transitions and other qualitative behaviour. For example, consider the Hamiltonian for a lithium atom with three electrons, including the spin dependent terms which we need not write down in detail. Since all electrons are alike, this Hamiltonian is invariant under the group $\mathfrak{P}_3$ of permutations (4.22) of the electron co-ordinates. The group $\mathfrak{P}_3$ was shown in § 4 to be isomorphic with the point-group 32, and therefore any representation of the group 32 is automatically also a representation of $\mathfrak{P}_3$. Thus $\mathfrak{P}_3$ has the three different (i.e. non-equivalent) irreducible representations Γ, $\mathscr{I}$ and $\mathscr{A}$ of Table 3 (cf. problem 5.6). From Theorem 1 the eigenfunctions associated with one energy level form an invariant vector space under the group $\mathfrak{P}_3$, and this vector space can be reduced into subspaces each of which transforms according to one of the representations Γ, $\mathscr{I}$, $\mathscr{A}$. A wave function transforming according to the representation $\mathscr{A}$ is antisymmetric in the usual quantum mechanical sense, i.e. it changes sign if we interchange the coordinates of any two electrons (cf. problem 6.1). Now it is known experimentally that these antisymmetric states corresponding to the representations $\mathscr{A}$ are the only ones ever found in nature, so that this sorting out of the wave functions according to the irreducible representations of the group $\mathfrak{P}_3$ is very important.

As another example consider the Hamiltonian (3.2) of a free atom. This is invariant under the group of transformations consisting of the identity transformation and the space-inversion Π (3.11). This group has only two one-dimensional irreducible representations, such that the corresponding function is multiplied by $+1$ or -1 by the inversion transformation, in which case the function is described as having even or odd parity respectively (cf. problem 5.21). As in the preceding example, the eigenfunctions associated with each energy level of an atom can be reduced until each eigenfunction has either even or odd parity. This, for instance, leads to the selection rule for optical transitions that the initial and final states must have opposite parity. Similarly, we shall see later that angular momentum quantum numbers L, S, J, m_L, m_S, m_J are simply labels for the irreducible representations of the rotation group.

Corollary to Theorem 1. *If a Hamiltonian is invariant under a group* $\mathfrak{G}$ *of transformations, then eigenfunctions of the Hamiltonian transforming according to one irreducible representation of* $\mathfrak{G}$ *belong to the same energy level.*

From Theorem 1, the vector space of eigenfunctions belonging to one level is either irreducible or can be reduced into subspaces each of which transforms according to an irreducible representation of $\mathfrak{G}$. It thus never happens that eigenfunctions belonging to the same irreducible vector space belong to different energy levels, which proves the corollary.

As a trivial example, consider the $2p$ wave functions of a hydrogen atom

$$\psi_x = xf(r),\ \psi_y = yf(r),\ \psi_z = zf(r), \tag{6.2}$$

where $f(r)$ is some function of r (Schiff 1955, p. 85). The Hamiltonian of a hydrogen atom is invariant under all rotations, and it can be shown that the functions (6.2) transform according to an irreducible representation of the rotation group (§ 8 or problem 5.23). They therefore belong to the same energy level.

The corollary tells us that eigenfunctions belonging to the same irreducible vector space necessarily belong to the same energy level. The question naturally arises, do eigenfunctions belonging to different irreducible vector spaces always belong to different energy levels? In general the answer is that they do not have to belong to different energy levels. However, if we find several irreducible vector spaces to be associated with each energy level in a systematic way, there must be some symmetry property that produces this degeneracy. Hence, if we have included *all* possible symmetry transformations in the group $\mathfrak{G}$, we would expect different irreducible vector spaces of eigenfunctions to have different energy, simply on the grounds that there is no symmetry property remaining to make them have the same energy. This conclusion is borne out by experience, though occasionally it has not been easy to discover all the symmetry transformations of a Hamiltonian, as already mentioned in § 4. Nevertheless, a few accidental degeneracies may still remain. For instance in a magnetic field two energy levels corresponding to different irreducible representations can cross one another as the field is varied, so that for a particular value of the field they are degenerate. Accidental degeneracies also occur among the energy levels in a crystal (problem 26.8). We can systematize our conclusion by defining an *accidental* degeneracy to be a degeneracy which is not brought about in the manner of Theorem 1 by any symmetry property of the Hamiltonian. With this definition we then have:

THEOREM 2. *If the group* $\mathfrak{G}$ *includes* all *possible symmetry transformations of the Hamiltonian, then the eigenfunctions of each energy level transform* irreducibly *under* $\mathfrak{G}$, *apart from accidental degeneracy.*

Proof: the eigenfunctions of one irreducible vector space are certainly degenerate by the corollary to Theorem 1. Furthermore they are transformed purely among themselves by $\mathfrak{G}$, and are not linked with any outside eigenfunction by any symmetry transformation. Thus by definition any remaining degeneracy is accidental.

The effect of perturbations

It is frequently convenient in quantum mechanics when discussing a complicated Hamiltonian $\mathscr{H}$, to split it up

$$\boxed{\mathscr{H} = \mathscr{H}_0 + \mathscr{H}_p} \tag{6.3}$$

into a relatively simple part $\mathscr{H}_0$ and a perturbation $\mathscr{H}_p$. Then the energy levels and eigenfunctions of $\mathscr{H}_0$ can be studied in detail, and hence the effect on them of adding $\mathscr{H}_p$ to the Hamiltonian can be calculated. This is particularly useful if it is possible to choose $\mathscr{H}_0$ and $\mathscr{H}_p$ so as to make the perturbation of the eigenfunctions and energy levels relatively small. Now $\mathscr{H}_0$ being simpler than $\mathscr{H}$ usually means that it has a higher degree of symmetry. For instance it is easier to calculate the energy levels of an electron in a spherically symmetrical potential than in a potential varying arbitrarily in any direction. If $\mathscr{H}_0$ has a higher degree of symmetry than $\mathscr{H}$, it implies in general that the energy levels of $\mathscr{H}_0$ are more degenerate because there are more symmetry transformations to make more eigenfunctions have the same energy in the manner of Theorem 1. Thus $\mathscr{H}_p$ tends to produce a splitting of these levels.

THEOREM 3. *In the notation of* (6.3), *if* $\mathscr{H}$, $\mathscr{H}_0$ *and* $\mathscr{H}_p$ *are all invariant under a group* $\mathfrak{G}$ *of symmetry transformations, and if the eigenfunctions of an energy level of* $\mathscr{H}_0$ *transform according to the representation* $D = D^{(1)} + D^{(2)} + \ldots + D^{(n)}$ *where the* $D^{(i)}$ *are the irreducible components of D, then the greatest splitting that the perturbation* $\mathscr{H}_p$ *can cause is into n levels. The eigenfunctions of each of these split levels transform according to a sum of the* $D^{(i)}$ *such that each* $D^{(i)}$ *is associated with one of the split levels.*

Consider the eigenfunctions and energy levels of the Hamiltonian

$$\mathscr{H}_\epsilon = \mathscr{H}_0 + \epsilon\mathscr{H}_p$$

where ϵ is varied continuously from 0 to 1. It may be possible to achieve this variation physically such as by the reduction of a magnetic field to zero, or it may be a purely mathematical device. For arbitrary $\epsilon \neq 0$ consider an energy level E_α. The eigenfunctions of this level transform according to some irreducible representation $D^{(\alpha)}$, or according to some reducible representation $D^{(\alpha)} + D^{(\alpha')} + \ldots$ if $\mathfrak{G}$ does not contain all the symmetry elements of $\mathscr{H}$. As ϵ is varied continuously, the energy levels and eigenfunctions vary continuously, and the representations $D^{(\alpha)}$ etc. cannot make a discontinuous change to some different (i.e. non-equivalent) representation. Thus as ϵ is varied continuously to zero, the only thing that can happen is that several energies coalesce into one, with the representation $D^{(1)} + D^{(2)} + \ldots + D^{(n)}$ of the composite level corresponding exactly to the components $D^{(\alpha)}$, $D^{(\beta)}$, $D^{(\gamma)}$ . . . $D^{(\nu)}$ which have coalesced. Looking at it the other way round now, we can say that the degenerate level is split by $\mathscr{H}_p$ into a maximum of n levels associated with the n irreducible components $D^{(1)}$ to $D^{(n)}$. This proves the theorem. By Theorem 2, if the group $\mathfrak{G}$ includes all the symmetry transformations of $\mathscr{H}$ (but not $\mathscr{H}_0$), the splitting would be the maximum allowed, except for possibly some accidental degeneracies of the types already mentioned. It should also be noted that the result of the theorem is precise and does not depend on $\mathscr{H}_p$ being small.

As an example, consider a free hydrogen atom with the electron in a $2p$ level, or any atom or ion with a single electron in a p level outside full quantum shells (Schiff 1955, p. 277). The Hamiltonian for the electron is

$$\mathscr{H}_0 = -\frac{\hbar^2}{2m}\nabla^2 - eV(r),$$

where $V(r)$ is the spherically symmetrical potential due to the proton or the closed shell ion core respectively. The p eigenfunctions have the form (6.2). If now the whole atom or ion is placed in an electric potential $V_{32}(\mathbf{r})$ with trigonal symmetry corresponding to the point-group 32, we have to include in the Hamiltonian $\mathscr{H}$ for the electron the term

$$\mathscr{H}_p = -eV_{32}(\mathbf{r}).$$

If the atom or ion is in a crystal, this potential could be due to the surrounding atoms (cf. problem 4.6). Thus $\mathscr{H}_p$, $\mathscr{H}_0$ and $\mathscr{H} = \mathscr{H}_0 + \mathscr{H}_p$ are all invariant under the point group 32. We have already seen above that in the free state in the absence of $\mathscr{H}_p$, the three eigenfunctions (6.2) all have the same energy. However, they transform according to the reducible representation $\Delta = \Gamma + \mathscr{I}$ (5.16) of the group 32, so that by Theorem 3 we may expect $\mathscr{H}_p$ to split the triply degenerate level of $\mathscr{H}_0$ into a non-degenerate level ($\mathscr{I}$) and a doubly degenerate level (Γ). We shall see in § 10 that the wave functions of a many electron atom or ion in a P state (Schiff 1955, p. 287) transform under rotations in exactly the same way as the functions x, y, z. Thus they also transform according to the representation Δ of the group 32, and when the atom or ion is placed in a trigonal crystalline electric field, we may expect the P level to be split as before into one singly and one doubly degenerate level. The $(2S + 1)$-fold electron spin degeneracy and the effects of spin-orbit coupling are superposed on this whole picture.

Summary

If a Hamiltonian $\mathscr{H}$ is invariant under a group $\mathfrak{G}$ of transformations, then the eigenfunctions belonging to one energy level E form the basis of a representation D of $\mathfrak{G}$. This representation can be used to characterize the level. If $\mathfrak{G}$ includes all symmetry transformations of $\mathscr{H}$, then D is irreducible, apart from accidental degeneracy. If $\mathscr{H}_0$ and $\mathscr{H}_p$ are both invariant under $\mathfrak{G}$ where $\mathscr{H}_p$ is a perturbation on $\mathscr{H}_0$, then $\mathscr{H}_p$ can split the energy level E_0 of $\mathscr{H}_0$ according to the number of irreducible components in the corresponding representation D.

These two theorems therefore substantiate the claims (i) and (ii) of § 1 for the uses of symmetry properties, and it remains to apply the method to particular systems by studying their symmetry transformations and the corresponding representations. It is convenient to defer until § 13 the third use of symmetry properties mentioned in § 1, namely the calculation of matrix elements and selection rules.

Problems

6.1 A three electron wave function ψ transforms according to the representation $\mathscr{A}$ (Table 3) of the group $\mathfrak{P}_3$ (cf. problem 5.6). Show in detail that it is antisymmetric in the usual quantum mechanical sense, i.e. it changes sign if we interchange the co-ordinates of two electrons or make a permutation corresponding to an odd number of such interchanges, and remains invariant under a permutation corresponding to an even number of such interchanges.

6.2 Verify that the five d wave functions

$$2(x \pm iy)zf(r),\ (x \pm iy)^2 f(r),\ \sqrt{(2/3)}(3z^2 - r^2)f(r) \qquad (6.4)$$

span an invariant vector space under proper and improper rotations about the x-, y- and z-axes. Note that this is not immediately obvious from the fact that they are all of the second degree in x, y, z, because there is a sixth such linear combination $x^2 + y^2 + z^2$. This transforms according to a different representation under rotations, and therefore does not get mixed in with the d functions.

6.3 From problem (6.2) and equations (5.20), (5.21) show that the five-fold degenerate level of an ion in a D state (neglecting spin degeneracy) splits into one singly and two doubly degenerate levels when the ion is placed in a crystalline electric field with trigonal point-group symmetry 32. This is the case of the Cu^{++} ion in copper fluorsilicate $CuSiF_6.6H_2O$ (Bleaney and Stevens 1953).

Chapter II

THE QUANTUM THEORY OF A FREE ATOM

The purpose of this chapter is to illustrate the use of group theory in quantum mechanics by giving a brief sketch of the quantum theory of a free atom or ion. In §§7, 8, and 9 we shall study the particular groups of symmetry transformations that we shall need, and the mechanics of handling their representations. In §§10, 11 and 12 we shall use the theorems of §6 to classify the energy levels and their wave functions, and to discuss the splitting of the levels by various perturbations. In §13 we shall develop the third use of group theory mentioned in §1, namely the calculation of matrix elements and selection rules. In fact we shall derive rigorously the usual features of what is known as the vector model of an atom (Pauling and Goudsmit 1930), and we shall lay the foundations of a complete treatment so that any energy level, transition probability, etc., can in principle be calculated. For references for this whole chapter, the reader is referred to the list of general references preceding the bibliography at the end of the book.

7. Some Simple Groups and Representations

Cyclic groups

The simplest type of group is the cyclic group $\mathfrak{C}_n$ of order n whose elements are $(E, A, A^2, A^3, \ldots A^{n-1})$ where $A^n = E$. Let $D_{ij}(A^r)$ be a representation of $\mathfrak{C}_n$. $D_{ij}(A)$ can always be reduced to the diagonal matrix

$$\Lambda = \text{diag}\,[\lambda_1, \lambda_2, \ldots] = \begin{bmatrix} \lambda_1 & \cdot & \cdot & \cdots \\ \cdot & \lambda_2 & \cdot & \cdots \\ \cdot & \cdot & \ddots & \\ \vdots & \vdots & & \end{bmatrix}$$

by an equivalence transformation (5.15) (Van der Waerden 1932, p. 26; Margenau and Murphy 1943, p. 316). Since A^r must always be represented by $[D(A)]^r$, the same transformation also transforms $D(A^r)$ into $\Lambda^r = \text{diag}\,[\lambda_1{}^r, \lambda_2{}^r, \ldots]$. Hence the representation D

has been reduced to a sum of one-dimensional representations, and we conclude all irreducible representations of $\mathfrak{C}_n$ are one-dimensional. Also since $A^n = E$, we must have $\lambda_m{}^n = 1$, whence

$$\boxed{\lambda_m = \exp \frac{2\pi i m}{n}.} \tag{7.1}$$

Thus $\mathfrak{C}_n$ has n different irreducible representations. An example of the group $\mathfrak{C}_n$ is the group of *n rotations about one axis* by $360/n$ degrees and multiples thereof. The *permutation group* $\mathfrak{P}_2$ *of order two* is a cyclic group of order two, and so is the point-group $\bar{1}$ consisting of the inversion $(x_i, y_i, z_i) = (-X_i, -Y_i, -Z_i)$ and the identity transformation, and likewise the point-group m consisting of a reflection like (3.12) and the identity transformation. From (7.1) each of these has two one-dimensional irreducible representations in which the identity element is represented by $+1$ and the other element by ± 1. These representations are called the symmetric or even one and antisymmetric or odd one respectively.

Abelian groups

As mentioned in § 4, an Abelian group is one in which any two elements commute. A cyclic group is automatically Abelian, but all Abelian groups are *not* isomorphic with one of the cyclic groups. For instance the translation group of a three-dimensional crystal lattice is not. However, it is shown in appendix E that all of a group of commuting matrices can be simultaneously reduced to diagonal form, so that *all the irreducible representations of an Abelian group are one-dimensional.*

The axial rotation group

The rotations (2.2) by all angles ϕ about a fixed axis form the axial rotation group. This is Abelian, since a rotation by ϕ_1 followed by a rotation by ϕ_2 is the same as a rotation by ϕ_2 followed by one by ϕ_1, as long as they are about the same axis. Thus the irreducible representations are all one-dimensional. Let $\chi(\phi)$ represent the rotation ϕ in one of these irreducible representations. Then we must have

$$\chi(\phi_1)\chi(\phi_2) = \chi(\phi_1 + \phi_2) \tag{7.2}$$

with $\chi(2\pi) = \chi(0)$. The solution of this equation is

$$\boxed{\chi(\phi) = e^{im\phi}, \; m = 0, \pm 1, \pm 2, \ldots} \tag{7.3}$$

and it will be convenient to refer to this as the mth or the $\exp(im\phi)$ representation, the latter notation being preferable when the sign of the exponent is important. Thus there are an infinite number of one-dimensional representations, and every representation is reducible into a sum of these.

The permutation group $\mathfrak{P}_n$

This group is not Abelian for $n > 2$ and the theory of its representations is not simple. We shall only prove here that for any n it always has two one-dimensional representations, namely, the symmetric and the antisymmetric ones. These two representations are the only ones in which we shall be interested, and all other irreducible representations have dimensions greater than one.

Consider two rows of digits (Fig. 4a) and the effect on the diagram

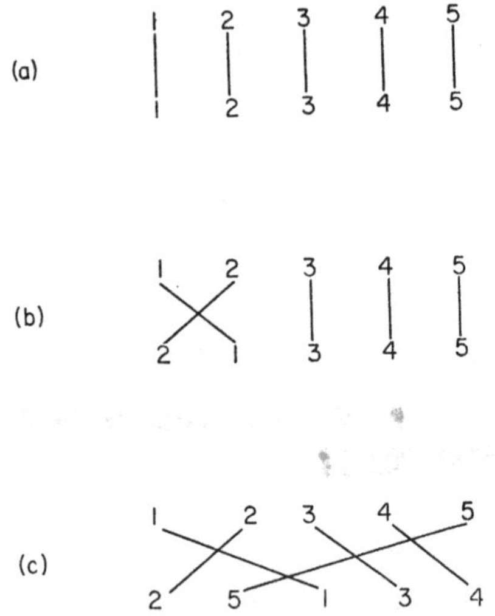

FIG. 4. Odd and even permutations.

of interchanging pairs of neighbouring digits in the lower row. Interchanging the digits 1 and 2, we obtain Fig. 4b, and thus introduce one intersection into the diagram of lines joining the same digits in the two rows. Further interchanges introduce further intersections, but the effect of one interchange is always to add an intersection or to remove one. For instance, in the diagram Fig. 4c, the interchange of 3 and 4 would add one intersection and the interchange of 5 and 1 would remove one. Hence starting with the permutation (12345) in the lower row, an odd (or even) number of consecutive interchanges results in a permutation giving a diagram with an odd (or even) number of intersections. A given permutation can accordingly be classed unambiguously as odd or even. Consider now the permutation transformation $P(ij \ldots)$

$$\begin{aligned} (x_1, y_1, z_1) &= (X_i, Y_i, Z_i) \\ (x_2, y_2, z_2) &= (X_j, Y_j, Z_j) \\ (x_3, y_3, z_3) &= \ldots\ldots \\ \ldots\ldots & \ \ \ldots\ldots \end{aligned} \tag{7.4}$$

of the n sets of co-ordinates (x_k, y_k, z_k), $k = 1$ to n. This transformation can be obtained as above by a sequence of elementary transformations in which only pairs of "neighbouring" co-ordinate sets are interchanged. As before, the transformation can be classed as odd or even according as the number of elementary interchanges is odd or even. Now consider the product P_2P_1 of two permutation transformations. If P_1 and P_2 are equivalent to sequences of n_1 and n_2 elementary interchanges respectively, then the product transformation P_2P_1 is equivalent to $n_1 + n_2$ elementary interchanges. Thus if n_1 and n_2 are both odd, P_2P_1 is an even permutation transformation. In this way we obtain the following skeleton multiplication table for permutations:

$$\begin{aligned} &(\text{odd}) \times (\text{odd}) = (\text{even}), \quad (\text{even}) \times (\text{even}) = (\text{even}), \\ &(\text{even}) \times (\text{odd}) = (\text{odd}), \quad\ \ (\text{odd}) \times (\text{even}) = (\text{odd}). \end{aligned} \tag{7.5}$$

This table shows immediately that there exists a one-dimensional representation of the group $\mathfrak{P}_n$ in which each even permutation is represented by $+1$ and each odd one by -1. This is known as the *antisymmetric representation*. As always, there is also the identity representation, here called the *symmetric representation*, in which each group element is represented by $+1$. When n is greater than 2, there are in addition irreducible representations with higher dimensions (cf. problem 5.6).

Summary

All the representations of Abelian groups are one-dimensional, the representations of the cyclic groups and the axial rotation group being particularly simple. The permutation group of any order always has the symmetric and antisymmetric representations.

Problems

7.1 Consider the group consisting of all rotations which are multiples of $2\pi/n$ radians, and write down the number representing a rotation by ϕ in the mth irreducible representation. Thus obtain the representations of the axial rotation group by letting n tend to infinity.

7.2 By putting $\phi_1 = \phi$, $\phi_2 = \phi + \delta\phi$ in (7.2), obtain a differential equation for $\chi(\phi)$, and hence show that (7.3) gives the only single-valued solutions of (7.2).

7.3 Classify all the permutation transformations of the groups $\mathfrak{P}_3$ and $\mathfrak{P}_4$ as odd or even, and verify the skeleton multiplication table (7.5).

7.4 Show that the point-group 222 (§ 16) is Abelian. Obtain all its irreducible representations by showing that each element can only be represented by ± 1.

8. The Irreducible Representations of the Full Rotation Group

The infinitesimal rotation operators

Before proceeding to derive the irreducible representations of the full rotation group, it is convenient to express all rotations in terms of three operators I_x, I_y, I_z known as infinitesimal rotation operators. We can then confine our attention in subsequent work simply to these three operators instead of having to deal with arbitrary rotations about arbitrary axes. Consider a rotation transformation $R(\alpha, \boldsymbol{\xi})$ by an angle α about the axis $\boldsymbol{\xi}$. From it we can define an infinitesimal rotation operator I_ξ given by

$$\underset{\alpha \to 0}{\text{Lt}} \frac{R(\alpha, \boldsymbol{\xi}) - 1}{\alpha} = iI_\xi, \tag{8.1}$$

or

$$\boxed{R(\alpha, \boldsymbol{\xi}) \approx 1 + i\alpha I_\xi \quad \text{when } \alpha \ll 1,} \tag{8.2}$$

where we have written 1 for the identity transformation E. Strictly speaking iI_ξ should be called the infinitesimal rotation operator, but it is more convenient and is becoming customary to work with I_ξ. $R(\alpha, \boldsymbol{\xi})$ can be expressed in terms of I_ξ for an arbitrary angle α which is not necessarily small, for a rotation by α is equal to n successive rotations by α/n. Thus from (8.2),

$$\begin{aligned} R(\alpha, \boldsymbol{\xi}) &= \underset{n \to \infty}{\text{Lt}} \left(1 + i\frac{\alpha}{n} I_\xi\right)^n \\ &= 1 + i\alpha I_\xi + \frac{(i\alpha I_\xi)^2}{2!} + \frac{(i\alpha I_\xi)^3}{3!} + \cdots \end{aligned} \tag{8.3a}$$

This series can be summed in a purely formal way, and we can write

$$R(\alpha, \xi) = \exp(i\alpha I_\xi). \tag{8.3b}$$

However, this should only be regarded as an abbreviated notation for the series (8.3a) and in any particular case the exponential has to be expanded before it can operate on a function.

Consider now $R(\alpha, \xi_1)$ when α is very small, where ξ_1 is an axis in the yz-plane making an angle θ with the z-axis (Fig. 5). This

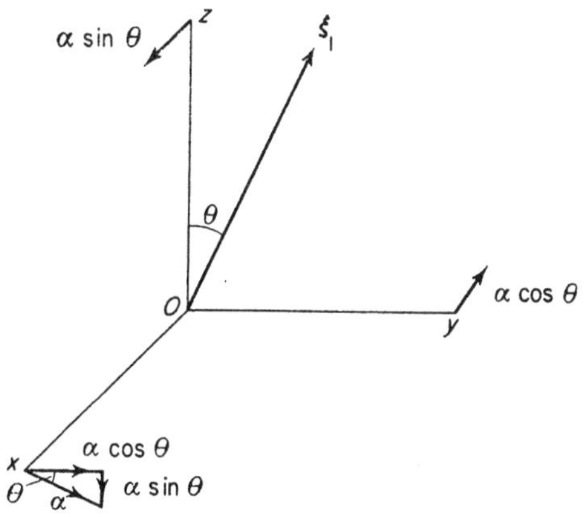

FIG. 5. Change of the co-ordinate axes in a small rotation $R(\alpha, \xi_1)$ where ξ_1 is in the yz-plane.

rotation is approximately equal to consecutive rotations $\alpha \cos\theta$ about the z-axis and $\alpha \sin\theta$ about the y-axis, correct to terms linear in α. This can be seen from Fig. 5 which shows the effect of $R(\alpha, \xi_1)$ on three points x, y, z at unit distances along the co-ordinate axes. $R(\alpha\cos\theta, z)$ displaces x and y each by $\alpha\cos\theta$, and $R(\alpha\sin\theta, y)$ displaces z and x each by $\alpha\sin\theta$. The total displacement of x is α in a plane perpendicular to ξ_1, and thus the effect on all three co-ordinate axes is equivalent to the single rotation $R(\alpha, \xi_1)$ when α is small. We have

$$R(\alpha, \xi_1) = R(\alpha \sin\theta, y)R(\alpha\cos\theta, z) + O(\alpha^2),$$

and hence from (8.1), (8.3)

$$I_{\xi_1} = I_y \sin\theta + I_z \cos\theta. \tag{8.4}$$

If now ξ is an axis making direction cosines $l = \sin\theta\cos\phi$, $m = \sin\theta\sin\phi$, $n = \cos\theta$ with the x-, y-, z-axes, then

$$I_\xi = I'\sin\theta + I_z\cos\theta \text{ where } I' = I_x\cos\phi + I_y\sin\phi,$$

i.e. we have

$$\boxed{I_\xi = lI_x + mI_y + nI_z.} \tag{8.5}$$

In fact infinitesimal operators add like unit vectors. This result corresponds to the fact that small rotations add vectorially to a first order of approximation, and that angular velocities in mechanics add vectorially (Milne 1948, p. 148; Goldstein 1950, p. 126). Of course finite rotations do not add like vectors, since they do not even commute (equation (4.5), problem 8.1). In conclusion we note that with the help of (8.5) and (8.3) any rotation can be expressed in terms of the infinitesimal rotation operators I_x, I_y, I_z.

Commutation relations

Consider now the physical rotation $\mathrm{Rot}(\alpha, \xi_1)$ when α is not small, where ξ_1 is the axis shown in Fig. 5. This rotation can also be achieved by first rotating by θ about Ox which brings the ξ_1 axis parallel to Oz, then rotating by α about Oz, and then returning ξ_1 to its original position by a rotation of $-\theta$ about Ox. Thus

$$\mathrm{Rot}(\alpha, \xi_1) = \mathrm{Rot}(-\theta, x)\mathrm{Rot}(\alpha, z)\mathrm{Rot}(\theta, x).$$

From § 2 a physical rotation by an angle $+\alpha$ is mathematically equivalent to a rotational transformation of co-ordinates by an angle $-\alpha$. Hence in terms of transformations of co-ordinates,†

$$R(-\alpha, \xi_1) = R(\theta, x)R(-\alpha, z)R(-\theta, x). \tag{8.6}$$

Expressing this in terms of infinitesimal rotation operators using (8.3) and (8.4), we obtain

$$\begin{aligned} 1 - i\alpha(I_y \sin\theta + I_z \cos\theta) + O(\alpha^2) \\ = [1 + i\theta I_x + O(\theta^2)][1 - i\alpha I_z + O(\alpha^2)][1 - i\theta I_x + O(\theta^2)]. \end{aligned} \tag{8.7}$$

Now (8.6) is valid for all α and θ, so that we may express (8.7) completely as a power series in θ and α and equate coefficients of $\theta\alpha$, which gives

$$-iI_y = I_x I_z - I_z I_x. \tag{8.8a}$$

By symmetry, we also obtain

$$\begin{aligned} -iI_x &= I_z I_y - I_y I_z, \\ -iI_z &= I_y I_x - I_x I_y. \end{aligned} \tag{8.8b}$$

† This relation could have been written down directly, but the author finds it easier to visualize the composition of physical rotations rather than that of co-ordinate changes.

In terms of

$$I_+ = I_x + iI_y, \qquad I_- = I_x - iI_y, \tag{8.9}$$

these commutation relations become

$$\boxed{\begin{aligned} I_zI_+ - I_+I_z &= I_+, \\ I_zI_- - I_-I_z &= -I_-, \\ I_+I_- - I_-I_+ &= 2I_z. \end{aligned}} \tag{8.10}$$

The irreducible representations†

It was mentioned in § 4 that all proper rotations about all axes through a point form the full rotation group, and there is in fact no difficulty in verifying that all the group requirements are satisfied (problems 8.2 and 8.3). We shall now take an arbitrary vector space $\mathfrak{R}$ which is invariant under the full rotation group, and start reducing it into its irreducible components. Since any rotation can be expressed in terms of I_x, I_y, I_z, it is not necessary to work with an arbitrary rotation but only with these three operators. More precisely, it follows from (8.3), (8.5) that if a space is invariant and irreducible under I_x, I_y, I_z, then it is also invariant and irreducible under all rotations and vice versa. Let us first reduce $\mathfrak{R}$ with respect to the axial rotation group about the z-axis, and let u_m be any vector transforming according to the mth representation (7.3).‡ Then

$$R(\phi, z)u_m = e^{im\phi}u_m,$$

whence from (8.1)

$$I_zu_m = mu_m. \tag{8.11}$$

If u_m transforms according to the mth representation of the axial rotation group, then I_+u_m, I_-u_m belong respectively to the $(m+1)^{\text{th}}$ and $(m-1)^{\text{th}}$ representations, for from (8.11), 8.10)

$$I_z(I_+u_m) = (I_+I_z + I_+)u_m = (m+1)I_+u_m \tag{8.12a}$$

and similarly

$$I_z(I_-u_m) = (m-1)I_-u_m. \tag{8.12b}$$

When $\mathfrak{R}$ is reduced according to the axial rotation group, let j be the highest value of m that occurs among the irreducible

† This derivation follows closely that given by Van der Waerden (1932).

‡ It is convenient to use u instead of ϕ for the vectors of $\mathfrak{R}$ so as to avoid confusion with the angle ϕ.

components, and let u_j be a vector corresponding to this value of m. Now since $\mathfrak{R}$ is invariant under rotations, I_+u_j also belongs to $\mathfrak{R}$ and from (8.12a) it has the m-value $j + 1$. However since j is already the largest m-value found in $\mathfrak{R}$, we must have $I_+u_j = 0$. On the other hand by repeated use of I_- we can *define* from u_j a sequence of vectors u_m with $m = j, j - 1, j - 2, \ldots$ such that

$$u_{m-1} = \alpha_m I_- u_m. \tag{8.13}$$

From (8.12b) u_m belongs to the eigenvalue m of I_z. α_m is a non-zero numerical constant which we shall determine later such that the u_m are all normalized. From (8.12a), I_+u_{m-1} belongs to the eigenvalue m of I_z, and we shall now prove from the definition (8.13) that it is actually a multiple of u_m. First suppose

$$I_+u_m = c_m\alpha_{m+1}u_{m+1}, \tag{8.14}$$

where c_m is an undetermined constant. Then from (8.13)

$$\begin{aligned} I_+u_{m-1} &= \alpha_m I_+I_-u_m = \alpha_m I_-I_+u_m + 2\alpha_m I_z u_m \\ &= \alpha_m I_- c_m\alpha_{m+1}u_{m+1} + 2m\alpha_m u_m \\ &= (c_m + 2m)\alpha_m u_m. \end{aligned} \tag{8.15}$$

Thus if (8.14) is true for one value of m, then by (8.15) it is also true for the value $m - 1$. But (8.14) is true for $m = j$ with $c_j = 0$, so that by induction it is true for all m. Further, it is possible to calculate the value of c_m, for from (8.15)

$$c_{m-1} = c_m + 2m.$$

The solution of this difference equation with the boundary condition $c_j = 0$ is†

$$c_m = j(j + 1) - m(m + 1). \tag{8.16}$$

Further if $\mathfrak{R}$ is of finite dimension, the sequence of vectors u_m must end at some point. I.e. we must have some $u_m = 0$ with $u_{m+1} \neq 0$. Hence $I_+u_m = 0$ and from (8.16) this only happens when $m = -j - 1$ (apart from $m = j$ already discussed). Thus the last of the sequence of vectors is u_{-j}, and the number of vectors is $2j + 1$. This is necessarily an integer so that j is an integer or half an odd integer.

The constants α_m can now be calculated so as to make all the u_m normalized. By appendix C, lemma 2, any rotation such as $R(\pm\theta, x)$

† The solution of difference equations is dealt with in most elementary algebra texts, e.g. Durell and Robson (1937).

leaves the scalar product of two arbitrary vectors u and v, or u and Rv, invariant. Thus

$$\begin{aligned}\int u^*R(\theta, x)v\,\mathrm{d}\tau &= \int R(-\theta, x)[u^*R(\theta, x)v]\,\mathrm{d}\tau \\ &= \int [R(-\theta, x)u]^*[R(-\theta, x)R(\theta, x)v]\,\mathrm{d}\tau \\ &= \int [R(-\theta, x)u]^*v\,\mathrm{d}\tau\end{aligned}$$

where $\mathrm{d}\tau$ is the volume element. This is true for any value of θ so that we may substitute (8.3) for R and equate coefficients of θ. Thus

$$\int u^*(I_xv)\,\mathrm{d}\tau = \int (I_xu)^*v\,\mathrm{d}\tau. \tag{8.17a}$$

A similar result holds for I_y, so that

$$\int u^*(I_+v)\,\mathrm{d}\tau = \int (I_-u)^*v\,\mathrm{d}\tau. \tag{8.17b}$$

Now from (8.13), (8.14), (8.17) we obtain

$$\begin{aligned}\int u_m{}^*u_m\,\mathrm{d}\tau &= \alpha_{m+1}\int (I_-u_{m+1})^*u_m\,\mathrm{d}\tau \\ &= \alpha_{m+1}\int u_{m+1}^*I_+u_m\,\mathrm{d}\tau \\ &= c_m(\alpha_{m+1})^2\int u_{m+1}^*u_{m+1}\,\mathrm{d}\tau.\end{aligned}$$

Hence, all the u_m are simultaneously normalized if we choose $\alpha_{m+1} = (c_m)^{-1/2}$, and (8.13), (8.14) become

$$\boxed{\begin{aligned}I_+u_m &= \sqrt{[j(j+1) - m(m+1)]}u_{m+1}, \\ &= \sqrt{[(j-m)(j+m+1)]}u_{m+1}, \\ I_-u_m &= \sqrt{[j(j+1) - m(m-1)]}u_{m-1}, \\ &= \sqrt{[(j+m)(j-m+1)]}u_{m-1}, \\ I_zu_m &= mu_m.\end{aligned}} \tag{8.18}$$

The u_m are also orthogonal to one another because they transform according to different irreducible representations of the axial rotation group. This follows immediately from appendix C, lemma 5, or from the fact that in the reduction of a vector space the different irreducible subspaces can always be made orthogonal to one another (§ 5). It can also be proved very simply directly, for by appendix C, lemma 2,

$$\int u_m{}^*u_\mu \, d\tau = \int R(\alpha, z)[u_m{}^*u_\mu] \, d\tau$$
$$= e^{i\alpha(\mu-m)} \int u_m{}^*u_\mu \, d\tau,$$

whence

$$\int u_m{}^*u_\mu \, d\tau = 0 \text{ if } m \neq \mu.$$

From (8.18) the vector space $\mathfrak{R}^{(j)}(u_j, u_{j-1}, \ldots u_{-j})$ is invariant under I_+, I_- and I_z, and therefore it is also invariant under all rotations. We now show that it is also irreducible. For, suppose $\mathfrak{R}^{(j)}$ contains an invariant subspace r, then r would also be invariant under the axial rotation group about the z-axis and would therefore be spanned by a set of the u_m. But from one u_m the operators I_+, I_- generate all the other ones, so that r can only be equal to the whole space $\mathfrak{R}^{(j)}$. *Thus the vectors* u_m, $m = j, j-1, \ldots -j$, *transforming according to* (8.18), *form the basis of an irreducible representation of the full rotation group. The different irreducible representations* $D^{(j)}$ *are given by the allowed values of* j, $j = 0, 1/2, 1, 3/2, 2, \ldots$, *and have dimensions* $2j + 1$. If a space $\mathfrak{R}^{(j)}$ transforms according to $D^{(j)}$, we shall refer to the particular set of base vectors satisfying (8.18) as the *standard* base vectors.

In this subsection we started with an arbitrary invariant vector space $\mathfrak{R}$, and set about reducing it. This led naturally to the above description and definition of the irreducible representations. But it should be noted that we have here in addition a systematic way of actually picking out an irreducible component from $\mathfrak{R}$, starting with a vector u_j with the highest m-value. Having picked out one irreducible subspace, we can then orthogonalize all remaining vectors to it and start the process again in the remaining vector space. Thus $\mathfrak{R}$ is gradually completely reduced. This scheme is described more fully in the next section where we actually employ it.

Examples

The spherical harmonics can be defined in various ways, but usually arise in quantum mechanics as the solutions of the equation

$$\left(\frac{1}{\sin\theta}\frac{\partial}{\partial\theta}\sin\theta\frac{\partial}{\partial\theta} + \frac{1}{\sin^2\theta}\frac{\partial^2}{\partial\phi^2}\right)Y = \lambda Y \tag{8.19}$$

belonging to the eigenvalue $\lambda = -l(l+1)$ (Schiff 1955, p. 70). Here θ, ϕ are spherical polar co-ordinates (problem 3.2). The spherical harmonics Y_{lm} are particular solutions of (8.19) having the form

$$\boxed{Y_{lm} = N_{lm}P_l^{|m|}(\cos\theta)e^{im\phi},} \tag{8.20}$$

where N_{lm} is a numerical normalizing factor, $P_l^{|m|}$ an associated Legendre polynomial, and m has the $2l+1$ integral values l, $l-1, \ldots -l$. The operator in (8.19) is invariant under rotations since it is just the angular part of the Laplacian ∇^2, and hence the Y_{lm} for given l span an invariant vector space $\mathfrak{R}^{(l)}$ (§ 6, Theorem 1). From (8.20) Y_{lm} belongs to the mth representation of the axial rotation group about the z-axis; the Y_{lm} are therefore orthogonal to one another and $\mathfrak{R}^{(l)}$ has the dimension $2l+1$. Now from one of them, Y_{ll} say, we can define using (8.18) $2l+1$ vectors Y'_{lm}. These span an invariant subspace of $\mathfrak{R}^{(l)}$ which must be equal to the whole space because it has the same dimension $2l+1$. *Thus the spherical harmonics Y_{lm} transform according to the irreducible representation $D^{(l)}$ of the full rotation group.* Moreover since Y_{lm} belongs to the mth representation of the axial rotation group, the Y_{lm} transform exactly like (8.18) if we give them the correct phase factors (Condon and Shortley 1935, p. 52). r^lY_{lm} can also be expressed as a polynomial of degree l in x, y, z. For instance

$$\begin{aligned}
rY_{1,1} &= N(x+iy), & rY_{1,0} &= -N\sqrt{2}z,\\
rY_{1,-1} &= -N(x-iy), & N &= (3/8\pi)^{1/2}:\\
r^2Y_{2,2} &= M\tfrac{1}{4}\sqrt{6}\,(x+iy)^2, & r^2Y_{2,1} &= -M\tfrac{1}{2}\sqrt{6}\,(x+iy)z,\\
r^2Y_{2,0} &= M\tfrac{1}{2}(3z^2-r^2), & r^2Y_{2,-1} &= M\tfrac{1}{2}\sqrt{6}\,(x-iy)z,\\
r^2Y_{2,-2} &= M\tfrac{1}{4}\sqrt{6}\,(x-iy)^2, & M &= (5/4\pi)^{1/2}.
\end{aligned} \tag{8.21}$$

These functions therefore transform according to the irreducible representation $D^{(1)}$ and $D^{(2)}$, and the signs have actually been chosen to make them standard base vectors. Thus x, y, z also transform according to $D^{(1)}$ but not as standard base vectors.

An interesting feature of the representations $D^{(j)}$ when j is *half an odd integer* is that they are *double-valued*, by which we mean the following. In § 7 we deduced the irreducible representations (7.3) of the axial rotation group, and concluded that m must be an integer because of the condition $\chi(2\pi) = \chi(0)$ and because $\chi(0)$

corresponding to no rotation at all must be equal to one. However, for the standard base vectors u_m (8.11), (8.18) transforming according to $D^{(j)}$, we have

$$R(\phi, z)u_m = \exp(im\phi)u_m$$

where m is half an odd integer if j is. In particular

$$R(2\pi, z)u_m = -u_m,$$

so that a rotation of 2π about the z-axis makes all the u_m change sign. Since $R(2\pi, z)$ is physically the same as no rotation at all, we have that the identity transformation is represented by two matrices, the unit matrix E and also $-E$. Similarly by compounding any rotation R with the identity transformation, R is represented by the two matrices $D^{(j)}(R)$ and $-D^{(j)}(R)$. This leads to no difficulties in quantum mechanics because the wave functions ψ and $-\psi$ always represent quantum mechanically the same physical state of a system, so that we can consider the matrices $\pm D^{(j)}(R)$ as inducing the same transformation among the base vectors. Clearly the u_m cannot be ordinary single-valued functions of x, y, z because

$$R(2\pi, z)f(x, y, z) = f(x, y, z).$$

In fact we shall see later that the representations $D^{(j)}$ where j is half an odd integer only arise in connection with spin functions.

We now show how we can calculate in principle the matrix $D^{(j)}(R)$ which represents any given rotation R in the representation $D^{(j)}$. If $R = R(\alpha, \xi)$ is a rotation by α about the axis ξ, the matrix $D^{(j)}(R)$ is completely determined by (8.18), (8.9), (8.5) and (8.3). When discussing such a general rotation it is usually convenient not to work in terms of the angle α and the direction cosines of the ξ axis, but to express the rotation parametrically in terms of three Eulerian angles ϕ, θ, χ defined in Fig. 6. Any rotation can be considered as shifting the z-axis to OP making polar angles ϕ, θ, plus a rotation by χ about this axis. Thus if the physical rotation $\mathrm{Rot}(\alpha, \xi)$ corresponds to the Eulerian angles (ϕ, θ, χ), we have from Fig. 6

$$\mathrm{Rot}(\alpha, \xi) = \mathrm{Rot}(\phi, z)\mathrm{Rot}(\theta, y)\mathrm{Rot}(\chi, z).$$

Note that OP is not the axis ξ; it is the direction into which Oz is rotated by $R(\alpha, \xi)$. In § 2 it was seen that a physical rotation by an angle α is mathematically equivalent to a rotational transformation of co-ordinates by $-\alpha$. Hence

$$\begin{aligned} R(-\alpha, \xi) &= R(-\phi, z)R(-\theta, y)R(-\chi, z), \\ R(\alpha, \xi) &= [R(-\alpha, \xi)]^{-1} \\ &= R(\chi, z)R(\theta, y)R(\phi, z). \end{aligned} \tag{8.22}$$

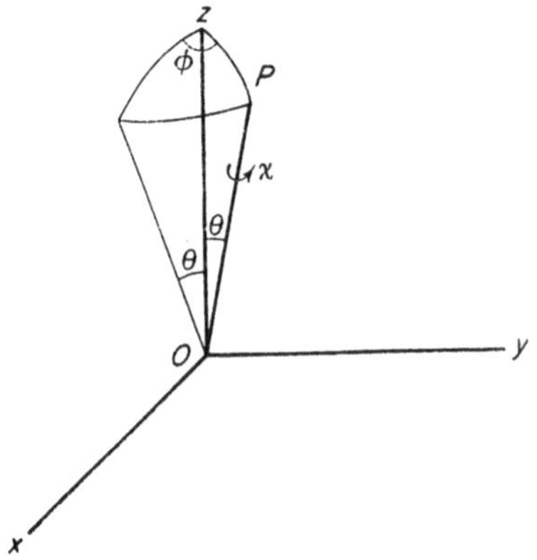

FIG. 6. The Eulerian angles χ, θ, ϕ.

We can now calculate the matrix representing $R(\alpha, \boldsymbol{\xi})$ in terms of the three parameters ϕ, θ, χ. As a definite example let us consider the representation $D^{(1/2)}$. From (8.18)

$$I_+ = \begin{bmatrix} \cdot & 1 \\ \cdot & \cdot \end{bmatrix}, \qquad I_- = \begin{bmatrix} \cdot & \cdot \\ 1 & \cdot \end{bmatrix}, \qquad I_z = \frac{1}{2}\begin{bmatrix} 1 & \cdot \\ \cdot & -1 \end{bmatrix}, \tag{8.23}$$

i.e. $$I_x = \frac{1}{2}\begin{bmatrix} \cdot & 1 \\ 1 & \cdot \end{bmatrix}, \qquad I_y = \frac{1}{2i}\begin{bmatrix} \cdot & 1 \\ -1 & \cdot \end{bmatrix},$$

$$\text{and } I_x^2 = I_y^2 = I_z^2 = \tfrac{1}{4}\,\mathbf{1},$$

where we have written $D^{(1/2)}(I_+)$ etc. as I_+ for short, and $\mathbf{1}$ for the unit matrix E. From (8.3) the matrix representing $R(\phi, z)$ is

$$\begin{aligned} D^{(1/2)}(\phi, z) &= \mathbf{1} + \sum_n \frac{(i\phi I_z)^n}{n!} \\ &= \mathbf{1} + \mathbf{1}\sum_{n\text{ even}} \frac{(\frac{1}{2}i\phi)^n}{n!} + 2I_z \sum_{n\text{ odd}} \frac{(\frac{1}{2}i\phi)^n}{n!} \\ &= \mathbf{1}\cos\tfrac{1}{2}\phi + 2I_z i \sin\tfrac{1}{2}\phi \\ &= \begin{bmatrix} e^{i\frac{1}{2}\phi} & \cdot \\ \cdot & e^{-i\frac{1}{2}\phi} \end{bmatrix}. \end{aligned}$$

Similarly

$$\begin{aligned} D^{(1/2)}(\theta, y) &= \mathbf{1}\cos\tfrac{1}{2}\theta + 2I_y i \sin\tfrac{1}{2}\theta \\ &= \begin{bmatrix} \cos\frac{1}{2}\theta & \sin\frac{1}{2}\theta \\ -\sin\frac{1}{2}\theta & \cos\frac{1}{2}\theta \end{bmatrix}. \end{aligned}$$

Hence from (8.22),

$$D^{(1/2)}(\alpha, \xi) = \pm\begin{bmatrix} e^{i\frac{1}{2}\chi} & \cdot \\ \cdot & e^{-i\frac{1}{2}\chi} \end{bmatrix} \begin{bmatrix} \cos\frac{1}{2}\theta & \sin\frac{1}{2}\theta \\ -\sin\frac{1}{2}\theta & \cos\frac{1}{2}\theta \end{bmatrix} \begin{bmatrix} e^{i\frac{1}{2}\phi} & \cdot \\ \cdot & e^{-i\frac{1}{2}\phi} \end{bmatrix}$$

$$= \pm\begin{bmatrix} e^{i\frac{1}{2}(\chi+\phi)}\cos\frac{1}{2}\theta & e^{i\frac{1}{2}(\chi-\phi)}\sin\frac{1}{2}\theta \\ -e^{-i\frac{1}{2}(\chi-\phi)}\sin\frac{1}{2}\theta & e^{-i\frac{1}{2}(\chi+\phi)}\cos\frac{1}{2}\theta \end{bmatrix}. \tag{8.24}$$

The $\pm$ has been included because $j = \frac{1}{2}$. It can in fact be seen explicitly that adding 2π to χ or ϕ changes the sign of the matrix.

Relationship to angular momentum

The infinitesimal rotation operators can be expressed directly in terms of co-ordinates. If f is a function of several sets of co-ordinates x_n, y_n, z_n, then from (2.2) and using the notation of (5.3),

$$\begin{aligned} R(\alpha, z)f(x_n, y_n, z_n) \\ &= f(x_n \cos\alpha - y_n \sin\alpha,\ x_n \sin\alpha + y_n \cos\alpha,\ z_n) \\ &= f(x_n, y_n, z_n) + \alpha \sum_n \left(x_n \frac{\partial f}{\partial y_n} - y_n \frac{\partial f}{\partial x_n}\right) + O(\alpha^2), \end{aligned}$$

whence from (8.1)

$$I_{z\ \mathrm{orb}} = -i \sum_n \left(x_n \frac{\partial}{\partial y_n} - y_n \frac{\partial}{\partial x_n}\right). \tag{8.25}$$

Here we have written $I_{z\ \mathrm{orb}}$ because as we shall see in § 11, I_z can also operate on the spin co-ordinate of an electron. Until we introduce spin co-ordinates, I_z and $I_{z\ \mathrm{orb}}$ can be considered as identical. In spherical polar co-ordinates

$$I_{z\ \mathrm{orb}} = -i \sum_n \frac{\partial}{\partial \phi_n}. \tag{8.26}$$

Now the quantum mechanical operator for the z-component of the orbital angular momentum $\mathbf{L}$ is (Schiff 1955, p. 74)

$$L_z = -i\hbar \sum_n \left(x_n \frac{\partial}{\partial y_n} - y_n \frac{\partial}{\partial x_n}\right) = -i\hbar \sum_n \frac{\partial}{\partial \phi_n}, \tag{8.27}$$

whence from (8.25), (8.26)

$$\boxed{L_z = \hbar I_{z\ \mathrm{orb}},\ \text{etc.}} \tag{8.28}$$

This result is a special case of the general relation due to Dirac

$$\boxed{p = \hbar I_q} \tag{8.29}$$

for the momentum operator p canonically conjugate to the co-ordinate q. The relation is proved in appendix F. The most common example of this relation is the usual linear momentum

$$p_x = \frac{\hbar}{i}\frac{\partial}{\partial x},$$

the i coming in through the definition (8.1) of an infinitesimal operator. Equation (8.29) is very important in discussing the angular momentum due to the electron spin. In the case of the momentum due to orbital motion, we have the classical expression $L_z = xp_y - yp_x$ from which (8.27) is derived, so that we have *verified* (8.28). However there is no classical analogue to spin angular momentum, and the only satisfactory way of defining it is via (8.29) (cf. Schiff 1955, p. 142; Dirac 1958, p. 142). Thus anticipating a little, we can define the total angular momentum vector J by

$$\boxed{J_x = \hbar I_x, \qquad J_y = \hbar I_y, \qquad J_z = \hbar I_z,} \tag{8.30}$$

where I_x, I_y, I_z apply to all orbital (or spatial) co-ordinates x_n, y_n, z_n and all spin co-ordinates (§ 11).

If we are transforming a function of many co-ordinates x_n, y_n, z_n, we can consider rotational transformations $R_n(\alpha, \xi)$ of the co-ordinates x_n, y_n, z_n, keeping all the other x_m, y_m, z_m fixed. These are not symmetry transformations of (3.2). Then

$$R(\alpha, \xi) = R_1(\alpha, \xi)R_2(\alpha, \xi) \ldots R_n(\alpha, \xi) \ldots \tag{8.31}$$

Correspondingly we can define infinitesimal transformations $I_{\xi n}$ and $I_{\xi n \text{ orb}}$ depending on whether we include or exclude the spin co-ordinate in the rotation. Then from (8.31), (8.2),

$$\boxed{I_\xi = \sum_n I_{\xi n}, \qquad I_{\xi \text{ orb}} = \sum_n I_{\xi n \text{ orb}}.} \tag{8.32}$$

The latter decomposition is clearly seen in (8.25), (8.26). Analogously to (8.28), (8.30) we define the angular momentum vectors $\mathbf{l}_n(l_{xn}, l_{yn}, l_{zn})$ and $\mathbf{j}_n(j_{xn}, j_{yn}, j_{zn})$ of the nth particle by

$$l_{xn} = \hbar I_{xn \text{ orb}}, \quad j_{xn} = \hbar I_{xn}, \text{ etc.} \tag{8.33}$$

These operators should not be confused with the quantum numbers l and j. From (8.32) we also have

$$\boxed{\mathbf{L} = \sum \mathbf{l}_n, \qquad \mathbf{J} = \sum \mathbf{j}_n.} \tag{8.34}$$

Let $\psi_m^{(j)}$ be a wave function transforming according to $D^{(j)}$ under rotations. Then from (8.18),

$$\boxed{\begin{aligned}(I_x^2 + I_y^2 + I_z^2)\psi_m^{(j)} &= (\tfrac{1}{2}I_+I_- + \tfrac{1}{2}I_-I_+ + I_z^2)\psi_m^{(j)} \\ &= j(j+1)\psi_m^{(j)}.\end{aligned}} \tag{8.35}$$

Thus $\psi_m^{(j)}$ describes an eigenstate of the total angular momentum with the eigenvalue $\sqrt{[j(j+1)]}\hbar$ and with z-component $m\hbar$. Similarly if a function transforms according to $D^{(l)}$ under rotational transformations of the orbital co-ordinates alone (leaving spin co-ordinates fixed), then it corresponds to a state with orbital angular momentum $\sqrt{[l(l+1)]}\hbar$.

Further references

A more detailed discussion of infinitesimal rotations, Eulerian angles, the representation $D^{(1/2)}$ and its relation to the Cayley-Klein parameters is given by Goldstein (1950, Chapter 4) from a somewhat different point of view. As regards the irreducible representations of the rotation group, there are also three main different approaches as exemplified by Van der Waerden (1932), Weyl (1931), Wigner (1931), Murnaghan (1938), Boerner (1955).

Summary

The irreducible representations of the full rotation group are called $D^{(j)}$ where j is an integer or half an odd integer. $D^{(j)}$ is of dimension $2j+1$, and the base vectors are conventionally chosen to transform according to (8.18) and labelled by $m = j, j-1, \ldots, -j$. A wave function transforming according to $D^{(j)}$ describes a state with total angular momentum $\sqrt{[j(j+1)]}\hbar$.

PROBLEMS

8.1 Start with a book in some definite position and apply successively the rotations Rot(90°, x) and Rot(90°, y) about two perpendicular axes Ox, Oy. Show that the final position depends on the order of applying the rotations, but that in either case it does not correspond to a single rotation by $90\sqrt{2}$ degrees about an axis bisecting Ox, Oy as would be expected if the rotations added like two vectors.

8.2 Prove that there always exists a unique single rotation $\mathrm{Rot}(\gamma, C)$ whose effect is the same as that of two consecutive given rotations $\mathrm{Rot}(\alpha, A)$ and $\mathrm{Rot}(\beta, B)$. Hint: suppose $\mathrm{Rot}(\alpha, A)$ and $\mathrm{Rot}(\beta, B)$ finally displace the arc PQ on the unit sphere to $P'Q'$ (Fig. 7). KC, LC are the right bisectors of PP', QQ'. Angle $PCQ = \angle P'CQ'$ and $\angle PCP' = \angle QCQ' = \gamma$.

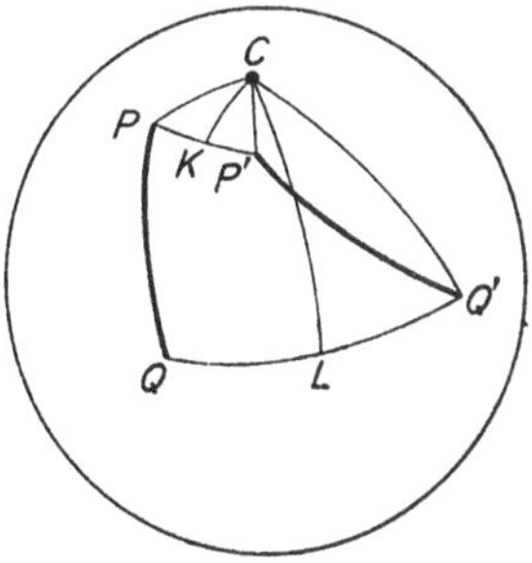

FIG. 7. Composition of rotations.

8.3 Using the result of problem 8.2, verify that all rotational transformations of co-ordinates satisfy the group postulates of § 4.

8.4 Let $\mathbf{r}$ be the position vector of a point P on a rigid body. If the body is rotated about the origin, show that $\mathbf{r}$ is carried into $\mathbf{r}' = \mathbf{T}.\mathbf{r}$, where $\mathbf{T}$ is the dyadic tensor,

$$\mathbf{vv} + (\mathbf{E} - \mathbf{vv}) \cos\theta + \mathbf{v} \wedge \mathbf{E} \sin\theta,$$

θ is the angle of rotation, $\mathbf{v}$ a unit vector along the axis of rotation, and $\mathbf{E}$ the unit tensor (Milne 1948, p. 36; Zachariasen 1945, p. 242). Hence show that $\mathrm{Rot}(\frac{1}{2}\pi, x)\ \mathrm{Rot}(\frac{1}{2}\pi, y) = \mathrm{Rot}(\frac{2}{3}\pi, \boldsymbol{\xi})$ where $\boldsymbol{\xi}$ is equally inclined to the positive x-, y- and z-axes. What single rotation is $\mathrm{Rot}(\frac{1}{2}\pi, y)\ \mathrm{Rot}(\frac{1}{2}\pi, x)$ equivalent to?

8.5 Draw a figure from which the truth of equation (8.5) may be derived directly in the same way as (8.4) follows from Fig. 5. Also prove (8.5) analytically using tensors of the form given in problem 8.4.

8.6 Calculate the coefficients of $\theta^2\ \alpha$ on both sides of equation (8.7) and show they are equal.

8.7 Starting with I_+, I_-, I_z expressed in Cartesian co-ordinates (cf. equation (8.25)), express them in spherical polar co-ordinates using the relations between the two co-ordinate systems (problem 3.2).

8.8 Verify directly that I_+, I_-, I_z expressed in Cartesian co-ordinates satisfy the commutation relations (cf. equation (8.25)). Also verify that the matrices (8.23) do.

8.9 Express I_+, I_-, I_z in Cartesian co-ordinates like (8.25) and hence verify that the functions (8.21) transform as standard base vectors (8.18) according to the irreducible representation $D^{(1)}$ of the rotation group. In particular verify that $I_+(x + iy) = I_-(x - iy) = 0$, and also (8.35).

8.10 From the definition (8.1) show that

$$I_\xi uv = (I_\xi u)v + u(I_\xi v),$$

so that I_ξ behaves like a single differentiation as regards its operation on a product function uv.

8.11 The vectors $u_{+1/2}$, $u_{-1/2}$ (abbreviated to u_+, u_-) transform as standard base vectors (8.18) according to the representation $D^{(1/2)}$ of the rotation group. Show from problem 8.10 that the vectors

$$u_m = \frac{u_+^{j+m} u_-^{j-m}}{[(j+m)!(j-m)!]^{1/2}}$$

transform as standard base vectors according to the irreducible representation $D^{(j)}$.

8.12 Apply the method used to obtain (8.24) to calculate $D^{(1)}(\theta, x)$ referred to standard base vectors. Check the result by writing down the matrix representing the effect of $R(\theta, x)$ on the functions x, y, z, and then transforming it to standard base vectors using (5.15).

8.13 Show that (8.24) is a unitary matrix with determinant $+1$, and conversely that any 2×2 unitary matrix can be written in the form (8.24).

8.14 Express the operator of (8.35) in spherical polar co-ordinates. What relation does it bear to the Laplacian ∇^2? Hence verify (8.35) for the spherical harmonics.

8.15 Consider

$$\begin{bmatrix} a & b \\ -b^* & a^* \end{bmatrix} \begin{bmatrix} -z & y+ix \\ y-ix & z \end{bmatrix} \begin{bmatrix} a^* & -b \\ b^* & a \end{bmatrix} = \begin{bmatrix} -z' & y'+ix' \\ y'-ix' & z' \end{bmatrix},$$

where $U_R = \begin{bmatrix} a & b \\ -b^* & a^* \end{bmatrix}$ is unitary with determinant $+1$, and where

x, y, z are real. Prove: (i) x', y', z' are real, (ii) $x'^2 + y'^2 + z'^2 = x^2 + y^2 + z^2$, (iii) $(\widetilde{U_1 U_2})^* = \tilde{U}_2^* \tilde{U}_1^*$. Hence show there is a homomorphism (appendix B) between the matrices U_R and the rotational transformations R of co-ordinates from (x, y, z) to (x', y', z').

8.16 In (8.1), note that iI_ξ is a real operator, in the sense that $iI_\xi\,\psi$ is real if ψ is real. Hence take the complex conjugate of (8.18), and show that the functions $u_m{}^*$ transform under rotations in exactly the same way as the functions $(-1)^{j-m}u_{-m}$.

8.17* Discuss the relationship between the full rotation group and the group of unitary 2×2 matrices with determinant $+1$. Show how this can be used to derive the irreducible representations of the full rotation group (Van der Waerden 1932, p. 57).

8.18 Derive a general expression for the elements $D_{\alpha\beta}{}^{(j)}(\chi, \theta, \phi)$ of the matrix representing a rotation by Eulerian angles χ, θ, ϕ (8.22) in the representation $D^{(j)}$. Particularize the answer to the rotation $R(\theta, x)$ and compare with the result of problem 8.12. Hint: use problem 8.11 and equation (8.24).

9. Reduction of the Product Representation $D^{(j)} \times D^{(j')}$

Product representations

In quantum mechanics it frequently happens that we wish to multiply wave functions or other functions by one another. For instance it may be convenient to express a two-electron wave function $\psi(\mathbf{r}_1, \mathbf{r}_2)$ in terms of products $\psi_1(\mathbf{r}_1)\psi_2(\mathbf{r}_2)$ of two one-electron functions ψ_1, ψ_2. Similarly when calculating a quantum mechanical matrix element $\int \psi_i{}^* x \psi_j \, d\tau$ we multiply three functions together. Let us therefore consider the vectors u_2, u_1, u_0, u_{-1}, u_{-2} and v_1, v_0, v_{-1} transforming respectively as standard base vectors according to the representations $D^{(2)}$ and $D^{(1)}$ of the rotation group. Then we can form the $5 \times 3 = 15$ different products $u_m v_\mu$, and these transform into one another under rotations so that they form the basis of a representation of the rotation group. This representation is written symbolically as $D^{(2)} \times D^{(1)}$.

Reduction of a product representation

The vector space of dimension 15 spanned by the $u_m v_\mu$ is called a product space. It and the representation $D^{(2)} \times D^{(1)}$ can now be reduced using the method of the last section which we may summarize as follows:

(i) Reduce the space according to the axial rotation group about the z-axis, so that each base vector belongs to a representation $\exp(iM\phi)$ for some value of M.

(ii) Pick out the vector (or one of the vectors) U_J with the largest value of M, this value being J.

(iii) Using the operator I_- and (8.18), define the vectors U_J, $U_{J-1}, \ldots U_{-J}$ transforming according to $D^{(J)}$.

(iv) Orthogonalize all other vectors to the U_M. This can be done without combining functions with different M since such functions are automatically orthogonal.

(v) Repeat the process (ii) to (iv) in the vector space remaining until the whole space has been reduced. At each stage the vector space remaining is still invariant under all rotations by appendix C, lemma 3.

In the present case the space $(\ldots, u_m v_\mu, \ldots)$ is already reduced according to the axial rotation group about the z-axis. For

$$R(\phi, z)u_m v_\mu = (e^{im\phi}u_m)(e^{i\mu\phi}v_\mu) = e^{i(m+\mu)\phi}u_m v_\mu, \tag{9.1a}$$

so that the base vector $u_m v_\mu$ belongs to the value

$$\boxed{M = m + \mu.} \tag{9.1b}$$

Thus all the base vectors with their values of M written underneath them are:

	u_2v_1	u_2v_0	u_2v_{-1}	u_1v_1	u_1v_0	u_1v_{-1}
$M =$	3	2	1	2	1	0
	u_0v_1	u_0v_0	u_0v_{-1}	$u_{-1}v_1$	$u_{-1}v_0$	$u_{-1}v_{-1}$
$M =$	1	0	−1	0	−1	−2
	$u_{-2}v_1$	$u_{-2}v_0$	$u_{-2}v_{-1}$			
$M =$	−1	−2	−3			

The vector $U_3 = u_2v_1$ has the largest value of M namely 3, and from it using (8.18) a vector U_2 with $M = 2$ may be constructed. This will not be u_2v_0 or u_1v_1 but a linear combination of them. At the present we are not interested in just what the correct linear combination is. The important fact is that whatever the linear combination, we obtain in step (iv) one other linearly independent vector, U'_2 say, orthogonal to U_2 because the value $M = 2$ occurs twice in our table. Similar considerations apply for the other values of M. Thus we can tick off on the list once each $M = 3$, 2, 1, 0, −1, −2 corresponding to a set of vectors U_M transforming according to $D^{(3)}$. There remains (step (iv) above) a set of vectors U'_M with the values $M = 2, 1, 1, 0, 0, -1, -1, -2$. Similarly from among these we can construct a set transforming according to $D^{(2)}$ and then a set according to $D^{(1)}$, which uses up all the 15 linearly independent vectors. Thus

$$D^{(2)} \times D^{(1)} = D^{(3)} + D^{(2)} + D^{(1)}.$$

Similarly if two sets of vectors transform according to $D^{(j)}$ and $D^{(j')}$, the product space transforming according to $D^{(j)} \times D^{(j')}$

contains the value $M = \pm(j+j')$ once, $M = \pm(j+j'-1)$ twice, $M = \pm(j+j'-2)$ three times, etc. The values M where $-|j-j'| \leqslant M \leqslant |j-j'|$ are contained $2j+1$ times each if $j \leqslant j'$ or $2j'+1$ times if $j' \leqslant j$. Proceeding in the same way as above, base vectors can be found transforming according to the representations $D^{(J)}$ with $J = j+j', j+j'-1, \ldots, |j-j'|$, and thus

$$\boxed{D^{(j)} \times D^{(j')} = D^{(j+j')} + D^{(j+j'-1)} + D^{(j+j'-2)} + \ldots + D^{|j-j'|}.} \tag{9.2}$$

For example

$$\begin{aligned} D^{(2)} \times D^{(1/2)} &= D^{(5/2)} + D^{(3/2)} \\ D^{(1)} \times D^{(1)} \times D^{(1)} &= D^{(1)} \times (D^{(2)} + D^{(1)} + D^{(0)}) \\ &= D^{(3)} + D^{(2)} + D^{(1)} + D^{(2)} + D^{(1)} \\ &\quad + D^{(0)} + D^{(1)}. \end{aligned} \tag{9.3}$$

The new base vectors

In simple cases there is no difficulty in carrying out the procedure of steps (i) to (v) above in complete detail and obtaining the actual base vectors transforming according to the different irreducible components $D^{(J)}$. For instance if u_1, u_0, u_{-1} and v_1, v_0, v_{-1} each transform according to $D^{(1)}$, the product representation is $D^{(1)} \times D^{(1)} = D^{(2)} + D^{(1)} + D^{(0)}$. There is only one vector $U_2^{(2)} = u_1 v_1$ with $M = 2$. From problem 8.10 and equation (8.18), we can define the next vector $U_1^{(2)}$ by operating on $U_2^{(2)}$ with I_-:

$$\begin{aligned} I_- U_2^{(2)} &= (I_- u_1)v_1 + u_1(I_- v_1) \\ &= 2^{1/2}(u_0 v_1 + u_1 v_0) \\ &= 2U_1^{(2)} \text{ by definition (8.18).} \end{aligned}$$

In this way we obtain the following base vectors $U_M^{(J)}$ transforming according to $D^{(2)}$, $D^{(1)}$ and $D^{(0)}$.

$$\begin{aligned} U_2^{(2)} &= N_2 u_1 v_1 \\ U_1^{(2)} &= (N_2/2^{1/2})(u_0 v_1 + u_1 v_0) \\ U_0^{(2)} &= (N_2/6^{1/2})(u_{-1} v_1 + 2u_0 v_0 + u_1 v_{-1}) \\ U_{-1}^{(2)} &= (N_2/2^{1/2})(u_{-1} v_0 + u_0 v_{-1}) \\ U_{-2}^{(2)} &= N_2 u_{-1} v_{-1} \end{aligned} \tag{9.4a}$$

$$\begin{aligned} U_1^{(1)} &= (N_1/2^{1/2})(u_0 v_1 - u_1 v_0) \\ U_0^{(1)} &= (N_1/2^{1/2})(u_{-1} v_1 - u_1 v_{-1}) \\ U_{-1}^{(1)} &= (N_1/2^{1/2})(u_{-1} v_0 - u_0 v_{-1}) \end{aligned} \tag{9.4b}$$

$$U_0^{(0)} = (N_0/3^{1/2})(u_{-1} v_1 + u_1 v_{-1} - u_0 v_0). \tag{9.4c}$$

Here N_2, N_1, N_0 are normalizing constants.

If (x_1, y_1, z_1) and (x_2, y_2, z_2) are the components of two ordinary vectors† $\mathbf{r}_1$ and $\mathbf{r}_2$, we can from (8.21) put

$$\begin{aligned} u_1 &= x_1+iy_1, \quad u_0=-2^{1/2}z_1, \quad u_{-1}=-(x_1-iy_1), \\ v_1 &= x_2+iy_2, \quad v_0=-2^{1/2}z_2, \quad v_{-1}=-(x_2-iy_2). \end{aligned} \tag{9.5}$$

The component (9.4c) transforming according to $D_0^{(0)}$ then becomes $-(2\sqrt{3}/3)(x_1x_2+y_1y_2+z_1z_2)$, which is proportional to the scalar product $\mathbf{r}_1 . \mathbf{r}_2$. The components $U_M^{(1)}$ become $-(z_1x_2-z_2x_1) \pm i(y_1z_2-y_2z_1)$ and $-i\sqrt{2}(x_1y_2-x_2y_1)$, so that from (8.21) the quantities

$$(y_1z_2-y_2z_1), \qquad (z_1x_2-z_2x_1), \qquad (x_1y_2-x_2y_1)$$

transform under rotations in the same way as the components (x, y, z) of an ordinary vector. As expected, they are the components of the vector product $\mathbf{r}_1 \wedge \mathbf{r}_2$. Consider now the components T_{ij} of a second rank tensor. By definition they transform in the same way as the products $r_{1i}r_{2j}$ where $r_{1i} = x_1, y_1, z_1$ for $i = 1, 2, 3$, etc. Hence the nine T_{ij} form a vector space of dimension nine transforming according to $D^{(1)} \times D^{(1)} = D^{(2)} + D^{(1)} + D^{(0)}$. The $D^{(0)}$ component is the scalar $T_{11}+T_{22}+T_{33}$, and the quantities $T_{23}-T_{32}$, $T_{31}-T_{13}$, $T_{12}-T_{21}$ transform like an ordinary vector according to $D^{(1)}$ (cf. Milne 1948, p. 46). The remaining five linearly independent combinations $T_{23}+T_{32}$, $T_{31}+T_{13}$, $T_{12}+T_{21}$, $T_{11}-T_{22}$, $T_{22}-T_{33}$ transform according to $D^{(2)}$, and form a symmetric second rank tensor with a zero sum of diagonal elements.

It is sometimes convenient to have the relations (9.4) expressed the other way round. We can solve them as simultaneous equations for the $u_m v_\mu$, and obtain

$$\begin{aligned} u_0v_1 &= (1/2^{1/2})(U_1^{(2)}/N_2 + U_1^{(1)}/N_1), \\ u_0v_0 &= (1/3^{1/2})(2^{1/2}U_0^{(2)}/N_2 - U_0^{(0)}/N_0), \\ u_0v_{-1} &= (1/2^{1/2})(U_{-1}^{(2)}/N_2 - U_{-1}^{(1)}/N_1), \text{ etc.} \end{aligned} \tag{9.6}$$

Wigner coefficients

The coefficients occurring in (9.4) and (9.6) are particular examples of certain general coefficients known as Wigner coefficients. Let

$$u_m, \; -j \leqslant m \leqslant j, \quad \text{and} \quad v_{m'}, \; -j' \leqslant m' \leqslant j',$$

† We shall refer to the common type of vector in three-dimensional space, such as the position vector $\mathbf{r}$ of a particle, as an *ordinary vector*, to distinguish it from the more general concept of a vector in a more general vector space.

be two sets of standard base vectors transforming respectively like (8.18) according to $D^{(j)}$ and $D^{(j')}$. Then by the procedure of steps (i) to (v) above, we can find in the product space $u_m v_{m'}$ several sets of vectors

$$U_M^{(J)} = N_J \sum_{m,m'} (jj'mm'|JM) u_m v_{m'} \tag{9.7}$$

transforming according to $D^{(J)}$, where from (9.2) $J = j+j', j+j'-1, \ldots |j-j'|$. The coefficients $(jj'mm'|JM)$ are the Wigner coefficients, also known as Clebsch–Gordon or vector coupling coefficients. Now our method of constructing $U_M^{(J)}$ is unique and, therefore, apart from the factor N_J, the Wigner coefficients as defined are uniquely determined. This is the most important fact about them: the coefficients do not depend on the detailed nature of the u_m and $v_{m'}$ which may be complicated many-electron wave functions. For instance we already showed from (9.1) that

$$(jj'mm'|JM) = 0 \text{ unless } M = m + m'.$$

Incidentally if we had chosen base vectors u_r, v_s, U_t not transforming in the standard way (8.18) but with some other definite transformation properties, the coefficients relating them would still be uniquely determined, apart from the N_J, by these transformation properties. The coefficients would not depend on the particular form of the functions, but just on their transformation properties. For instance, in the example (9.4) we might have used $x_1, y_1, z_1, x_2, y_2, z_2$ as base vectors instead of the standard linear combinations (9.5), and this would have given somewhat different but still uniquely determined coefficients.

As before, the equations (9.7) can be inverted to give

$$u_m v_{m'} = \sum_{J,M} (jj'mm'|JM)(1/N_J) U_M^{(J)}. \tag{9.8}$$

Actually it is not obvious that the coefficients in (9.8) are the same as those in (9.7), but we shall prove in § 20 that with appropriate choice of N_J this always is so: compare for instance (9.4) and (9.6). We shall also derive a general formula for $(jj'mm'|JM)$ in § 20, but it is too cumbersome for general use so that numerical tables of Wigner coefficients are given in appendix I.

Summary

If two sets of vectors u_m and $v_{m'}$ transform according to $D^{(j)}$ and $D^{(j')}$, then the product space spanned by the vectors $u_m v_{m'}$

transforms according to a representation $D^{(j)} \times D^{(j')}$ which is reducible into the components $D^{(J)}$ given by (9.2). If all base vectors are chosen in the standard way according to (8.18), the vectors transforming according to the different $D^{(J)}$ are given in terms of the $u_m v_{m'}$ by the Wigner coefficients.

PROBLEMS

9.1 From (9.2) write down the irreducible components of $D^{(j)} \times D^{(0)}$, $D^{(l)} \times D^{(1/2)}$, $D^{(1/2)} \times D^{(1/2)}$, $D^{(1/2)} \times D^{(1/2)} \times D^{(1/2)}$.

9.2 Derive all the vectors (9.4) and (9.6) in detail, and verify that the same coefficients are involved whichever way round ((9.7) or (9.8)) one writes the relationship between them.

9.3 Find the vectors $U_M{}^{(J)}$ in the space $u_m v_{m'}$ where the u_m and $v_{1/2}$, $v_{-1/2}$ transform as standard base vectors according to $D^{(l)}$ and $D^{(1/2)}$ respectively.

9.4 What happens to the vectors (9.4), (9.5) if we put $x_1, y_1, z_1 = x_2, y_2, z_2$? Show that this situation is an example of the following general theorem, and construct another illustration. *Theorem*: if a set of linearly *independent* vectors ϕ_i and another set of linearly *dependent* vectors ψ_i (i.e. linearly dependent among themselves, not linearly dependent on the ϕ_i) transform with the same matrices under a group of transformations, and if the ϕ_i transform according to a reducible representation $D^{(\alpha)} + D^{(\beta)} + D^{(\gamma)} + \ldots + D^{(\nu)}$, then the ψ_i transform according to a representation $\sum D^{(r)}$, where the $D^{(r)}$ are some of the irreducible components out of the set $D^{(\alpha)}, D^{(\beta)}, D^{(\gamma)}, \ldots D^{(\nu)}$. For a proof of this theorem, see Van der Waerden 1932, p. 74.

9.5 Show that the Wigner coefficients $(jj'mm'|JM) = 1$ when $M = m + m' = \pm J = \pm (j + j')$.

9.6 Using the operators I_+, I_-, I_z, show that $\sum\limits_m (-1)^m u_{-m} v_m$ is invariant under rotations, where u_m, v_m are two sets of functions each transforming according to $D^{(j)}$. Note that (9.4c) is a special case of this.

9.7 Show algebraically that

$$\sum_{|j-j'|}^{j+j'} (2J + 1) = (2j + 1)(2j' + 1).$$

What is the significance of this result in connection with equations (9.2) and (9.7)?

9.8 T_{ij} is a symmetric second rank tensor ($T_{ij} = T_{ji}$) such that the sum of the diagonal elements is zero. Show that the components transform under rotations according to the irreducible representation $D^{(2)}$.

10. Quantum Mechanics of a Free Atom; Orbital Degeneracy

The total Hamiltonian $\mathscr{H}$ of a free atom can conveniently be divided into three parts

$$\boxed{\mathscr{H} = \mathscr{H}_{\text{orb}} + \mathscr{H}_{\text{spin}} + \mathscr{H}_{\text{nucl}}} \tag{10.1}$$

where

$$\boxed{\mathscr{H}_{\text{orb}} = -\frac{\hbar^2}{2m}\sum_i^n \nabla_i^2 - \sum_i^n \frac{Ze^2}{r_i} + \sum_{i<j}^n\sum^n \frac{e^2}{r_{ij}}.} \tag{10.2}$$

Here Ze is the charge of the nucleus, and n the number of electrons which is not equal to Z if the atom is ionized. $\mathscr{H}_{\text{spin}}$ depends on the electron spin and will be discussed in detail in the next section. $\mathscr{H}_{\text{nucl}}$ takes into account the motion of the nucleus, its finite size and the deviation of the potential from a pure Coulomb field near and inside the nucleus, and also its spin, magnetic moment, quadrupole moment, etc. It is small so that we shall neglect it completely in this chapter, but return to consider it in § 21.

Now even the simplest part of the Hamiltonian, $\mathscr{H}_{\text{orb}}$ (10.2), does not have eigenfunctions that can be expressed exactly in a convenient closed form, though with present day computing machinery it is possible to get numerical solutions with a high degree of accuracy. We shall therefore of necessity be concerned with approximation procedures and perturbation theory. In particular the wave functions we derive will usually be only the first terms in a complete expansion of the wave function, but for many purposes this is quite adequate. However, it is important to realize that this does not mean the whole theory will be approximate. The theorems of § 6 allow us to make precise statements about the transformation properties of wave functions, even though these wave functions are unknown, very complicated, many-electron functions. Consequently anything that depends purely on transformation properties can be discussed exactly. Examples of such things are the degeneracy of energy levels, qualitative splittings by perturbations, the introduction of spin angular momentum, selection rules for transitions between various levels, and the relative strengths of a group of spectral lines. In this context the wave functions we shall write down are just aids to the imagination. On the other hand, if one wants to calculate such things as the actual energies of certain levels or the absolute intensity of a spectral line, one does require a knowledge of the wave function. In this connection, therefore, we shall need to write down approximate wave functions and to have systematic ways of obtaining more accurate ones when required.

In short the main features of our group-theoretical discussion in the remainder of this chapter are the following: (i) the rigorous arguments depending on symmetry properties are clearly separated from the approximations inherent in any particular wave function; (ii) in particular the treatment of spin angular momentum is made rigorous; (iii) the argument applies generally to an atom with any number of electrons.

Self-consistent field

We shall now start to consider (10.1) in the usual quantum mechanical manner, first the largest terms of $\mathscr{H}_{\text{orb}}$, then the next smaller ones as a perturbation, and then still smaller ones in a definite sequence. As a first approximation we regard each electron moving independently in the average potential, considered as fixed, of the other electrons. Such independent motion corresponds to a single product wave function ψ of one-electron *orbitals* ϕ_i;

$$\psi = \phi_1(\mathbf{r}_1)\,\phi_2(\mathbf{r}_2)\,\ldots\,\phi_n(\mathbf{r}_n). \tag{10.3}$$

We now use the variational principle to calculate the ϕ_i that makes ψ come closest to the correct wave function. This principle† states that the lower the energy E of the wave function ψ

$$E = \frac{\int \psi^* \mathscr{H}_{\text{orb}}\,\psi\,\mathrm{d}\tau}{\int \psi^*\psi\,\mathrm{d}\tau}, \tag{10.4}$$

the closer ψ is to the true eigenfunction of $\mathscr{H}_{\text{orb}}$. The lowest energy obtainable with ψ having the form (10.3) can be calculated, and it is found that the ϕ_i are given by the set of n Hartree equations (Schiff 1955, p. 284)

$$\left[-\frac{\hbar^2}{2m}\nabla_i^2 + V_i(\mathbf{r}_i)\right]\phi_i(\mathbf{r}_i) = E_i\phi_i(\mathbf{r}_i), \tag{10.5a}$$

$$V_i(\mathbf{r}_i) = -\frac{Ze^2}{r_i} + \sum_{j\neq i}\int \phi_j^*(\mathbf{r}_j)\phi_j(\mathbf{r}_j)\frac{e^2}{r_{ij}}\,\mathrm{d}\tau_j. \tag{10.5b}$$

As anticipated, the potential energy V_i is the expectation value (10.5b) of the electrostatic potential energy due to the other electrons and the nucleus. It is called the self-consistent field because through it each ϕ_i depends on every ϕ_j, so that we have to obtain simultaneously a self-consistent set of solutions ϕ_i of all the equations.

† This is a very rough formulation of the variational principle, but it suffices for the present purposes (Schiff 1955, p. 171).

$V_i(\mathbf{r}_i)$ as defined by (10.5b) is not actually spherically symmetrical, but for practical purposes of computation it is made so by averaging over all directions for fixed r_i, and we shall assume that this has been done. (10.5) is then invariant under rotations so that we obtain degenerate sets of eigenfunctions transforming according to $D^{(l_i)}$. These in fact have the form (Schiff 1955, p. 69)

$$\boxed{\phi_i(\mathbf{r}_i) = f_{n_i l_i}(r_i)\, Y_{l_i m_{li}}(\theta_i, \theta_i),} \tag{10.6}$$

where the spherical harmonic Y_{lm} shows that they transform according to $D^{(l_i)}$. If the radial function f has $n_i - l_i - 1$ nodes in it, it is designated by the *principal quantum number* n_i and the *orbital angular momentum quantum number* l_i. The latter derives its name from the fact that from (8.35), ϕ_i has angular momentum $\sqrt{[l_i(l_i+1)]}\hbar$. Instead of using l, an orbital is usually specified using the *spectroscopic notation* by a symbol nx, where n is the principal quantum number and x is the letter

$$s, p, d, f, g, h, \ldots \text{ for } l = 0, 1, 2. 3, 4, 5, \ldots \tag{10.7}$$

A state of the form (10.3) is specified by a symbol such as $(1s)^2 2s(2p)^2$ where the indices denote the number of $1s$ etc. orbitals occurring. In general the orbital eigenvalues E_i of (10.5) in order of increasing energy are $1s$, $2s$, $2p$, $3s$, $3p$, $3d$, $4s$, . . . , the $1s$ orbital having the lowest energy because it lies closest to the nucleus. Thus to obtain the lowest energy for (10.3), we would expect to take each of the ϕ_i as a $1s$ orbital. However as we shall see in § 12, the exclusion principle prevents the atom from collapsing in this way.

The spectroscopic notation does not specify an orbital or a set of orbitals (10.3) completely, for it leaves each of the m_l arbitrary. For instance a symbol such as $(1s)^2 2s(2p)^2$ corresponds to a whole set of

$$(2l_1+1)(2l_2+1)\ldots(2l_i+1)\ldots \tag{10.8}$$

wave functions (10.3) given by all possible combinations of the m_{li}. Since the orbitals (10.6) are degenerate, these wave functions all have the same energy (10.4). They are known collectively as a *configuration*, and this name is sometimes also applied to their energy. All the wave functions (10.3) of one configuration are eigenfunctions of a central self-consistent field Hamiltonian

$$\mathscr{H}_{\text{cscf}} = -\frac{\hbar^2}{2m}\sum_i \nabla_i^2 + \sum_i V_i(r_i). \tag{10.9}$$

They all belong to the same eigenvalue of $\mathscr{H}_{\text{cscf}}$, as follows from the fact that (10.9) is invariant under separate rotations of the individual electron co-ordinates x_i, y_i, z_i alone. All the wave functions can be generated from the one with every $m_{li} = l_i$ by using the operators $(I_-)_{i\ \mathrm{orb}}$ of (8.32) and the relation (8.18).

Energy levels of $\mathscr{H}_{\mathrm{orb}}$

Since the configuration of wave functions (10.3) are degenerate eigenfunctions of $\mathscr{H}_{\mathrm{cscf}}$, we obtain the energy levels of $\mathscr{H}_{\mathrm{orb}}$ by applying

$$\mathscr{H}_{\mathrm{es}} = \mathscr{H}_{\mathrm{orb}} - \mathscr{H}_{\mathrm{cscf}} \tag{10.10}$$

as a perturbation. This represents the difference between the true electrostatic interaction between the electrons and the self-consistent field. It will produce a splitting of each configuration, because the total Hamiltonian $\mathscr{H}_{\mathrm{orb}}$ is now no longer invariant under *separate* rotations of the co-ordinates x_i, y_i, z_i alone, so that the degeneracy (10.8) no longer applies. However, $\mathscr{H}_{\mathrm{orb}}$ is invariant under *simultaneous* rotations of all the co-ordinates x_i, y_i, z_i. Under these rotations the wave functions (10.3) transform according to the reducible representation

$$D^{(l_1)} \times D^{(l_2)} \times \ldots \times D^{(l_n)} = \sum D^{(L)}, \tag{10.11}$$

so that by Theorem 3 of § 6 the configuration is split into a set of levels according to the different values of L occurring in (10.11). For example the configuration $(1s)^2 2s2p3p$ would give levels with $L = 2$, 1 and 0.

Using the Wigner coefficients (9.7), we can write down the actual linear combinations transforming according to $D^{(L)}$. We first combine the $\phi_1(\mathbf{r}_1)$ and $\phi_2(\mathbf{r}_2)$ to form the functions

$$\sum_{m_{l1}m_{l2}} (l_1 l_2 m_{l1} m_{l2} | L_2 M_2)\, \phi_{n_1 l_1 m_{l1}}(\mathbf{r}_1)\, \phi_{n_2 l_2 m_{l2}}(\mathbf{r}_2)$$

transforming according to $D^{(L_2)}$ where $D^{(l_1)} \times D^{(l_2)} = \sum D^{(L_2)}$ (9.2). These are then combined with the $\phi_3(\mathbf{r}_3)$ to form functions transforming according to $D^{(L_3)}$ where $D^{(L_2)} \times D^{(l_3)} = \sum D^{(L_3)}$. These are then combined with the $\phi_4(\mathbf{r}_4)$ etc. until we reach $L_n = L$. The wave functions then are

$$\begin{aligned} \psi_{LM_L} = \sum (l_1 l_2 \ldots | L_2 .)(L_2 l_3 \ldots | L_3 .)(L_3 l_4 \ldots | L_4 .) \ldots \\ (L_{n-1} l_n \ldots | LM_L)\, \phi_{m_1}(\mathbf{r}_1)\, \phi_{m_2}(\mathbf{r}_2) \ldots \phi_{m_n}(\mathbf{r}_n). \end{aligned} \tag{10.12}$$

Here we have dropped for simplicity all but the most important suffices. The summation is over all combinations of the m_i such that $\Sigma m_i = M_L$. Now, although the wave functions (10.12) have the correct symmetry properties, there is no reason why they should be eigenfunctions of $\mathscr{H}_{\text{orb}}$, and in general, we have to write the correct eigenfunctions in the form

$$\psi_{LM_L}(\mathbf{r}_1,\mathbf{r}_2,\ldots\mathbf{r}_n)$$
$$= \alpha_0(\psi_{LM_L}\text{ given by (10.12))}$$
$$+\sum\alpha_n(\psi_{LM_L}\text{ from other configurations}). \quad (10.13)$$

Here $\alpha_0 \approx 1$ and the other coefficients α_n are usually small. The mixing in of wave functions from other configurations is called *configurational interaction*. Since the whole wave function has to transform according to $D^{(L)}$, only terms with the same L can give contributions to (10.13). When the energy difference between various configurations is comparable with the splittings produced by $\mathscr{H}_{\text{es}}$, then the corresponding coefficients in (10.13) are all large and the configurational mixing is important in calculating the energies of the various levels. However, their number and L values cannot be affected, since by Theorem 3 of §6 these depend only on symmetry properties. In (10.12), it happens occasionally that we obtain two unrelated sets of functions with the same L from one configuration. In this case we have to take linear combinations between them to obtain even approximate eigenfunctions, and both sets will appear in (10.13) with large coefficients.

Summary

The Hartree self-consistent field equation (10.5) is invariant under rotation of the co-ordinates x_i, y_i, z_i, so that its eigenfunctions transform according to $D^{(l_i)}$ and its energy levels are $(2l_i + 1)$-fold degenerate. The central self-consistent field Hamiltonian $\mathscr{H}_{\text{cscf}}$ (10.9) is invariant under separate rotations of the co-ordinates of the different electrons, so that its energy levels have the degeneracy (10.8). The Hamiltonian $\mathscr{H}_{\text{orb}}$ (10.2) is invariant under simultaneous rotation of all electron co-ordinates, its eigenfunctions transform according to $D^{(L)}$ and its energy levels are $(2L + 1)$-fold degenerate where L is given by (10.11).

PROBLEMS

10.1 What values of L could the following configurations give rise to: $(1s)^2 2p3s3d$, $(1s)^2(2s)^2(2p)^2$, $(1s)^2(2s)^2 2p3p$, $(1s)^2(2s)^2(2p)^3$.

10.2 Using the Wigner coefficients of (9.4), write down the wave function (10.12) transforming according to $D^{(0)}$ in the configuration $(1s)^2(2s)^2(2p)^3$.

10.3 Using problem 9.5, write down the wave function ψ_{LM_L} with $L = M = 3$ in the configuration $(1s)^2(2s)^2(2p)^3$. By operating with I_- and using (8.18), generate the other wave functions with $L = 3$.

10.4 An atom is in a state with $L = 0$. Show that it has a spherically symmetrical charge distribution even when configurational interaction is fully taken into account.

10.5* Discuss the degeneracy of the states of a hydrogen atom. Note that the pure Coulomb $1/r$ nature of the potential introduces an extra symmetry into the Hamiltonian, which in turn is responsible for the extra degeneracy (Fock 1935).

11. Quantum Mechanics of a Free Atom including Spin

The electron spin

In the 1920's experimental evidence accumulated which indicated that electrons cannot be described correctly by the Hamiltonian $\mathscr{H}_{\text{orb}}$ (10.2), or rather its equivalent in the old quantum theory of the time. It was found necessary to make the following additional assumptions, which have no classical analogue, about the nature of electrons.

I. Each electron has an intrinsic (or internal) angular momentum $\mathbf{s}$ of magnitude $\sqrt{\frac{3}{4}}\hbar$. This is called its spin angular momentum, or spin for short.

II. The component of $\mathbf{s}$ along any direction can only have the values $\pm\frac{1}{2}\hbar$. It is usual to choose a definite z-axis and to describe the state of the electron using an additional or intrinsic co-ordinate σ_z, where $\sigma_z = \pm 1$ corresponds to the component s_z being $\pm\frac{1}{2}\hbar$. This is called the spin co-ordinate.

III. In general the effects of spin are small compared with the Coulomb repulsion between electrons, or more precisely with the self-consistent field of (10.5).

IV. The electron has a permanent magnetic moment $(-e/mc)\mathbf{s}$.

An electron circling in an orbit with orbital angular momentum $\mathbf{l}$ has an effective magnetic moment $(-e/2mc)\mathbf{l}$ due to the purely classical circulation of the current. We note therefore that the ratio of spin magnetic moment to spin angular momentum is twice as large as the ratio of orbital magnetic moment to orbital angular momentum. We shall not discuss here the actual experiments, such as the Stern-Gerlach experiment and the Zeeman effect in the spectra of alkali atoms (see for instance Born 1933), that originally led Uhlenbeck and Goudsmit (1925) and others to these assumptions.

Suffice it to say that in the course of time the assumptions have been found to give a description of the nature of electrons which is in complete accordance with experiment apart from some relativistic and radiative corrections.

Actually the assumptions I to IV above are not logically independent, and we shall only make the much more limited assumption :

> *An electron has some internal (or intrinsic or spin) degree of freedom which gives it direction-dependent properties. The co-ordinate, σ_z say, describing the internal degree of freedom can only have two values, $\sigma_z = \pm 1$ say.* (11.1)

From this we shall derive in the present section the existence of the intrinsic spin angular momentum and its magnitude (items I and II above). Later in § 32 we shall look for a relativistic description of particles having such a spin momentum in ordinary space, and shall arrive at the Dirac equation from which the magnetic moment and its interactions follow (items III and IV above). Though logically sounder, it would be very inconvenient to defer all discussion of spin-orbit coupling, etc., till after § 32, so that in the present section we shall anticipate the results III and IV above, and base our discussion of spin-dependent forces on them.

Spin operators

From our assumption, we must write a wave function

$$\psi(x, y, z, \sigma_z) \tag{11.2}$$

for one electron as depending, not only on the orbital (or positional or spatial) co-ordinates x, y, z, but also on the spin co-ordinate σ_z. Let us now operate on ψ with a rotation $R(\alpha, \xi)$. Going back to our fundamental definition of what this means (§ 5), we first refer ψ to new axes OX, OY, OZ. Now the co-ordinate σ_z is defined with respect to a definite axis Oz, and will therefore be expressed also in terms of a new co-ordinate σ_Z with respect to the new axis OZ. We then replace X, Y, Z, σ_Z by x, y, z, σ_z again throughout to obtain $R\psi(x, y, z, \sigma_z)$. From this description, it is seen that R can be divided into two independent transformations

$$R(\alpha, \xi) = R_{\text{orb}}(\alpha, \xi)R_{\text{spin}}(\alpha, \xi), \tag{11.3}$$

where R_{orb} expresses only x, y, z in terms of the new axes and R_{spin} only σ_z. As in (8.2) the rotations can be expanded in powers of α with the infinitesimal rotation operators, and (11.3) becomes

$$1 + i\alpha I_\xi + O(\alpha^2) = (1 + i\alpha I_{\xi\ \text{orb}} + \ldots)(1 + i\alpha I_{\xi\ \text{spin}} + \ \ldots).$$

Equating coefficients of α, we obtain

$$I_\xi = I_{\xi\,\mathrm{orb}} + I_{\xi\,\mathrm{spin}}, \tag{11.4}$$

where $I_{\xi\,\mathrm{orb}}$ operates only on x, y, z and $I_{\xi\,\mathrm{spin}}$ only on σ_z.

We can now show that an electron has a spin (or intrinsic or internal) angular momentum. In the absence of a classical analogue to spin, the angular momentum has to be defined by the fundamental Dirac relation

$$P_q = \hbar I_q. \quad \text{[see (8.29) or (F.2)]}$$

Thus the total angular momentum of an electron is given by

$$\boxed{j_\xi = \hbar I_\xi = \hbar I_{\xi\,\mathrm{orb}} + \hbar I_{\xi\,\mathrm{spin}} = l_\xi + s_\xi, \qquad (\xi = x, y, z)} \tag{11.5}$$

where the l_ξ are the components of the orbital momentum

$$l_x = -i\hbar\left(y\frac{\partial}{\partial z} - z\frac{\partial}{\partial y}\right), \text{ etc.} \tag{8.27}$$

as already defined in § 8, and where

$$\boxed{s_\xi = \hbar I_{\xi\,\mathrm{spin}} \quad (\xi = x, y, z)} \tag{11.6}$$

are the components of an additional, non-classical, spin angular momentum **s**. The magnitude of the spin momentum will be derived in (11.14). Similarly for several electrons, their combined total angular momentum **J** is the sum of the combined orbital and spin angular momenta **L** and **S**:

$$J_\xi = \hbar I_\xi = \hbar I_{\xi\,\mathrm{orb}} + \hbar I_{\xi\,\mathrm{spin}} = L_\xi + S_\xi, \tag{11.7a}$$

$$\text{where } L_\xi = \hbar I_{\xi\,\mathrm{orb}} = \sum_i l_{\xi i} = \sum_i \hbar I_{\xi i\,\mathrm{orb}},$$

$$S_\xi = \hbar I_{\xi\,\mathrm{spin}} = \sum_i s_{\xi i} = \sum_i \hbar I_{\xi i\,\mathrm{spin}}. \quad (\xi = x, y, z) \tag{11.7b}$$

Alternatively **J** can be split up in terms of the total orbital plus spin angular momenta $\mathbf{j}_i$ of the individual electrons:

$$J_\xi = \hbar I_\xi = \sum_i \hbar I_{\xi i} = \sum_i j_{\xi i} = \sum_i (l_{\xi i} + s_{\xi i}). \quad (\xi = x, y, z) \tag{11.7c}$$

Spin-orbit coupling

We can now calculate $\mathscr{H}_{\text{spin}}$, the spin-dependent part of the total Hamiltonian (10.1), expressing it in terms of the above operators $\mathbf{s}_i$. Consider an electron in the field of the nucleus, the electron being at the position $\mathbf{r}$ relative to the nucleus which is considered to be at rest. An observer moving with the electron would see the nucleus moving past him with velocity $-\mathbf{v}$, generating because of its motion a magnetic field $\mathbf{H} = (Ze/cr^3)(-\mathbf{v}) \wedge \mathbf{r}$ at the electron. This interacts with the magnetic moment $\mathbf{m} = -e\mathbf{s}/mc$ of the electron, giving a contribution

$$-\mathbf{m}\cdot\mathbf{H} = \frac{Ze^2}{m^2c^2r^3}\,\mathbf{l}\cdot\mathbf{s}$$

to the total energy, i.e. to the Hamiltonian. Here we have put $\mathbf{l} = m\mathbf{r} \wedge \mathbf{v}$. The fact that we want the Hamiltonian expressed in a frame of reference in which the atom is at rest gives an additional factor of $\frac{1}{2}$ (Thomas 1926). We also obtain corresponding terms for the interaction of the electrons among themselves, and have therefore the following contribution to the Hamiltonian (Condon and Shortley 1951, p. 211; Heisenberg 1926)

$$\mathscr{H}_{LS} = \frac{Ze^2}{2m^2c^2}\sum_i \frac{\mathbf{l}_i\cdot\mathbf{s}_i}{r_i^3} - \frac{e^2}{mc^2}\sum_{i \neq j}\sum \left[\frac{(\mathbf{r}_i - \mathbf{r}_j) \wedge (\mathbf{v}_i - \mathbf{v}_j)}{r_{ij}^3} - \frac{1}{2}\frac{(\mathbf{r}_i - \mathbf{r}_j) \wedge \mathbf{v}_i}{r_{ij}^3}\right]\cdot\mathbf{s}_i. \tag{11.8}$$

This contribution to $\mathscr{H}_{\text{spin}}$ is known as the spin-orbit coupling because it depends on the orbital motion of the electrons as well as the spin. We shall see in § 32 how the spin-orbit coupling for one electron comes out of the relativistic Dirac equation without putting in the value of the spin magnetic moment. In addition there is the spin-spin coupling which is the direct magnetostatic interaction of the electron magnetic moments

$$\mathscr{H}_{SS} = \frac{e^2}{m^2c^2}\sum_{i < j}\sum \mathbf{s}_i\cdot\mathbf{s}_j/r_{ij}^3 - 3(\mathbf{s}_i\cdot\mathbf{r}_{ij})(\mathbf{s}_j\cdot\mathbf{r}_{ij})/r_{ij}^5. \tag{11.9}$$

Comparable in magnitude with (11.8) and (11.9) is also the classical magnetic interaction between the electron orbits viewed as currents, but we shall neglect this, since it leads to no very important effects.

Spin-dependent wave functions

In ψ (11.2) the variable σ_z is only allowed to have the discrete values $\sigma_z = \pm 1$, in distinction from the other variables x, y, z which vary continuously. There are three common ways of writing a function of a discrete variable like σ_z. They are all three completely equivalent and we shall use them interchangeably according to convenience. Let us introduce them by examples. Let $N(a)$ be the number N of children in a certain class whose age was a years last birthday. Then $N(a)$ is mathematically a well defined function of the discrete variable a which by its definition only takes integer values. Our writing $N(a)$ and thinking of it as a function of a is the first way of writing such a function and corresponds to the form (11.2) for ψ which we have already used. The second way consists of thinking of $N(a)$ as a table of values, for instance:

$a =$	$\leqslant 6$	7	8	9	10	11	$\geqslant 12$
$N(a) =$	0	2	10	13	6	1	0.

This also specifies the function completely, and similarly we can express ψ (11.2) in terms of two components in a little table:

$\sigma_z =$	$+1$	-1
$\psi(x, y, z, \sigma_z) =$	$\psi_+(x, y, z)$	$\psi_-(x, y, z)$.

Here the entries ψ_+, ψ_- are still functions of the continuous variables x, y, z, but they are two ordinary functions independent of one another and of σ_z. Thus the second way of writing ψ is in component form

$$\boxed{\psi(x, y, z, \sigma_z) = [\psi_+(x, y, z), \psi_-(x, y, z)],} \tag{11.10}$$

remembering that the two components refer to $\sigma_z = \pm 1$. Consider now the position vector $\mathbf{r}$ of a point in ordinary space. We write it as (x, y, z) in terms of three components analogously to (11.10). However we sometimes consider the vector $\mathbf{r}$ as a single entity in itself and then it may be written $x\mathbf{i} + y\mathbf{j} + z\mathbf{k}$ in terms of the unit vectors $\mathbf{i}$, $\mathbf{j}$, $\mathbf{k}$ along the three co-ordinate axes. Similarly we write

$$\boxed{\psi(x, y, z, \sigma_z) = \psi_+(x, y, z)u_+ + \psi_-(x, y, z)u_-,} \tag{11.11a}$$

where u_+, u_- are functions of σ_z only, defined by

$$\boxed{u_+(+1) = 1,\ u_+(-1) = 0; \qquad u_-(+1) = 0,\ u_-(-1) = 1.} \tag{11.12}$$

Clearly (11.11a), (11.12) is the same function ψ defined by our table, and this is our third way of writing ψ (11.2). u_+ and u_- are called spin functions or spinors. An n-electron wave function ψ is usually written in this third way as

$$\boxed{\psi = \sum_{\alpha\beta\ldots\nu} \psi_{\alpha\beta\ldots\nu}(x_1, y_1, z_1, \ldots x_n, y_n, z_n) u_{\alpha 1} u_{\beta 2} \ldots u_{\nu n}} \tag{11.11b}$$

where we have to sum over all the 2^n arrangements $\alpha\beta \ldots \nu$ of the $n+$ and $-$ signs, and where $u_+(\sigma_{z1})$ is abbreviated to u_{+1} etc.

Transformation properties of spin functions

Since all our spin operators s_x, etc., are defined in terms of infinitesimal rotations (11.6), before we can operate with them on a wave function we have to know how u_+, u_- transform under rotations R. Now any function of σ_z can be expressed in terms of u_+, u_-. In particular this applies to $R\psi$ and also Ru_+ and Ru_-. Thus we can always write

$$\begin{aligned} Ru_+ &= au_+ + bu_-, \\ Ru_- &= cu_+ + du_-, \end{aligned} \tag{11.13}$$

so that u_+, u_- transform into one another by rotations. They therefore form a basis of a representation of the rotation group, which can only be $D^{(1/2)}$ or $D^{(0)} + D^{(0)}$ since it is two-dimensional. In the latter case u_+ and u_- would be invariant under rotations and we would have no directional properties contrary to our assumption (11.1). *Hence u_+, u_- transform according to $D^{(1/2)}$*, and u_+, u_- are often distinguished by their m quantum number $m_s = \pm\frac{1}{2}$.

From this conclusion and (8.18), (8.35), we have

$$\boxed{\begin{aligned} s_z u_+ &= \tfrac{1}{2}\hbar u_+, & s_z u_- &= -\tfrac{1}{2}\hbar u_-, \\ s^2 u_+ &= \tfrac{3}{4}\hbar^2 u_+, & s^2 u_- &= \tfrac{3}{4}\hbar^2 u_-, \end{aligned}} \tag{11.14}$$

so that *u_+, u_- correspond to states with a spin angular momentum of magnitude $\sqrt{\tfrac{3}{4}}\hbar$ with components $\pm\tfrac{1}{2}\hbar$ in the z-direction* as stated at the beginning of the section.

Term wave functions

We first generalize the discussion of § 10 by including spin in all the wave functions. Since $\mathscr{H}_{\text{csef}}$ (10.9) and $\mathscr{H}_{\text{orb}}$ (10.2) do not depend on spin, their spin dependent eigenfunctions can be obtained by multiplying each of their orbital eigenfunctions by any one of the 2^n products $u_{\alpha 1}u_{\beta 2} \ldots u_{\nu n}$, where $\alpha\beta \ldots \nu$ is a set of $+$ and $-$ signs as in (11.11b). Thus the eigenfunctions of $\mathscr{H}_{\text{orb}}$ are

$$\psi = \psi_{LM_L \text{ orb}}(\mathbf{r}_1, \mathbf{r}_2, \ldots \mathbf{r}_n)u_{\alpha 1}u_{\beta 2} \ldots u_{\nu n}. \tag{11.15}$$

These 2^n functions are all degenerate, corresponding to the fact that $\mathscr{H}_{\text{orb}}$ does not depend on σ_{zi} and is invariant under a rotation $R_{\text{spin } i}$ (11.3) of the ith spin co-ordinate alone. (Compare the degeneracy (10.8) of a configuration.) However, instead of using in (11.15) the simple products $u_{\alpha 1}u_{\beta 2} \ldots u_{\nu n}$ which transform according to the reducible representation

$$D^{(1/2)} \times D^{(1/2)} \times \ldots \times D^{(1/2)} = \sum D^{(S)}, \tag{11.16}$$

we can first pick out the linear combinations U_{SM_S} transforming according to the irreducible components $D^{(S)}$. These can be written down using the Wigner coefficients as in (10.12). Then instead of (11.15), the eigenfunctions of $\mathscr{H}_{\text{orb}}$ are written

$$\psi(L, M_L, S, M_S) = \psi_{LM_L \text{ orb}} U_{SM_S}, \tag{11.17}$$

and for given L and S they form a degenerate set of $(2L+1)(2S+1)$ states called a *term*. This degeneracy corresponds to the fact that $\mathscr{H}_{\text{orb}}$ is invariant under a rotation of all spatial co-ordinates or of all spin co-ordinates simultaneously. If we combined a given $\psi_{LM_L \text{ orb}}$ with each of the U_{SM_S}, we would get 2^n wave functions (11.17) which are just linear combinations of the functions (11.15). The reason for preferring the form (11.17) is that in fact the exclusion principle does not allow us to use in (11.17) for each L all values of S occurring in (11.16). In fact we shall see in the next section that a given value of L gets combined with only *one* value of S in (11.16) and thus gives rise to only one term, not to several. In the spectroscopic notation, a term is specified by a capital letter S, P, D, F, G, etc., corresponding to $L = 0, 1, 2, 3, 4$, analogously to (10.7). The value of S is indicated by the superscript $2S+1$. For instance a term with $S = 1$, $L = 2$ is written ^3D. The superscript is pronounced "singlet", "doublet", "triplet", etc.

Splitting due to $\mathscr{H}_{\text{spin}}$

What splittings occur if we now include $\mathscr{H}_{\text{spin}}$ in the Hamiltonian? We first have to derive the symmetry transformations of $\mathscr{H}_{\text{spin}}$. In § 8 it was shown that if

$$\boldsymbol{\xi} = l\mathbf{i} + m\mathbf{j} + n\mathbf{k}$$

is a unit vector along some axis 0ξ, then I_ξ is given by

$$I_\xi = lI_x + mI_y + nI_z. \tag{11.18}$$

Thus if R is a transformation to new co-ordinate axes 0ξ, 0η, 0ζ where

$$\xi = lx + my + nz,$$

R transforms I_x into I_ξ (11.18), and we see that I_x, I_y, I_z behave under rotations like the components of an ordinary vector. It follows from the definitions (8.28), (8.30), (8.33), (11.5), etc., that the components of the angular momenta $\mathbf{s}_i$, $\mathbf{l}_i$, $\mathbf{J}$, etc., all transform under rotations like the components of vectors, as we have anticipated by our notation. Now $\mathscr{H}_{\text{spin}}$ contains products of vectors, for instance $\mathbf{l}_i \cdot \mathbf{s}_j$ (cf. (11.8) and (11.9)). The Hamiltonian is therefore no longer invariant under orbital and spin rotations separately. However, $\mathbf{l}_i \cdot \mathbf{s}_j$ is invariant under a simultaneous rotation in both orbital and spin space, because it is the scalar product of two vectors and transforms according to the identity representation $D^{(0)}$, and the same applies to all the parts of $\mathscr{H}_{\text{spin}}$ (11.8) and (11.9). Hence the eigenfunctions of $\mathscr{H}_{\text{orb}} + \mathscr{H}_{\text{spin}}$ can be sorted out according to the representations $D^{(J)}$ of the combined-orbital-and-spin rotation group.

In particular the term functions (11.17) transform according to

$$D^{(L)} \times D^{(S)} = \sum D^{(J)} \tag{11.19a}$$

under simultaneous orbital and spin rotations, where the values

$$J = L + S,\ L + S - 1, \ldots |L - S| \tag{11.19b}$$

are given by (9.2). By Theorem 3 of § 6 we therefore expect the term to be split into separate levels, one for each value of J, and these are still $(2J + 1)$-fold degenerate. In the spectroscopic notation the value of J is indicated by a subscript. For example the spin-orbit coupling splits the term ^3D into the levels $^3D_3$, $^3D_2$ and $^3D_1$. If we want to specify the level even more completely we write in front the configuration from which the term arises.

The procedure just outlined is particularly profitable when the splittings produced by $\mathscr{H}_{\text{spin}}$ are small compared with energy separations between terms. For then we obtain good approximations to the eigenfunctions by using the Wigner coefficients to pick out from the term functions $\psi(L, M_L, S, M_S)$ the linear combinations

$$\psi(L, S, J, M_J) = \sum_{M_L M_S} (LSM_L M_S | JM_J)\, \psi(L, M_L, S, M_S) \tag{11.20}$$

transforming according to $D^{(J)}$. This situation is known as *Russell–Saunders coupling* and applies particularly to the light atoms.

However $\mathscr{H}_{\text{orb}} + \mathscr{H}_{\text{spin}}$ is not invariant under separate orbital and spin rotations, so that L, M_L, S, M_S, are no longer good quantum numbers for characterizing a state. As regards the wave functions this means that the exact eigenfunctions are not just (11.20) but contain, in addition, contributions with the same J and M_J from other terms with different L and S. In fact in the very heavy atoms (and likewise for nuclear forces), it happens that the effects of $\mathscr{H}_{\text{spin}}$ are larger than the splittings due to $\mathscr{H}_{\text{es}}$, so that the wave functions are very mixed. One then obtains better zero order approximations to the correct eigenfunctions by following the scheme known as *jj coupling*. In the central self-consistent field approximation, all terms in (11.8) vanish except the ones which reduce to

$$\boxed{\begin{aligned} \mathscr{H}'_{\text{spin}} &= \sum \xi(r_i)\mathbf{l}_i\cdot\mathbf{s}_i, \\ \text{where}\quad \xi(r) &= \frac{1}{2m^2c^2}\frac{1}{r}\frac{\mathrm{d}V_i}{\mathrm{d}r}. \end{aligned}} \tag{11.21}$$

This can be deduced directly from (11.8) or more simply as in equation (32.26) from the Dirac equation. We also neglect $\mathscr{H}_{SS}$ (11.9). With this approximation $\mathscr{H}_{\text{cscf}} + \mathscr{H}'_{\text{spin}}$ is invariant under a simultaneous rotation of the space and spin co-ordinates of the *separate* electrons, and its eigenfunctions are just products

$$\psi_{j_1 m_{j1}}(\mathbf{r}_1, \sigma_{z1})\,\psi_{j_2 m_{j2}}(\mathbf{r}_2, \sigma_{z2}) \ldots \psi_{j_n m_{jn}}(\mathbf{r}_n, \sigma_{zn}). \tag{11.22}$$

Here the spin-dependent one-electron orbitals ψ_{jm_j} transform according to $D^{(l)} \times D^{(1/2)} = \sum D^{(j)}$. The products (11.22) transform according to

$$D^{(j_1)} \times D^{(j_2)} \times \ldots \times D^{(j_n)} = \sum D^{(J)},$$

and we can use the Wigner coefficients as in (10.12) to pick out the linear combinations $\psi(j_1, j_2, \ldots j_n, J, M_J)$ transforming according to the $D^{(J)}$. These will be the zero order wave functions when we split the configuration (11.22) by applying the remainder of the Hamiltonian, namely $\mathscr{H}_{\text{es}} + \mathscr{H}_{\text{spin}} - \mathscr{H}'_{\text{spin}}$, as a perturbation. There will of course be again some configurational mixing as in (10.13).

Parity

As already noted in § 3, $\mathscr{H}_{\text{orb}}$ is invariant under the space inversion Π (3.11). Now Π commutes with every rotation R, $\Pi R = R\Pi$. Hence if $(\psi_1, \ldots, \psi_n)$ is a vector space invariant under Π

and transforming irreducibly under rotations, the matrix $D(\Pi)$ representing Π commutes with all the irreducible matrices $D(R)$ representing the rotations. By Schur's lemma, appendix D, we have $D(\Pi) = wE$ where E is the unit matrix. Since Π^2 is the identity transformation, we have $w = 1$ or $w = -1$ when the representation is single-valued, and corresponding functions ψ such that $\Pi\psi = \psi$ or $\Pi\psi = -\psi$ are said to have *even or odd parity*. As regards the functions u_+ and u_-, Πu_+ and Πu_- must be expressible again in terms of u_+, u_- as in (11.13), so that the vector space (u_+, u_-) is invariant under Π. However in this case it transforms according to the double-valued representation $D^{(1/2)}$ under rotations. The identity transformation is represented by the two matrices E and $-E$, so that

$$\Pi u_+ = wu_+, \qquad \Pi u_- = wu_- \tag{11.23a}$$

where $w = 1, -1, i$, or $-i$. It may be verified *a posteriori* that all the results of physical significance which are derived using Π, e.g. selection rules, are quite independent of which value of w is chosen in (11.23). Briefly the reason for this is that physically observable quantities contain products like $\psi_i^*\psi_j$ so that $\Pi\psi_i^*\psi_j$ contains the factor $(w^*w)^n$ where n is the number of electrons. This factor is always unity for each value of w (11.23). For simplicity therefore we choose

> the spin functions u_+, u_- to have even parity $w = 1$.

(11.23b)

To determine the parity of the various wave functions we have written down, we first note that the spherical harmonic Y_{lm} can be expressed as a polynomial of degree l in x/r, y/r, z/r (§ 8). It therefore has parity $(-1)^l$. Consequently the configurational wave functions (10.3) have parity

$$\boxed{w = (-1)^{\Sigma l_i}} \tag{11.24}$$

and the same applies to their linear combinations, the ψ_{LM_L} (10.12). Now since $\mathscr{H}_{\text{orb}}$ is invariant under Π, its eigenfunctions have definite parity, so that the configurational mixing in (10.13) can only be between configurations with the same parity. When we include the spin functions U_{SM_S} in (11.17), the parity is still (11.24) since we have already made the convention that all spin functions have even parity. $\mathscr{H}_{\text{spin}}$ is also invariant under Π, since all the $\mathbf{l}_i$ and $\mathbf{s}_i$ are (appendix F). Thus $\mathscr{H}_{\text{orb}} + \mathscr{H}_{\text{spin}}$ is invariant under Π, and its eigenfunctions $\psi(L, S, J, M_J)$ (11.20) can be assigned a definite parity (11.24) which is determined by the configuration from which the level arises. This is true no matter how much mixing between configurations and terms there is in the wave functions.

Summary

The electron spin has been introduced and the property of spin angular momentum deduced. $\mathscr{H}_{\mathrm{orb}}$ is invariant under separate rotations of all orbital co-ordinates and of all spin co-ordinates, so that its eigenfunctions transform according to $D^{(L)}$ and $D^{(S)}$ under these rotations. Hence its energy levels, called terms, are labelled by quantum numbers L and S, and have a degeneracy of $(2L+1)(2S+1)$. $\mathscr{H}_{\mathrm{orb}} + \mathscr{H}_{\mathrm{spin}}$ is invariant under simultaneous rotations of all orbital and spin co-ordinates, and we get sets of eigenfunctions transforming according to $D^{(J)}$ given by (11.19). A level characterized by J is $(2J+1)$-fold degenerate. The total Hamiltonian is invariant under the space inversion Π, so that the parity given by (11.24) is also an exact quantum number.

PROBLEMS

11.1 Prove that $s_x u_+ = \frac{1}{2}\hbar u_-$, $s_x u_- = \frac{1}{2}\hbar u_+$. Show that if we write the wave function (11.10) as a column matrix $\begin{bmatrix}\psi_+\\ \psi_-\end{bmatrix}$ then we have to write any one-electron operator as a 2×2 matrix. In particular show that the matrix for a one-electron Hamiltonian is obtained by multiplying the spin-independent part by the 2×2 unit matrix **1** and by putting in the spin-dependent part

$$s_x = \frac{1}{2}\hbar\begin{bmatrix} \cdot & 1 \\ 1 & \cdot \end{bmatrix}, \qquad s_y = \frac{1}{2}\hbar\begin{bmatrix} \cdot & -i \\ i & \cdot \end{bmatrix}, \qquad s_z = \frac{1}{2}\hbar\begin{bmatrix} 1 & \cdot \\ \cdot & -1 \end{bmatrix}.$$

These matrices, without the factor $\frac{1}{2}\hbar$, are known as the Pauli spin matrices $\sigma_x, \sigma_y, \sigma_z$. Show that they have the property

$$\sigma_x\sigma_y = -\sigma_y\sigma_x = i\sigma_z, \text{ etc.,} \qquad \sigma_x^2 = \sigma_y^2 = \sigma_z^2 = 1.$$

11.2 Express the function $\sigma_z(\psi_+u_+ + \psi_-u_-)$ in component form, (i) when σ_z is considered as the matrix $\begin{bmatrix} 1 & \cdot \\ \cdot & -1 \end{bmatrix}$ in the sense of problem 11.1, and (ii) when σ_z is regarded as a co-ordinate as in (11.11a). Hence show that the double meaning of σ_z can lead to no confusion.

11.3 The normalized pure spin function u is such that $u = 1$ when the ξ component of **s** along the ξ-axis is $\frac{1}{2}\hbar$, where 0ξ makes spherical polar angles θ, ϕ with the z-axis. Express u in terms of u_+, u_-.

11.4 It is found that an electron in a state ψ has equal probabilities of having its spin parallel or antiparallel to the z-axis at any particular point $\mathbf{r}$. What form does ψ have, expressed in terms of u_+ and u_-?

11.5 What values of S can we have for an atom with three electrons? Write down the spin functions U_{SM_S} for $S = \frac{3}{2}$ (use the method of problem 10.3).

11.6 Show that $\sum_{|L-S|}^{L+S} (2J+1) = (2L+1)(2S+1)$, and interpret this with respect to the degeneracy of a term.

11.7 Show that an S term ($L = 0$) does not necessarily have even parity.

11.8 Starting with a particular configuration, show that one obtains the same J values using the Russell–Saunders coupling and the jj coupling schemes.

11.9 Show that the vectors $\mathbf{r}_1$ and $\mathbf{r}_1 \wedge \mathbf{r}_2$ respectively do and do not change sign under the space inversion Π. Such vectors are called regular vectors and pseudovectors respectively. Classify the following vectors into regular and pseudo: velocity, momentum, grad, angular momentum, spin, force, electric field, magnetic field, magnetic vector potential, magnetic moment. (Note that we define the vector product $\mathbf{r}_1 \wedge \mathbf{r}_2$ to have the components $y_1z_2 - y_2z_1$, etc., in either a left-handed or a right-handed reference frame. Similarly, to obtain the transformation properties of electric and magnetic fields, we require that Maxwell's equations be the same written for a left- or right-handed set of axes.) Show that the Hamiltonian for an atom is invariant under Π when the interactions among the electrons and with the nucleus (§ 21) are included, and the interaction with a uniform magnetic field. Hence, deduce that parity is an exact quantum number for describing the state of an atom under these conditions, but that it would not be if the spin-orbit coupling had the form $\lambda\Sigma\mathbf{r}_i\cdot\mathbf{s}_i$ (Lee and Yang 1956). Consider also an atom interacting with a general time-dependent electromagnetic field and show that the parity of the whole system (atom, field and sources of the field) does not change with time (problem 30.2). In particular verify this when the atom undergoes a radiative dipole transition (Schiff 1955, pp. 224, 246, 257).

12. The Effect of the Exclusion Principle

Antisymmetric wave functions

So far we have not taken account of the fact that $\mathscr{H}_{\mathrm{orb}} + \mathscr{H}_{\mathrm{spin}}$ is invariant under the group $\mathfrak{P}_n$ of permutations of the electron co-ordinates. Let ψ be an eigenfunction of $\mathscr{H}_{\text{orb}}$ or $\mathscr{H}_{\text{orb}} + \mathscr{H}_{\text{spin}}$. If P is any permutation of the co-ordinates $\mathbf{r}_i$, σ_{zi}, then by § 6 $P\psi$ is also an eigenfunction with the same energy. This increases the degeneracies deduced in the last two sections by a factor of up to $n!$ (factorial n) and we can generate an enormous number of degenerate eigenfunctions. For instance, suppose we have obtained, using the Hartree equations, an approximate eigenfunction

$$\psi = \phi_{1s}(\mathbf{r}_1)\phi_{2p,m}(\mathbf{r}_2)u_{+1}u_{+2}$$

of $\mathscr{H}_{\text{orb}}$, corresponding to the configuration $1s2p$ of the helium atom. We have really a degenerate set of functions with the three allowed values of $m = 1, 0, -1$ and two values $\pm\frac{1}{2}$ each for m_{s1}, m_{s2}. By operating with P_{12}, the transformation that interchanges x_1, y_1, z_1, σ_{z1} and x_2, y_2, z_2, σ_{z2}, we generate a new function $P_{12}\psi$ from each one of these, so that we have a total of $3 \times 2^2 \times 2 = 24$ degenerate functions. These can be sorted out according to the irreducible representations of $\mathfrak{P}_2$. Clearly the linear combinations

$$\Psi_s = \psi + P_{12}\psi, \qquad \Psi_a = \psi - P_{12}\psi \tag{12.1}$$

transform according to the symmetric and antisymmetric representations respectively (§ 7).

Now in § 7 it was shown that for any number n of electrons, $\mathfrak{P}_n$ always has an antisymmetric representation. *The Pauli exclusion principle says that the only states occurring in nature are the ones with wave functions transforming according to these antisymmetric representations.* This can be derived from the fact that electrons have spin $\frac{1}{2}$ [see discussion above equation (32.27)]. In our example with two electrons, the exclusion principle eliminates half the possible number of wave functions, namely those of the form Ψ_s (12.1). When $n > 2$, the reduction in the number of wave functions is even more drastic, a fraction $1/n!$ being in general antisymmetric. Hence in making calculations for an atom with several electrons, it would appear to be a waste of effort to set up a large number of wave functions, only to reject most of them at the end because of the exclusion principle. Following Slater (1929), it is much better to restrict oneself to antisymmetric wave functions from the beginning. From any wave function ψ, we can always construct an antisymmetric one Ψ analogously to (12.1), namely

$$\boxed{\Psi = A\psi = \frac{1}{\sqrt{n!}}\sum_{P}^{n!} \delta_P P\psi.} \tag{12.2}$$

Here the antisymmetrizing operator A means applying a permutation P to ψ, multiplying by $\delta_P = \pm 1$ as the permutation is even or odd (§ 7), summing over all P in the group $\mathfrak{P}_n$, and multiplying by the normalizing factor. When ψ is a simple product of one-electron functions, Ψ can be written as a determinant: more generally when ψ is a sum of products, Ψ is a sum of determinants.

Configuration wave functions

Let us now modify the discussion of §§ 10 and 11, and express it entirely in terms of antisymmetric wave functions. It is essential to include spin from the beginning because the exclusion principle applies to permutations of all co-ordinates, not of the orbital ones alone. Thus analogously to (10.3), we write as our first approximation to the wave function a single determinant

$$\boxed{\begin{aligned}\Psi &= A\,\phi_1(\mathbf{r}_1)u_{\alpha 1}\,\phi_2(\mathbf{r}_2)u_{\beta 2}\ldots\phi_n(\mathbf{r}_n)u_{\nu n}\\ &= \frac{1}{\sqrt{n!}}\begin{vmatrix}\phi_1(\mathbf{r}_1)u_{\alpha 1} & \phi_2(\mathbf{r}_1)u_{\beta 1}\ldots\\ \phi_1(\mathbf{r}_2)u_{\alpha 2} & \phi_2(\mathbf{r}_2)u_{\beta 2}\ldots\\ \vdots & \vdots\end{vmatrix}.\end{aligned}} \tag{12.3}$$

As before $\alpha\beta\ldots\nu$ is some arrangement of $\pm$ signs. We first neglect $\mathscr{H}_{\text{spin}}$. Our Ψ (12.3) is too simple to be an eigenfunction of $\mathscr{H}_{\text{orb}}$, but we can define the expectation value of the energy E as in (10.4). We again obtain the best wave function subject to its having the form (12.3) by varying the ϕ_n so as to minimize E. This gives a set of equations analogous to (10.5) from which the ϕ_n can be determined. These are the Hartree–Fock self-consistent field equations (Hartree 1957)

$$\begin{aligned}\left[-\frac{\hbar^2}{2m}\nabla_i^2-\frac{Ze^2}{r_i}+\sum_j\int\phi_j^*(\mathbf{r}_j)\phi_j(\mathbf{r}_j)\frac{e^2}{r_{ij}}\,d\tau_j\right]\phi_i(\mathbf{r}_i)\\ -\sum_{j\parallel}\left[\int\phi_j^*(\mathbf{r}_j)\phi_i(\mathbf{r}_j)\frac{e^2}{r_{ij}}\,d\tau_j\right]\phi_j(\mathbf{r}_i)=E_i\phi_i(\mathbf{r}_i).\end{aligned} \tag{12.4a}$$

The electrostatic average potential appears as before. The last term on the left is the so-called exchange term which has no classical analogue, and in it the summation is only over electrons having a parallel spin direction, i.e. $\sigma_{zi} = \sigma_{zj}$. A clear discussion of its physical significance has been given by Slater (1951). In making actual calculations with these equations, it is usual to average the potential and exchange terms over all directions if they are not already spherically symmetrical. Then (12.4a) is invariant under rotations of the x_i, y_i, z_i alone, and the wave functions have the form (10.6) transforming according to $D^{(l_i)}$. In this way the individual energies E_i and the total energy E of Ψ become independent of the m_{li} and m_{si} though they depend through the self-consistent field on the l_i and the principal quantum numbers n_i. It follows that the total energy of the wave function (12.3) for the whole atom is completely specified by the quantum numbers n_i, l_i. Such a level is called a *configuration* and has degeneracy

$$\boxed{2^n(2l_1+1)(2l_2+1)\ldots(2l_n+1).} \tag{12.4b}$$

There is one important difference between the determinant Ψ (12.3) and the simple product Hartree wave function (10.3). In the latter there is no restriction on the quantum numbers n_i, l_i, m_{li} chosen, but we notice from the determinantal form that Ψ vanishes if we put two electrons say i and j into states with exactly the same quantum numbers n, l, m_l, m_s, i.e. if for instance $\phi_1(\mathbf{r})u_\alpha = \phi_2(\mathbf{r})u_\beta$. This happens because the two columns i and j of the determinant become identical so that the determinant is zero. Hence applied to determinantal wave functions of the type (12.3), the exclusion principle takes the form that *only one electron can occupy a particular orbital, designated by a particular set of quantum numbers* n, l, m_l, m_s. This drastically limits the number of wave functions in a configuration. For example in the configuration $(1s)^2 2s(2p)^2$ any arrangement with $m_{s1}=m_{s2}=\frac{1}{2}$ or $m_{s1}=m_{s2}=-\frac{1}{2}$ is excluded. Furthermore, from the antisymmetry of the wave function we have

$$\Psi(m_{s1}=\tfrac{1}{2}, m_{s2}=-\tfrac{1}{2}) = -\Psi(m_{s1}=-\tfrac{1}{2}, m_{s2}=\tfrac{1}{2}),$$

(where all other quantum numbers are supposed the same on both sides) since these two wave functions are just the same determinant except for an interchange of the first two columns. Thus the two combinations of quantum numbers do not give linearly independent functions. Similarly $m_{l4}=m_{l5}=1$, $m_{s4}=m_{s5}=\frac{1}{2}$ is excluded, but $m_{l4}=1$, $m_{l5}=0$, $m_{s4}=m_{s5}=\frac{1}{2}$ is allowed.

The periodic table

The exclusion principle not only limits the number of states in a configuration, but also limits the number of allowed configurations. For instance $(1s)^3$ is excluded because there are only two possible $1s$ functions corresponding to $m_l=0$, $m_s=\pm\frac{1}{2}$. Similarly there are two $2s$ functions ($l=0$), six $2p$ functions ($l=1$), and, in general,

$2(2l + 1)$ functions with principal quantum number n and orbital quantum number l. These make up respectively the $1s$ shell, $2s$ shell, $2p$ shell, etc., as they are called. We can now write down the configurations corresponding to the states of lowest energy of the lightest atoms. In hydrogen the single electron goes into the state with lowest energy, namely $1s$. In helium we have a second $1s$ electron, which completes the $1s$ shell. In lithium the third electron must go into the orbital with the next lowest energy, namely $2s$, and in this way we build up the configurations shown in Table 4.

TABLE 4

Configurations of the Elements

H	$1s$	Li	$1s^2\ 2s$	Na	$1s^2\ 2s^2\ 2p^6\ 3s$
He	$1s^2$	Be	$1s^2\ 2s^2$	Mg	$1s^2\ 2s^2\ 2p^6\ 3s^2$
		B	$1s^2\ 2s^2\ 2p$	Al	$1s^2\ 2s^2\ 2p^6\ 3s^2\ 3p$
		C	$1s^2\ 2s^2\ 2p^2$	Si	$1s^2\ 2s^2\ 2p^6\ 3s^2\ 2p^2$
		N	$1s^2\ 2s^2\ 2p^3$	P	$1s^2\ 2s^2\ 2p^6\ 3s^2\ 3p^3$
		O	$1s^2\ 2s^2\ 2p^4$	S	$1s^2\ 2s^2\ 2p^6\ 3s^2\ 3p^4$
		F	$1s^2\ 2s^2\ 2p^5$	Cl	$1s^2\ 2s^2\ 2p^6\ 3s^2\ 3p^5$
		Ne	$1s^2\ 2s^2\ 2p^6$	A	$1s^2\ 2s^2\ 2p^6\ 3s^2\ 3p^6$

We notice the similarity between the configurations of the sequences Li, Be, . . . , F, Ne and Na, Mg, . . . Cl, A. They differ only by the closed shells $(2s)^2(2p)^6$ which is a very stable structure, as shown by the inertness of neon, compared with the more loosely bound additional electrons which have to go into orbitals with not nearly such a low energy. The latter give rise to the chemical properties of the elements and are called valence electrons. In this way the whole periodic classification of the elements can be accounted for (see for instance Slater 1939, p. 344).

Enumeration of terms

As in § 11, all the determinantal wave functions Ψ (12.3) belonging to one configuration transform under orbital rotations according to

$$\boxed{D^{(l_1)} \times D^{(l_2)} \times \ldots \times D^{(l_n)} = \sum D^{(L)}} \tag{12.5a}$$

and under rotations in spin space according to

$$\boxed{D^{(1/2)} \times D^{(1/2)} \times \ldots \times D^{(1/2)} = \sum D^{(S)}.} \tag{12.5b}$$

Using the Wigner coefficients we can reduce the vector space with respect to both these groups and construct a set of $(2L + 1)(2S + 1)$ linear combinations $\Psi(L, M_L, S, M_S)$. These transform according to $D^{(L)}$ and $D^{(S)}$ under orbital and spin rotations respectively, and one such set (or its energy) is called a *term* as in § 11. Taking linear combinations of determinants keeps the wave function antisymmetric of course.

Following Slater (1929) we can enumerate all the terms from a given configuration which are allowed by the exclusion principle. This is done by counting in a way similar to § 9 all the values of M_L and M_S that occur, and then grouping them into complete terms. Applying a rotation to all orbital variables keeping the σ_{zi} fixed, we have from (12.3), (10.6)

$$R_{\text{orb}}(\phi, z)\,\Psi = \exp(i\phi \sum m_{li})\,\Psi = \exp(iM_L\phi)\,\Psi.$$

Thus

$$\boxed{M_L = \sum m_{li}, \qquad M_S = \sum m_{si}.} \tag{12.5c}$$

In a configuration such as $(1s)^2 2s(2p)^2$, a full shell like $(1s)^2$ never gives any contribution to M_L and M_S. This is because the states in one shell can be grouped in pairs with quantum numbers m_l, m_s and $-m_l$, $-m_s$, and in a full shell they are all occupied. Let us therefore designate a state of the configuration $(1s)^2 2s(2p)^2$ by writing down $(n_i, l_i, m_{li}, m_{si})$ for each of the three electrons (i.e. $2s(2p)^2$) outside the closed shell. In this way we can systematically write down all the states allowed by the exclusion principle. They are shown in Table 5, where $\pm$ denotes $\pm\frac{1}{2}$ and where all states with negative M_L or M_S are omitted. These are simply given by symmetry from the states with both M_L and M_S positive. To group the values of M_L, M_S into terms, we pick out the highest value of M_L with the highest M_S accompanying it, namely $M_L = 2$, $M_S = \frac{1}{2}$. From the corresponding wave function, we can generate with $I_{-\text{orb}}$, $I_{-\text{spin}}$ all the wave functions belonging to a term with $L = 2$, $S = \frac{1}{2}$, i.e. a ${}^2D$ term. This uses up the M_L, M_S values marked (a) in Table 5 and corresponding negative values. Continuing the process among the remaining ones, we obtain also the other terms given in Table 5.

Two orbitals having the same l are said to be *equivalent* if they also have the same n. If in the above example we had taken the two p orbitals to be inequivalent, e.g. the configuration $(1s)^2 2s2p3p$, there would have been some extra allowed entries in Table 5. For instance the determinant $(2\,0\,0\,+)(2\,1\,1\,+)(3\,1\,1\,+)$ in the configuration $2s2p3p$ is non-zero, whereas the determinant $(2\,0\,0\,+)(2\,1\,1\,+)(2\,1\,1\,+)$ in the configuration $2s(2p)^2$ vanishes.

As another example consider the configuration $(2p)^4$. Corresponding to any given set of quantum numbers, e.g. $(2\,1\,1\,+)$

TABLE 5

Enumeration of Wavefunctions in the Configuration $(1s)^2 2s(2p)^2$

	M_L	M_S	Term†
(200+)(211+)(211−)	2	1/2	a
(200+)(211+)(210+)	1	3/2	b
(200+)(211+)(210−)	1	1/2	a
(200+)(211+)(21−1+)	0	3/2	b
(200+)(211+)(21−1−)	0	1/2	a
(200+)(211−)(210+)	1	1/2	b
(200+)(211−)(21−1+)	0	1/2	b
(200+)(210+)(210−)	0	1/2	d
(200−)(211+)(210+)	1	1/2	c
(200−)(211+)(21−1+)	0	1/2	c

† (a) ^2D; $L = 2$, $S = 1/2$. (b) ^4P; $L = 1$, $S = 3/2$.
(c) ^2P; $L = 1$, $S = 1/2$. (d) ^2S; $L = 0$, $S = 1/2$.

(2 1 1 −)(2 1 0 +)(2 1 −1 −), with total M quantum numbers M_{LA}, M_{SA}, we can write down the remaining two orbitals (2 1 0 −) (2 1 −1 +) required to fill the shell. These have total quantum numbers $M_{LB} = -M_{LA}$, $M_{SB} = -M_{SA}$, because the full shell has $M_L = M_S = 0$. Since in any term the values $\pm M_L$ and $\pm M_S$ occur in pairs, the terms that can be formed from the configuration $(2p)^4$ are the same as those from $(2p)^2$. Similarly $(2p)^5$ and $2p$ give the same terms, and so do $(nd)^x$ and $(nd)^{10-x}$, etc.

Table 6 lists the terms from some configurations which consist only of equivalent electrons outside a full shell. A configuration composed only of closed shells has only the one state with $M_L = M_S = 0$, i.e. it gives a ^1S term.

The discussion of parity in § 11 is unaffected by the exclusion principle, and the parity (11.24) applies to all states derived from one configuration.

TABLE 6

Terms from Simple Configurations

s	;	^2S.
s^2, p^6 or d^{10}	;	^1S.
p or p^5	;	^2P.
p^2 or p^4	;	^3P, ^1D, ^1S.
p^3	;	^4S, ^2D, ^2P.
d or d^9	;	^2D.
d^2 or d^8	;	^3F, ^3P, ^1G, ^1D, ^1S.

Energy levels

As already mentioned, with some approximations in the Hartree–Fock equations (12.4), all the determinantal wave functions (12.3) belonging to one configuration have the same energy.† Now $\mathscr{H}_{\text{orb}}$ is invariant under separate rotations of all orbital co-ordinates and of all spin co-ordinates. Hence the eigenfunctions can be grouped into terms transforming according to $D^{(L)}$ and $D^{(S)}$ (12.5), different terms in this approximation having different energies. The wave functions are approximately linear combinations of determinants (12.3) from the configuration to which the terms belong, but strictly they include also small contributions from other configurations. These contributions must come from terms with the same L and S so that the whole wave functions still transform precisely according to $D^{(L)}$ and $D^{(S)}$. This limits the amount of configurational interaction considerably, and there are further restrictions, e.g. the configurations must have the same parity.

For a detailed discussion of the calculation of the term energies we refer to Condon and Shortley (1951, p. 191). However, we can see qualitatively why different terms belonging to the same configuration have different energies. Terms with different L have different spatial arrangements of their charge density, and so differ in electrostatic interaction energy. But how do terms with the same L but different S differ in energy when $\mathscr{H}_{\text{orb}}$ does not depend on spin co-ordinates, e.g. ^4P and ^2P of Table 5? The answer is that the exclusion principle is not spin-independent; it demands antisymmetry in all the co-ordinates including the σ_{zi}. Consider a determinant Ψ (12.3) in which two electrons i and j have parallel spin, i.e. $m_{si} = m_{sj}$. Then for both electrons at the same point $\mathbf{r}_i = \mathbf{r}_j$ we have

$$P_{ij}\Psi(\mathbf{r}_i = \mathbf{r}_j) = \Psi(\mathbf{r}_i = \mathbf{r}_j). \tag{12.6}$$

However by the exclusion principle $P_{ij}\Psi = -\Psi$, so that we must have $\Psi(\mathbf{r}_i = \mathbf{r}_j) = 0$. I.e. *there is zero probability of finding the two electrons with the same spin simultaneously at the same point.* Clearly

† There is actually no single Hamiltonian analogous to $\mathscr{H}_{\text{cscf}}$ (10.9), of which the Hartree–Fock determinants are the eigenfunctions. However by further approximating to the exchange terms in (12.4), Slater has derived a central field Hamiltonian of the form

$$\mathscr{H}_{\text{Slater}} = -\frac{\hbar^2}{2m}\sum \nabla_i^2 + \sum V(r_i)$$

with determinant eigenfunctions. These approximate closely to the Hartree–Fock wave functions, and for one configuration all have the same energy eigenvalue (Slater 1951, Pratt 1952).

(12.6) does not hold for electrons with different spin directions, and for these $\Psi \neq 0$ for $\mathbf{r}_i = \mathbf{r}_j$. Thus the exclusion principle automatically keeps electrons with parallel spins out of each other's way, which lowers their electrostatic interaction energy. Of all the states in Table 5, only the one with $M_S = 3/2$ has all three electron spins parallel, and we expect it therefore to have the lowest energy. Of course, the other states belonging to the same term must have the same low energy. More generally, *in any configuration, the term with the lowest energy* has the maximum number of parallel spins, i.e. it *has the largest S. Subject to this condition it also has the largest L possible.* This is Hund's rule. The L rule follows by considering the state with maximum M_L. It consists of determinants with mainly positive values of m_{li}. The angular momenta of the electrons tend to be lined up so that the electrons are going round the nucleus in the same direction. This allows them to keep as far apart as possible on opposite sides of the nucleus as they revolve with a consequent lowering of their electrostatic energy. On the other hand electrons going round the nucleus in opposite directions (opposite m_l) meet one another every revolution, so that they are close together for a considerable fraction of the time, and this raises the electrostatic energy. These rules only apply rigorously to the lowest term of a configuration.

As in § 11, when we include $\mathscr{H}_{\text{spin}}$ in the Hamiltonian, each term is split according to

$$\boxed{D^{(L)} \times D^{(S)} = \sum D^{(J)}} \tag{12.7}$$

into levels characterized by the quantum numbers J, M_J and parity. In light atoms the effect of $\mathscr{H}_{\text{spin}}$ is small compared with the differences between terms (Russell–Saunders coupling case). Using Wigner coefficients the zero order wave functions $\Psi(L, S, J, M_J)$ can be written down, but the exact eigenfunctions contain admixtures from other terms. *The level with the minimum J has lowest energy, unless the configuration contains a partly filled shell more than half full, in which case the lowest energy is given by the maximum value of J.* Fig. 8 gives a sketch of the levels of the configuration $(1s)^2 2s(2p)^2$, the levels from other configurations having too high or too low energy to be included in the diagram.

Summary

The exclusion principle requires all physically applicable wave functions to be antisymmetric with respect to permutations of the orbital and spin co-ordinates together. This drastically limits the

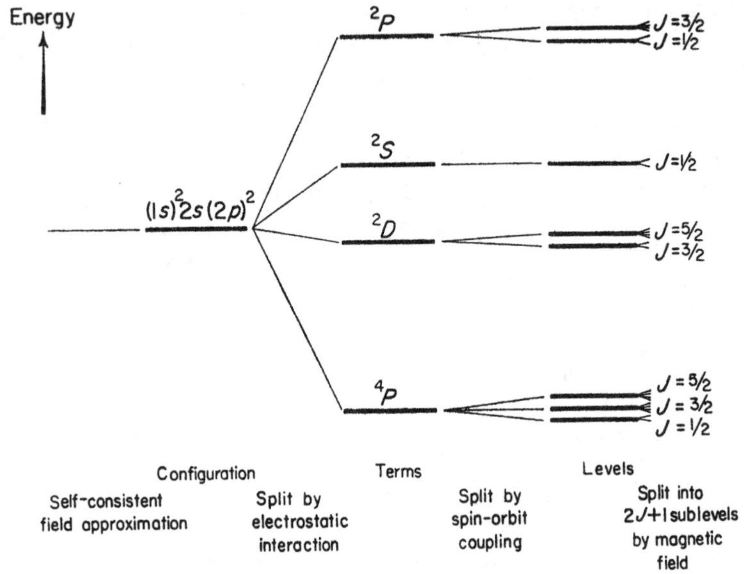

FIG. 8. Energy levels of the configuration $(1s)^2\ 2s\ (2p)^2$.

number of allowed configurations and the degeneracy of each configuration. Systematic ways of deriving the allowed configurations and terms have been developed.

PROBLEMS

12.1 What is the lowest level of the Mn^{++} ion? What configurations could be admixed with it? The ground state configuration is $(3d)^5$.

12.2 What terms arise from the configurations $(2p)^3$, $(2p)^2 3p$, and $2p3p4p$?

12.3 A configuration consists of two partly filled shells. Show how its terms may be derived by first obtaining the terms from the two shells separately and then combining them.

12.4 Write down the spin functions U_{SM_S} (11.17) for two electrons, and show that the $S = 1$ and $S = 0$ ones are respectively symmetric and antisymmetric with respect to permutations. Hence show that all wave functions $\Psi(L, M_L, S, M_S)$ of helium terms can be written as a symmetric (or antisymmetric) orbital wave function times an antisymmetric (or symmetric) spin function. Hence set up all the terms arising from low lying configurations of helium and draw a rough energy level diagram (Van der Waerden 1932, p. 111).

Discuss how an analogous (but more complicated!) procedure can be followed for more than two electrons (see § 28; also Wigner 1931, p. 270).

12.5 Illustrate the argument of equation (12.6) with a simple two-electron wave function.

12.6 Write down the Hamiltonian $\mathscr{H}$ for an atom in a uniform magnetic field in the z-direction (Schiff 1955, pp. 292, 293). Show that $\mathscr{H}$ is invariant under rotations about the z-axis, and that we would expect a level characterized by the quantum number J to be split into $2J+1$ levels. Are J and M_J exact quantum numbers in the presence of the field?

13. Calculating Matrix Elements and Selection Rules

Matrix elements

In quantum mechanics all physically measurable quantities appear as matrix elements or are expressible in terms of them. A matrix element is an expression of the form†

$$M_{ij} = \int \psi_i^* Q \psi_j \, \mathrm{d}\tau \tag{13.1}$$

involving an operator Q and two wave functions ψ_i, ψ_j. For example, if we are calculating energy levels, the operator is the Hamiltonian or some part of the Hamiltonian that is being applied as a perturbation. Also the transition probabilities between energy levels with the emission or absorption of radiation, and the expansions for wave functions, all involve matrix elements. In this section we shall establish *selection rules* which state which matrix elements are non-zero. We shall also show how non-zero matrix elements for different i and j can be related to one another and calculated.

We may assume without loss of generality that ψ_i and ψ_j belong to an orthogonal normalized set of functions ψ_n which transform in a definite way according to the irreducible representations of some group. That is,

$$\int \psi_m^* \psi_n \, \mathrm{d}\tau = \delta_{mn}, \tag{13.2}$$

where the Kronecker δ_{mn} is defined by (A.15). We now expand $Q\psi_j$ in terms of the complete set ψ_n:

$$Q\psi_j = \sum_n M_{nj}\, \psi_n. \tag{13.3}$$

† If ψ_i, ψ_j are spin-dependent functions, we shall define the symbol $\int \ldots \mathrm{d}\tau$ to include also a summation over the values ± 1 of all spin co-ordinates σ_z, i.e. setting $u_\alpha^* u_\beta = \delta_{\alpha\beta}$ for each electron.

Multiplying by ψ_i^* and integrating, we obtain using (13.2) *that the expansion coefficient* M_{ij} *is just the matrix element* (13.1). Conversely if we have an arbitrary wave function ϕ_j and expand it in terms of a complete set of functions ψ_i, then the expansion coefficients are just the matrix elements $\int \psi_i^* \phi_j \, d\tau$. In what follows we shall be constantly alternating between the two points of view (matrix element and expansion coefficient), according to which is more helpful for the purpose at hand.

Fundamental theorem

Instead of specifying the ψ_n by the single suffix n, let us sort them out to transform according to the irreducible representations $D^{(\mu)}$ of some group $\mathfrak{G}$. Now $D^{(\mu)}$ will occur several times, for example among the eigenfunctions of the hydrogen atom we have $2p$, $3p$, $4p$, . . . functions all transforming according to the representation $D^{(1)}$ of the rotation group, and we shall use the index r to distinguish the rth set of functions transforming according to $D^{(\mu)}$. The different functions inside one such set will be distinguished by a subscript such as i or j, and we shall choose the functions always in some conventionally chosen standard way so that the particular representations $D^{(\mu, r)}$ with different r are not only equivalent but completely identical. Instead of the $Q\psi_i$ of (13.3), we start by considering some set of functions $\phi_j^{(\lambda)}$ transforming according to $D^{(\lambda)}$ under $\mathfrak{G}$, and expand them in terms of the ψ's.

$$\phi_j^{(\lambda)} = \sum_r \sum_\mu \sum_i M_{ij}^{(\mu r)} \psi_i^{(\mu r)}. \tag{13.4}$$

Here the M's are now the matrix elements

$$M_{ij}^{(\mu r)} = \int \psi_i^{(\mu r)*} \phi_j^{(\lambda)} \, d\tau. \tag{13.5}$$

Now the right-hand side of (13.4) has to transform in the same way as the left-hand side, and this suggests that *the only* $\psi_i^{(\mu r)}$ *which can appear on the right-hand side of* (13.4) *are those that transform according to the same row* j *of the same irreducible representation* $D^{(\lambda)}$ *as* $\phi_j^{(\lambda)}$ *itself*. I.e. (13.4) reduces to

$$\boxed{\phi_j^{(\lambda)} = \sum_r a_{\lambda r} \psi_j^{(\lambda r)}.} \tag{13.6}$$

This result is indeed so plausible and well known in particular instances (e.g. in Fourier analysis an even function can be expanded in terms of cosines alone without any sine terms), that we have already used it implicitly in (10.13) etc. The result can be re-expressed in terms of the $M_{ij}^{(\mu r)}$, to give what we shall refer to as the *fundamental theorem* of this section.

IA. *When* $\mu \neq \lambda$, $M_{ij}^{(\mu r)} = 0$. *I.e. there can be no contribution to* (13.4) *from sets of functions not transforming according to* $D^{(\lambda)}$.

IB. *When* $\mu = \lambda$, $M_{ij}^{(\lambda r)} = a_{\lambda r}\sigma_{ij}$. *I.e. inside one set* $\psi_i^{(\lambda r)}$, *the matrix* M_{ij} *is a multiple* $a_{\lambda r}$ *of the unit matrix.*

Proof. So far the reasoning has been purely heuristic. The proper proof of the theorem is as follows. If T is any transformation of $\mathfrak{G}$, $T\phi_j^{(\lambda)}$ can be written in two ways,

$$T\phi_j^{(\lambda)} = \sum_{\mu r}\sum_{ik} M_{ij}^{(\mu r)}\, D_{ki}^{(\mu)}(T)\, \psi_k^{(\mu r)},$$

and

$$T\phi_j^{(\lambda)} = \sum_i D_{ij}^{(\lambda)}(T)\, \phi_i^{(\lambda)}$$

$$= \sum_{\mu r}\sum_{ik} D_{ij}^{(\lambda)}(T)\, M_{ki}^{(\mu r)}\, \psi_k^{(\mu r)}.$$

Comparing these two, we see that the matrix $M_{ij}^{(\mu r)}$ for each μ, r satisfies Schur's lemma (appendix D), from which the results IA and IB above follow immediately. Q.E.D.

For most applications, we require two extensions of the theorem. Suppose that instead of the $\phi_j^{(\lambda)}$ in (13.4) and (13.6), we have linear combinations

$$\Phi_j^{(\lambda)} = \sum_i P_{ij}\, \phi_i^{(\lambda)}$$

which do not transform in the standard manner adopted for the $\phi_i^{(\lambda)}$ and $\psi_i^{(\lambda r)}$. From (13.4) and (13.6) we have

$$\Phi_j^{(\lambda)} = \sum_r\sum_\mu\sum_i M_{ij}^{(\mu r)}\, \psi_i^{(\mu r)}$$

$$= \sum_r\sum_i a_{\lambda r}\, P_{ij}\, \psi_i^{(\lambda r)},$$

i.e.

$$M_{ij}^{(\mu r)} = \delta_{\lambda\mu} a_{\lambda r} P_{ij}.$$

Thus the matrix elements are determined uniquely by the P_{ij}, apart from the factor $a_{\lambda r}\delta_{\lambda\mu}$ which is a constant for each set of ψ's. (By a set of ψ's we mean the $\psi_i^{(\mu r)}$, $i = 1, 2, \ldots$ forming a set of base vectors for the irreducible representation $D^{(\mu)}$.) Furthermore the P_{ij} determine the transformation properties of the $\Phi_j^{(\lambda)}$ as shown explicitly by (5.15), and conversely the transformation properties uniquely determine the coefficients P_{ij} as can easily be proved in the manner of problem D.2. We therefore have symbolically:

$$\left.\begin{array}{l}\text{transformation}\\ \text{properties of}\\ \text{the } \Phi_j^{(\lambda)}\end{array}\right\} \rightarrow \quad P_{ij} \quad \rightarrow \left\{\begin{array}{l}\text{matrix elements } M_{ij}^{(\mu r)}\\ \text{apart from constant}\\ \text{factor.}\end{array}\right.$$

Similarly if the ψ's being used do not transform in the standard way, then this can be taken into account in the same manner. This gives the first extension of the fundamental theorem:

IIA. *If the* $\Phi_j^{(\lambda)}$ *and* $\psi_i^{(\mu r)}$ *have not been chosen to transform in the standard way but in some other manner, then* IA *still holds.* IIB. *Also the* $M_{ij}^{(\mu r)}$ *are completely determined as regards their dependence on* i *and* j *by the transformation properties of the* $\Phi_j^{(\lambda)}$ *and* $\psi_i^{(\mu r)}$, *except for a constant factor* $a_{\lambda r}$.

A further extension is obtained by taking linear combinations

$$\Phi_j = \sum_i \sum_\mu P_{ij\mu}\, \phi_i^{(\mu)}$$

of ϕ's from several irreducible representations. These are again expanded in terms of the ψ's;

$$\Phi_j = \sum_r \sum_\mu \sum_i M_{ij}^{(\mu r)}\, \psi_i^{(\mu r)}. \tag{13.7}$$

By using the substitution (13.6) and arguments analogous to the ones above, we easily deduce:

IIIA. *If in* (13.7) *the* Φ_j *on the left transform according to a reducible representation* $D^{(\alpha)} + D^{(\beta)} + \ldots + D^{(\epsilon)}$, *then* $M_{ij}^{(\mu r)} = 0$ *unless* $\mu = \alpha, \beta, \ldots$ *or* ϵ. IIIB. *When* $\mu = \alpha, \beta, \ldots$ *or* ϵ, *then the matrix elements* $M_{ij}^{(\mu r)}$ *are completely determined as regards dependence on* i *and* j *by the transformation properties of the* Φ's *and* ψ's, *except for a constant factor* $a_{\mu r}$ *if* $\alpha, \beta \ldots \epsilon$ *are all different. If* $D^{(\alpha)}, D^{(\beta)} \ldots D^{(\epsilon)}$ *are not all different and* $D^{(\mu)}$ *occurs among them* n *times, then the* $M_{ij}^{(\mu r)}$ *involve* n *undetermined constants* $a_{\mu r}, a'_{\mu r}, a''_{\mu r}, \ldots$. The latter result comes about because the linear combinations Φ_j contain terms from the n sets of functions $\phi_i^{(\mu)}, \phi'^{(\mu)}_i, \phi''^{(\mu)}_i, \ldots$, so that from (13.6) the matrix elements involve the n constants $a_{\mu r}, a'_{\mu r}, a''_{\mu r}, \ldots$. When we describe these constants as "undetermined", we mean that they are not determined by symmetry properties alone, but depend on the detailed form of the functions involved.

As stated at the beginning of this section, the most common application of these theorems is to derive selection rules for the matrix elements

$$\boxed{M_{ij} = \int \psi_i^{(\lambda)*}\, Q_k^{(\mu)}\, \psi_j^{(\nu)}\, d\tau,} \tag{13.8a}$$

where we suppose $\psi_i^{(\lambda)}$, $Q_k^{(\mu)}$ and $\psi_j^{(\nu)}$ transform respectively according to $D^{(\lambda)}$, $D^{(\mu)}$ and $D^{(\nu)}$. From (13.3), these matrix elements are obtained by expanding $Q_k^{(\mu)}\psi_j^{(\nu)}$ which transforms according to $D^{(\mu)} \times D^{(\nu)}$, in terms of the $\psi_i^{(\lambda)}$. Theorem IIIA and B therefore gives:

A. *In* (13.8a) M_{ij} *is non-zero only if the reduction of* $D^{(\mu)} \times D^{(\nu)}$ *contains* $D^{(\lambda)}$. B. *If* $D^{(\lambda)}$ *is contained once* (*or n times*), *then the* M_{ij} *are uniquely determined by the symmetry properties of the* ψ's *and* Q's, *apart from a constant factor* (*or n undetermined constants*).	(13.8b)

In (13.8b) the representation $D^{(\lambda)}$ must be irreducible, but $D^{(\mu)}$ and $D^{(\nu)}$ need not be. Alternatively the matrix elements (13.8a) are given by expanding the whole integrand $\psi_i^{(\lambda)*}Q_k^{(\mu)}\psi_j^{(\nu)}$ in terms of any complete set of functions, the first of which is the constant unity which of course always transforms according to the identity representation. Theorem IIIA and B then gives the selection rule

A. *In* (13.8a) M_{ij} *is non-zero only if the reduction of* $D^{(\lambda)*} \times D^{(\mu)} \times D^{(\nu)}$ *contains the identity representation.*

B. *If the identity representation is contained once* (*or n times*), *then the* M_{ij} *are uniquely determined by the symmetry properties of the* ψ's *and* Q's, *except for a constant factor* (*or n undetermined constants*). (13.8c)

In the latter formulation, none of the representations $D^{(\lambda)}$, $D^{(\mu)}$, $D^{(\nu)}$ need necessarily be irreducible, though usually they will be. An alternative proof of the important results above will be given in § 14 using group characters.

Electric-dipole transition selection rules

An atom in one quantum state can make a transition to another quantum state either spontaneously with the emission of radiation, or under the influence of externally applied radiation. We shall restrict ourselves to the type of transition described as electric-dipole, and the spontaneous and induced transition probabilities are then proportional to the modulus squared of the matrix element (Schiff 1949, p. 247)

$$\boxed{\langle i|r_\alpha|j\rangle = \int \psi_i^* \, r_\alpha \, \psi_j \, d\tau.} \tag{13.9}$$

Here

$$r_1 = -\sqrt{\tfrac{1}{2}}\sum_n (x_n + iy_n), \qquad r_{-1} = \sqrt{\tfrac{1}{2}}\sum_n (x_n - iy_n),$$

$$r_0 = \sum_n z_n, \tag{13.10}$$

and ψ_i and ψ_j are the final and initial states. The summations are over all electrons in the atom. Also $\alpha = 0$ applies to radiation linearly polarized along the z-direction, and $\alpha = 1, -1$ give the probabilities for radiation circularly polarized about the z-axis. In the Zeeman effect (problem 13.6) the linearly and circularly polarized radiation are known as the π and σ components respectively.

It is now simple to write down selection rules stating when $\langle i|r_\alpha|j\rangle$ can be non-zero. In § 12 it was shown that degenerate eigenfunctions of an atom transform according to the representations $D^{(J)}$ of the rotation group. Suppose ψ_i, ψ_j belong to sets with $J = J_1, J_2$ respectively. Then the $r_\alpha\psi_j$ transform according to

$$D^{(1)} \times D^{(J_2)} = D^{(J_2+1)} + D^{(J_2)} + D^{(J_2-1)} \text{ if } J_2 \geqq 1. \tag{13.11}$$

From (13.8b) above, $\langle i|r_\alpha|j\rangle = 0$ unless $J_1 = J_2 + 1$, J_2 or $J_2 - 1$. However, when $J_2 = 0$, then $D^{(1)} \times D^{(0)} = D^{(1)}$, so that only $J_1 = 1$ is allowed and the transition $J_1 = 0$ to $J_2 = 0$ is impossible. We can summarize the results by the selection rule:

$$\boxed{\Delta J = 0, 1 \text{ or } -1, \text{ except } 0 \to 0.} \tag{13.12}$$

Suppose now that ψ_i, ψ_j in (13.9) transform according to the representations M_{J1}, M_{J2} of the axial rotation group about the z-axis. The r_α transform with $M = 1, 0, -1$ for $\alpha = 1, 0, -1$ and the whole matrix element transforms with $M = -M_{J1} + \alpha + M_{J2}$. We have the selection rule from (13.8c)

$$\boxed{\begin{aligned} &\Delta M_J = \pm 1, \text{ circularly polarized radiation, } (\alpha = \pm 1), \\ &\Delta M_J = 0, \text{ linearly polarized radiation, } (\alpha = 0). \end{aligned}} \tag{13.13}$$

Similarly for $\langle i|r_\alpha|j\rangle$ to be non-zero, the initial and final levels must have opposite parity (11.24).

In the case of Russell–Saunders coupling (§§ 11 and 12), we derived in § 12 approximate eigenfunctions† $\psi(L, S, J, M_J)$ trans-

† We shall not follow any more the notation of § 12 as regards the use of a capital Ψ for antisymmetric wave functions. All wave functions are now assumed to be antisymmetric.

forming according to $D^{(L)}$ and $D^{(S)}$ under orbital and spin rotations. The r_α have $L = 1$ and $S = 0$, so that analogously to (13.12) we have

$$\boxed{\begin{array}{ll} \Delta L = 0, 1 \text{ or } -1 & \text{except } 0 \to 0, \\ \Delta S = 0. & \end{array}} \tag{13.14}$$

However, since L and S are not good quantum numbers, this selection rule only distinguishes large from small matrix elements.

Transition probabilities

So far we have used what amounts to part A of our theorem to sort out the non-zero matrix elements. We can now use part B to calculate the relative intensities of several related transitions. Consider the matrix elements between two energy levels with eigenfunctions $\psi_i = \psi(J = 2, M_2)$ and $\psi_j = \psi(J = 1, M_1)$. From (13.8), all matrix elements between these two levels are completely determined apart from one common numerical factor by the transformation properties of the $\psi(J = 2, M_2)$, the r_α and the $\psi(J = 1, M_1)$. Hence to calculate the relative values of these matrix elements, we can replace these functions by the functions $U_{M_2}{}^{(2)}$, u_α, v_{M_1} of (9.4) which transform in exactly the same way. From (9.6)

$$\int U_1{}^{(2)*} u_0 v_1 \, \mathrm{d}\tau = \sqrt{\tfrac{1}{2}}/N_2,$$

$$\int U_0{}^{(2)*} u_0 v_1 \, \mathrm{d}\tau = 0,$$

$$\int U_0{}^{(2)*} u_0 v_0 \, \mathrm{d}\tau = \sqrt{\tfrac{2}{3}}/N_2, \text{ etc.}$$

whence we obtain the relative magnitudes of all the matrix elements for $\alpha = 0$ (Table 7). Those for $\alpha = \pm 1$ may be obtained similarly. More generally when ψ_i, ψ_j transform according to $D^{(J_1)}$, $D^{(J_2)}$,

TABLE 7

Relative Values of $\langle J = 2, M_2 | r_0 | J = 1, M_1 \rangle$

M_1 \ M_2	2	1	0	−1	−2
1	0	$\sqrt{\frac{1}{2}}$	0	0	0
0	0	0	$\sqrt{\frac{2}{3}}$	0	0
−1	0	0	0	$\sqrt{\frac{1}{2}}$	0

the matrix elements $\langle J_1 M_{J1}|r_\alpha|J_2 M_{J2}\rangle$ between a particular pair of levels are proportional to the Wigner coefficients

$$\langle J_1 M_{J1}|r_\alpha|J_2 M_{J2}\rangle \propto (1 J_2 \alpha M_{J2}|J_1 M_{J1}), \tag{13.15}$$

as follows from (9.8) by the same argument.

Although (13.15) determines the relative values of the matrix elements completely, we shall now calculate the ones for $\Delta J = 0$ transitions (13.12) in an alternative way purely to illustrate the use of the fundamental theorem. The ψ_i, ψ_j are now two sets of functions ψ_{M_1} and ϕ_{M_2} say, both transforming according to $D^{(J)}$ but belonging to different energy levels. In § 11 it was shown from (11.18) that the infinitesimal rotation operators I_x, I_y, I_z transform under rotations in the same way as x, y, z. Consequently from (8.21) the I_α,

$$I_1 = -(I_x + iI_y), \qquad I_0 = \sqrt{2}I_z, \qquad I_{-1} = I_x - iI_y, \tag{13.16}$$

transform in the same way as the r_α (13.10). For purposes of calculating the matrix elements, we therefore replace ψ_{M_1}, r_α, ϕ_{M_2} by ϕ_{M_1}, I_α, ϕ_{M_2}. Taking

$$|\langle JM_1|r_\alpha|JM_2\rangle|^2 \propto \left|\int \phi_{M1}{}^* \, I_\alpha \, \phi_{M2} \, \mathrm{d}\tau\right|^2,$$

we obtain from (8.18) the relative transition probabilities:

$$\begin{aligned}
&\alpha = 1, M+1 \rightarrow M; && \text{rel. prob.} = J(J+1) - M(M+1);\\
&\alpha = 0, M \rightarrow M; && \text{rel. prob.} = 2M^2;\\
&\alpha = -1, M-1 \rightarrow M; && \text{rel. prob.} = J(J+1) - M(M-1).
\end{aligned}$$

Perturbation theory

As in § 6, let $\mathscr{H} = \mathscr{H}_0 + \mathscr{H}_p$ be a Hamiltonian consisting of a major or unperturbed part $\mathscr{H}_0$ and a small perturbation $\mathscr{H}_p$. If the energy levels E_{or} of $\mathscr{H}_0$ are all non-degenerate, the energy levels of $\mathscr{H}$ are (Schiff 1955, p. 153)

$$\boxed{E_r = E_{or} + \langle r|\mathscr{H}_p|r\rangle + \sum_{s\neq r} \frac{\langle r|\mathscr{H}_p|s\rangle\langle s|\mathscr{H}_p|r\rangle}{E_{or} - E_{os}} + \ldots} \tag{13.17}$$

Here we use the Dirac notation (13.9) and $\langle r|\mathscr{H}_p|s\rangle$ is a matrix element of $\mathscr{H}_p$ with respect to the eigenfunctions ψ_r, ψ_s of $\mathscr{H}_0$. The three terms give the energy correct to the zeroth, first and second order of perturbation respectively.

Now Schiff 1955 (p. 155) shows that (13.17) does not apply and perturbation theory becomes at first sight very complicated, when the unperturbed energy levels of $\mathscr{H}_0$ are degenerate as they usually are. The reason is that one in general gets zero energy denominators in (13.17). To avoid these, it is necessary to choose such linear combinations ϕ_r, ϕ_s of the eigenfunctions ψ of $\mathscr{H}_0$, that

$$\int \phi_r^* \mathscr{H}_p \phi_s \, d\tau = 0 \qquad \text{when } E_{or} = E_{os}. \tag{13.18}$$

Then formula (13.17) becomes correct again using matrix elements with respect to the ϕ_r. Fortunately group theory and our fundamental theorem help us considerably to sort out the eigenfunctions of $\mathscr{H}_0$ such that (13.18) is satisfied. Let $\mathfrak{G}$ be the group of all symmetry transformations of $\mathscr{H}$. $\mathscr{H}_p$ is invariant under $\mathfrak{G}$ so that $\mathscr{H}_p\psi_s$ transforms in exactly the same way as ψ_s. Hence if we sort out the ψ's according to the irreducible representations of $\mathfrak{G}$ and use the indices μ, s, i, etc. of (13.4), we have from results IA and IB that

$$\langle \lambda ri | \mathscr{H}_p | \mu sj \rangle = \delta_{\lambda\mu}\, \delta_{ij}\, a_{rs}. \tag{13.19}$$

Thus (13.19) is automatically satisfied unless an irreducible component $D^{(\lambda)}$ occurs more than once in the vector space of one degenerate energy level of $\mathscr{H}_0$. This happens infrequently: if it does, one has no option but to solve an $n \times n$ secular equation, where n is the number of times $D^{(\lambda)}$ occurs in a particular unperturbed level. Hence we can usually obtain the correct energy levels by applying the ordinary non-degenerate perturbation theory (13.17) to each set of states $\psi_i^{(\lambda r)}$ having the same λ and i.

Relative splittings due to spin-orbit coupling

We shall now calculate to first order the energies of the split levels when we apply the spin-orbit coupling $\mathscr{H}_{LS}$ (11.8) as a perturbation to the degenerate states of one term. The eigenfunctions of a term can be sorted out according to either the quantum numbers L, S, M_L, M_S or L, S, J, M_J, where L and S are fixed for one term and the others take on several sets of values. For the present we shall use the former classification. Now the angular momenta l_{ix} and L_x transform in the same way under orbital rotations and spin rotations (they are invariant under the latter). Hence (13.8) implies that the matrix elements of l_{ix} and L_x inside one term are proportional:

$$\langle LSM_LM_S | l_{ix} | LSM'_LM'_S \rangle \propto \langle LSM_LM_S | L_x | LSM'_LM'_S \rangle.$$

Similarly the matrix elements of $(\mathbf{r}_i \wedge \mathbf{v}_j)_x$ and L_x are proportional;

so are those of s_{kx} and S_x, and those of $\mathbf{r}_i \wedge \mathbf{v}_j \cdot \mathbf{s}_k$ and $\mathbf{L} \cdot \mathbf{S}$. Thus the whole of $\mathscr{H}_{LS}$ (11.8) can be written effectively as

$$\boxed{\mathscr{H}_{LS} = \zeta(LS)\, \mathbf{L} \cdot \mathbf{S}} \tag{13.20}$$

as regards the vector space of one term, where ζ is a constant. Since $\mathbf{J} = \mathbf{L} + \mathbf{S}$, (13.20) can also be written

$$\mathscr{H}_{LS} = \tfrac{1}{2}\zeta(LS)(\mathbf{J}^2 - \mathbf{L}^2 - \mathbf{S}^2). \tag{13.21}$$

In order to apply the perturbation (13.21) to our term wave functions, we first have to sort out the eigenfunctions $\psi(LSJM_J)$ which transform according to $D^{(J)}$ under rotations, so that (13.18) is satisfied. Then from (13.17) and (8.35),

$$\boxed{E(LSJM_J) = E_0 + \tfrac{1}{2}\zeta(LS)[J(J+1) - L(L+1) - S(S+1)],} \tag{13.22}$$

where E_0 is the energy of the term without spin-orbit splitting.

Absolute values of spin-orbit splittings

The fundamental theorem only establishes the *relative* magnitudes of all matrix elements between one pair of vector spaces. For instance in the case of the spin-orbit coupling, to calculate the *absolute* values of the splittings (13.22) we still have to determine the $\zeta(LS)$. This is not really a group theoretical problem, but since it occurs in any application of the theory, we shall show as an example how the $\zeta(LS)$ are calculated. In principle there is no difficulty. In § 12 we have shown how to write down (neglecting configurational and inter-term mixing) the wave functions $\psi(LSJM_J)$ in terms of Wigner coefficients and the one-electron orbitals ϕ_i. It would only be necessary to calculate the energy of one such state in each term to fix $\zeta(LS)$. In practice these wave functions are very complicated and it is easier to proceed slightly differently. We calculate the value of *any* convenient matrix element or *sum of matrix elements* in two ways, once from (13.20) and once from the original Hamiltonian (11.21),

$$\mathscr{H}'_{\text{spin}} = \sum \xi(r_i)\mathbf{l}_i \cdot \mathbf{s}_i, \tag{11.21}$$

and equate them. This gives some simultaneous equations for the required quantities, in our case the $\zeta(LS)$.

To apply this method in detail, (13.20) is written

$$\mathscr{H}_{LS} = \zeta(LS)\mathbf{L} \cdot \mathbf{S} = \zeta(LS)(\tfrac{1}{2}L_+S_- + \tfrac{1}{2}L_-S_+ + L_zS_z). \tag{13.23}$$

Let us calculate its diagonal matrix element† with respect to $\psi(LSM_LM_S)$. Operating on this wave function with L_-S_+, we generate using (8.18) the wave function with $M'_S = M_S + 1$, $M'_L = M_L - 1$ which is orthogonal to $\psi(LSM_LM_S)$. Hence in (13.23) only the last term contributes to the matrix element and

$$\langle LSM_LM_S|\mathscr{H}_{LS}|LSM_LM_S\rangle = \zeta(LS)M_LM_S. \tag{13.24}$$

Similarly from (11.21), (12.2), (12.3), we obtain for the matrix element with respect to a single determinant wave function

$$\langle n_il_im_{li}m_{si}|\mathscr{H}_{LS}|n_il_im_{li}m_{si}\rangle = \sum_i \zeta_{n_il_i}m_{li}m_{si}, \tag{13.25}$$

$$\zeta_{n_il_i} = 4\pi \int_0^\infty f_{n_il_i}(r)\xi_i(r)f_{n_il_i}(r)r^2\,\mathrm{d}r. \tag{13.26}$$

Here f is the radial part of a one-electron orbital as in (10.6), so that (13.26) can be calculated once the Hartree–Fock solutions are known. We note from (13.25) that a closed shell of electrons does not contribute to the matrix elements of $\mathscr{H}_{LS}$.

Consider now the specific configuration $2p3p$. Its terms may be found analogously to Table 5, and they are ^3D, ^3P, ^1D, ^1P, ^3S, ^1S. In the ^3D term, the wave function $\psi(L = 2,\ S = 1,\ M_L = 2,\ M_S = 1)$ consists of a single determinant $(2\,1\,1\,+)(3\,1\,1\,+)$ in the notation of Table 5. Thus from (13.24), (13.25), we have

$$2\zeta(^3D) = \tfrac{1}{2}\zeta_{2p} + \tfrac{1}{2}\zeta_{3p}. \tag{13.27}$$

We cannot determine the other $\zeta(LS)$ so simply, and have to make use of a theorem called the sum rule. Let ψ_i, $i = 1$ to n, be a set of wave functions and let $\phi_\alpha = \Sigma C_{i\alpha}\psi_i$ be n mutually orthogonal linear combinations. Further let M_{ij}, $M'_{\alpha\beta}$ be matrix elements of any operator with respect to the ψ_i, ϕ_α. Then in matrix notation $M' = \tilde{C}^*MC$. If ψ_i, ϕ_α are normalized, $\int\psi_i^*\psi_i\,\mathrm{d}\tau = \int\phi_\alpha^*\phi_\alpha\,\mathrm{d}\tau = 1$, so that $\tilde{C}^*C = E$ and C is unitary. Applying the result of problem A.7 (appendix A), we have that the *sum of the diagonal matrix elements $M'_{\alpha\alpha}$ is equal to the sum of the M_{ii}*.

In our case the two wave functions with $M_L = 1$, $M_S = 1$ of the ^3D and ^3P terms are linear combinations of the two determinants $(2\,1\,1\,+)(3\,1\,0\,+)$ and $(2\,1\,0\,+)(3\,1\,1\,+)$. Hence calculating the diagonal matrix elements (13.24) and (13.25) and using the sum rule, we obtain

$$\zeta(^3D) + \zeta(^3P) = \tfrac{1}{2}\zeta_{2p} + \tfrac{1}{2}\zeta_{3p}. \tag{13.28}$$

† The matrix element $\int\psi_i^*Q\psi_j\,\mathrm{d}\tau$ is said to be diagonal if $\psi_i = \psi_j$.

Similarly, simultaneous equations for all the $\zeta(LS)$ are derived. In particular from (13.27), (13.28) we have

$$\zeta(^3D) = \zeta(^3P) = \tfrac{1}{4}\zeta_{2p} + \tfrac{1}{4}\zeta_{3p}$$

Hence from (13.22) the total splitting of the ^3D levels is 5/3 times the spread of the ^3P levels, a conclusion which is very amenable to experimental confirmation.

Summary

A fundamental theorem has been established, which in essence states the following. If some functions ϕ_i are expanded in terms of a complete set of functions ψ_j, and if each set has been sorted out according to the irreducible representations of a group, then each of the terms appearing in the expansion must transform in the same way as the ϕ_i being expanded. The expansion coefficients are matrix elements, many of which are therefore zero. Selection rules indicate which matrix elements are non-zero, and their relative magnitudes have also been calculated. This has been applied to electric-dipole transitions, degenerate perturbation theory, and level splittings due to spin-orbit coupling. The absolute magnitudes of the matrix elements can only be determined by evaluating one of them, or a sum of several, explicitly.

PROBLEMS

13.1 Complete Table 7 for r_1 and r_{-1}. Hence verify that the total intensity radiated in all directions is the same from each of the initial levels $J = 2, M_2$. Prove this result in general for transitions from the $2J + 1$ different states of one level. The result is not unexpected because the various M_J states only differ in the *orientation* of the angular momentum, which should not affect the total intensity.

13.2 In (13.9), (13.10), $-e\Sigma x_n$, $-e\Sigma y_n$, $-e\Sigma z_n$ is the electric dipole moment operator. To obtain magnetic-dipole transition probabilities, it has to be replaced by the total magnetic moment $(-e/2mc)(\mathbf{L} + 2\mathbf{S})$. Calculate the selection rules for magnetic dipole transitions, and also those for electric quadrupole transitions (Schiff 1955, pp. 253, 292; Condon and Shortley 1951, p. 93).

13.3 If we use an electric or magnetic field to split the levels of a term, show that there can be no electric dipole transitions between the levels of one term or even one configuration, but that magnetic dipole transitions can occur. In paramagnetic resonance at microwave frequencies the quanta of energy are so small that one is dealing with transitions inside one term. The samples are therefore placed in a resonant cavity at a position of maximum $\mathbf{H}_{rf}$ to induce magnetic dipole transitions.

13.4 Show that all electric and magnetic multipole transitions are forbidden for the transition $J = 0$ to $J = 0$.

13.5 Write down the Hamiltonian for an atom (not hydrogen) in a uniform electric field. Show that it produces no splitting of the levels in the first order of perturbation (Schiff 1955, p. 159).

13.6 Show that the linear term in the Hamiltonian for an atom in a uniform magnetic field H along the z-axis can be written $\beta H(I_{z\text{ orb}} + 2I_{z\text{ spin}})$, where the Bohr magneton β is $e\hbar/2mc$. Show that this is effectively $(1 + \alpha)\beta H I_z$ as regards the vector space of eigenfunctions of one energy level. Use the relation $\mathbf{L}^2 = (\mathbf{J} - \mathbf{S})^2 = \mathbf{J}^2 + \mathbf{S}^2 - \mathbf{J}\cdot\mathbf{S} - \mathbf{S}\cdot\mathbf{J}$ to calculate α in the case of Russell–Saunders coupling. Hence determine the energy levels for small fields, and express them in the usual form $E_0 + g_L\beta H M_J$, where g_L is the Landé splitting factor

$$\boxed{g_L = 1 + \frac{J(J+1) + S(S+1) - L(L+1)}{2J(J+1)}.}$$

This splitting is known as the anomalous Zeeman effect (Schiff 1955, p. 294).

13.7 Using the perturbation of problem 13.6, calculate the energy levels of an atom in a large field H, for which the magnetic energy is much greater than the splittings due to spin-orbit coupling but is small compared with the separations between terms (Paschen–Back effect; Schiff 1955, p. 294). Sketch schematically how the energy levels of a ^2P term change as the field is reduced to zero.

13.8 In dipole transitions, how do the intensities of the π and σ radiation vary with direction (Schiff 1955, p. 253)? Calculate the relative intensities of all possible π and σ Zeeman components (in a weak magnetic field in the z-direction) for the transitions between a $^2P_{1/2}$ and a $^2P_{3/2}$ state. Using the g-factors of problem 13.6 to determine the energy differences, prepare a diagram showing the splitting, polarization and intensity pattern that would be observed for the $^2P_{3/2} \to {}^2P_{1/2}$ line in the z-direction and the x-direction (Condon and Shortley 1951, p. 378).

13.9 The centre of gravity of a set of levels is defined to be their mean energy weighted according to the degeneracy of each level. Using (13.24) and the sum rule, show that the centre of gravity of the levels J of one term is equal to the unsplit term energy E_0 of (13.22).

13.10 Calculate all the $\zeta(LS)$ for the $2p3p$ configuration.

13.11 Consider the $(nd)^2$ configuration.

(i) Work out the allowed terms of this configuration.

(ii)* Express the energies of the allowed terms in terms of the standard radial integrals $F_0(nd^2)$, $F_2(nd^2)$, $F_4(nd^2)$ (Condon and Shortley 1951, p. 193). Find the condition that these F's must satisfy if Hund's rule is to apply.

(iii) Find the spin-orbit coupling parameter $\zeta(LS)$ for all terms in terms of ζ_{nd}.

(iv) Work out the eigenfunctions of the $M_L = 3$, $M_S = 0$ states belonging to the ^1G and ^3F terms in terms of simple determinants. Using the tables of Wigner coefficients (appendix I), find the $M_J = 3$ eigenfunctions of both the $J = 3$ and $J = 4$ levels of the ^3F term, again as linear combinations of determinant wave functions.

13.12 An atom in a state transforming according to $D^{(J)}$ is placed inside a crystal in an electric field due to the surrounding ions. Show that the centre of gravity of the $2J+1$ levels is not shifted in first order if the crystal potential V contains no constant term. Hint: show $\sum_M \psi_M{}^*\psi_M$ transforms according to $D^{(0)}$; expand V in spherical harmonics; show $\int V \sum \psi_M{}^*\psi_M \, d\tau = 0$.

13.13 $D^{(\lambda)}$ and $D^{(\nu)}$ are two irreducible representations. By putting $Q = 1$ in (13.8) and comparing (13.8b) and (13.8c), prove that $D^{(\lambda)*} \times D^{(\nu)}$ contains the identity representation if and only if $D^{(\lambda)} = D^{(\nu)}$.

13.14 Verify directly that (13.8b) and (13.8c) give identical results for the selection rule (13.12).

Chapter III

THE REPRESENTATIONS OF FINITE GROUPS

In the preceding two chapters we developed and illustrated the use of group theory in quantum mechanics. In the remaining chapters we shall apply these methods to a selection of problems from various branches of physics. The present chapter develops the mechanical aspects of handling group representations in general. At a first reading, only pp. 113–117 and a look at pp. 125–131 are vital.

14. Group Characters

Before we can use group theory in quantum mechanics, we have to have systematic procedures, applicable to an arbitrary group, for:
(i) labelling and describing the irreducible representations;
(ii) reducing a given representation;
(iii) deriving all the different irreducible representations.
We shall not concern ourselves very much with this last point apart from just indicating how it is done, because for a simple group the irreducible representations are almost sure to be tabulated, or at least deducible by inspection from those of a tabulated related group. Complicated groups are usually treated by special methods, cf. the rotation group (§ 8), the permutation group (Wigner 1931), and the Lorentz group (§ 31).

In §§ 8 and 9 we developed a method for doing the three things listed above in the case of the rotation group by using the infinitesimal rotation operators. This method can be extended more or less to continuous groups in general, but certainly not to groups consisting of a finite number of elements because the infinitesimal transformations cannot be defined. We shall therefore develop a new method, that of group characters, which is particularly suited to finite groups but which can also be extended to continuous groups (Wigner 1931, p. 103).

Classes

Let us consider again the point-group 32 of elements E, A, B,

K, L, M defined in (4.1) or (4.13). From its multiplication table (Table 1), we have

$$\begin{aligned} KAK^{-1} &= B, & KBK^{-1} &= A, \\ LAL^{-1} &= B, & LBL^{-1} &= A, \\ MAM^{-1} &= B, & MBM^{-1} &= A, \\ BAB^{-1} &= A, & BBB^{-1} &= B, \\ AAA^{-1} &= A, & ABA^{-1} &= B, \\ EAE^{-1} &= A, & EBE^{-1} &= B, \end{aligned}$$

and we note that SAS^{-1} and SBS^{-1} are equal to either A or B for every group element S. A and B are said to form a class. Definition: *a class is a set of elements out of a group, such that if T is one element of the class, then STS^{-1} also belongs to that class for all group elements S.* It may easily be verified by direct multiplication as above that the three classes of the group 32 are (E), (A, B), (K, L, M). In general, to sort the elements of a finite group into classes, we would start with one element T_1 and form all products ST_1S^{-1}. This gives a set of elements $T_1, T_2, \ldots T_i$. We then form $ST_2S^{-1}, \ldots, ST_iS^{-1}$ and in general this gives the old elements $T_1, T_2, \ldots T_i$ with possibly some new ones $T_{i+1}, \ldots$. If there are no new ones, then $T_1, T_2, \ldots T_i$ form a class; if there are additional elements $T_{i+1}, \ldots$, then we repeat the process until we have found a closed class. It is clear that in the process of forming classes, we cannot have an element left over at the end which does not fit into any class: for let T be such an element. Then either all the products STS^{-1} are equal to T, in which case T forms a class by itself, or one of the products, $S_1TS_1^{-1}$ say, is equal to some other element R say. In the latter case we have

$$(S_1^{-1})R(S_1^{-1})^{-1} = T,$$

and since S_1^{-1} is some group element this equation shows that T would already have been included in the class of R in the procedure described above. Thus all the group elements can be sorted into classes with each element belonging to one class.

Characters

Let $D_{ij}(T)$ be any representation of a group of transformations T. The sum of the diagonal elements of $D_{ij}(T)$ is called the character χ of T in this representation.

$$\boxed{\chi(T) = \sum_i D_{ii}(T).} \tag{14.1}$$

If S is any matrix, we have using the summation convention,

$$[SD(T)S^{-1}]_{ii} = S_{ij}D_{jk}(T)S_{ki}^{-1} = S_{ki}^{-1}S_{ij}D_{jk}(T)$$
$$= \delta_{kj}D_{jk}(T) = \chi(T).$$

From this we have two results. *The characters of all elements in one class are the same.* Also from the definition of equivalence (5.15), *the characters of equivalent representations are the same.* We shall later prove the converse of this, namely that two representations having the same set of characters are equivalent. Here a *set* of characters means the characters for all the classes in a group. Thus such a set of characters is a very useful way of specifying a representation. It does not distinguish between equivalent representations, which is usually a convenience rather than a hindrance, as discussed in § 5. However it does distinguish a given representation from all inequivalent ones. Moreover the number of classes in a group is considerably smaller than the number of elements, and hence so is the number of different characters. In particular the irreducible representations of a group can be tabulated easily in this way. For instance from Table 3, we have for the irreducible representations of the group 32 the characters shown in Table 8.

TABLE 8

Character Table for the Point-group 32

	E	A	K
χ^{Γ}	2	−1	0
$\chi^{\mathscr{A}}$	1	1	−1
$\chi^{\mathscr{I}}$	1	1	1

This is called the character table of the group, and the character of only one element in each class is usually listed. Clearly the identity element E always forms a class of its own and $\chi(E)$ is the dimension of the representation.

Reduction of a representation

Suppose a representation D is reducible into irreducible components $D^{(\alpha)} + D^{(\beta)} + \ldots D^{(\epsilon)}$. Then we can bring D into reduced form by a transformation $P^{-1}D(T)P$, and hence the characters are

$$\boxed{\begin{aligned}\chi(T) &= \chi^{(\alpha)}(T) + \chi^{(\beta)}(T) + \ldots + \chi^{(\epsilon)}(T)\\ &= \sum_{\lambda} c_{\lambda}\,\chi^{(\lambda)}(T).\end{aligned}} \qquad (14.2)$$

Here c_λ is the number of times the irreducible component $D^{(\lambda)}$ occurs in the reduction of D. Hence after setting up $\chi(T)$ for each of the classes, a representation can be reduced by determining the c_λ by inspection from (14.2), or if necessary by solving the (14.2) as simultaneous equations.† For instance consider the representation of the group 32 in the vector space $(x^2, y^2, z^2, yz, zx, xy)$. These vectors are neither normalized nor orthogonal but they still form the basis for a representation in this space, with $D(A)$ being given above equation (5.20). By inspection the characters are

$$\chi(E) = 6, \qquad \chi(A) = 0, \qquad \chi(K) = 2.$$

Hence from the character table of the group (Table 8) and solving the simultaneous equations (14.2) by inspection, we have

$$c_\Gamma = 2, \qquad c_{\mathscr{A}} = 0, \qquad c_{\mathscr{I}} = 2,$$

and $$D = \Gamma + \Gamma + \mathscr{I} + \mathscr{I},$$

as was found before in (5.21).

This method can be applied easily to product representations (§ 9). Let u_i, v_j be base vectors transforming according to the representations $D^{(\lambda)}$ and $D^{(\mu)}$ of a group, and T some transformation of the group. Suppose

$$Tu_i = T_{ii}^{(\lambda)}\, u_i + \text{other terms in } u_k,$$
$$Tv_j = T_{jj}^{(\mu)}\, v_j + \text{other terms in } v_l.$$

Then $T(u_i v_j) = T_{ii}^{(\lambda)} T_{jj}^{(\mu)}(u_i v_j) + \text{terms in } (u_k v_l)$. Hence if we take $(u_i v_j)$ as the nth base vector in the product space, the diagonal element $D_{nn}(T)$ of the matrix $D(T)$ of the product representation $D = D^{(\lambda)} \times D^{(\mu)}$ is just $T_{ii}^{(\lambda)} T_{jj}^{(\mu)}$. Therefore

$$\boxed{\chi(T) = \sum_n D_{nn}(T) = \sum_{ij} T_{ii}^{(\lambda)} T_{jj}^{(\mu)} = \chi^{(\lambda)}(T)\, \chi^{(\mu)}(T).} \tag{14.3}$$

I.e. *the characters of a product representation are the products of the constituent characters.* For example from Table 8 we obtain for the product representation $\Gamma \times \Gamma$ of the group 32 the characters

$$\chi(E) = 4, \qquad \chi(A) = 1, \qquad \chi(K) = 0,$$

whence from (14.2)

$$\Gamma \times \Gamma = \Gamma + \mathscr{A} + \mathscr{I}.$$

† We shall prove later that these equations have in fact always a unique solution (see below equation (14.18)).

Another application of (14.2) is to determine according to what representation a set of spherical harmonics transforms under a group consisting of a finite number of rotations. Let $\mathrm{Rot}(\phi, \mathfrak{x}_1)$ be a physical rotation in the sense of § 2 about the axis $\mathfrak{x}_1$, and $\mathrm{Rot}(S)$ any other rotation. Suppose $\mathrm{Rot}(S)$ rotates the axis $\mathfrak{x}_2$ into $\mathfrak{x}_1$. Then

$$[\mathrm{Rot}(S)]^{-1}\,\mathrm{Rot}(\phi, \mathfrak{x}_1)\,\mathrm{Rot}(S) = \mathrm{Rot}(\phi, \mathfrak{x}_2).$$

For $\mathrm{Rot}(S)$ rotates $\mathfrak{x}_2$ into the position $\mathfrak{x}_1$, $\mathrm{Rot}(\phi, \mathfrak{x}_1)$ then applies a rotation of ϕ about it, and $[\mathrm{Rot}(S)]^{-1}$ moves it back to its original position $\mathfrak{x}_2$. Thus *all rotations by one angle ϕ about different axes belong to the same class of the full rotation group.* By considering $\mathrm{Rot}(\phi, z)u_m = e^{im\phi}u_m$, we obtain for the characters of the irreducible representation $D^{(j)}$ of the full rotation group

$$\boxed{\chi(\phi) = \sum_{m=-j}^{j} \exp(im\phi) = \frac{\sin\,(j+\frac{1}{2})\phi}{\sin\frac{1}{2}\,\phi}.} \tag{14.4}$$

The spherical harmonics Y_{lm} transform according to $D^{(l)}$ under the full rotation group, and therefore they also form the basis of a representation D of some finite point-group. Now the matrices of D are just the appropriate matrices taken from $D^{(l)}$, so that the characters of D are given by (14.4) with $j = l$. The representation can now be reduced by (14.2). Consider for instance the spherical harmonics with $l = 2$ transforming according to a representation D under the group 32. For A and K, ϕ is $-120°$ and $180°$ respectively and

$$\chi(E) = 5, \qquad \chi(A) = -1, \qquad \chi(K) = 1,$$

whence from (14.2) and Table 8,

$$D = \Gamma + \Gamma + \mathscr{I}.$$

Thus if we have the character table of some group—and this can usually be looked up, for instance in appendix K or L—we can find the irreducible components from (14.2) of any given representation. We have therefore achieved the main purpose of this section. The remainder of the section is devoted to deducing some of the other important properties of characters which are required for other applications.

Orthogonality relations

Consider the matrix

$$P = \sum_{T} D^{(\lambda)}(T^{-1})\,C\,D^{(\mu)}(T), \tag{14.5}$$

where $D^{(\lambda)}$, $D^{(\mu)}$ are two particular irreducible representations of a group of h elements. C is any matrix, and the summation is over all group elements T. We have

$$PD^{(\mu)}(S) = D^{(\lambda)}(S) \sum_T D^{(\lambda)}(S^{-1}T^{-1})\, CD^{(\mu)}(TS).$$

Here S is any group element, and $S^{-1}T^{-1} = (TS)^{-1}$ (problem A.5). Now the elements TS are just the group elements rearranged, for there are h of them and we cannot have $TS = RS$ unless $T = R$. Hence the summation over T can be replaced by a summation over TS and we have

$$PD^{(\mu)}(S) = D^{(\lambda)}(S)P.$$

P therefore satisfies the conditions for Schur's lemma (appendix D), and we have $P = 0$ if $D^{(\lambda)}$ and $D^{(\mu)}$ are inequivalent, and $P = a_c E$ if they are identical, where a_c depends on the particular matrix C. Since C is arbitrary, let us choose all elements zero except one, $C_{\alpha i} = 1$. This gives

$$\sum_T D_{\beta\alpha}{}^{(\lambda)}(T^{-1}) D_{ij}{}^{(\mu)}(T) = a_{\alpha i}\delta_{\beta j}\delta_{\lambda\mu}, \tag{14.6}$$

where we have used $\delta_{\lambda\mu}$ to denote 1 when $D^{(\lambda)}$ and $D^{(\mu)}$ are identical, and zero when they are inequivalent. When they are identical, we evaluate $a_{\alpha i}$ by putting $\beta = j$ and summing over the repeated suffix. Thus

$$\sum_T D_{\beta\alpha}{}^{(\lambda)}(T^{-1}) D_{i\beta}{}^{(\lambda)}(T) = n_\lambda a_{\alpha i} = h\delta_{\alpha i}$$

since $D^{(\lambda)}(T^{-1})D^{(\lambda)}(T) = E$. Here n_λ is the dimension of $D^{(\lambda)}$. Hence $a_{\alpha i} = h\delta_{\alpha i}/n_\lambda$. Also

$$D^{(\lambda)}(T^{-1}) = [D^{(\lambda)}(T)]^{-1} = [\tilde{D}^{(\lambda)}(T)]^* \tag{14.7}$$

if we suppose the representation is unitary (problem C.2). Thus (14.6) becomes

$$\sum_T D_{\alpha\beta}{}^{(\lambda)*}(T)\, D_{ij}{}^{(\mu)}(T) = \frac{h}{n_\lambda}\, \delta_{\alpha i}\delta_{\beta j}\delta_{\lambda\mu}. \tag{14.8}$$

This is the fundamental *orthogonality relation of the first kind* for irreducible representations. By putting $i = j$, $\alpha = \beta$ and summing, we obtain the orthogonality relation for the characters

$$\boxed{\sum_T \chi^{(\lambda)*}(T)\, \chi^{(\mu)}(T) = h\delta_{\lambda\mu}.} \tag{14.9}$$

Projection operators

We now discuss the reduction of a given vector space and the corresponding representation and vector space in further detail. First we can use (14.9) to obtain the explicit solution of equations (14.2) determining the irreducible components of any representation. Multiplying (14.2) by $\chi^{(\lambda)*}(T)$ and summing, we obtain

$$\boxed{c_\lambda = \frac{1}{h}\sum_T \chi^{(\lambda)*}(T)\,\chi(T).} \tag{14.10}$$

The uniqueness of the reduction shows that two representations with the same characters are reducible into the same irreducible components and hence are equivalent.

Let $\mathfrak{R}$ be a vector space transforming according to a representation, which can be reduced into irreducible components by (14.10). It remains to pick out the actual vectors in $\mathfrak{R}$ transforming according to the different components. Let $D_{ij}{}^{(\lambda)}(T)$ be one of the irreducible component representations, and ϕ any vector of $\mathfrak{R}$. Consider the vectors

$$\boxed{\phi_i{}^{(\lambda)} = \sum_T D_{ij}{}^{(\lambda)*}(T)\, T\phi} \tag{14.11a}$$

for any fixed value of j. Then from (14.11a) and (14.7), we obtain

$$\begin{aligned} S\phi_i{}^{(\lambda)} &= \sum_T D_{ij}{}^{(\lambda)*}(T) ST\phi \\ &= D_{i\alpha}{}^{(\lambda)*}(S^{-1}) \sum_T D_{\alpha j}{}^{(\lambda)*}(ST)ST\phi \\ &= D_{\alpha i}{}^{(\lambda)}(S)\phi_\alpha{}^{(\lambda)}, \end{aligned}$$

which shows that the $\phi_i{}^{(\lambda)}$ are vectors in $\mathfrak{R}$ transforming according to the particular irreducible representation $D_{ij}{}^{(\lambda)}(T)$. Putting $i = j$ in (14.11a) and summing, we have that

$$\boxed{\phi^{(\lambda)} = \sum_T \chi^{(\lambda)*}(T)\, T\phi} \tag{14.11b}$$

belongs to an irreducible subspace transforming according to $D^{(\lambda)}$. This relation is useful if we only have the characters and not the particular matrices $D_{ij}{}^{(\lambda)}(T)$, or if we do not require a proper set of base vectors in the subspace. In vector analysis, $\mathbf{i}\cdot\mathbf{v}$ is the projection of the vector $\mathbf{v}$ on $\mathbf{i}$ and is the part of $\mathbf{v}$ that lies in the direction $\mathbf{i}$.

By analogy the operators in (14.11) are called projection operators, because they pick out from ϕ the part lying in a particular subspace.

Matrix elements

We can now give an alternative derivation of the fundamental theorem of § 13 on the calculation of matrix elements. Let $\phi_j^{(\lambda)}$ be a set of functions transforming according to the irreducible representation $D^{(\lambda)}$. Since an integral is invariant under transformations (lemma 2, appendix C), we have

$$\begin{aligned} h\int \phi_j^{(\lambda)}\,\mathrm{d}\tau &= \sum_T \int T\phi_j^{(\lambda)}\,\mathrm{d}\tau \\ &= \sum_T D_{ij}^{(\lambda)}(T)\int \phi_i^{(\lambda)}\,\mathrm{d}\tau \\ &= \Big[\sum_T 1\cdot D_{ij}^{(\lambda)}(T)\Big]\int \phi_i^{(\lambda)}\,\mathrm{d}\tau. \qquad (14.12) \end{aligned}$$

(i summed)

Here the 1 in the square bracket denotes the element $D_{\alpha\beta}(T)$ of the identity representation. Thus from the orthogonality relation (14.8), (14.12) is zero unless $D^{(\lambda)}$ is the identity representation, i.e. we have $\int\phi_j^{(\lambda)}\,\mathrm{d}\tau = 0$. If $D^{(\lambda)}$ is the identity representation, then the i and j suffices become superfluous and (14.12) becomes

$$h\int\phi^{(\lambda)}\,\mathrm{d}\tau = h\int\phi^{(\lambda)}\,\mathrm{d}\tau$$

and the integral in general has some arbitrary non-zero value. If we now take ϕ to be some function in a vector space $\mathfrak{R}$ transforming according to a reducible representation D, then we can express ϕ in terms of the irreducible base vectors of $\mathfrak{R}$, and hence $\int\phi\,\mathrm{d}\tau = 0$ if D does not contain the identity representation. If D contains the identity representation n times, $\int\phi\,\mathrm{d}\tau$ is a linear combination of n undetermined integrals. We have therefore proved the theorem of § 13 on the evaluation of matrix elements in the particular form of result (13.8c).

Let us consider the reduction of

$$D = D^{(\lambda)*} \times D^{(\mu)},$$

where $D^{(\lambda)}$ and $D^{(\mu)}$ are irreducible representations. The characters of D are given by (14.3). Suppose D contains the identity representation c times. The characters of the identity representation are all unity, so that from (14.10) c is given by

$$\begin{aligned} c &= \frac{1}{h}\sum_T 1\cdot\chi^{(\lambda)*}(T)\chi^{(\mu)}(T) \\ &= \delta_{\lambda\mu} \text{ from (14.9).} \qquad (14.13) \end{aligned}$$

In words, $D^{(\lambda)*} \times D^{(\mu)}$ *contains the identity representation once, if and only if* $D^{(\lambda)} = D^{(\mu)}$. This is just the result of problem 13.13. With this and (13.8c) already proved, (13.8b) and part III of the theorem of § 13 follow immediately, and parts I and II then as special cases.

The regular representation D^{reg} of a group is defined as follows. First write out the group multiplication table, rearranging the rows so that all E's occur on the diagonal, as shown in Table 9 for the

TABLE 9

Construction of Regular Representation

$= T_j^{-1}$ applied first \ $= T_i$ applied second	E	A	B	K	L	M
$E^{-1} = E$	E	A				
$A^{-1} = B$		E	A			
$B^{-1} = A$	A		E			
$K^{-1} = K$				E		A
$L^{-1} = L$				A	E	
$M^{-1} = M$					A	E

point-group 32 (4.1). The columns are labelled by the elements T_i and the rows by T_j^{-1}. Then the matrix $D^{\text{reg}}(A)$ is obtained by putting all A's in the table equal to one and all other elements zero. Thus $D^{\text{reg}}_{ij}(A) = 1$ when $T_i T_j^{-1} = A$, i.e. when

$$AT_j = T_i = D^{\text{reg}}_{ij}(A) T_i.$$

Thus the regular representation is a representation among the group elements considered as base vectors. Now by construction, $\chi^{\text{reg}}(E) = h$ and $\chi^{\text{reg}}(T) = 0$ for $T \neq E$. We may reduce D^{reg}, and applying (14.10) we find $c_\lambda = n_\lambda$. I.e. *every irreducible representation* $D^{(\lambda)}$ *is contained in the regular representation* n_λ *times, where* n_λ *is the dimension of* $D^{(\lambda)}$. Counting up the total number of base vectors, we obtain as a by-product the relation

$$\boxed{\sum_\lambda n_\lambda^2 = h.} \tag{14.14}$$

In the reduction, let $Y_i^{(\lambda r)}$, $i = 1, 2, \ldots$ denote the rth set of base vectors transforming according to $D^{(\lambda)}$.

We shall now construct all transformations P that commute with every element of the group, and that are linear combinations of the group. By the latter condition P transforms every vector space $Y_i^{(\lambda r)}$, $i = 1, 2, \ldots$ into itself. Hence P is represented with respect to the $Y_i^{(\lambda r)}$ by a matrix in reduced form. Since we also require it to commute with all group elements, it must, by Schur's lemma (appendix D), have the particular form diag $\{a_{\lambda 1}E, a_{\lambda 2}E, \ldots, a_{\mu 1}E, a_{\mu 2}E, \ldots\}$ where the E's are little unit matrices along the diagonal. Further it must retain this form if we choose new base vectors $Y_i^{(\lambda r)} \pm Y_i^{(\lambda s)}$, whence $a_{\lambda r} = a_{\lambda s} = a_\lambda$ say. Thus we can write P in the form

$$P = \sum_\lambda a_\lambda E_\lambda, \tag{14.15a}$$

where the E_λ are operators such that $E_\lambda Y_i^{(\lambda r)} = Y_i^{(\lambda r)}$, $E_\lambda Y_i^{(\mu r)} = 0$, and where the a_λ are arbitrary constants. E_λ is the unit operator inside all the subspaces transforming according to $D^{(\lambda)}$. Now we can also construct P in a different way. Since P commutes with any element S of the group, it must contain the elements T and STS^{-1} an equal number of times. Hence P can be expressed in the form

$$P = \sum_k b_k M_k, \tag{14.15b}$$

where M_k is the sum of all elements in the kth class. Thus P can be considered as a vector in a space spanned by the E_λ or the M_k. Either way the dimension of this space has to be the same, whence comparing (14.15a) and (14.15b) we have

number of irreducible representations = number of classes.

(14.16)

Let the number of elements in the kth class be h_k. Then (14.9) can be written

$$\sum_k \left(\frac{h_k}{h}\right)^{1/2} \chi_k^{(\lambda)*} \left(\frac{h_k}{h}\right)^{1/2} \chi_k^{(\mu)} = \delta_{\lambda\mu}. \tag{14.17}$$

If we define the matrix U such that

$$U_{k\lambda} = \left(\frac{h_k}{h}\right)^{1/2} \chi_k^{(\lambda)},$$

then (14.16) shows that U is square. Also (14.17) becomes $\tilde{U}^*U = E$. Hence U is unitary and we also have $U^*\tilde{U} = E$, which when written out becomes

$$\sum_{\lambda} \left(\frac{h_j}{h}\right)^{1/2} \chi_j^{(\lambda)*} \left(\frac{h_k}{h}\right)^{1/2} \chi_k^{(\lambda)} = \delta_{jk}. \tag{14.18}$$

These are the *orthogonality relations of the second kind*. The unitary property of U also ensures that the simultaneous equations (14.2) always have a unique solution [namely (14.10)]. Thus we have an alternative proof of Theorem 3, appendix C, namely that the reduction of a representation is unique, apart from equivalence.

As an application of the above formulae, we shall set up from first principles the character table for the group 32. As already deduced, there are three classes, (E), (A, B), (K, L, M). From (14.16) there are three irreducible representations, and (14.13) becomes $n_1^2 + n_2^2 + n_3^2 = 6$. The only solution is $n_1 = n_2 = 1$, $n_3 = 2$. We always have the identity representation with characters 1, 1, 1. Since $K^2 = E$, $A^3 = E$, the only possibilities for the characters of the second representation are 1; 1, ω or ω^2; ± 1; where $\omega^3 = 1$. Relation (14.17) with $\lambda = 1$, $\mu = 2$ gives 1, 1, -1 as the only possibility. The third representation has $\chi(E) = 2$. Then (14.18) with $j = 1$, $k = 2$ and $j = 1$, $k = 3$ determines the other two characters, namely -1 and 0. This gives just the character table as shown in Table 8. For more complicated groups the result of problem 14.9 can be very helpful.

References

More detailed accounts are given of the properties of characters in Speiser (1937, pp. 166–175), of the method of characters applied to continuous groups in Wigner (1931), and of characters in general in Littlewood (1940). A systematic method of setting up the character table is given by Bhagavantam and Venkatarayudu (1951, p. 254).

Summary

Group characters have been defined, and used to label representations, particularly the irreducible representations. Any representation may be reduced using (14.2) and (14.10). Functions with given transformation properties may be determined using the projection operators (14.11). The two types of orthogonality relations and other results may be used to derive the characters of all the irreducible representations of a group.

PROBLEMS

14.1 The spherical harmonics of order 3 transform according to D under the point-group 32. Reduce D.

14.2 A rare-earth ion in the free state is in a state with $J = 1$ with energy E_0. In a crystal it is in an electric potential with symmetry 32 and mean zero. Show that the level is split into two levels E_1, E_2, and determine the ratio $(E_1 - E_0)/(E_2 - E_0)$ (cf. § 6, also problem 13.12).

14.3 Prove equation (9.2) using group characters.

14.4 u_+, u_- transform according to $D^{(1/2)}$ of the full rotation group and according to D under the point-group 32. Try to reduce D. What has gone wrong?

14.5 Use (14.11) to find base vectors in the space (x^2, y^2, z^2, yz, zx, xy) transforming under the group 32 according to the irreducible component representations (cf. equation (5.20)).

14.6 Show that in an Abelian group each element forms a class by itself. Hence deduce that all the irreducible representations are one-dimensional.

14.7 From (14.10) show that $\sum_T |\chi(T)|^2 = h \sum_\lambda c_\lambda^2$. Hence show that a necessary and sufficient condition for a representation to be irreducible is that $\sum |\chi(T)|^2 = h$.

14.8 Show (14.14) is a particular case of (14.18). Also deduce that $\sum_\lambda n_\lambda \chi^{(\lambda)}(T) = h$ if $T = E$, and is zero otherwise.

14.9 Let C_i be the ith class of a group. The product C_iC_j denotes all elements obtained by multiplying those of C_i by those of C_j. Show that these can be grouped into classes. Let the class C_k appear c_{ijk} times. This is written $C_iC_j = \sum_k c_{ijk}C_k$. With this notation, show that

$$h_i h_j \chi_i^{(\lambda)} \chi_j^{(\lambda)} = \chi^{(\lambda)}(E) \sum_k c_{ijk} h_k \chi_k^{(\lambda)}.$$

Hint: show $M_k = \eta_k E$ (14.15b) as regards its operating on irreducible vector spaces; $\eta_i \eta_j = \sum c_{ijk} \eta_k$; the character of M_k is $\eta_k \chi^{(\lambda)}(E) = h_i \chi_i^{(\lambda)}$ (Speiser 1937).

14.10 Set up the character table for the point-group 422. This is the group of proper rotations that leaves a square invariant. Use any of the relations of this chapter and of problems 14.7, 14.8, 14.9, and trial and error to set up the character table for this group. Verify the result of problem 14.9 in a few cases. Also reduce all the product representations $D^{(\lambda)} \times D^{(\mu)}$.

14.11 $\phi_i^{(\lambda)}$ and $\psi_j^{(\mu)}$ transform irreducibly according to $D^{(\lambda)}$ and $D^{(\mu)}$. Show that $M_{ij} = \int \phi_i^* \psi_j \, d\tau = \int (T\phi_i)^*(T\psi_j) \, d\tau$. Hence using the orthogonality relations show that $M_{ij} = 0$ when $D^{(\lambda)}$ and $D^{(\mu)}$ are inequivalent. Deduce also the result IB of the fundamental theorem of § 13.

14.12 In the fundamental theorem of § 13, in (13.8) and problem 13.13, in (14.12), (14.13) and problem 14.11, we have a series of closely related results with at least two independent possible lines of proof. Summarize how these are all logically inter-related.

14.13 Suppose you are given the characters of an irreducible representation. How would you construct in a systematic way an actual set of matrices to constitute the representation?

14.14* Discuss the application of group character techniques to continuous groups, in particular to the rotation group (Wigner 1931, p. 97).

14.15 A set of base vectors ϕ_i transform according to the irreducible representation $D_{ij}{}^{(\lambda)}(T)$ of some group of transformations T. Show from the orthogonality relation (14.8) that

$$\phi_i = \sum_T D_{ij}{}^{(\lambda)*}(T)\, T\phi_j,$$

where the suffix j is *not* summed over. Thus having once obtained one vector ϕ_j of a representation, we can define all the other ϕ_i in terms of it.

14.16 $D^{(\lambda)}(T)$ is an irreducible representation of a finite group, and $D^{(\lambda)*}(T)$ the representation which is obtained by taking the complex conjugate of every matrix of $D^{(\lambda)}(T)$. If $D^{(\lambda)}$ and $D^{(\lambda)*}$ are not equivalent, show from the orthogonality relation (14.8) that

$$\sum_T \chi^{(\lambda)}(T^2) = 0.$$

If $D^{(\lambda)}$ and $D^{(\lambda)*}$ are identical, i.e. real, show similarly that

$$\sum_T \chi^{(\lambda)}(T^2) = h$$

where h is the number of group elements. For an application of these results, see equation (19.22). Note: $D(T^2) = D(T)D(T)$ in any representation.

14.17 Prove just from the orthogonality relation of the first kind (14.9), that the number of irreducible representations of a group must be less than or equal to the number of classes.

15. Product Groups

Irreducible representations of product groups

Let us first consider the full rotation and reflection group, which consists of all proper and improper rotations about a point. In particular it contains the inversion Π (3.11), and any improper rotation can be written $R\Pi$ where R is a proper one. Let $\mathfrak{R}$ be a vector space invariant under this group. In $\mathfrak{R}$ we can use the proper rotations to pick out a set of base vectors ψ_m transforming according to $D^{(l)}$ under the rotation group, where for the present we shall consider l to be an integer. Then we can form the functions

$$\psi_{m+} = \psi_m + \Pi\psi_m, \qquad \psi_{m-} = \psi_m - \Pi\psi_m, \tag{15.1}$$

so that

$$\Pi\psi_{m+} = \psi_{m+}, \qquad \Pi\psi_{m-} = -\psi_{m-}. \tag{15.2}$$

Now Π commutes with every proper rotation. Hence operating on (15.1) with some rotation R we obtain

$$\begin{aligned} R\psi_{m+} &= R\psi_m + R\Pi\psi_m = R\psi_m + \Pi R\psi_m \\ &= D^{(l)}_{nm}(R)(\psi_n + \Pi\psi_n) = D^{(l)}_{nm}(R)\psi_{n+}. \end{aligned} \tag{15.3}$$

The ψ_{m+} transform therefore according to $D^{(l)}$ under proper rotations. Thus from (15.2) and (15.3), the ψ_{m+} form a vector space which is invariant under both proper and improper rotations, and which forms the basis of an irreducible representation $D^{(l;+1)}$ of the rotation and reflection group. Similarly the ψ_{m-} give an irreducible representation which we shall denote by $D^{(l;-1)}$. When the ψ_m transform according to $D^{(j)}$ under proper rotations, where j is half an odd integer, the representation is double-valued. Since $\Pi^2 = E$ and E is represented by plus or minus the unit matrix, we can similarly construct functions ψ_{m1}, ψ_{mi} such that

$$\Pi\psi_{m;1} = \pm\psi_{m;1}, \qquad \Pi\psi_{m;i} = \pm i\psi_{m;i}. \tag{15.4}$$

These give representations which we shall denote by $D^{(j;\pm 1)}$ and $D^{(j;\pm i)}$.

We shall not prove here that the above representations are the only irreducible ones, but proceed to the general theory. Consider two groups $\mathfrak{g}_1$ and $\mathfrak{g}_2$, which are such that every element P_1 of $\mathfrak{g}_1$ commutes with every element T_2 of $\mathfrak{g}_2$, i.e.

$$\boxed{P_1T_2 = T_2P_1.} \tag{15.5}$$

Then we can form a group $\mathfrak{G}$ consisting of all products P_1T_2 or T_2P_1 of elements from $\mathfrak{g}_1$ and $\mathfrak{g}_2$. Clearly these products satisfy the group postulates, e.g. $(P_1T_2)(Q_1S_2) = (P_1Q_1)(T_2S_2)$ is always another such product. For instance if $\mathfrak{g}_1$ is the full rotation group and $\mathfrak{g}_2$ is (E, Π), we have the rotation and reflection group. Other examples are the following pairs:

$\mathfrak{g}_1$	$\mathfrak{g}_2$
rotation of all co-ordinates r_i, σ_{zi}	permutation of all co-ordinates r_i, σ_{zi}
rotation of all orbital co-ordinates r_i	rotation of all spin co-ordinates σ_{zi}
rotation of co-ordinates r_1, σ_{z1}	rotation of r_2, σ_{z2}

The group $\mathfrak{G}$ is called the *direct product* of $\mathfrak{g}_1$ and $\mathfrak{g}_2$, written

$$\boxed{\mathfrak{G} = \mathfrak{g}_1 \times \mathfrak{g}_2.} \tag{15.6}$$

We shall now construct all the irreducible representations of $\mathfrak{G} = \mathfrak{g}_1 \times \mathfrak{g}_2$. In any invariant vector space we can construct a set ψ_i, $i = 1, 2, \ldots$ transforming according to the irreducible representation $D^{(1\lambda)}$ of $\mathfrak{g}_1$. Let T, S be any elements of $\mathfrak{g}_2$. Then from ψ_1 we can generate a set of functions $T\psi_1$, $S\psi_1$, etc., which clearly transform into one another under $\mathfrak{g}_2$, and we can reduce them and form a set of linear combinations

$$\psi_{1\alpha} = \sum_T b_{T\alpha} T\psi_1$$

transforming according to some irreducible representation $D^{(2\mu)}$ of $\mathfrak{g}_2$. Using the other ψ_i, let us now define the functions

$$\psi_{i\alpha} = (\sum_T b_{T\alpha} T)\psi_i. \tag{15.7}$$

Since any element of $\mathfrak{g}_1$ commutes with every T, it commutes with the operator $\sum b_{T\alpha} T$ in (15.7), and the $\psi_{i\alpha}$, $i = 1, 2, \ldots$ transform under $\mathfrak{g}_1$ in exactly the same way as the ψ_i, i.e. according to $D^{(1\lambda)}$. What is now the effect of operating on $\psi_{i\alpha}$ with an element S of $\mathfrak{g}_2$? Since S commutes with any element of $\mathfrak{g}_1$, $S\psi_{i\alpha}$, $i = 1, 2, \ldots$ transform according to $D^{(1\lambda)}$. Hence from part I of the fundamental theorem of § 13, we have

$$S\psi_{i\alpha} = \sum_\beta a_\beta(\alpha)\psi_{i\beta}$$

where $a_\beta(\alpha)$ does not depend on i. Thus the $\psi_{i\alpha}$, $\alpha = 1, 2, \ldots$ (i fixed) transform into one another under $\mathfrak{g}_2$ in the same way as the $\psi_{1\alpha}$, i.e. according to $D^{(2\mu)}$. We have therefore obtained a rectangle of functions $\psi_{i\alpha}$ such that the functions in each column (α fixed) transform under $\mathfrak{g}_1$ according to the irreducible representation $D^{(1\lambda)}$, and the functions in each row (i fixed) transform irreducibly according to $D^{(2\mu)}$ under $\mathcal{G}_2$. Together they give the representation $D^{(1\lambda;2\mu)}$ of the product group $\mathcal{G}_1 \times \mathcal{G}_2$, and this representation is clearly irreducible. *In this way we can construct systematically all irreducible representations of* $\mathcal{G}_1 \times \mathcal{G}_2$ *by combining each* $D^{(1\lambda)}$ *with each* $D^{(2\mu)}$.

Applying this to the earlier example, the irreducible representations of E, Π are

$$\chi(E) = 1,\ 1,\ \pm 1,\ \pm 1;$$
$$\chi(\Pi) = 1,\ -1,\ \pm 1,\ \pm i.$$

When j is half an odd integer, $D^{(j;+)}$ and $D^{(j;-)}$ are identical with $D^{(j;\pm 1)}$ because $D^{(j)}$ is already double-valued and $\Pi = \Pi E$ is represented by $\pm$ the unit matrix. This gives just the two irreducible representations $D^{(j;\pm 1)}$, $D^{(j;\pm i)}$ already obtained before. When l is an integer, we have the four representations $D^{(l;+)}$, $D^{(l;-)}$, $D^{(l;\pm 1)}$, $D^{(l;\pm i)}$, but the last two have such unphysical properties, (single-valuedness under proper rotations, double-valuedness under Π) that they have found no application (as yet!).

Summary

We have defined what is meant by a product group and shown how to construct all its irreducible representations out of those of the component groups.

PROBLEMS

15.1 In the notation of (15.5), verify that all the products $P_1 T_2$ satisfy the group requirements of § 4.

15.2 The functions u_{+i}, u_{-i}, $i = 1, 2, 3$, are spin functions for three electrons. Sort out linear combinations of the products $u_{\pm 1}\ u_{\pm 2}\ u_{\pm 3}$ into rectangles that transform irreducibly under the product group (rotation of all σ_{zi}) $\times$ (permutation of σ_{zi}). Hint: the correct linear combinations can be obtained using problems 11.5 and 5.17. The answer is given in (28.16).

15.3 What are the *complete* symmetry groups of $\mathscr{H}_{\text{cscf}}$, $\mathscr{H}_{\text{orb}}$, $\mathscr{H}_{\text{orb}} + \mathscr{H}_{\text{spin}}$ defined by (10.9), (10.2), (11.8), (11.9)? Interpret some of the degeneracies of §§ 10, 11 and 12 in terms of representations of product groups, particularly (10.8), (11.17). In terms of the complete symmetry group of $\mathscr{H}_{\text{orb}} + \mathscr{H}_{\text{spin}}$, what representations are allowed by the exclusion principle?

16. Point-groups

A point-group is a group of rotations about a point, i.e. of rotations whose axes intersect at a point. This distinguishes them from the more general space-groups which involve translations and rotations about non-intersecting axes. In this section we first describe and tabulate the thirty-two point-groups of crystallographic interest. Then follows a more detailed discussion of selected aspects of point-groups, namely: how all the crystallographic point-groups are derived; other point-groups such as apply to molecules, including those derived from the axial rotation group; how the character tables are derived; and derivation of the double-valued spin representations. The complete character tables of the most important point-groups are tabulated in appendices K and L.

Crystal point-groups

In a crystal the atoms are arranged in a regular periodic array, and the fact that it is periodic severely restricts the kind of rotational symmetries that the arrangement can have. The rotations by all multiples of $360°/n$ are referred to as an n-fold rotation axis, and it will be shown below that crystals can only have 1-fold, 2-fold, 3-fold, 4-fold and 6-fold axes. For instance it is impossible to have a periodic array with five-fold symmetry. These proper rotations are given the symbols 1, 2, 3, 4, 6, and in addition there can be improper ones. A roto-inversion axis $\bar{n}$ consists of all multiples of a rotation by $360°/n$ followed by the inversion through the origin Π (3.11). In particular $\bar{1}$ is just E and the inversion itself, and the group $\bar{2}$ is E plus a reflection m in a mirror plane perpendicular to the axis. In fact the axis $\bar{2}$ is usually denoted by m. There are also roto-reflection axes $\tilde{n}$, whose basic element is a rotation by $360°/n$ followed by a reflection in a mirror plane perpendicular to the axis. In these improper axes n again can only have the values 1, 2, 3, 4, 6. The two types of improper axis are not independent. In fact the roto-inversion axes are used almost exclusively now although in the older literature, particularly in the Schoenflies notation, the roto-reflection axes were adopted as the basic improper axes. We have the relations between them $\tilde{1} = \bar{2} = \mathrm{m}$, $\tilde{2} = \bar{1}$, $\tilde{3} = \bar{6}$, $\tilde{4} = \bar{4}$, $\tilde{6} = \bar{3}$.

These and other facts about point-groups can be seen mostly easily by drawing stereograms. Consider a sphere with a spot marked on it at some arbitrary point. If we apply all the rotations of a point-group to the sphere, we can represent each element of the group by the position taken up by the spot. The position can be plotted by taking a bird's-eye view of the sphere from above the north pole, a dot meaning a position in the northern hemisphere and a circle one in the southern hemisphere. Such a diagram is called a stereogram.† Thus if we plot a dot or circle for each element

† This description of a stereogram is technically not quite correct (Phillips 1946), but it is near enough for our present discussion.

of a point-group, we obtain a pattern of dots and circles representing a set of equivalent directions in space which are rotated into one another under the action of the group. It can now be seen, for example, that the stereogram for the point-group $\bar{3}$ (Fig. 9) is

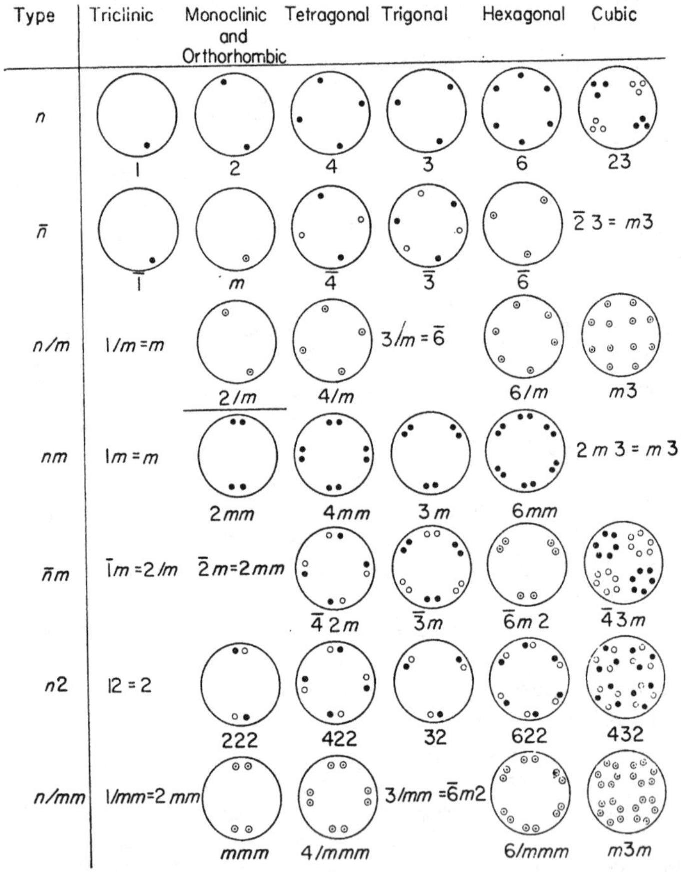

FIG. 9. Stereograms of crystal point-groups.

invariant under a 60° rotation followed by a horizontal reflection, so that $\bar{3}$ is equivalent to $\tilde{6}$.

We can form more complicated point-groups by combining two or more axes. For instance the stereogram for 422 (Fig. 9) shows that this point-group contains a four-fold axis (perpendicular to the page) and four two-fold axes in the plane of the paper. Fig. 9 shows all the thirty-two point-groups that can apply to crystals. They are labelled in the abbreviated international notation. First a complete symmetry symbol is constructed as follows. The main axis is given first. Then $\frac{n}{\mathrm{m}}$, usually written n/m, denotes a mirror plane perpendicular to this axis. Then follow other non-equivalent symmetry axes and planes. For instance 4mm has two pairs of vertical mirror planes which are at 45° to one another and hence are not equivalent to one another under the action of the group. The complete symmetry symbols are then abbreviated, just enough being retained to characterize the group completely. Appendix J shows the complete symmetry symbols, the abbreviated (international) symbols, and also the older Schoenflies symbols for all the point-groups.

In the point-groups of the cubic system, the 3-fold axes are along the body diagonals of a cube, the 4-fold or main 2-fold axes perpendicular to the cube faces, and other 2-fold axes parallel to the face diagonals. For a fuller description of the point-groups, we refer to any book on crystallography (e.g. Phillips 1946, Buerger 1956).

The character tables of all the crystal point-groups are given in appendix K, where the notation for the irreducible representations is also explained. The derivation of the character tables is discussed later in this section.

Derivation of the crystal point-groups

We shall now show how the point-groups are derived and why there are only a finite number. Very complete derivations of this and also of the space-groups have been given by Seitz (1934, etc.), Zachariasen (1945) and Murnaghan (1938). To bring out the main ideas, we shall consider in detail the point-groups which consist of proper rotations only. There are eleven of these (Table 10). We shall derive them from first principles and show no others can exist. The improper point-groups may easily be derived from these as shown in the above references and in problem 16.4.

We first prove that only n-fold axes can exist, where $n = 1, 2, 3, 4$ or 6. Because of the periodicity of the crystal structure, any translation

$$\mathbf{t} = n_1\mathbf{a}_1 + n_2\mathbf{a}_2 + n_3\mathbf{a}_3 \tag{16.1}$$

is a symmetry element and conversely any purely translational symmetry is of the form (16.1). Here the $\mathbf{a}_i$ are the primitive translation vectors, which in general do not have the same length and may make skew angles with one another. The n_i are integers. Let $R(\phi)$ be a purely rotational symmetry. Then $R(\phi)tR(-\phi)$ is another pure translation, say

$$\mathbf{t}' = n'_1\mathbf{a}_1 + n'_2\mathbf{a}_2 + n'_3\mathbf{a}_3,$$

which is just the translation $\mathbf{t}$ rotated by an angle ϕ as follows directly from (5.4†) or from a diagram. If we write

$$n'_j = D_{ij}(\phi)n_i, \tag{16.2}$$

then $D_{ij}(\phi)$ forms a representation of the rotation. By choosing new base vectors $\mathbf{i}_1$, $\mathbf{i}_2$, $\mathbf{i}_3$ along the co-ordinate axes, we see that the representation (16.2) is equivalent to the representation $D^{(1)}$ based on the functions x, y, z. The character of $D_{ij}(\phi)$ is therefore from (14.4)

$$\chi(\phi) = 1 + 2\cos\phi. \tag{16.3}$$

Now if we put $n_1 = 1$, $n_2 = n_3 = 0$, then since the n'_j have to be integers, (16.2) shows that the matrix elements $D_{1j}(\phi)$ are integers and similarly so are all the $D_{ij}(\phi)$. Consequently the character is an integer since it is just the sum of the diagonal elements. Comparing with (16.3), we have $1 + 2\cos\phi$ is an integer, whence the allowed values of ϕ are given by

$$\begin{aligned} 1 + 2\cos\phi &= -1, \quad 0, \quad 1, \quad 2, \quad 3; \\ \phi &= 180°, \quad \pm 120°, \quad \pm 90°, \quad \pm 60°, \quad 0°. \end{aligned}$$

I.e. ϕ is a multiple of 60° or 90°, *which limits the rotational symmetries to* $n = 1, 2, 3, 4$ *or* 6. The allowed angles can conveniently be written

$$\boxed{\begin{aligned} \phi = 2\pi k/n,\ n &= 1, 2, 3, 4 \text{ or } 6, \\ k &= 1, 2, \ldots, n. \end{aligned}} \tag{16.4}$$

We now deduce a relation between the numbers s_n of n-fold axes in one point-group $\mathfrak{G}$. Let S be the sum of all the elements of $\mathfrak{G}$ which we will assume has at least two rotation axes. We can write

$$S = E + \sum_{n,r} S_{nr}, \tag{16.5}$$

where S_{nr} is the sum of all elements except E belonging to the

r^{th} n-fold axis. E belongs simultaneously to all axes. Summing (16.3) over all angles (16.4) with $k = 1, 2, \ldots, n$-1 gives the character of a single S_{nr},

$$\chi(S_{nr}) = n + 2\sum_{k=1}^{n} \cos 2\pi k/n - \chi(E) = n - 3.$$

Thus from (16.5),

$$\chi(S) = 3 + \sum_{2}^{6} (n-3)s_n = 3 - s_2 + s_4 + 3s_6. \qquad (16.6)$$

We now calculate $\chi(S)$ in another way. Let R_1 and R_2 be rotations about two different axes. The collection of elements $R_1\mathfrak{G}$ is just the group $\mathfrak{G}$ itself (cf. below equation (14.5), also problem 4.10). Hence $R_1S = R_2S = S$. From (16.2) we may regard a rotation as a tensor that turns the vector $\mathbf{t}$ into $\mathbf{t}'$ and write $\mathbf{t}' = \mathbf{R}(\phi)\cdot\mathbf{t}$. In this notation, we have

$$\mathbf{R}_1 \cdot \mathbf{St} = \mathbf{R}_2 \cdot \mathbf{St} = \mathbf{St}.$$

Now $\mathbf{R}_1 \cdot \mathbf{St} = \mathbf{St}$ implies that the vector $\mathbf{St}$ is parallel to the rotation axis of $\mathbf{R}_1$. Likewise it is parallel to the second axis, and since these were assumed to be different, $\mathbf{St} = 0$ for all $\mathbf{t}$, whence $\mathbf{S} = 0$ and $\chi(S) = 0$. Substituting into (16.6) we obtain

$$\boxed{s_2 = 3 + s_4 + 3s_6,} \qquad (16.7)$$

which is the required result. This relation does not apply to point-groups having only one rotation axis.

We now consider the angles between rotation axes. Suppose $R(\phi_1, \xi_1)$ and $R(\phi_2, \xi_2)$ are two rotations belonging to a point-group. Then

$$R(\phi_3, \xi_3) = R(\phi_2, \xi_2)R(\phi_1, \xi_1)$$

has to be another allowed rotation, so that ϕ_3 is one of the angles (16.4). Now ϕ_3 is given by†

$$\cos \tfrac{1}{2}\phi_3 = \cos \tfrac{1}{2}\phi_1 \cos \tfrac{1}{2}\phi_2 - \sin \tfrac{1}{2}\phi_1 \sin \tfrac{1}{2}\phi_2 \cos \gamma, \qquad (16.8)$$

where γ is the angle between the ξ_1 and ξ_2 axes. Consider the case

† The reader can prove this formula as follows. Let $\mathbf{T}_1$ and $\mathbf{T}_2$ be the dyadics representing the first and second rotations (see problem 8.4). The dyadic $\mathbf{T}_2\cdot\mathbf{T}_1$ then represents the resultant rotation, and its scalar (sum of diagonal elements) is therefore $1 + 2\cos\phi_3$; i.e. $\mathbf{T}_2:\mathbf{T}_1 = 1 + 2\cos\phi_3$. This equation reduces to (16.8) after some manipulation. For a more geometrical proof, see Buerger (1956, p. 38).

when ξ_1 and ξ_2 are both 6-fold axes. Putting $\phi_1 = \phi_2 = 60°$ and $\phi_1 = 60°$, $\phi_2 = 120°$ we obtain respectively

$$\cos \tfrac{1}{2}\phi_3 = \tfrac{3}{4} - \tfrac{1}{4}\cos \gamma,$$

and

$$\cos \tfrac{1}{2}\phi'_3 = \tfrac{1}{4}\sqrt{3} - \tfrac{1}{4}\sqrt{3}\cos \gamma, \tag{16.9a}$$

whence eliminating $\cos \gamma$ we have

$$\cos \tfrac{1}{2}\phi_3 - \sqrt{\tfrac{1}{3}}\cos \tfrac{1}{2}\phi'_3 = \tfrac{1}{2}. \tag{16.9b}$$

From (16.4) the allowed values of $\cos \frac{1}{2}\phi_3$ and $\cos \frac{1}{2}\phi'_3$ are $\pm\frac{1}{2}$, $\pm\frac{1}{2}\sqrt{3}$, $\pm\sqrt{\frac{1}{2}}$, 0, ± 1, and we see by inspection that the only allowed solutions of (16.9b) are

(i) $\cos \frac{1}{2}\phi_3 = \frac{1}{2}$, $\quad\cos \frac{1}{2}\phi'_3 = 0$;
(ii) $\cos \frac{1}{2}\phi_3 = 1$, $\quad\cos \frac{1}{2}\phi'_3 = \frac{1}{2}\sqrt{3}$;
(iii) $\cos \frac{1}{2}\phi_3 = 0$, $\quad\cos \frac{1}{2}\phi'_3 = -\frac{1}{2}\sqrt{3}$.

From (16.9a), these solutions give respectively

(i) $\cos \gamma = 1$, $\gamma = 0°$; (ii) $\cos \gamma = -1$, $\gamma = 180°$; (iii) $\cos \gamma = 3$.

Case (iii) is meaningless, and in either of the other cases the axes ξ_1, ξ_2 coincide. Thus there can be at most one 6-fold axis in a point-group.

TABLE 10

Number of Axes in Proper Point-groups

	s_2	s_3	s_4	s_6	point-group
(i)	0	0	0	0	1
(ii)	1	0	0	0	2
(iii)	0	1	0	0	3
(iv)	0	0	1	0	4
(v)	0	0	0	1	6
(vi)	6	0	0	1	622
(vii)	4	0	1	0	422
(viii)	6	4	3	0	432
(ix)	3	1	0	0	32
(x)	3	4	0	0	23
(xi)	3	0	0	0	222

We can now tabulate all the possible combinations of axes. If there is only one axis we have the point-groups (i) to (v) (Table 10). We next consider the case when there is a 6-fold axis; there can only be one as we have just proved. From this it follows that there can be no separate 3-fold axis since this would require at least three 6-fold axes to give a 3-fold symmetry. Likewise there can be no four-fold axes. Then (16.7) gives only case (vi) (Table 10) with $s_6 > 0$, and in enumerating the other possibilities we shall assume without further mention that $s_6 = 0$. Consider the case of a single 4-fold axis. There can be no 3-fold axes because these would lead to more than one 4-fold axis, but from (16.7) there are four 2-fold axes, giving case (vii). If there are two or more 4-fold axes, it can easily be shown from (16.8) that they intersect at 90°. Now from (16.8) or problem 8.4, $R(90°, x)\ R(90°, y)$ is a 120° rotation. Hence we have a 3-fold axis and a third 4-fold axis in the z-direction. There can be no more 4-fold axes because there are no further directions at right angles to each of the x-, y- and z-axes. Further there must be four 3-fold axes to preserve the four-fold symmetry, which gives case (viii). This exhausts the cases with 4-fold axes. If there is one 3-fold axis, any 2-fold axes must be perpendicular to it and from (16.7) there are three of them, giving case (ix). If there are more than one 3-fold axes, (16.8) gives $\cos\gamma = \pm\frac{1}{3}$. Further $\phi_1 = \phi_2 = 120°$, $\cos\gamma = \frac{1}{3}$ gives $\phi_3 = 180°$, i.e. a 2-fold axis. Then from (16.7) there must be three 2-fold axes, which in turn generate four 3-fold axes. The latter are oriented with respect to one another as the body diagonals of a cube, and the $\cos\gamma = \pm\frac{1}{3}$ leaves no room for more. Hence we have only case (x). From (16.7) there remains only the possibility of three 2-fold axes, case (xi). Having derived all the possible numbers of axes, it can now be shown that for each case of Table 10 there is essentially only one way of arranging the axes in space. This follows from elementary geometrical reasoning. For instance in case (x), the restriction that the three-fold axes intersect at $\cos^{-1}(\pm\frac{1}{3})$ allows only one arrangement of them. The two-fold axes must lie either perpendicular to each three-fold axis or bisecting the angle between two of them. For there to be only three two-fold axes it can easily be seen that the latter alternative must hold, which fixes their directions uniquely. This completes the derivation of all proper point-groups, and the ones with improper axes may be derived from them as already mentioned (Zachariasen 1945, p. 40; problem 16.4).

Character tables and other point-groups

Besides the crystallographic point-groups, there are other finite subgroups of the full rotation group. Using the same notation as before, an n-fold rotation axis n is a *cyclic* point-group for any value of n. We can add 2-fold axes perpendicular to it to obtain the *dihedral* group $n2$. Similarly by adding mirror planes (improper

2-fold axes) we obtain the groups n, $n2$, n/m, $n\mathrm{m}$, n/mm, $\bar{n}$, $\bar{n}\ 2\mathrm{m}$, etc. These are not all different, but which ones are the same depends on whether we have $n = 2q + 1$, $4q$, or $4q + 2$ where q is an integer (cf. Fig. 9). Apart from these there is only one other proper point-group, and that corresponds to the symmetry of a regular icosahedron (Murnaghan 1938, p. 336).

In addition to the above finite point-groups, there are several subgroups of the full rotation group that are associated with the axial rotation group. The latter has the symbol ∞ in the present notation, and the other derived groups are ∞/m, $\infty 2$, $\infty\mathrm{m}$ and ∞/mm. These are very important for describing diatomic molecules and we shall now derive their character tables. Firstly ∞ is a cyclic group and its irreducible representations have been derived in § 7. Each rotation forms a class of its own (cf. problem 14.6), and the character table is as shown in Table 11 and appendix L.

TABLE 11

Character Tables of ∞ and $\infty\mathrm{m}$

Irred. rep.	The group ∞ $\chi(E)$	$\chi[R(\phi, z)]$
A_0	1	1
A_k	1	$\exp(ik\phi)$
A_{-k}	1	$\exp(-ik\phi)$
$(k \geqq 1)$		

Irred. rep.	$\chi(E)$	The group $\infty\mathrm{m}$ $\chi[R(\phi, z)]$	$\chi(\mathrm{m}_x)$
A_+	1	1	1
A_-	1	1	-1
E_k	2	$2\cos k\phi$	0
$(k \geqq 1)$			

Note: The notation for the irreducible representations is not quite the standard notation of appendices K and L, but is more suited to the present discussion.

Since there are an infinite number of classes we expect an infinite number of irreducible representations. Consider next the group $\infty\mathrm{m}$, operating on some invariant vector space. This space can first be reduced according to the axial rotation group, and let ψ_1 be a vector in it which transforms according to the representation

$\exp(ik\phi)$ of Table 11. In spherical polar co-ordinates, the reflection m_x perpendicular to the x-axis induces the transformation

$$\Theta = \theta, \qquad \Phi = -\phi.$$

If we put

$$\psi_2 = \mathrm{m}_x\psi_1,$$

then we have

$$\begin{aligned} R(\phi, z)\psi_2 &= R(\phi, z)\mathrm{m}_x\psi_1 \\ &= \mathrm{m}_x R(-\phi, z)\psi_1 \\ &= \mathrm{m}_x \exp(-ik\phi)\psi_1 \\ &= \exp(-ik\phi)\psi_2. \end{aligned} \tag{16.10}$$

Hence ψ_2 belongs to the representation $\exp(-ik\phi)$ of ∞, so that if $k \neq 0$ it is linearly independent of ψ_1. However, from (16.10) ψ_2 belongs to the same irreducible vector space as ψ_1, and we have a two-dimensional irreducible representation E_k. The character of $R(\phi, z)$ is

$$\exp(ik\phi) + \exp(-ik\phi) = 2\cos k\phi,$$

and the character of m_x is 0 because it just interchanges ψ_1 and ψ_2. For $k = 0$, ψ_1 and ψ_2 need not be linearly independent. We can always form the symmetric and antisymmetric linear combinations $\psi_1 \pm \psi_2$, which give two one-dimensional representations. In this way we obtain the complete character table shown in Table 11. The theory for the other groups derived from ∞ is similar. The group $\infty 2$ is isomorphic with ∞m and hence has the same character table. Also ∞/m is the direct product $\infty \times (E, \mathrm{m}_z)$, and ∞/mm is the direct product of ∞m with the inversion group $\bar{1} = (E, \Pi)$. Hence their character tables can be written down from the prescription of § 15 or appendix K.

Similar considerations apply to the finite groups n, $n2$, n/m, nm, n/mm, etc. However there is the following difference. For n even, k in (16.10) may have the value $\frac{1}{2}n$, so that $\exp(-ik\phi) = \pm 1$ for all possible values $2\pi r/n$ of ϕ. Hence ψ_1 and ψ_2 in (16.10) need not be linearly independent and we obtain two one-dimensional representations instead of a two-dimensional one. In this way the irreducible representations of all the point-groups except the cubic ones (and the icosahedral ones) can easily be obtained. The character tables for these remaining ones can be determined from the relations of § 14.

Double-valued (or spin) representations

Let u_+, u_- be the spin functions transforming according to $D^{(1/2)}$ under the full rotation group. Now this representation is double-valued, with the base vectors changing sign under a rotation of

360° about any axis (§ 8), so that u_+, u_- must also transform according to a double-valued representation under any point-group. The double-valued irreducible representations of all the crystal point-groups have been tabulated by Koster (1957).

We shall now show how the double-valued representations may be found, following the method of Opechowski (1940) based on the earlier work of Bethe (1929). These representations can be found for the cyclic groups n, the dihedral groups $n2$ and associated groups by using half integer values of k in (16.10) and proceeding as before. However the following device is more general and convenient. Consider a point-group $\mathfrak{g}$ of proper rotations, and the corresponding group $\mathfrak{G}$ of matrices taken from the representation $D^{(1/2)}$ (8.24) of the full rotation group. Clearly these matrices satisfy all the group requirements, because (8.24) forms a representation. However because of the + or − sign in (8.24), i.e. the double-valuedness of $D^{(1/2)}$, $\mathfrak{G}$ contains twice as many elements as $\mathfrak{g}$ does. Consequently any single-valued representation of $\mathfrak{G}$ automatically furnishes a double-valued representation of $\mathfrak{g}$. The single-valued representations of $\mathfrak{G}$ may of course be found by the methods of § 14 in a straightforward way. The group of matrices $\mathfrak{G}$ is called the *double-group* of $\mathfrak{g}$.

The procedure becomes clearer by considering as a specific example the point-group 32 of transformations (4.13). Let B be the 2×2 matrix (8.24), taken with a plus sign, which represents the rotation $R(\frac{2}{3}\pi, z)$, i.e. the transformation B' of (4.13). Let $\bar{B}$ be the same matrix taken with a minus sign. Similarly let E, $\bar{E}$, A, $\bar{A}$, K, $\bar{K}$, L, $\bar{L}$, M, $\bar{M}$ be the other elements of $\mathfrak{G}$. Now suppose we have relations $K\bar{A} = \bar{L}$, etc., among the elements of $\mathfrak{G}$. These must correspond to the relations $K'A' = L'$, etc., among the rotations of $\mathfrak{g}$. Thus if we have a representation D of the double group $\mathfrak{G}$ so that $D(K)D(\bar{A}) = D(\bar{L})$, etc., then this also gives a representation of the original point-group $\mathfrak{g}$, which however will be double-valued because we must associate both $D(A)$ and $D(\bar{A})$ with the same rotation A'. We shall not write out the whole multiplication table of $\mathfrak{G}$, but typical relations are

$$B^2 = \bar{A}, \qquad B^3 = \bar{E}, \qquad B^4 = A^2 = \bar{B}, \qquad \bar{A}^3 = B^6 = E,$$
$$K^2 = \bar{E} = \bar{K}^2, \qquad K\bar{K} = E = K^4. \tag{16.11}$$

Notice in particular that if successive rotations like B^3 or K^2 add up to a rotation of 360° about one axis, they are always equal to $\bar{E}$. The elements can easily be collected into their six classes

$$(E), (\bar{E}), (A, B), (\bar{A}, \bar{B}), (K, L, M), (\bar{K}, \bar{L}, \bar{M}).$$

The classes are very much as before, e.g. A and B form one class. Now $\mathfrak{G}$ contains 12 elements and 6 classes, so that from (14.13), (14.6) there are four one-dimensional and two two-dimensional irreducible representations. Some of these can be written down immediately. Suppose we have an ordinary single-valued irreducible representation Δ of $\mathfrak{g}$. Then we can obtain an irreducible representation of $\mathfrak{G}$ by associating the matrix $\Delta(R')$ of Δ with each of the elements R and $\bar{R}$ of $\mathfrak{G}$. Thus from Table 8 we obtain the representations Γ_1, Γ_2, Γ_3 of Table 12. Consider now the element $\bar{E}$. It commutes with every element (matrix) of $\mathfrak{G}$, and hence by

TABLE 12

Character Table for the Double-group of the Point-group 32

	E	$\bar{E}$	A,B	$\bar{A},\bar{B}$	K,L,M	$\bar{K},\bar{L},\bar{M}$
Γ_1	1	1	1	1	1	1
Γ_2	1	1	1	1	-1	-1
Γ_3	2	2	-1	-1	0	0
Γ_4	1	-1	1	-1	i	$-i$
Γ_5	1	-1	1	-1	$-i$	i
Γ_6	2	-2	-1	1	0	0

Γ_1, Γ_2, Γ_3 give the single-valued representations of the group 32 which are identical with $\mathscr{I}$, $\mathscr{A}$ and Γ of Tables 3 and 8 and with A_1, A_2 and E of appendix K. Γ_4, Γ_5, Γ_6 give extra double-valued representations.

Schur's lemma (appendix D) it is represented in any irreducible representation by a multiple cE of the unit matrix E. Since $\bar{E}^2 = E$, we have $c = \pm 1$. Since $\bar{R} = R\bar{E}$ for any R, choosing $c = 1$ gives a representation of the type already considered with $\chi(R) = \chi(\bar{R})$ for all R. Thus to obtain further representations we must choose $c = -1$. Consider first the one-dimensional representations. Since $K^2 = \bar{E}$, we must have $\chi(K) = -\chi(\bar{K}) = \pm i$. The relation $K\bar{A} = \bar{L}$ then gives $\chi(\bar{A}) = -\chi(A) = -1$, whence we obtain the representations Γ_4, Γ_5 of Table 12. Γ_6 may then be obtained from the relation of problem 14.8. This then gives all the double-valued and single-valued irreducible representations of the point-group 32 by taking $\chi(R)$, $\chi(\bar{R})$ together. No very systematic or generally adopted method of labelling the double-valued ones exists at present, and they are usually just numbered consecutively.

We now formulate some general rules which aid considerably in finding the double-valued irreducible representations of any particular point-group or other finite group $\mathfrak{g}$. We first note that the device of constructing the double-group $\mathfrak{G}$ of matrices (8.24) is quite general. Let R' be any rotation of $\mathfrak{g}$. Then (8.24) gives the two matrices R and $\bar{R}$ of $\mathfrak{G}$, and *any irreducible representation D of $\mathfrak{G}$ gives a double-valued irreducible representation $D(R)$, $D(\bar{R})$ of $\mathfrak{g}$.* In working out a character table, the first job is to collect the elements into classes. Now since the matrices R, $\bar{R}$ of $\mathfrak{G}$ form a representation of $\mathfrak{g}$, they multiply in exactly the same way as the rotations R' of $\mathfrak{g}$, except for the possible introduction of minus signs. For instance a relation

$$P'R'(P')^{-1} = S' \text{ for } \mathfrak{g}$$

which shows that R' and S' belong to the same class, becomes

$$P(R \text{ or } \bar{R})P^{-1} = S \text{ or } \bar{S} \text{ for } \mathfrak{G}.$$

Thus the class structure of $\mathfrak{G}$ is very similar to that of $\mathfrak{g}$. To be precise, if a set of rotations R' form a class of $\mathfrak{g}$, then the matrices R, $\bar{R}$ form one or two classes of $\mathfrak{G}$. It is not clear from the above discussion whether the set R, $\bar{R}$ makes up one class or breaks up into two classes. This can be decided as follows. From (8.24), the matrices R and $\bar{R}$ have characters

$$\chi(R) = -\chi(\bar{R}) = 2 \cos \tfrac{1}{2}\theta \cos \tfrac{1}{2}(\chi + \phi). \tag{16.12}$$

Since they have different characters, they cannot belong to the same class, i.e. we obtain two classes. By suitable choice of angles in (16.12), e.g. choosing $\phi = -120°$ or $+240°$, we can always arrange that the R matrices form the one class and the $\bar{R}$ the other, which gives the first rule:

I. *If a set of rotations R' form a class of $\mathfrak{g}$, then the matrices R form one class and the matrices $\bar{R}$ another class of $\mathfrak{G}$.*

The above argument does not work when the angle of rotation of R' is 180°, because then the character (16.12) becomes zero. In this case R and $\bar{R}$ may or may not belong to the same class. Suppose that they do. Then there exists some element S such that $\bar{R} = S^{-1}RS$, i.e.

$$S\bar{R} = RS. \tag{16.13}$$

Let us choose the rotation axis of R' as the z-axis. Then from (8.24) R takes the form

$$R = \begin{bmatrix} i & \cdot \\ \cdot & -i \end{bmatrix}.$$

If we take the general form (8.24) for S and substitute in (16.13), we find $\theta = 180°$. This means that the z-axis (Fig. 6) gets inverted, so that S is a rotation by 180° about an axis perpendicular to the z-axis. The argument can also be applied in reverse, and we have as our second rule:

II. *There is one exception to rule* I. *If the rotations are through* 180°, *then* R *and* $\bar{R}$ *belong to the same class of the double-group if, and only if, there is also in the group another rotation by* 180° *about an axis perpendicular to the axis of* R.

Any representation $D(R')$ of $\mathfrak{g}$ gives a representation of $\mathfrak{G}$ by associating $D(R')$ with both R and $\bar{R}$, but this does not give a double-valued representation of $\mathfrak{g}$ except in a very trivial sense. To obtain the extra irreducible representations of $\mathfrak{G}$ which give genuine double-valued representations of $\mathfrak{g}$ we must have the element $\bar{E}$ represented by minus the unit matrix as already proved in connection with the group 32. We therefore have from $\bar{R} = R\bar{E}$ the third rule:

III. *For the extra irreducible representations of the double-group, we have*

$$\chi(R) = -\chi(\bar{R}),$$

and in the exceptional case of rule II, $\chi(R) = 0$.

The above procedure can only be applied as it stands to proper point-groups. However any point-group containing improper rotations is either a direct product of a proper group and the inversion group $\bar{1}$, or it is isomorphic with a proper group which is obtained by replacing each improper rotation by the corresponding proper one. The former case can be handled using § 15 and equation (11.23), and the latter case by considering the isomorphic proper group. We can therefore reduce the representation given by any set of functions, including spin-dependent functions, with respect to any point-group.

References

Bethe (1929) and Opechowski (1940) discuss the derivation of the double-valued irreducible representations. Koster (1957) has tabulated these for the crystal point-groups. A derivation and description of the crystal point-groups are given by Seitz (1934), Zachariasen (1945) and Murnaghan (1938). Murnaghan (1938) also discusses the icosahedral group.

Summary

We have tabulated all the thirty-two crystal point-groups (Fig. 9 and appendix J), and the characters of their single-valued irreducible representations (appendix K). We have shown how the single- and double-valued irreducible representations of any point-group may be obtained. We have discussed the groups derived from the axial rotation group and derived their character tables (appendix L). We have also shown how the crystal point-groups may be derived from first principles.

Problems

16.1 Determine the classes of the point-groups 32, 422, 622, and hence calculate their character tables by the method used for ∞2 in the text.

16.2 Show that 422 and $\bar{4}$m2 must have the same character tables.

16.3 Write down four functions which transform respectively according to the four one-dimensional representations of ∞/mm. Also write down five functions whose *maximum* symmetry groups are respectively ∞, ∞/m, ∞2, ∞m and ∞/mm.

16.4 An improper point-group contains the inversion element Π. Show that the group is a direct product $\mathfrak{g} \times \bar{1}$, where $\mathfrak{g}$ is a proper point-group. If an improper point-group does not contain the inversion Π as an element, show that an isomorphic proper point-group can be constructed from it by replacing every improper axis $\bar{n}$ by the corresponding proper axis n. Hence derive all the improper tetragonal point-groups of Fig. 9 from the proper ones of Table 10.

16.5 The functions ϕ_i, $i = 1, 2, 3$ and u_+, u_- transform according to $D^{(1)}$ and $D^{(1/2)}$ respectively under the full rotation group. According to what representation (in irreducible components) do the products $\phi_i u_+$ transform under the point-group 32?

16.6 A set of functions transform according to $D^{(J)}$ under the full rotation group. Let us also use $D^{(J)}$ for the representation of the group 32 which they form. With this notation, show that

$$D^{(J)} = D^{(J-6)} + 2\mathscr{I} + 2\mathscr{A} + 4\Gamma$$

for $J =$ an integer $\geqslant 6$ in the notation of Tables 3 and 8, and that

$$D^{(J)} = D^{(J-3)} + \Gamma_4 + \Gamma_5 + 2\Gamma_6$$

for $J =$ half an odd integer >3 in the notation of Table 12.

16.7 Construct the characters of the extra double-valued representations of the group 422.

16.8 R and S are any two elements of a group in the same class. Show that there exists an element T such that $T^{-1}ST = R$.

16.9 Show that in any representation $\chi(R)^* = \chi(R^{-1})$. What are the consequences of this for the characters of 180° rotations in single-valued and double-valued irreducible representations (Opechowshi 1940)?

16.10 Show how to construct polynomials of degree n which transform according to a given representation of a point-group (Olson and Rodrigues, 1957).

17. The Relationship Between Group Theory and the Dirac Method

In this book we have developed the group theoretical method for sorting out and labelling a complete set of functions, usually the eigenfunctions of a Hamiltonian. This contrasts at first sight with the more usual procedure, developed by Dirac (1958), in which one uses as a complete set of functions the simultaneous eigenfunctions of a set of commuting operators (see for example Schiff 1955, p. 143). The purpose of the present section is to relate these two approaches,† and show that they are completely equivalent.

In brief, Dirac proceeds as follows. Let ψ be an eigenfunction of two operators A and B belonging to eigenvalues a, b;

$$A\psi = a\psi, \qquad B\psi = b\psi.$$

It follows that

$$AB\psi = Ab\psi = ba\psi = aB\psi = Ba\psi = BA\psi,$$

i.e.

$$(AB - BA)\psi = 0.$$

This suggests that simultaneous eigenfunctions like ψ are most likely to exist if $AB - BA = 0$, i.e. if the two operators commute. In fact it can be shown (Dirac 1958, p. 49) that the simultaneous eigenfunctions of two commuting operators form a complete set of functions. By using several commuting operators, we can arrange it that the sets of eigenvalues of two different functions are always different. This then gives a definite way of achieving a sorted and labelled complete set of functions. One of the operators is usually chosen to be the Hamiltonian, so that the eigenfunctions are sorted out according to their energies. In this case the other operators are

† The results of this section are not used elsewhere in the book. It has been included for the benefit of those readers whose original introduction to quantum mechanics was through Dirac's book, or who for other reasons like to think in terms of complete sets of commuting operators.

constants of the motion in the quantum mechanical sense. In detail, for any operator A not depending explicitly on time, we have (Schiff 1955, p. 134)

$$\frac{\mathrm{d}}{\mathrm{d}t}\langle A\rangle = \frac{1}{i\hbar}\langle A\mathscr{H} - \mathscr{H}A\rangle, \qquad (17.1)$$

where $\langle A\rangle$ is the expectation value $\int\psi^* A\psi\,\mathrm{d}\tau$ of A. Clearly $\langle A\rangle$ is constant for any state $\psi(t)$ if A commutes with $\mathscr{H}$, and A is called a constant of the motion accordingly.

Continuous groups

The relationship between the group theoretical and the Dirac methods of labelling eigenfunctions is very simple when the Hamiltonian $\mathscr{H}$ is invariant under a group of transformations forming a continuous sequence in terms of some co-ordinates q_i. Suppose $\mathscr{H}$ does not depend on q_1. Then if we regard $\mathscr{H}$ as a classical Hamiltonian the momentum p_1 conjugate to q_1 is a constant of the motion because

$$\frac{\mathrm{d}p_1}{\mathrm{d}t} = -\frac{\partial\mathscr{H}}{\partial q_1} = 0.$$

Quantum mechanically, p_1 commutes with all p_i and all q_i except q_1. Since $\mathscr{H}$ does not involve q_1, p_1 also commutes with $\mathscr{H}$. From (17.1) it is a constant of the motion just as in the classical analysis, and may be taken as one of the set of commuting operators. As a simple example we may cite the case of the electron of a hydrogen atom in a uniform magnetic field H in the z-direction. The Hamiltonian (neglecting spin) is

$$-\frac{\hbar^2}{2m}\nabla^2 + \frac{e^2}{r} + \frac{ei\hbar}{2mc}H\frac{\partial}{\partial\phi} + \frac{e^2}{8mc^2}H^2 r^2 \sin^2\theta. \qquad (17.2)$$

This is not exactly in canonical form, but it does not depend explicitly on ϕ. Thus the conjugate momentum, namely the angular momentum L_z about the z-axis, is a constant of the motion as we would expect classically. The eigenfunctions therefore have a definite value of L_z, namely $m\hbar$ (Schiff 1955, p. 75). This illustrates the Dirac viewpoint.

Now group theoretically we would proceed as follows. Since $\mathscr{H}$ is independent of q_1, it is invariant under all transformations $q_1 = Q_1 + \Delta q_1$. Thus $\mathscr{H}$ is also invariant under the infinitesimal transformation I_1. Moreover all the transformations form a continuous group. Then if we have a vector space transforming irreducibly under the group, we may use I_1 to pick out in it the functions invariant under I_1 and use these as base vectors. Now these eigenfunctions of I_1 are just the same ones as we would have found using the constant of the motion p_1 above, since there exists the relation $p_1 = \hbar I_1$ (8.29) between them. There is thus a very close relationship between the classical result, the Dirac approach and the group theoretical procedure.

Finite groups

The above discussion completely breaks down when the symmetry group of the Hamiltonian is a finite group $\mathfrak{G}$, for in this case the infinitesimal transformations do not exist. Consider for instance the Hamiltonian (4.12) of an electron moving in the field of three protons arranged in an equilateral triangle. What classical constant of the motion is there that corresponds to the fact that the potential has trigonal symmetry? We could introduce a co-ordinate $q_1 = 1$, 2, 3, 4, 5, 6 corresponding to the six equivalent positions of the triangle. Then $\mathscr{H}$ is certainly independent of q_1, but there is no conjugate momentum because q_1 is not a continuous variable and differentiations such as $\partial q_1/\partial t$, $\partial \mathscr{H}/\partial q_1$ become meaningless. In fact it is not at all obvious what complete set of commuting operators to use in this problem to obtain a convenient set of simultaneous eigenfunctions. Let us therefore adopt the group theoretical method and assume that we have the eigenfunctions $\psi_i^{(\lambda r)}$ of the Hamiltonian $\mathscr{H}$ sorted out to transform according to the irreducible representations $D^{(\lambda)}$ of the symmetry group $\mathfrak{G}$ of $\mathscr{H}$. We shall now construct a set of operators P_k of which the $\psi_i^{(\lambda r)}$ are simultaneous eigenfunctions.

Consider the operator

$$P_k = (1/h_k) \sum_{\text{all } T \text{ in class } k} T_k \tag{17.3a}$$

$$= (1/h) \sum_{\text{all } S \text{ in } \mathfrak{G}} S T_k S^{-1}, \tag{17.3b}$$

where S is any element of $\mathfrak{G}$ and T_k any element of the kth class. As in § 14, h is the number of elements of $\mathfrak{G}$ and h_k the number in the kth class. From (17.3b), P_k commutes with every element of $\mathfrak{G}$. Hence by Schur's lemma (appendix D) P_k is represented by a_k times the unit matrix with respect to an irreducible set of base vectors $\psi_i^{(\lambda r)}$, $i = 1, 2, \ldots$. Hence using (17.3a) and taking the sum of diagonal matrix elements

$$a_k n_\lambda = (1/h_k) \sum_{T_k} \chi^{(\lambda)}(T_k),$$

i.e.

$$a_k = \chi_k^{(\lambda)}/n_\lambda, \tag{17.4}$$

where n_λ is the dimension of $D^{(\lambda)}$. Thus $\psi_i^{(\lambda r)}$ is an eigenfunction of P_k with the eigenvalue (17.4). Since the characters $\chi_k^{(\lambda)}$, $k = 1, 2, \ldots$ characterize an irreducible representation completely, the set of eigenvalues a_k (17.4) serve to label uniquely which irreducible representation a given ψ belongs to. Furthermore since P_k commutes with any T, from (17.3a) it commutes with any P_l. Now any element T of $\mathfrak{G}$ is a symmetry transformation of $\mathscr{H}$. Hence operating on $\mathscr{H}\psi$, it affects only the wave function ψ and we have $T\mathscr{H}\psi = \mathscr{H}T\psi$, whence

$$T\mathscr{H} = \mathscr{H}T. \tag{17.5}$$

Thus T is a constant of the motion, and from (17.3a) so is P_k. Our functions $\psi_i^{(\lambda r)}$ transforming irreducibly according to $D^{(\lambda)}$ are therefore simultaneous eigenfunctions of the set $\mathscr{H}$, P_1, $P_2, \ldots$ of commuting operators. The eigenvalues of the P_k distinguish different irreducible representations, and the energies (eigenvalues of $\mathscr{H}$) differentiate between different values of r, e.g. between the $2p$, $3p$, $4p$, . . . wave functions in an atom since these all transform according to $D^{(1)}$ under rotations. The operators P_k cannot distinguish between the different $\psi_i^{(\lambda r)}$, $i = 1, 2, \ldots$ of one irreducible representation, but some more operators can easily be devised to do this too (problem 17.3).

Summary

The group theoretical method consists of sorting out the eigenfunctions of the Hamiltonian $\mathscr{H}$ according to the irreducible representations of the symmetry group of $\mathscr{H}$. We have shown that this is completely equivalent to the Dirac method, in which we set up a complete set of simultaneous eigenfunctions of $\mathscr{H}$ and other commuting constants of the motion.

Problems

17.1 In the notation of the text, show that I_1 is a constant of the motion without using the relation $P_1 = \hbar I_1$.

17.2 Show that the two expressions in (17.3) are equal.

17.3 With the notation of § 14, consider the operators

$$Q_{\lambda i} = \sum_T D_{ii}^{(\lambda)*}(T)T,$$

where the summation is over all T in the group, and i is not summed. Show that $Q_{\lambda i}$ commutes with any $Q_{\mu j}$, any P_k (17.3), and with the Hamiltonian. What is the effect of $Q_{\lambda i}$ on an eigenfunction $\psi_j^{(\mu r)}$ of the Hamiltonian? Hence discuss the use of the $Q_{\lambda i}$, with or without the P_k, as a complete set of commuting constants of the motion.

17.4 An electron is moving in a potential whose maximum symmetry is the point-group 32. Set up the minimum number of commuting operators forming a complete set so that a simultaneous eigenfunction is uniquely determined by its set of eigenvalues. Give the operators and eigenvalues in detail.

Chapter IV

FURTHER ASPECTS OF THE THEORY OF FREE ATOMS AND IONS

In Chapter II we outlined the quantum theory of free atoms, the main purpose there being to illustrate the use of group theory in quantum mechanics. In the present chapter we shall apply group theory to some more specialized aspects of the theory of atoms, and then in the remaining chapters to a selection of topics from other branches of physics.

18. Paramagnetic Ions in Crystalline Fields

Introduction

Most atoms when they form ions, such as in a crystalline salt, achieve a closed shell configuration either by losing their valence electrons (metallic ions) or building up an incomplete shell (negative ions). Such closed shell ions are therefore in a ^1S state, which is diamagnetic, non-degenerate and generally does not give rise to interesting effects. However, the transition and rare earth metals are exceptions to this rule because of their inner, incomplete $3d$, $4d$, $5d$, $4f$, $5f$ shells. As free ions these in general have paramagnetic, degenerate ground states which in crystals get split by the electric fields. In a crystalline salt, such an ion finds itself surrounded by some regular arrangement of other ions and water molecules. Now strictly one should consider the ion and its neighbours as a big molecule, and discuss their interaction in terms of covalent and other bonds, etc. However it has been found a good approximation to consider the neighbours as simply giving rise to an electrostatic potential, the *crystalline field*, which acts on the paramagnetic ion. This field, combined with the spin-orbit coupling, splits the ground term into a sequence of levels. Since 1946 paramagnetic resonance has been used to study in great detail the lowest of these energy levels and their variation in an applied magnetic field. Consequently there is considerable interest in making accurate calculations of these levels. Comparison between the calculated and observed levels has then determined the magnitude of various parameters such as the strength of the crystalline field, and has also brought to light some refinements that are required in the quantum theory of atoms. Reviews of the subject have been given by Bleaney and Stevens (1953) and by Bowers and Owen (1955).

In Chapter II, the use of symmetry properties led to very broad general results, applicable to any free atom or ion, but in this section we shall use group theory in quite a different role. Because of the different ground states of the paramagnetic ions and the different crystal structures, each salt almost has to be discussed separately. Group theory is used to help calculate particular matrix elements and splittings as required. In the present section we shall therefore not attempt at any general presentation but consider two particular salts, cerium ethylsulphate and chromous sulphate, which suffice to illustrate most of the ideas. The energy levels are determined by setting up the Hamiltonian matrix and solving the secular equation, often using perturbation theory (Schiff 1955, p. 128). In detail this involves a long and complicated calculation, so that we shall only show in outline how group theory is used to calculate the type of matrix elements that are required during the calculation, and what the kind of splitting is at each stage.

The crystalline field in cerium ethylsulphate

The composition of this salt is $Ce(C_2H_5SO_4)_3.9H_2O$, and it crystallizes in a hexagonal form with two molecules per unit cell. The two cerium ions are situated at equivalent special points in the cell lying on a three-fold rotation axis, with the water molecules

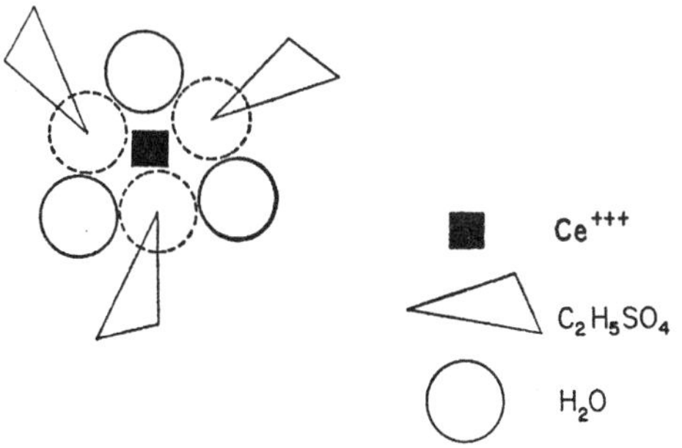

FIG. 10. A diagram of one cerium ethyl sulphate molecule $Ce(C_2H_5SO_4)_3.9H_2O$. The full circles represent one water molecule each in the plane of the paper, and the broken circles each two water molecules with one above and one below the plane of the paper. The Ce^{+++} and $C_2H_5SO_4^-$ ions are also in the middle plane.

and ethylsulphate ions arranged round them as shown in Fig. 10. The cerium ion and water molecules have symmetry $\bar{6}m2$, but the arrangement of atoms in the ethylsulphate radical does not have a vertical mirror plane, so that the overall symmetry is $\bar{6}$. Let us consider the potential energy $V_c(\mathbf{r})$ of an electron belonging to the cerium ion, due to the electrostatic field of all the water molecules and other ions. Since the charge density of the other ions is zero over the region of the Ce^{+++} ion, we can write V_c in terms of spherical harmonics† Y_l^m,

$$V_c = \sum_{l,m} A_l^m r^l Y_l^m(\theta, \phi). \tag{18.1}$$

Since $R(\phi, z) Y_l^m = e^{im\phi} Y_l^m$, we must have

$$A_l^m = 0 \text{ unless } m = 0, \pm 3, \pm 6, \ldots \tag{18.2}$$

for V_c to have three-fold symmetry. Since the group $\bar{6}$ contains the mirror plane m_z, (18.1) must contain only even powers of z, and we have

$$A_l^m = 0 \text{ unless } l - m = \text{even}. \tag{18.3}$$

The cerium ion has the configuration $4f$, and if we neglect mixing with other configurations having some 40,000 cm^{-1} higher energy, all matrix elements of V_c involve the integrals

$$\int \phi_{4f}^* r^l Y_l^m \phi_{4f}\, d\tau \tag{18.4}$$

where ϕ_{4f} is a $4f$ orbital. Since the ϕ_{4f}'s transform according to $D^{(3)}$ under rotations, the $Y_l^m \phi_{4f}$ transform according to

$$\sum D^{(j)}, \; |l - 3| \leqq j \leqq l + 3;$$

and from the fundamental theorem (13.8b), the integral (18.4) is zero unless the values of j include $j = 3$, i.e. unless $l \leqq 6$. Similarly Y_l^m must have even parity, i.e. we have $l = $ even. It follows from these two restrictions and (18.2), (18.3) that we need only retain the constants A_0^0, A_2^0, A_4^0, A_6^0, A_6^6, A_6^{-6}, with the condition $A_6^{-6} = A_6^{6*}$ so that (18.1) is real. The origin of the ϕ co-ordinate can now be chosen to make A_6^6 real, and expressing the $r^l Y_l^m$ in terms of x, y, z, we have

$$\begin{aligned} V_c = {} & B_0^0 + B_2^0(3z^2 - r^2) + B_4^0(35z^4 - 30r^2z^2 + 3r^4) + \\ & + B_6^0(231z^6 - 315r^2z^4 + 105r^4z^2 - 5r^6) + \\ & + B_6^6(x^6 - 15x^4y^2 + 15x^2y^4 - y^6), \end{aligned} \tag{18.5}$$

where the normalization constants have been absorbed into the B's. Because we are restricting ourselves to the lowest configuration,

† In this section we write the spherical harmonics Y_l^m and other quantities with m as a superscript instead of a subscript as in § 8, in accordance with customary usage in this particular subject.

(18.5) actually has the higher symmetry $\bar{6}m2$, rather than the required $\bar{6}$. This $\bar{6}m2$ is therefore the symmetry of the effective Hamiltonian for the cerium ion.

Energy level splittings in cerium ethylsulphate

The lowest configuration of Ce^{+++}, namely $4f$, contains only one term ^2F, and in the free ion this is split by the spin orbit coupling into two levels $J = 5/2$ and $J = 7/2$. By Hund's rule (§ 12) the $J = 5/2$ level lies lowest. The separation between the levels is about 2000 cm^{-1}, whereas the splittings produced by V_c are about 200 cm^{-1}, so that it is a fair approximation to treat V_c as a perturbation. We shall therefore consider the splitting of the $J = 5/2$ level caused by V_c using first order perturbation theory. A more accurate theory is possible, and indeed necessary to interpret the observed data completely: this involves considering the spin-orbit coupling and V_c simultaneously, and solving the secular equation for the whole 14 states together without using perturbation theory (Elliott and Stevens 1952). Considering the perturbation (18.5), we note that $B_0{}^0$ shifts all levels equally so that we shall neglect it. Also the $B_6{}^0$ and $B_6{}^6$ give only zero matrix elements among the $J = 5/2$ manifold of states, because the products $Y_6{}^m\psi(J = 5/2, M_J)$ transform according to

$$\sum D^{(J)}, J = 7/2, 9/2, \ldots, 17/2,$$

which does not contain $J = 5/2$. We are therefore left to consider the perturbation

$$\sum_i V'_c(\mathbf{r}_i) = \sum_i [V_2{}^0(\mathbf{r}_i) + V_4{}^0(\mathbf{r}_i)] \qquad (18.6)$$

to the Hamiltonian, where the summation is over all the electrons of the Ce^{+++} ion, and where

$$V_2{}^0 = B_2{}^0(3z^2 - r^2), \qquad (18.7)$$

$$V_4{}^0 = B_4{}^0(35z^4 - 30r^2z^2 + 3r^4). \qquad (18.8)$$

The perturbation (18.6) is invariant under rotations about the z-axis, so that the quantum number $M_J = M$ can still be used to label the states. Also (18.6) is invariant under a two-fold rotation about the x-axis, so that the states $\psi(M)$, $\psi(-M)$ have the same energy (cf. § 16 and appendix L). The relative energies of the three doublets are therefore

$$\langle M | V'_c | M \rangle \text{ with } M = \pm\tfrac{1}{2}, \pm\tfrac{3}{2}, \pm\tfrac{5}{2}, \qquad (18.9)$$

where we have used the Dirac notation (13.9) for the matrix elements. Since the Ce^{+++} ion has an odd number of electrons, Kramers' theorem (§ 19) shows that the splitting into doublets is the greatest splitting that can occur, and higher order effects from (18.5) can only shift the levels about.

Let us consider the first term in (18.6). Since $\sum 3z_i^2 - r_i^2$ and $3J_z^2 - J^2$ transform† in the same way under rotations, their matrix elements are proportional inside a given irreducible manifold of states by the argument of § 13 and

$$\langle M|\sum V_2{}^0|M'\rangle = \langle M|\alpha[3J_z{}^2 - J(J+1)]|M'\rangle. \tag{18.10}$$

Hence the contribution of $V_2{}^0$ to the energies (18.9) becomes

$$-8\alpha(M = \pm\tfrac{1}{2}),\ -2\alpha(M = \pm\tfrac{3}{2}),\ 10\alpha(M = \pm\tfrac{5}{2}). \tag{18.11}$$

It remains to evaluate α. Let $\psi(m_l, m_s)$ be a single determinant wave function for all the electrons in a Ce^{+++} ion, where m_l, m_s denote the quantum numbers of the single $4f$ electron outside closed shells. We shall make the convention that when we use a single quantum number it refers to M_J, and a pair of numbers to m_l, m_s. The functions $\psi(m_l, m_s)$ transform according to $D^{(3)} \times D^{(1/2)}$ under rotations, and we can pick out the linear combination

$$\psi(M_J = 5/2) = a\psi(3, -\tfrac{1}{2}) + b\psi(2, \tfrac{1}{2}),$$

where a and b are Wigner coefficients as in (9.7). They are (appendix I) $a = \sqrt{(6/7)}$, $b = -\sqrt{(1/7)}$. Thus

$$\langle\tfrac{5}{2}|\sum V_2{}^0|\tfrac{5}{2}\rangle = a^2\langle 3, -\tfrac{1}{2}|\sum V_2{}^0|3, -\tfrac{1}{2}\rangle + b^2\langle 2, \tfrac{1}{2}|\sum V_2{}^0|2, \tfrac{1}{2}\rangle. \tag{18.12}$$

Since $3z^2 - r^2$ and $3l_z{}^2 - l(l+1)$ transform in the same way under rotations, we have

$$\begin{aligned}\langle 3, -\tfrac{1}{2}|\sum V_2{}^0|3, -\tfrac{1}{2}\rangle &= \langle 3, -\tfrac{1}{2}|\beta\sum[3l_{zi}{}^2 - l_i(l_i+1)]|3, -\tfrac{1}{2}\rangle = 15\beta,\\ \langle 2, \tfrac{1}{2}|\sum V_2{}^0|2, \tfrac{1}{2}\rangle &= \langle 2, \tfrac{1}{2}|\beta\sum[3l_{zi}{}^2 - l_i(l_i+1)]|2, \tfrac{1}{2}\rangle = 0,\end{aligned} \tag{18.13}$$

where the closed shells do not contribute (problem 13.12). Further if

$$\phi(3, -\tfrac{1}{2}) = F_{4f}(r)\sin^3\theta\, e^{i3\phi}u_-$$

† Throughout this section we shall drop the factor $\hbar$ from all angular momenta for the sake of custom and convenience. Thus we use $J_z, l_{z,i}$ for $I_{z\ \mathrm{tot}}, I_{z,\ \mathrm{orb},\ i}$, etc.

is the $m_l = 3$, $m_s = -\frac{1}{2}$ orbital of the $4f$ electron, we have

$$\langle 3, -\tfrac{1}{2}|\sum V_2^0|3, -\tfrac{1}{2}\rangle = \frac{B_2^0 \int_0^\infty F^2 r^4 \, dr \int_0^\pi \sin^6 \theta (3\cos^2\theta - 1) \sin\theta \, d\theta}{\int_0^\infty F^2 r^2 \, dr \int_0^\pi \sin^6 \theta \sin\theta \, d\theta} = -\tfrac{2}{3} B_2^0 \overline{r^2}, \tag{18.14}$$

where $\overline{r^2}$ is the expectation value of r^2 for any $4f$ state. Now combining (18.11), (18.12), (18.13) and (18.14), we have $\alpha = -(2/35) B_2^0 \overline{r^2}$. Similarly the matrix elements of V_4^0 (18.8) may be calculated, and the final energy levels are shown in Table 13 (problems

TABLE 13

Energies and g-values of Lowest Doublets

Doublet M	Energy	g_z	g_x, g_y
$\pm\frac{1}{2}$	$-8\alpha + 2\gamma$	$\frac{6}{7}$	$\frac{18}{7}$
$\pm\frac{3}{2}$	$-2\alpha - 3\gamma$	$\frac{18}{7}$	0
$\pm\frac{5}{2}$	$10\alpha + \gamma$	$\frac{30}{7}$	0

$\alpha = -(2/35)B_2^0 r^2$ and $\gamma = (8/21)B_4^0 r^4$.

18.1, 18.2). The method we have used to calculate such elements by replacing x, y, z by J_x, J_y, J_z or l_x, l_y, l_z, is called the method of operator equivalents. It has been expounded in detail by Stevens (1952), who also gives some useful tables.

Cerium ethylsulphate g-values

We consider now the effect of a small magnetic field $\mathbf{H}$ on the energy levels. The perturbation to the Hamiltonian is

$$\beta \, \mathbf{H} \cdot (\mathbf{L} + 2\mathbf{S}) \tag{18.15}$$

where β is the Bohr magneton $e\hbar/2mc$ (Schiff 1955, p. 293). This usually splits each doublet, and the splitting ΔE is usually expressed in the form

$$\Delta E = g(M)\beta H, \tag{18.16}$$

where g is a factor which we shall now calculate. The operators $\mathbf{L}$, $\mathbf{S}$ and $\mathbf{J}$ all transform as vectors under rotations, so that the matrix elements of their components are proportional to one another inside an irreducible manifold of states belonging to one value of J. Thus the perturbation (18.15) can be written $g_L \beta \mathbf{H} \cdot \mathbf{J}$ where $g_L = 6/7$ is the Landé splitting factor calculated in problem

13.6. With **H** in the z-direction, each energy level is increased by $g_L \beta H M$, whence comparison with (18.16) gives

$$g_z(M) = 2g_L |M|.$$

For **H** along the x-axis, we have

$$\langle \pm 3/2 | J_x | \pm 3/2 \rangle = \langle \pm 5/2 | J_x | \pm 5/2 \rangle = 0$$

and no first order splitting of the $M = 3/2$ and $M = 5/2$ doublets. From (8.18) we have

$$\begin{aligned} \langle -\tfrac{1}{2} | J_x | \tfrac{1}{2} \rangle &= \tfrac{1}{2} \langle -\tfrac{1}{2} | J_+ + J_- | \tfrac{1}{2} \rangle, \\ &= \tfrac{1}{2}[J(J+1) - M(M-1)]^{1/2} \text{ with } M = \tfrac{1}{2}, \\ &= \tfrac{3}{2}. \end{aligned}$$

Thus for the $\pm \frac{1}{2}$ doublet the matrix of the perturbation (18.15) is

$$\frac{9}{7} \beta H \begin{bmatrix} \cdot & 1 \\ 1 & \cdot \end{bmatrix},$$

and the energy levels are $\pm (9/7) \beta H$, whence $g_x = 18/7$. From symmetry, $g_y = g_x$, and we have the numerical values shown in Table 13.

Paramagnetic resonance experiments at low temperatures show that in cerium ethylsulphate the lowest doublet has $g_z = 0{\cdot}955$, $g_x = g_y = 2{\cdot}185$, and that there is another doublet about 3 cm^{-1} higher with $g_z = 3{\cdot}72$, $g_x = g_y = 0{\cdot}2$. Comparison with the g-values of Table 13 shows that we must identify these levels with the $M = \pm \frac{1}{2}$ and $M = \pm 5/2$ doublets respectively. The discrepancies between the calculated and observed g-values are due to the perturbing influence of the $J = 7/2$ level some 2000 cm^{-1} higher and the $B_6{}^0$, $B_6{}^6$ terms in (18.5). Magnetic susceptibility measurements indicate the presence of another level about 130 cm^{-1} above the two low lying doublets, and this must be the $M = \pm 3/2$ level. Comparing these figures with the relative energies of the levels shown in Table 13, we obtain

$$B_2{}^0 \overline{r^2} = 25 \text{ cm}^{-1}, \; B_4{}^0 \overline{r^4} = -74 \text{ cm}^{-1}.$$

These values are again rather rough, but they illustrate the kind of way in which the magnitudes of the crystal field parameters can be obtained from experiment. Elliott and Stevens (1952) have given a much better fit to the data, and determined the values of all the constants in (18.5) (except $B_0{}^0$). In particular $B_6{}^6$ turns out to be large and important, as we might expect from the fact that it alone reflects the trigonal arrangement of the water molecules, all the other terms being axially symmetrical.

Crystal field splittings in chromous sulphate

This salt $CrSO_4.5H_2O$ appears to have the same structure as copper sulphate. The Cr^{++} ion is at the centre of a square of four water molecules with two oxygen ions centrally located above and below the square. The six oxygen atoms form approximately a regular octahedron, and the crystalline field consists predominantly of a large cubic component with symmetry m3m, and a smaller tetragonal component with symmetry 4/mmm. In addition a distortion of the square of water molecules gives a still smaller orthorhombic (usually abbreviated to "rhombic") field with symmetry mmm, whose effect is comparable with that of the spin-orbit coupling. We shall now discuss qualitatively the energy level splittings on the basis of these symmetries and orders of magnitudes.

The chromous ion Cr^{++} is in the configuration $(3d)^4$ whose lowest term is ^5D (§ 12). Under the influence of the cubic field, the Hamiltonian is invariant under the group of rotations m3m of the orbital variables, and under all spin rotations. The $2S+1=5$-fold spin degeneracy of each orbital level therefore remains. The unperturbed orbital functions transform according to $D^{(2)}$ under rotations, and this is reducible into the representations E_g and T_{2g} of the cubic group (Table 14). The characters of this representation with respect

TABLE 14

Irreducible Representations for Splitting of ^5D Term

Cubic field, point-group m3m. $D^{(2)}$ (rotation group) $= E_g + T_{2g}$.

	$\chi(E)$	$\chi(3)$	$\chi(2_z)$	$\chi(2_d)$	$\chi(4_z)$	parity
rotation group $D^{(2)}$	5	-1	1	1	-1	even
E_g	2	-1	2	0	0	even
T_{2g}	3	0	-1	1	-1	even

Tetragonal field, point-group 4/mmm. E_g(cubic group) $= A_{1g} + B_{1g}$.

	$\chi(E)$	$\chi(2_z)$	$\chi(2_x)$	$\chi(2_d)$	$\chi(4_z)$	parity
cubic group E_g	2	2	2	0	0	even
A_{1g}	1	1	1	1	1	even
B_{1g}	1	1	1	-1	-1	even

Rhombic field, point-group mmm.

Orbital function: B_{1g} (tetragonal group) $= A_g$ (rhombic).
Spin functions: $D^{(2)} = A_g + A_g + B_{1g} + B_{2g} + B_{3g}$.
Complete functions: $A_g \times D^{(2)} = A_g + A_g + B_{1g} + B_{2g} + B_{3g}$.

to the group m3m are obtained from (14.4), the irreducible representations of m3m from appendix K, and the reduction of the representation from (14.2). The $L = 2$ term is therefore split into an orbital doublet and an orbital triplet, and it so happens that the doublet lies lowest (Fig. 11). Under the action of the tetragonal field the lower doublet splits into two orbital singlet states transforming according to the representations A_{1g} and B_{1g} of the group 4/mmm (Table 14). We shall suppose that the B_{1g} level lies lowest (Fig. 11). The states of this level transform under orbital rotations

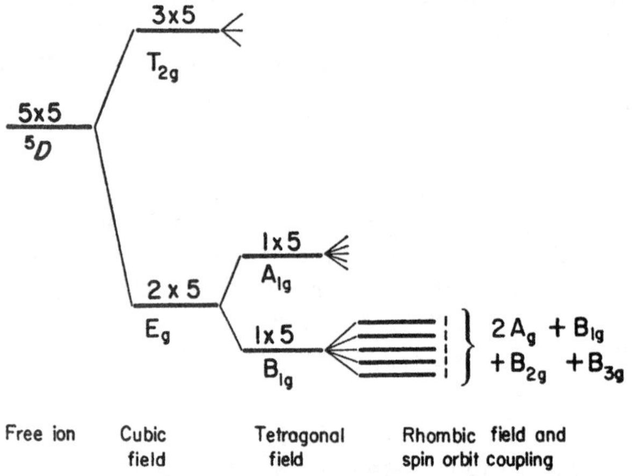

FIG. 11. Splitting of ^5D term by crystalline fields. The degeneracy and symmetry of each level is indicated.

according to the representation B_{1g} of the group 4/mmm, which becomes the representation A_g of the group mmm. Under spin rotations the states transform according to $D^{(S)}$ with $S = 2$, and under orbital and spin rotations together according to $A_g \times D^{(2)}$, which is reducible into five one-dimensional representations (Table 14). Thus under the combined influence of the rhombic field and spin-orbit coupling, we finally obtain five non-degenerate levels (Fig. 11).

The spin-Hamiltonian for chromous sulphate

The splittings produced by the cubic and tetragonal fields can easily be calculated in the same general way as the energies in Table 13, using the method of operator equivalents as discussed by Stevens (1952). We shall only concern ourselves here with the relative energies of the five lowest levels shown in Fig. 11, because it is these energies that can be studied experimentally as functions of the strength and direction of an applied magnetic field. Now it is not usually possible to give analytic expressions for these energies as functions of the field, so that instead the theoretical and experimental results are conveniently interpreted in terms of a 5×5 matrix whose eigenvalues are the required energy values (Schiff 1955, p. 128). This matrix, which is called the *spin-Hamiltonian*, we shall now proceed to calculate.

We start with the 1×5-fold degenerate level transforming according to $B_{1g} \times D^{(2)}$ under orbital and spin rotations of the group 4/mmm (Fig. 11). We apply the rhombic field, the spin-orbit coupling and the magnetic field as a perturbation. Since all three produce effects of the same order of magnitude (1 cm^{-1}), they must be treated together as a single perturbation. To be precise, the rhombic field and $\lambda\mathbf{L} \cdot \mathbf{S}$ produce no splitting of the level in a first order of approximation, but they are intrinsically large enough for their second and higher order terms to be comparable with the first order magnetic field splittings. All other levels in the tetragonal field approximation lie considerably higher, so that perturbation theory can be used. The rhombic field can be expanded in terms of spherical harmonics (18.1), and we shall consider only the typical term

$$V_{\text{rh}} = \sum_i A(x_i^2 - y_i^2).$$

The spin-orbit coupling takes the form $\lambda\mathbf{L} \cdot \mathbf{S}$ (13.20), since we shall neglect the effect of all levels not arising from the ^5D term. With the magnetic interaction (18.15) the whole perturbation becomes

$$\mathscr{H}_p = V_{\text{rh}} + \lambda\mathbf{L} \cdot \mathbf{S} + \beta\mathbf{H} \cdot (\mathbf{L} + 2\mathbf{S}). \tag{18.17}$$

We shall denote the unperturbed states, i.e. the states in the tetragonal field approximation of Fig. 11, by two quantum numbers n, M. Here n denotes the orbital state, numbered from the lowest one (B_{1g}) as zero, and M is the spin quantum number M_S, $-2 \leqq M \leqq 2$. As a further piece of notation the matrix elements of some operators can be written in a simplified form. For example in the operator L_xS_x, L_x operates only on orbital variables and S_x on spin variables. Thus (problem 18.5)

$$\boxed{\begin{aligned} &\langle nM|L_xS_x|n'M'\rangle \\ &\quad = \langle nM''|L_x|n'M''\rangle\langle n''M|S_x|n''M'\rangle \\ &\quad = \langle n|L_x|n'\rangle\langle M|S_x|M'\rangle, \end{aligned}} \tag{18.18}$$

where in the last line we have dropped the quantum numbers n'', M'' since the matrix elements are independent of them. We shall follow this convention throughout.

We now apply the perturbation (18.17) to first order, the energy levels being obtained from the 5×5 matrix

$$\langle 0M \mid \mathscr{H}_p \mid 0M' \rangle. \tag{18.19}$$

Since V_{rh} is spin-independent, it gives a constant diagonal contribution to (18.19) which shifts all energies equally. This we shall neglect. The spin-orbit coupling gives

$$\lambda \sum_{x,y,z} \langle 0|L_i|0\rangle\langle M|S_i|M\rangle.$$

This is zero because $\langle 0|L_i|0\rangle$ is zero, as we shall now prove. Suppose that $\psi = \phi_{\text{orb}} U_{\text{spin}}$, or an antisymmetric linear combination of such functions, satisfies a spin-independent Schrödinger equation as our unperturbed states do. In the absence of magnetic fields, all terms in the Hamiltonian are real. Thus if $\phi_{\text{orb}} U_{\text{spin}}$ satisfies the Schrödinger equation, then so does $\phi_{\text{orb}}{}^* U_{\text{spin}}$. Now our ground state is orbitally non-degenerate, so that we must have $\phi^* = \alpha\phi$ where α is some constant. By incorporating in ϕ a suitable phase factor, we can choose ϕ real. Hence

$$\langle 0|L_z|0\rangle = i\hbar \int \phi\left(-x\frac{\partial}{\partial y} + y\frac{\partial}{\partial x}\right)\phi \, \mathrm{d}\tau = ia, \tag{18.20}$$

where a is real. But from (8.17a), we have

$$\begin{aligned}\langle 0|L_z|0\rangle^* &= \left[\int \phi^* l_z \phi \, \mathrm{d}\tau\right]^* = \left[\int (l_z\phi)^*\phi \, \mathrm{d}\tau\right]^* \\ &= \int \phi^*(l_z\phi)\, \mathrm{d}\tau = \langle 0|L_z|0\rangle,\end{aligned}$$

and hence $\langle 0|L_z|0\rangle$ is real. Comparing this with (18.20), we conclude

$$\langle 0|L_i|0\rangle = 0, \qquad i = x, y, z. \tag{18.21}$$

We next calculate the magnetic term in (18.19). The contribution from $\beta\mathbf{H} \cdot \mathbf{L}$ is zero because of (18.21), and we are left with

$$\mathscr{H}_s^{(1)} = \sum_i 2\beta H_i \langle M|S_i|M'\rangle. \tag{18.22}$$

If $\mathbf{H}$ is in the z-direction, (18.22) is diagonal and the energy levels to this order of approximation are

$$E_{\text{M}} = 2\beta HM. \tag{18.23}$$

Now a system with only orbital angular momentum has magnetic energy levels $\beta H M_L$ and gives a contribution to the susceptibility the same as that of a classical magnetic moment of strength $\mu = [L(L+1)]^{1/2}\beta$. Similarly a free ion with spin has a magnetic moment

$$\boxed{\mu = g_L[J(J+1)]^{1/2}\beta,} \tag{18.24}$$

where g_L is the Landé splitting factor (problem 13.6). However in the crystalline field, from (18.23) the Cr^{++} ion has a magnetic moment of

$$\boxed{\mu = 2[S(S+1)]^{1/2}\beta,} \tag{18.25}$$

as if only the spin and not the orbital angular momentum contributed to the magnetic moment. This situation is described by saying "the orbital moment is quenched". Table 15 compares the

TABLE 15

Magnetic Moments of Paramagnetic Ions

Ion	Configuration	Lowest level	μ(calc) free ion	μ(calc) spin only	μ(exp)
Ti^{+++}, V^{4+}	$3d^1$	$^2D_{3/2}$	1·55	1·73	1·8
V^{+++}	$3d^2$	$^3F_2$	1·63	2·83	2·8
Cr^{+++}, V^{++}	$3d^3$	$^4F_{3/2}$	0·77	3·87	3·8
Mn^{+++}, Cr^{++}	$3d^4$	$^5D_0$	0·00	4·90	4·9
Fe^{+++}, Mn^{++}	$3d^5$	$^6S_{5/2}$	5·92	5·92	5·9
Fe^{++}	$3d^6$	$^5D_4$	6·70	4·90	5·4
Co^{++}	$3d^7$	$^4F_{9/2}$	6·63	3·87	4·8
Ni^{++}	$3d^8$	$^3F_4$	5·59	2·83	3·2
Cu^{++}	$3d^9$	$^2D_{5/2}$	3·55	1·73	1·9

The values of μ are given in units of the Bohr magneton $\beta = e\hbar/2mc$. The free-ion values are calculated from (18.24) and the spin-only ones from (18.25). The experimental values are for representative salts (Kittel 1956).

observed magnetic moments of transition metal ions in salts with the calculated free-ion value (18.24) and the spin-only value (18.25). The quenching of the orbital moment is seen to be a general phenomenon for such ions, and the physical reason for it is as follows. In a free ion the states of different M_L have the same energy, and in a magnetic field the atom tries to line up its magnetic moment with the field and hence to take up the state with minimum M_L. However, in a crystal this is impossible, because the states with different M_L have quite different energies. Their charge distributions differ, and there is one lowest orbital state whose charge distribution avoids as much as possible the electrons from the nearest neighbouring atoms. In this way the Coulomb repulsion between the two sets of electrons is lowest. The ion is therefore not free to take up the state of minimum M_L and the orbital angular momentum gives no contribution to the magnetic moment in a first order of approximation.

We continue with the perturbation calculation of (18.17). Since the lowest five unperturbed states $\psi(0, M)$ are degenerate, their energies have to be obtained by first calculating a 5×5 matrix

$$\mathscr{H}_s = \mathscr{H}_s^{(1)} + \mathscr{H}_s^{(2)} + \mathscr{H}_s^{(3)} + \dots \qquad (18.26)$$

where $\mathscr{H}_s^{(1)}$, $\mathscr{H}_s^{(2)}$ etc. are obtained from first order, second order, etc., perturbation theory. This matrix is called the spin-Hamiltonian, and the energies of the states are its eigenvalues (Schiff 1955, p. 158; Pryce 1950). $\mathscr{H}_s^{(1)}$ has already been calculated (18.22). We shall continue to express the whole of $\mathscr{H}_s$ like $\mathscr{H}_s^{(1)}$ in terms of the matrices $\langle M|S_i|M'\rangle$, $i = x, y, z$. These matrices can easily be calculated from (18.18): for instance $\langle M|S_x|M'\rangle$ is

$$\begin{bmatrix} \cdot & 1 & \cdot & \cdot & \cdot \\ 1 & \cdot & \frac{1}{2}\sqrt{6} & \cdot & \cdot \\ \cdot & \frac{1}{2}\sqrt{6} & \cdot & \frac{1}{2}\sqrt{6} & \cdot \\ \cdot & \cdot & \frac{1}{2}\sqrt{6} & \cdot & 1 \\ \cdot & \cdot & \cdot & 1 & \cdot \end{bmatrix}. \qquad (18.27)$$

Also for the sake of simplicity we shall from now on use S_x for the matrix $\langle M|S_x|M'\rangle$ like (18.27), as well as for the more general spin operator $S_x = I_{x\ \text{spin}}$. With this notation (18.22) can be written

$$\mathscr{H}_s^{(1)} = 2\beta\mathbf{H}\cdot\mathbf{S} = 2\beta(H_xS_x + H_yS_y + H_zS_z). \qquad (18.22')$$

The second order contribution to the spin-Hamiltonian is

$$\boxed{\mathscr{H}_s^{(2)} = -\sum_{n,m} \frac{\langle OM|\mathscr{H}_p|nm\rangle\langle nm|\mathscr{H}_p|OM'\rangle}{E_n - E_0}.} \qquad (18.28)$$

On substituting (18.17) for $\mathscr{H}_p$ and multiplying out the numerator, we obtain a number of terms, one of which is

$$-\sum_{n,\,m} \frac{\langle OM|\lambda L_zS_z|nm\rangle\langle nm|\beta H_zL_z|OM'\rangle}{E_n - E_0}. \qquad (18.29a)$$

This can be expressed in the spin-Hamiltonian form

$$\tfrac{1}{2}\Delta g_z\beta H_zS_z \qquad (18.29b)$$

where
$$\Delta g_z = -2\lambda \sum_n \frac{|\langle 0|L_z|n\rangle|^2}{E_n - E_0}. \quad (18.29c)$$

This term and $\mathscr{H}_s^{(1)}$ (18.22′) can therefore be collected together and written
$$\beta(g_x H_x S_x + g_y H_y S_y + g_z H_z S_z) \quad (18.30)$$
where
$$g_i = 2 + \Delta g_i.$$

Since we have not used V_{rh} so far, the system has tetragonal symmetry and $g_x = g_y$. Further, for an incomplete shell less than half full of electrons, it can easily be shown from (11.8), (13.16) that λ is positive in accordance with Hund's rule (§ 12). Thus from (18.29c) and (18.30), the values of g_i tend to be less than 2. For instance in chromous sulphate $g_z = 1{\cdot}95$. For the same reason the observed magnetic moments of ions with less than a half full 3d shell tend to be smaller than the spin-only values (18.25). On the other hand a shell more than half full behaves like a few positive holes, which changes the sign of λ. This explains why in the lower half of Table 15 the observed magnetic moments are larger than the spin-only values. Furthermore selection rules for $\langle 0|L_i|n\rangle$ show that there is no contribution to (18.29), (18.30) from the level $n = 1$, so that the energy denominator $E_n - E_0$ is at least the large splitting produced by the cubic field (Fig. 11). Thus the Δg_i in chromous sulphate are small, of the order of 0·05.

Similarly, it follows that the term $\langle|\mathbf{L}\cdot\mathbf{S}|\rangle\langle|\mathbf{L}\cdot\mathbf{S}|\rangle$ in (18.28) contributes
$$\sum_{i,j} d_{ij} S_i S_j$$
to the spin-Hamiltonian, where d_{ij} is a tensor. Because of the tetragonal symmetry of the unperturbed system, the tensor must have the form
$$aS_x^2 + aS_y^2 + bS_z^2 = D[S_z^2 - \tfrac{1}{3}S(S+1)] + \text{const.} \quad (18.31)$$

There is a similar term $\Lambda\beta^2 H_z^2$ due to $\langle|\beta\mathbf{H}\cdot\mathbf{L}|\rangle\langle|\beta\mathbf{H}\cdot\mathbf{L}|\rangle$ in (18.28), which is very small and which we shall neglect. Further, selection rules show that all terms in (18.28) involving V_{rh} are zero (problem 18.6). In fact it is only in the fourth order that terms involving V_{rh} like
$$\sum_{\substack{n,m,n',m'\\ n'',m''}} \frac{\langle OM|V_{\text{rh}}|nm\rangle\langle nm|V_{\text{rh}}|n'm'\rangle \langle n'm'|\mathbf{L}\cdot\mathbf{S}|n''m''\rangle\langle n''m''|\mathbf{L}\cdot\mathbf{S}|OM'\rangle}{(E_n - E_0)(E_{n'} - E_0)(E_{n''} - E_0)}$$

give non-zero contributions to the spin-Hamiltonian. Because of the rhombic symmetry, we expect some of these to have the form (18.31) and some the form $E(S_x^2 - S_y^2)$. Similarly, there are very small contributions of the form (18.30) with $g_x \neq g_y$. Still higher orders of perturbation will give other rhombic components like $S_x^4 - S_y^4$ which are very small. Thus we may write the total spin-Hamiltonian approximately as

$$\mathcal{H}_s = D[S_z^2 - \tfrac{1}{3}S(S+1)] + E(S_x^2 - S_y^2) + \\ + \beta(g_x H_x S_x + g_y H_y S_y + g_z H_z S_z). \qquad (18.32)$$

The constants are not entirely independent of one another because of the way they depend on λ, $\langle 0|L_i|n\rangle$, etc. Ono *et al.* (1954) have calculated the energy levels on the basis of this Hamiltonian, and compared them with their observed paramagnetic resonance lines. From this they obtain the following values of the constants:

$$|D| = 2{\cdot}24 \text{ cm}^{-1}, \qquad |E| = 0{\cdot}10 \text{ cm}^{-1},$$
$$g_x \approx g_y = 1{\cdot}99, \qquad g_z = 1{\cdot}95,$$

which they show to be compatible with the expected values of λ, the cubic field splitting, etc.

References

The reviews by Bleaney and Stevens (1953) and Bowers and Owen (1955) have already been mentioned. The method of spin-Hamiltonians was first developed by Pryce (1950), and extended by Abragam and Pryce (1951a) to hyperfine interactions in crystals. An illuminating derivation of the spin-Hamiltonian has been given by Nierenberg (1957). The operator equivalents of crystal fields are developed by Stevens (1952), who gives some useful tables.

Summary

We have shown how symmetry properties are used to determine the qualitative nature of the splittings due to crystalline electric fields in paramagnetic salts. In setting up a quantitative theory, it is necessary to calculate a large number of matrix elements of various kinds. Here we have shown by discussing two particular salts how symmetry properties can be useful in a wide variety of ways in calculating these matrix elements, so much so that they often enable one to set up a complete theory and derive a spin-Hamiltonian involving just a few constants which then have to be calculated in detail or found by experiment.

Problems

18.1 In writing down an operator equivalent for the potential (18.8) $35z^4 - 30r^2z^2 + 3r^4$ analogously to (18.10), show that r^2z^2 must not be replaced by $J_z^2J(J+1)$, but by the expression

$$\tfrac{1}{6}\sum_{i=x,y,z} (J_z^2J_i^2 + J_zJ_iJ_zJ_i + J_zJ_i^2J_z + J_i^2J_z^2 + J_iJ_zJ_iJ_z + J_iJ_z^2J_i).$$

Using the commutation relations among the J_i, show that the complete operator equivalent for (18.8) is

$$\delta[35J_z^4 - 30J(J+1)J_z^2 + 25J_z^2 - 6J(J+1) + 3J^2(J+1)^2].$$

18.2 Calculate δ in problem 18.1 for a Ce^{+++} ion, and hence derive the V_4^0 contribution to the energy levels of Table 13.

18.3 Discuss the splitting of the upper level in Fig. 11 under a tetragonal field, and then under an additional rhombic field and spin-orbit coupling.

18.4 Show that L_x, L_y, L_z have zero matrix elements between the levels marked A_{1g} and B_{1g} in the tetragonal field approximation in Fig. 11.

18.5 Prove the first step in equation (18.18) analytically.

18.6 Show that the second order terms (18.28) which involve V_{rh} are all zero or a constant.

18.7 Calculate the eigenvalues of the spin-Hamiltonian (18.32) when the magnetic field is zero. Associate each level with the correct irreducible representation (Table 14) of the group mmm.

18.8 Calculate the selection rules for magnetic dipole transitions (problem 13.2) among states transforming under the various irreducible representations of the group mmm.

18.9 A sample of chromous sulphate is placed in a microwave cavity such that the oscillating magnetic field is in the x-direction, referred to the same axes as the spin-Hamiltonian (18.32). What magnetic dipole transitions are allowed among the levels (a) in zero magnetic field, (b) when the steady applied magnetic field is in the z-direction, and (c) when the applied field is in the y-direction?

18.10* Outline the theory of the paramagnetic resonance spectra of hydrated cobalt salts (Abragam and Pryce 1951b).

18.11 In a dilute alloy of cobalt in copper, three cobalt atoms are arranged as a cluster in a straight line at the nearest neighbour distance apart. Assume that they have a spin of one, and that their interaction energy is

$$\mathscr{H} = a\mathbf{S}_1 \cdot \mathbf{S}_2 + a\mathbf{S}_2 \cdot \mathbf{S}_3 + b\mathbf{S}_1 \cdot \mathbf{S}_3,$$

where the $\mathbf{S}_i$ are the spins of the three atoms with atom 2 being the central one, and a is positive (antiferromagnetic coupling). Show that the ground state of the cluster may be expected to have $S = 0$ or 1 for physical reasons. Hence calculate the ground state energy by writing down the wave functions using Wigner coefficients (§ 9) and all the symmetries of $\mathscr{H}$. Discuss how you would do the calculation if $\mathscr{H}$ contained higher order terms like $(\mathbf{S}_1 \cdot \mathbf{S}_2)^2 + (\mathbf{S}_2 \cdot \mathbf{S}_3)^2$ or anisotropic exchange $(S_{1z} + S_{3z})S_{2z}$.

19. Time-Reversal and Kramers' Theorem

Introduction

In addition to the types of symmetry transformation listed in § 3 and studied so far, almost all† quantum mechanical systems have an additional symmetry of a rather different nature, called *time-reversal*. Consider the time-dependent Schrödinger equation

$$\mathscr{H}\Psi = i\hbar \frac{\partial \Psi}{\partial t}.$$

If we apply the simple *time-inversion* substitution $t \to -t$ (denoted by τ), we obtain

$$\mathscr{H}\Psi = -i\hbar \frac{\partial \Psi}{\partial t},$$

and we note that the Schrödinger equation is clearly not invariant under τ because of the minus sign. However we can easily fix this up by taking the complex conjugate as well, and the combined operation of time-inversion and complex conjugation is called the *time-reversal transformation* T. We shall first exhibit some of its features in a simplified form by an example.

An electron is moving, say in an atom, in a spherically symmetrical potential on which is superposed a uniform electric field $\mathscr{E}$ in the z-direction. The time-independent Schrödinger equation is

$$\left[-\frac{\hbar^2}{2m}\nabla^2 + V(r) + e\mathscr{E}z\right]\psi = E\psi. \tag{19.1}$$

The Hamiltonian is invariant under rotations about the z-axis, and we can sort out the eigenfunctions to transform according to

† The only known exception is a system in an externally applied magnetic field. There is also reason to suspect that the nuclear interaction involved in beta decay (§ 33) may not be invariant under time-reversal (Jackson *et al.*, 1957).

the representations $\exp(im\phi)$ (7.3). Since we are considering only one electron, the wave function can in fact be written

$$\psi = f(r, \theta) \exp(im\phi). \tag{19.2}$$

These representations are all one-dimensional, and at first sight one might suppose that they all have different energies so that all levels are non-degenerate. However, this is not so, as can be seen as follows. We take the complex conjugate of (19.1), obtaining

$$\left[-\frac{\hbar^2}{2m}\nabla^2 + V(r) + e\mathscr{E}z\right]\psi^* = E\psi^*. \tag{19.3}$$

This shows that ψ^* is also an eigenfunction belonging to the same energy as ψ. Now (19.2) shows that ψ^* belongs to the representation with $-m$, so that if $m \neq 0$, ψ and ψ^* must be linearly independent. Thus all the levels with $m \neq 0$ are at least two-fold degenerate; and when they are n-fold degenerate then n must be even. More generally, the wave function including spin for a many-electron atom satisfies a Schrödinger equation which is much more complicated than (19.1). However, the two-fold degeneracy between wavefunctions with quantum numbers M_J and $-M_J$ remains. In particular for an odd number of electrons

$$M_J = \text{half an odd integer} \neq 0,$$

so that all levels have even degeneracy. This is called the Kramers' degeneracy.

In the present example, the degeneracy between wave functions belonging to representations m and $-m$ could have been proved using the reflection symmetry of (19.1) in the plane $x = 0$ (cf. for instance the character table for the group ∞m, appendix L). The use of the time-reversal symmetry $\psi \rightarrow \psi^*$ is not really necessary.

Summarizing, we have illustrated in a vague kind of way the following essential aspects of time-reversal symmetry.

(i) Time-reversal symmetry involves taking the complex conjugate $\psi \rightarrow \psi^*$. It is, therefore, not a simple transformation of coordinates as discussed in § 3. Consequently the representation theory of § 5 and appendix C do not apply, and the time-reversal symmetry has to be considered from first principles.

(ii) To prove the existence of a degeneracy due to time-reversal, we have to prove firstly that ψ and $T\psi$ belong to the same energy level, where T is the time-reversal operator (19.5) below, and secondly that ψ and $T\psi$ are linearly independent. In this way it can be shown that an atom, with an odd number of electrons placed in any type of electric potential but in the absence of a magnetic field, has all energy levels n-fold degenerate where n is *even*. This is Kramers' theorem.

(iii) It is usually only necessary to consider time-reversal symmetry explicitly for systems having a low spatial symmetry and having spin dependent wave functions. In other cases the degeneracy produced by time-reversal is usually produced also by some spatial symmetry. These points will become clearer as we proceed with the detailed development.

The time-reversal operator

The time-dependent Schrödinger equation is

$$\left(\mathscr{H} - i\hbar \frac{\partial}{\partial t}\right)\Psi(t) = 0. \tag{19.4}$$

Recapitulating briefly the initial argument, the operator in this equation is clearly not invariant under the simple time-*inversion* substitution $t \to -t$. Thus if there is to be a time-reversal transformation, it must have a more complicated form, and we define the time-*reversal* operator T by the relation

$$\boxed{T\Psi(\mathbf{r}_i, \sigma_{zi}, t) = \Psi^*(\mathbf{r}_i, \sigma_{zi}, -t).} \tag{19.5a}$$

Here we have abbreviated all the electron co-ordinates $\mathbf{r}_1, \sigma_{z1}, \mathbf{r}_2, \sigma_{z2}, \ldots \mathbf{r}_n, \sigma_{zn}$ to $\mathbf{r}_i, \sigma_{zi}$.

This operator T has some important properties. Firstly suppose Ψ is an energy eigenfunction, i.e. it has the form

$$\Psi(\mathbf{r}_i, \sigma_{zi}, t) = \psi(\mathbf{r}_i, \sigma_{zi}) \exp(-iEt/\hbar). \tag{19.6}$$

In this section we shall always use Ψ and ψ to denote time-dependent and time-independent wave functions respectively. We have

$$T\Psi = \psi^*(\mathbf{r}_i, \sigma_{zi}) \exp(-iEt/\hbar), \tag{19.7}$$

so that *if* $T\Psi$ is also an eigenfunction of the Hamiltonian, then it belongs to the same energy level.

From (19.5) we have immediately the second property

$$T(a\Psi_1 + b\Psi_2) = a^*T\Psi_1 + b^*T\Psi_2. \tag{19.8}$$

Thus T is not a linear operator, and much of the representation theory of § 5 and appendix C does not hold, for instance the equations leading to the equivalence relation (5.15). We shall just note one consequence of this to avoid confusion later. In § 7 it was shown that ordinarily a cyclic group has only one-dimensional irreducible representations, but it is shown below in proving Kramers' theorem that the cyclic group E, T, $T^2 = -E$, $-T$ (cf. equation (19.13b)) has two-dimensional irreducible representations. Thus time-reversal can lead to degeneracy.

In § 11 it was shown that the spin functions u_+, u_- transform under rotations according to the representation $D^{(1/2)}$ with matrices (8.24). Since these matrices are complex, the functions u_+, u_- have been regarded quite properly as complex quantities. However this notion requires some care because it is possible to interpret the complex conjugates u_+^*, u_-^* in two different ways.† To obtain the first interpretation, we recall the definition (11.12) of u_+, u_-, and taking the complex conjugate of (11.12) we obtain $u_+^*(1) = 1$, $u_+^*(-1) = 0$, $u_-^*(1) = 0$, $u_-^*(-1) = 1$. Since u_+^* is still a function of the variable σ_z, it must be expressible as a linear combination of u_+ and u_-. Comparing with (11.12), we have

$$u_+^* = u_+, \qquad u_-^* = u_-. \tag{19.9a}$$

This is the interpretation one uses in calculating matrix elements: for instance we have $u_+^*(1)u_-(1) = 0 = u_+^*(-1)u_-(-1)$ and $\sum_{\sigma_z} u_+^* u_- = 0$. More generally

$$\sum_{\sigma_z} u_\alpha^* u_\beta = \delta_{\alpha\beta}, \tag{19.9b}$$

as in the footnote to equation (13.1).

The second interpretation of u_+^*, u_-^* is obtained from the transformation properties. The complex conjugate of (11.13) is

$$\begin{aligned} Ru_+^* &= a^*u_+^* + b^*u_-^*, \\ Ru_-^* &= c^*u_+^* + d^*u_-^*, \end{aligned} \tag{19.10}$$

where R is a real transformation of the real co-ordinate axes Ox, Oy, Oz. From the form (8.24) of the matrices of $D^{(1/2)}$, we have

$$a^* = d, \qquad b^* = -c, \qquad c^* = -b, \qquad d^* = a,$$

and

$$\begin{aligned} R(u_-^*) &= a(u_-^*) + b(-u_+^*), \\ R(-u_+^*) &= c(u_-^*) + d(-u_+^*). \end{aligned}$$

† This confusion arises because our discussion of spinors is somewhat inadequate, in particular because we have not distinguished between row and column spinors. In fact (19.9b) should be written $\Sigma \tilde{u}_\alpha^* u_\beta = \delta_{\alpha\beta}$ using the row spinors $\tilde{u}_\alpha^*$, and (19.10) is correct as it stands with the u_α^* being column spinors like u_α.

Thus u_-^*, $-u_+^*$ transform like u_+, u_-, as follows alternatively from problem 8.16. As before, u_+^*, u_-^* must be expressible as linear combinations of u_+, u_-, so that by Schurs' lemma (problem D.2) u_-^*, $-u_+^*$ are proportional to u_+, u_-. We may choose the arbitrary phase factor as unity, and obtain

$$u_+^* = -u_-, \qquad u_-^* = u_+. \tag{19.11}$$

It is this latter interpretation that is required in connection with (19.5a), so to be explicit we shall complete the definition of time-reversal (19.5a) by the relations

$$\boxed{Tu_+ = -u_-, \qquad Tu_- = u_+.} \tag{19.5b}$$

For instance for one electron we have

$$T[f_+(\mathbf{r})u_+ + f_-(\mathbf{r})u_-] = f_-^*(\mathbf{r})u_+ - f_+^*(\mathbf{r})u_-,$$

and for a many-electron function (11.11b) we obtain

$$\begin{aligned} T\psi_{\text{orb}}\,&(\mathbf{r}_1, \mathbf{r}_2, \ldots)u_{\alpha 1}u_{\beta 2}\ldots \\ &= \psi^*_{\text{orb}}\,(\mathbf{r}_1, \mathbf{r}_2, \ldots)(Tu_{\alpha 1})(Tu_{\beta 2})\ldots \end{aligned} \tag{19.12}$$

For an ordinary function $f(\mathbf{r})$ we have $(f^*)^* = f$, but note that from (19.11) $(u_+^*)^* = -u_+$, $(u_-^*)^* = -u_-$. It follows then from (19.12) that

$$\boxed{\begin{aligned} T^2\Psi &= \Psi && \text{for an even number of electrons,} \\ &= -\Psi && \text{for an odd number of electrons.} \end{aligned}} \quad \begin{aligned} &(19.13\text{a}) \\ &(19.13\text{b}) \end{aligned}$$

Transformation of the Hamiltonian

Operating on the Schrödinger equation (19.4) with T, we obtain from (19.5a), (19.8)

$$\begin{aligned} 0 &= T(\mathscr{H} - i\hbar\partial/\partial t)\Psi \\ &= T\mathscr{H}T^{-1}T\Psi - (-i\hbar)\partial/\partial(-t)T\Psi \\ &= (T\mathscr{H}T^{-1} - i\hbar\partial/\partial t)T\Psi, \end{aligned} \tag{19.14}$$

where $T\mathscr{H}T^{-1}$ is the transformed Hamiltonian (cf. footnote to equation (5.4)). Since (19.5) does not state how to calculate $T\mathscr{H}T^{-1}$, this product is defined in terms of its effect on an arbitrary function. Let ϕ be arbitrary; then $T\phi$ is also arbitrary. Suppose $\mathscr{H} = \mathscr{H}_r + i\mathscr{H}_i$, where $\mathscr{H}_r$ and $\mathscr{H}_i$ are real operators. By a real operator we mean an operator which operating on a real function gives a real function.† Then

$$
\begin{aligned}
[T(\mathscr{H}_r + i\mathscr{H}_i)T^{-1}]T\phi &= T\mathscr{H}_r\phi + Ti\mathscr{H}_i\phi \\
&= \mathscr{H}_r T\phi - i\mathscr{H}_i T\phi \\
&= (\mathscr{H}_r - i\mathscr{H}_i)T\phi,
\end{aligned}
$$

as follows as an extension of (19.8) or alternatively by writing $\phi = \phi_r + i\phi_i$. Thus we have

$$T\mathscr{H}T^{-1} = \mathscr{H}^*,$$

i.e. we obtain the time-reversed operator $T\mathscr{H}T^{-1}$ by replacing every operator in $\mathscr{H}$ by its complex conjugate. For instance a potential $V(\mathbf{r})$ is real and remains unchanged. The momentum $p_x = -i\hbar\partial/\partial x$ is pure imaginary and changes sign, $\mathbf{p} \to -\mathbf{p}$. Also the angular momentum operators change sign. For the orbital angular momentum this follows from the definition $\mathbf{l} = \mathbf{r} \wedge \mathbf{p}$, but for the spin angular momentum it is necessary to go back to its definition (11.6) as follows. A rotation R is a real operator,‡ and it follows from (8.1) that iI_ξ is real, and that all angular momenta $\hbar I_\xi$ are pure imaginary operators. Since $\mathscr{H}$ is a real function of $\mathbf{r}$, $\mathbf{p}$, $\mathbf{s}$, we have

$$\boxed{T\mathscr{H}T^{-1} = \mathscr{H}^* = \mathscr{H}(\mathbf{r}_i, -\mathbf{p}_i, -\mathbf{s}_i).} \tag{19.15}$$

Kramers' theorem

This theorem states: in the presence of any electric potential but in the absence of an external magnetic field, every energy level of a system with an *odd* number of electrons is n-fold degenerate, where n is an *even* number (not necessarily the same for each level).

We first note that in the absence of an external magnetic field, the Hamiltonian contains only even powers of the momenta. This is

† This notation differs from the one used for instance by Dirac (1958) who uses "real" to describe an operator with all real eigenvalues.

‡ This is obvious as regards its action on an ordinary function $f(x, y, z)$ because it is a real transformation of the real co-ordinates x, y, z. It is also true with respect to its effect on spin functions, for consider it operating on the real function $(u_+ + u_+{}^*)$:

$$
\begin{aligned}
R(u_+ + u_+{}^*) &= (au_+ + a^*u_+{}^*) + (bu_- + b^*u_-{}^*) \\
&= \text{real.}
\end{aligned}
$$

Incidentally, note that in deriving this result we have used (19.10), which explains why we use the definition (19.11) of complex conjugation derived from (19.10), rather than using (19.9).

so for the kinetic energy $\mathbf{p}^2/2m$ and for all the spin-orbit and spin-spin coupling terms of (11.8), (11.9) like $\mathbf{l}_i \cdot \mathbf{s}_i$, $\mathbf{r}_i \wedge \mathbf{p}_j \cdot \mathbf{s}_i$, $\mathbf{s}_i \cdot \mathbf{s}_j$, etc. Thus

$$\mathscr{H}(\mathbf{r}_i, \mathbf{p}_i, \mathbf{s}_i) = \mathscr{H}(\mathbf{r}_i, -\mathbf{p}_i, -\mathbf{s}_i), \tag{19.16}$$

or from (19.15)

$$T\mathscr{H}T^{-1} = \mathscr{H}. \tag{19.17}$$

In the presence of an external magnetic field $\mathbf{H}$, the Hamiltonian contains the term $(e/2mc)\mathbf{H}\cdot(\mathbf{L} + 2\mathbf{S})$ (Schiff 1955, p. 292) which is linear in the angular momenta, so that (19.16) does not hold. This shows why the theorem must be restricted to systems not in an external magnetic field. Note, however, that interactions such as the spin-orbit coupling which depend on the *internal* magnetic fields of the system generated by the moving electrons are invariant under time-reversal, since these internal fields change sign if the momenta of all the particles are reversed. This is all completely analogous to the situation in classical mechanics (problem 19.4).

It follows from (19.17) that if $\mathscr{H}\psi = E\psi$, then $\mathscr{H}T\psi = ET\psi$, so that ψ and $T\psi$ are eigenfunctions belonging to the same energy level. For this to give a degeneracy we have to show that they are linearly independent. Suppose

$$T\psi = \alpha\psi \tag{19.18}$$

where α is some constant. Then

$$T^2\psi = T\alpha\psi = \alpha^* T\psi = \alpha^*\alpha\psi.$$

For a system with an odd number of electrons, this gives a contradiction to $T^2\psi = -\psi$ (19.13b) since $\alpha^*\alpha$ is positive and cannot possibly equal –1. Thus (19.18) is false, and ψ and $T\psi$ are linearly independent. Since $T^2\psi = -\psi$, the degeneracy of every energy level is *even*, which proves the theorem.

Energy level degeneracies

Let us consider an atom in a $J = 3/2$ state placed in an electric field with point-group symmetry 32 due to its neighbours in a crystal or molecule, and let us determine the possible splitting of the level in the manner of §§ 6 and 14. The $J = 3/2$ states transform according to a double-valued representation Γ under rotations, and their characters for the double-group of 32 are (§ 16 and equation (14.4))

$\chi(E)$	$\chi(\bar{E})$	$\chi(B)$	$\chi(A)$	$\chi(K)$	$\chi(\bar{K})$
4	−4	−1	1	0	0

From Table 12 we have

$$\Gamma = \Gamma_4 + \Gamma_5 + \Gamma_6, \tag{19.19}$$

so that we might expect the level to split into two singlets (Γ_4, Γ_5) and one doublet (Γ_6). However we have not considered the full symmetry group of the Hamiltonian because this also includes time reversal, and Kramers' theorem shows that the greatest splitting possible is into doublets. Hence in this example time-reversal leads to a degeneracy, at least between Γ_4 and Γ_5, in addition to the degeneracies (Γ_6) due to rotational and other symmetries.

We shall now investigate the question of degeneracy more generally, and consider a degenerate set of states ψ_i transforming according to an irreducible representation D under some group $\mathfrak{G}$ of rotational or other symmetry transformations R.

$$R\psi_j = D_{ij}(R)\psi_i. \tag{19.20}$$

There are three cases to consider.

(a) The representation $D_{ij}(R)$ is real, or can be made completely real by an equivalence transformation (5.15) of base vectors.

(b) The representations $D_{ij}(R)$ and $D_{ij}^*(R)$ are not equivalent.

(c) The representations $D_{ij}(R)$ and $D_{ij}^*(R)$ are equivalent, but they cannot be made completely real by an equivalence transformation (5.15) of base vectors.

We have to subdivide these further and discuss separately systems with even and odd numbers of electrons. We also assume that the Hamiltonian is invariant under time-reversal, and shall determine in which cases this leads to some degeneracy in addition to that expected from the symmetry group $\mathfrak{G}$ alone.

Consider first a system with an *even* number of electrons. *Case (a)*. Suppose the ψ_i have already been so chosen that $D_{ij}(R)$ is real. Then

$$RT\psi_j = D_{ij}^*(R)T\psi_i = D_{ij}(R)T\psi_i, \tag{19.21}$$

and the functions $\phi_i = \psi_i + T\psi_i$ also transform under $\mathfrak{G}$ according to $D_{ij}(R)$. Further from (19.13a), $T\phi_i = \phi_i$ so that the set ϕ_i transforms into itself under *all* symmetry transformations, both those of $\mathfrak{G}$ and time-reversal. Thus there are no symmetry properties connecting the set ϕ_i with any other wave functions, and hence there is no additional degeneracy (§ 6, Theorem 2). An example of this case is the representation $D^{(l)}$ of the rotation group when l is an integer. The functions $Y_{lm} + (-1)^m Y_{l,-m}$ and $i[Y_{lm} - (-1)^m Y_{l,-m}]$ are real base vectors transforming according to $D^{(l)}$ so that the matrices of the representation are all real. Here the Y_{lm} are spherical harmonics transforming in the standard way (8.18). Other examples are the single-valued representations of the group 32. *Case* (*b*). The $T\psi_i$ transform according to $D_{ij}^*(R)$ which is not equivalent to $D_{ij}(R)$, so that the ψ_i and $T\psi_i$ are linearly independent (lemma 5, appendix C). The representations $D_{ij}(R)$ and $D_{ij}^*(R)$ therefore always occur together as a pair. Since $\mathscr{H}$ is invariant under T, they belong to the same energy level and we have an additional degeneracy. There are several examples of this among the representations of the crystal point-groups (appendix K), for instance the two complex representations of the group 4 which are bracketed together and labelled E. Since the two representations are always degenerate when there is time-reversal symmetry, they behave for elementary purposes as a single doubly-degenerate irreducible representation. This is the reason for bracketing them together and labelling the pair with a single letter. *Case* (*c*). In this case time-reversal leads to additional degeneracy, as the reader should have no difficulty in proving by following the hints in problem 19.11. The only example of this case which the author has come across is given in problem 26.10.—Note that if we are neglecting the electron spin completely, an orbital wave function always satisfies (19.13a) and the effect of time-reversal symmetry is exactly the same as in the even-number-of-electrons case.

Consider now a system with an odd number of electrons. The difference is that we have to use $T^2\psi = -\psi$ now instead of $T^2\psi = \psi$ (19.13). *Case* (*a*). The functions ψ_i and $T\psi_i$ again transform in the same way (19.20), (19.21) under $\mathfrak{G}$. As in the proof of Kramers' theorem (below 19.18)), we cannot have $T\psi_i = \alpha\psi_i$, which is the only possibility if they belong to the same irreducible vector space. The representation D therefore occurs twice and we have an extra doubling of the degeneracy because of time-reversal. *Case* (*b*). Since in the discussion of case (b) for an even number of electrons we did not use (19.13), the situation is exactly the same for even and odd numbers of electrons. *Case* (*c*). In case (c) for an odd number of electrons, it can be shown that we may always choose base vectors ψ_i such that $T\psi_j$ is a linear combination of the ψ_i (Wigner 1932, p. 557; Johnston 1958). There is therefore no extra degeneracy. An example of this case is the representation $D^{(1/2)}$ of the rotation group (problem 19.8). In fact all the representations $D^{(j)}$ of the rotation group come under this case when j is half an odd integer (problem 19.9). Also it can easily be shown (cf. 19.19)) that u_+, u_- transform according to Γ_6 under the point-group 32 (Table 14), which forms another example.

We summarize these results as follows. The three cases are:

> (a) the representation D can be transformed to real form;
> (b) D and D^* are inequivalent;
> (c) D and D^* are equivalent but cannot be made real.
>
> For an *even* number of electrons or *neglecting spin*, this implies respectively the following consequences:
> (a) there is no additional degeneracy;
> (b) D and D^* occur together and there is an additional degeneracy between them;
> (c) there is an additional degeneracy.
>
> For an *odd* number of electrons the consequences of (a) and (c) above are reversed.

There now remains the question of deciding which category a given irreducible representation comes into, particularly deciding between (a) and (c) since case (b) can be picked out by inspection of the character table. We state without proof the following test which depends only on the group characters:

$$\begin{aligned} \sum_R \chi(R^2) &= h, & &\text{case (a)} \\ &= 0, & &\text{case (b)} \\ &= -h. & &\text{case (c)} \end{aligned} \tag{19.22}$$

Here the summation is over all the h group elements R of the group $\mathfrak{G}$. Parts (a) and (b) of this relation can be proved easily by elementary methods using the hints in problem 14.16; case (c) is more difficult (Wigner 1932; Frobenius and Schur 1906; Johnston 1958). For example, from Table 12 it follows that for the group 32, Γ_1, Γ_2, Γ_3 belong to case (a), Γ_4 and Γ_5 to case (b), and Γ_6 to case (c).

Returning to the example of a $J = 3/2$ level being split by a crystal field of symmetry 32, we have that the states transform under the group 32 according to Γ (19.19). Since J is half an odd integer, we are dealing with an odd number of electrons. The representations Γ_4 and Γ_5 come under case (b) and give a doubly degenerate level. Γ_6 is two dimensional and comes under case (c); this gives no additional degeneracy due to time-reversal and we have another doublet. Similarly the states of a $J = 5/2$ level transform according to

$$\Gamma_4 + \Gamma_5 + \Gamma_6 + \Gamma_6$$

under the group 32. Γ_4 and Γ_5 together give a doublet. The two representations Γ_6 each give a doublet, and there is no reason to expect a degeneracy between them.

References

More detailed discussions have been given by the following: general introductory treatment, Klein (1952); time-reversal and the irreducible representations of point-groups, Bethe (1929), Opechowski (1940); Kramers' degeneracy, Kramers (1930); explicit introduction of time-reversal symmetry and discussion of cases (a), (b) and (c), Wigner (1932); application to Brillouin zone theory, Herring (1937a), Elliott (1954); mathematical treatment of cases (a), (b) and (c), Johnston (1958); relativistic treatment, Jauch and Rohrlich (1955), Racah (1937), Pauli (1941).

Summary

We have discussed the transformation known as time-reversal. It is a symmetry transformation of the Hamiltonian for any electronic system in any electrostatic field provided there is no externally applied magnetic field. It leads to various additional degeneracies beyond those expected from rotational and other symmetries, particularly in the case of systems with a relatively low symmetry. In particular Kramers' theorem states that for an odd number of electrons all states have an even degeneracy.

Problems

19.1 The Hamiltonian for a system of electrons is known to be invariant under space inversion Π and under time reversal T. Consider the following interactions between two particles i and j where $\mathbf{p} = \mathbf{p}_i - \mathbf{p}_j$ and $\mathbf{r} = \mathbf{r}_i - \mathbf{r}_j$: $(\mathbf{s}_i \cdot \mathbf{s}_j)(\mathbf{p} \cdot \mathbf{r})$, $\mathbf{r} \cdot \mathbf{p} \wedge (\mathbf{s}_i + \mathbf{s}_j)$, $\mathbf{r} \cdot \mathbf{p} \wedge (\mathbf{s}_i - \mathbf{s}_j)$, $\mathbf{r} \cdot \mathbf{p} \wedge (\mathbf{s}_i \wedge \mathbf{s}_j)$, $\mathbf{r} \cdot (\mathbf{s}_i - \mathbf{s}_j)$, $(\mathbf{r} \cdot \mathbf{s}_i)(\mathbf{r} \cdot \mathbf{s}_j)$. Show that only the second and the last can occur in the Hamiltonian.

19.2 Consider a single-electron time-independent wave function ψ. Show that the operator T can be written $i\sigma_y C$ as regards its effect on ψ, where C takes the complex conjugate of the *orbital* part of ψ, and σ_y is the Pauli spin operator of problems 11.1, 11.2. Show that for n electrons T becomes $(i)^n \sigma_{y1}\sigma_{y2} \cdots \sigma_{yn} C$. Use this form to establish (19.13). For one electron write the Hamiltonian in the form $\mathscr{H}_0 + \mathscr{H}_1\sigma_x + \mathscr{H}_2\sigma_y + \mathscr{H}_3\sigma_z$ and hence prove (19.15) (Klein 1952).

19.3 The function ψ_M belongs to the representation $\exp(iM\phi)$ of the axial rotation group. Show that $T\psi_M$ belongs to $\exp(-iM\phi)$.

19.4 A *classical* system of particles is moving under velocity independent forces. Show that the Hamiltonian $\mathscr{H}(q_i, p_i) = \mathscr{H}(q_i, -p_i)$, where the q_i, p_i are the generalized co-ordinates and momenta. If $q_i(t)$, $p_i(t)$ gives a particular motion of the system, show that $q_i(-t)$, $-p_i(-t)$ gives another possible motion, which is simply the original motion reversed. Discuss physically and mathematically how this situation breaks down in the presence of an external magnetic field.

19.5 Discuss the splitting of a free-atom level with quantum number J under the action of an electric field with point-group symmetry 32, when $J = 0, \frac{1}{2}, 1, \frac{3}{2}, \ldots 6$.

19.6 Discuss the splitting of the $J = 5/2$ ground state of a cerium ion in a field of symmetry $\bar{6}$2m (§ 18). The double-valued irreducible representations of this group can be calculated in the manner of § 16 or found in Bethe (1929) or Koster (1957).

19.7 Show that $\int (T\psi)^*(T\phi)\,d\tau = \int \phi^*\psi\,d\tau$, where ϕ^* is used in the sense of (13.1) and (19.9). L is any angular momentum operator. Prove its Hermitian property $\int \psi^* L\phi\,d\tau = \int (L\psi)^*\phi\,d\tau$, and that it is a pure imaginary operator $TL\psi = -LT\psi$. Hence show $\int \psi^* L\psi\,d\tau = -\int (T\psi)^* L(T\psi)\,d\tau$. Also prove that the expectation value for the magnetic moment $\beta(\mathbf{L} + 2\mathbf{S})$ is zero for any non-degenerate level.

19.8 Show from first principles that there are no linear combinations w_1, w_2 of the vectors u_+, u_- which transform with real coefficients under all rotations. Hint: suppose that w_1, w_2 exist. Using problem D.2 show that $Tw_i = \alpha w_i$, which, however, leads to a contradiction as shown in connection with (19.18).

19.9 The functions ψ_m transform according to $D^{(j)}$ under rotations, where j is half an odd integer. Using problem 8.16, show that the functions $\phi_m = \psi_m + i^{-2m} T\psi_{-m}$ transform into one another under rotations and time-reversal. Show the same for the functions $\chi_m = i(\psi_m - i^{-2m} T\psi_{-m})$. Prove that this representation always comes under case (c) of the text. Hint: show that cases (a) and (b) do not apply by using the method of problem 19.8.

19.10* If the ground state of a molecule or molecular complex is degenerate, then the molecule spontaneously distorts itself so as to split the ground state and lower the degeneracy. This is the Jahn–Teller effect. Exceptions occur for linear molecules, or when the degeneracy is the two-fold Kramers' degeneracy which cannot be split. Illustrate the origin of this effect, and discuss the importance in it of time-reversal symmetry, particularly when taking spin into account (Jahn and Teller 1937, Jahn 1938).

19.11 Prove that in case (c) for an even number of electrons, time-reversal symmetry leads to a doubling of the degeneracy.

Hints: show that $T\psi_j$ cannot be linearly dependent on the ψ_i for all j, because if it were we would have functions $\phi_j = \psi_j + T\psi_j$ in the vector space $(\psi_1, \ldots \psi_n)$ satisfying $T\phi_j = \phi_j$ giving case (a). Thus at least one $T\psi_j$ is not a linear combination of the ψ_i, and by orthogonalizing and then using problem 14.15 we can construct from it a whole set of functions orthogonal to the ψ_i but degenerate with them.

20. Wigner and Racah Coefficients

Coupling of angular momenta

Suppose that $U_m^{(j)}$ and $V_{m'}^{(j')}$ are two sets of normalized functions transforming respectively according to $D^{(j)}$ and $D^{(j')}$ under rotations. For the present we shall also assume that the two sets are functions of two quite different lots of variables, such as orbital variables and spin variables or the co-ordinates of two different particles. Then the products $U_m^{(j)}V_{m'}^{(j')}$ are also normalized. Using the method of § 9, we can in principle pick out normalized linear combinations

$$W_M^{(J)} = \sum_{m,m'} (jj'mm'|JM)U_m^{(j)}V_{m'}^{(j')} \tag{20.1}$$

which transform according to $D^{(J)}$, where

$$|j-j'| \leqslant J \leqslant j+j'. \tag{20.2}$$

The phase of the coefficients $(jj'mm'|JM)$ is fixed by making

$$(jj'j, J-j|JJ) = \text{real and positive}, \tag{20.3}$$

for each value of J. This defines the vectors (20.1) completely and hence also the coefficients $(jj'mm'|JM)$, which are known as Wigner coefficients or alternatively as vector-coupling or Clebsch–Gordon coefficients. Now since in (20.1) the vectors $W_M^{(J)}$ and $U_m^{(j)}V_{m'}^{(j')}$ are all normalized, (20.1) can be regarded as a unitary transformation in the vector space $\{\ldots, U_m^{(j)}V_{m'}^{(j')}, \ldots\}$ so that the matrix A of Wigner coefficients† is unitary $A^{-1} = \tilde{A}^*$ (problem A.10). Furthermore since (8.18) contains only real coefficients, the process described in § 9 for picking out the vectors $W_M^{(J)}$ (20.1) leads only to real numbers for the Wigner coefficients, so that

$$A^{-1} = \tilde{A}. \tag{20.4}$$

† In this matrix each row is labelled by a pair of numbers m, m', and each column by J, M. The matrix is clearly square of order equal to the dimension of the vector space (cf. problem 9.7).

Hence we can invert the equations (20.1) and write

$$U_m^{(j)} V_{m'}^{(j')} = \sum_{J,M} (jj'mm'|JM) W_M^{(J)}. \tag{20.5}$$

The situation becomes slightly more complicated when the U's and V's are not functions of two completely different sets of variables, for now the products $U_m^{(j)} V_{m'}^{(j')}$ are in general not normalized. However the vectors

$$\sum_{m,m'} (jj'mm'|JM) U_m^{(j)} V_{m'}^{(j')} \tag{20.6}$$

still have the same transformation properties (8.18) as in the previous case of (20.1). Hence (§ 8) all the vectors with the same J have the same amplitude, and we only need to include a factor N_J in (20.6) to obtain normalized vectors

$$\boxed{W_M^{(J)} = N_J \sum_{m,m'} (jj'mm'|JM) U_m^{(j)} V_{m'}^{(j')},} \tag{20.7}$$

where N_J is independent of M. Equation (20.4) can again be used to invert (20.7).

$$\boxed{U_m^{(j)} V_{m'}^{(j')} = \sum_{M,J} (jj'mm'|JM) W_M^{(J)}/N_J.} \tag{20.8}$$

This justifies the equations (9.7), (9.8) given in § 9. As already pointed out in § 9, we have

$$(jj'mm'|JM) = 0 \text{ unless } M = m + m', \tag{20.9}$$

so that the summation in (20.8) reduces to a summation over J only.

Spinor invariants

We shall now carry out the first part† of the calculation of a general formula for the Wigner coefficients. This will also serve to introduce and illustrate the technique known as the method of spinor invariants (Brinkman 1956). Consider a sum

$$\Phi = c_j \phi_j \qquad (j \text{ summed}), \tag{20.10}$$

to which we apply a unitary transformation T. With the nomenclature explained below equation (5.14), we suppose that the ϕ_j transform as base vectors

$$T\phi_j = T_{ij}\phi_i \tag{20.11a}$$

† This essentially follows the discussion of Van der Waerden (1933).

and the c_j as coefficients

$$Tc_j = c_i[T^{-1}]_{ji} = T_{ij}^* c_i, \tag{20.11b}$$

where the last equality follows from the unitary property (problem C.2). Then the sum (20.10) remains invariant under T, and is called an invariant. In order to compound angular momenta in the manner of (20.7), it is convenient not to consider arbitrary sets of U's and V's, but to use the specific sets of vectors

$$U_m^{(j)} = \frac{u_+^{j+m} u_-^{j-m}}{[(j+m)!(j-m)!]^{1/2}}, \qquad V_{m'}^{(j')} = \frac{v_+^{j'+m'} v_-^{j'-m'}}{[(j'+m')!(j'-m')!]^{1/2}}. \tag{20.12}$$

Here u_+, u_- and v_+, v_- are spinors transforming according to $D^{(1/2)}$ under rotations, and we recall from problem 8.11 that the vectors (20.12) do transform in the standard way (8.18) according to the representations $D^{(j)}$ and $D^{(j')}$. The method of spinor invariants derives its name from the use of invariants (20.10) in which the functions are of the form (20.12).

Step 1. We first suppose that x_+, x_- are two quantities that transform as coefficients according to $D^{(1/2)}$. Thus

$$x_+ u_+ + x_- u_- \tag{20.13}$$

is an invariant. *Step 2.* Since (20.13) is invariant, so is the quantity

$$(x_+ u_+ + x_- u_-)^{2J} = (2J)! \sum_M X_M^{(J)} U_M^{(J)}, \tag{20.14}$$

where the $U_M^{(J)}$ are defined in the manner of (20.12). The coefficients $X_M^{(J)}$ therefore transform according to $D^{(J)}$ as coefficients. By expanding the binomial in (20.14) we obtain explicitly

$$X_M^{(J)} = \frac{x_+^{J+M} x_-^{J-M}}{[(J+M)!(J-M)!]^{1/2}}. \tag{20.15}$$

Step 3. Under a rotation R the vectors v_+, v_- transform according to (8.24), which may be written

$$\begin{aligned} Rv_+ &= av_+ + bv_-, \\ Rv_- &= -b^* v_+ + a^* v_-, \end{aligned} \tag{20.16a}$$

or

$$\begin{aligned} R(v_-) &= a^*(v_-) + b^*(-v_+), \\ R(-v_+) &= -b(v_-) + a(-v_+). \end{aligned} \tag{20.16b}$$

Comparison of (20.16) with (20.11) shows that v_-, $-v_+$ transform as coefficients, and hence

$$v_- u_+ - v_+ u_- \tag{20.17}$$

is an invariant.

Step 4. We now consider the expression

$$Z = (v_- u_+ - v_+ u_-)^\lambda (x_+ u_+ + x_- u_-)^{2j-\lambda} (x_+ v_+ + x_- v_-)^{2j'-\lambda} \tag{20.18}$$

where $$\lambda = j + j' - J.$$

From (20.13), (20.17), Z is an invariant. Each factor can be expanded by the binomial theorem

$$(v_- u_+ - v_+ u_-)^\lambda = \sum_r (-1)^r \binom{\lambda}{r} (v_- u_+)^{\lambda - r} (v_+ u_-)^r,$$

where the binomial coefficient $\binom{\lambda}{r}$ is defined by (20.25) below. Collecting the powers of $u_\pm$, $v_\pm$, $x_\pm$ together, we obtain

$$Z = \sum_m \sum_{m'} [(j+m)!(j-m)!(j'+m')!(j'-m')!(J+M)!(J-M)!]^{1/2}$$
$$\sum_r (-1)^r \binom{\lambda}{r} \binom{2j-\lambda}{j-m-r} \binom{2j'-\lambda}{j'+m'-r} U_m{}^{(j)} V_{m'}{}^{(j')} X_{m+m'}^{(J)}$$
$$= \lambda!(2j-\lambda)!(2j'-\lambda)! \sum_m \sum_{m'} a_{mm'}^J U_m{}^{(j)} V_{m'}{}^{(j')} X_{m+m'}^{(J)}, \tag{20.19}$$

where†

$$a_{mm'}^J = \sum_r (-1)^r \times$$
$$\times \frac{[(j+m)!(j-m)!(j'+m')! \times (j'-m')!(J+M)!(J-M)!]^{1/2}}{(j-m-r)!(j+m-\lambda+r)!(j'+m'-r)! \times (j'-m'-\lambda+r)!(r)!(\lambda-r)!}. \tag{20.20}$$

Since the $X_M{}^{(J)}$ transform as coefficients according to $D^{(J)}$, the quantities

$$w_M{}^{(J)} = \sum_{m,m'} a_{mm'}^J U_m{}^{(j)} V_{m'}{}^{(j')} \tag{20.21}$$

with $M = m + m'$

† The summation is over all terms such that none of the brackets becomes negative. Also $0! = 1$.

which multiply the $X_M^{(J)}$ in (20.19) must transform as base vectors according to $D^{(J)}$. This is the coupling of angular momenta which we require, and comparison with (20.7) gives the proportionality

$$(jj'mm'|JM) = b_{Jjj'}a^J_{mm'} \tag{20.22}$$

where $b_{Jjj'}$ is a constant.

A general formula for Wigner coefficients

A consideration of the result (20.22) illustrates both the power and the limitation of group theory. We have used transformational properties to relate the vectors $w_M^{(J)}$ with the same value of J to one another, and this gives the relationship (20.20), (20.22) between the appropriate Wigner coefficients. However we have not obtained the actual values of the Wigner coefficients, or what is the same thing, we cannot so far compare coefficients with different J's in any systematic way. To do this we have to take account of some algebraic considerations (as distinct from the effects of rotational transformations above), namely (20.3) and the fact that the definition (20.1) of the Wigner coefficients requires vectors *normalized to unity*. Now there is no direct way of normalizing the $U_m^{(j)}$ and $V_{m'}^{(j')}$ defined by (20.12), so that we have to use instead the algebraic relation (20.4)†. This gives

$$\tilde{A}A = E \text{ or}$$

$$\sum_{m,m'} (jj'mm'|JM)^2 = 1, \tag{20.23}$$

where the summation over m, m' is subject to $m + m' = M$.

Following E. R. Cohen (1949), we now calculate $b_{Jjj'}$ by substituting (20.22) in (20.23) for one convenient value of M, namely $M = J$. Equation (20.22) then gives all the Wigner coefficients. For $M = J$ we have $m' = J - m$, and the summation (20.20) reduces to the single term $r = j - m$. Hence from (20.22)

$$(jj'm, J - m|JJ)$$

$$= \frac{b_{Jjj'}[(2J)!]^{1/2}(-1)^{j-m}}{(j + J - j')!(j' + J - j)!}\left[\frac{(j' + J - m)!(j + m)!}{(j' - J + m)!(j - m)!}\right]^{1/2}.$$

† By interpreting the u_+, u_- as creation and annihilation operators for spin $\frac{1}{2}$ particles, Schwinger has been able to utilize the symmetry properties of expressions like (20.12) and at the same time to work with normalized wave functions. This then gives the Wigner coefficients directly.

Substituting this into (20.23), we obtain

$$\frac{b^2{}_{Jjj'}(2J)!}{(j+J-j')!(j'+J-j)!}\sum_m \binom{j'+J-m}{j-m}\binom{j+m}{j'-J+m} = 1. \tag{20.24}$$

We now use the formula

$$\begin{aligned}\binom{\alpha}{\beta} &= \frac{\alpha(\alpha-1)(\alpha-2)\ \ldots\ (\alpha-\beta+1)}{\beta!} \\ &= (-1)^\beta \frac{(\beta-\alpha-1)(\beta-\alpha-2)\ \ldots\ (1-\alpha)(-\alpha)}{\beta!} \\ &= (-1)^\beta \binom{\beta-\alpha-1}{\beta}.\end{aligned} \tag{20.25}$$

The sum in (20.24) becomes

$$\begin{aligned}&\sum_m \binom{j'+J-m}{j-m}\binom{j+m}{j'-J+m} \\ &= \sum_m (-1)^{j+j'-J}\binom{j-j'-J-1}{j-m}\binom{j'-j-J-1}{j'-J+m} \\ &= (-1)^{j+j'-J}\binom{-2J-2}{j+j'-J} = \binom{j+j'+J+1}{j+j'-J}.\end{aligned}$$

Here the summation of the binomial coefficients is obtained by picking out the coefficient of $x^{j+j'-J}$ on each side of the identity

$$(1+x)^{j-j'-J-1}(1+x)^{j'-j-J-1} = (1+x)^{-2J-2}.$$

Thus (20.24) together with (20.20) and (20.22) gives

$$\boxed{\begin{aligned}(jj'mm'|JM) &= \left[\frac{(j+j'-J)!(j+J-j')! \times (j'+J-j)!(2J+1)}{(j+j'+J+1)!}\right]^{1/2}\delta^M_{m+m'} \\ &\times \sum_r (-1)^r \frac{[(j+m)!(j-m)!(j'+m')! \times (j'-m')!(J+M)!(J-M)!]^{1/2}}{(j-m-r)!(J-j'+m+r)!(j'+m'-r)! \times (J-j-m'+r)!r!(j+j'-J-r)!}.\end{aligned}} \tag{20.26}$$

Numerical values of some Wigner coefficients are given in appendix I, and algebraic tables for the cases $j' = \frac{1}{2}, 1, \frac{3}{2}$ are given by Condon and Shortley (1951, p. 76). For references to further tables, see appendix I.

Symmetry properties

The Wigner coefficients have several symmetry properties obtained by permuting the quantum numbers. These symmetry properties could be obtained directly from (20.26) with some manipulation. However, it is more instructive to proceed as in the derivation of the general formula (20.26), i.e. to use first the transformational properties to establish a proportionality, and then to calculate the proportionality constant algebraically by considering a special case. We shall from now on use a, b, c instead of j, j', J, and denote the corresponding magnetic quantum numbers by α, β, γ.

Let $U_\alpha{}^a$, $V_\beta{}^b$, $W_\gamma{}^c$ be functions transforming according to $D^{(a)}$, $D^{(b)}$, $D^{(c)}$. From (20.8) we obtain the matrix element

$$\int (W_\gamma{}^c)^* U_\alpha{}^a V_\beta{}^b \,\mathrm{d}\tau = (\text{const})(ab\alpha\beta|c\gamma). \tag{20.27}$$

This matrix element can also be calculated in another way by recalling (problem 8.16) that

$$(-1)^{\pm(c-\gamma)}(W_\gamma{}^c)^* \tag{20.28}$$

transforms under rotations in exactly the same way as $W_{-\gamma}{}^c$. Hence analogously to (20.27) we have

$$\int [(-1)^{-a+\alpha} U_\alpha{}^{a*}]^* \, [(-1)^{c-\gamma} W_\gamma{}^{c*}] V_\beta{}^b \,\mathrm{d}\tau$$
$$= (\text{another constant})(c, b, -\gamma, \beta|a, -\alpha).$$

Comparing this with (20.27) and putting $\gamma = \alpha + \beta$, we obtain

$$(ab\alpha\beta|c\gamma) = (-1)^{b+\beta} A_{abc}(cb\, -\gamma\beta|a\, -\alpha). \tag{20.29}$$

The two Wigner coefficients in (20.29) can easily be calculated from (20.26) for the special case $\alpha = a$, $\beta = c - a$, $\gamma = c$, and hence the constant A_{abc} evaluated. We obtain

$$(ab\alpha\beta|c\gamma) = (-1)^{b+\beta} \left(\frac{2c+1}{2a+1}\right)^{1/2} (cb\, -\gamma\beta|a\, -\alpha). \tag{20.30}$$

All of the following basic symmetry properties can be proved in a similar way, and further symmetries derived by combining them.

$$\begin{aligned}(ab\alpha\beta|c\gamma) &= (ba\,-\beta\,-\alpha|c\,-\gamma),\\ &= (-1)^{a+b-c}(ba\beta\alpha|c\gamma),\\ &= (-1)^{a+b-c}(ab\,-\alpha\,-\beta|c\,-\gamma),\\ &= (-1)^{a-\alpha}\left(\frac{2c+1}{2b+1}\right)^{1/2}(ac\alpha\,-\gamma|b\,-\beta),\\ &= (-1)^{b+\beta}\left(\frac{2c+1}{2a+1}\right)^{1/2}(cb\,-\gamma\beta|a\,-\alpha).\end{aligned} \tag{20.31}$$

Racah coefficients

It happens in many calculations that sums of products of Wigner coefficients occur which are very difficult and tedious to sum by using the general formula (20.26) or a table of numerical values. By the use of the symmetry properties (20.31) and the unitary property (20.4), (20.23), it is usually possible to relate these sums to the Racah coefficients $W(abcdef)$ which are defined by the relation

$$\begin{aligned}W(abcdef) = \sum_{\alpha,\beta,\gamma,\delta,\phi} & \frac{(-1)^{e+f-b-c}}{[(2b+1)(2c+1)]^{1/2}}\\ & \times (af\alpha\phi|c\gamma)(cd\gamma\delta|e\epsilon)(fd\phi\delta|b\beta)(ab\alpha\beta|e\epsilon).\end{aligned} \tag{20.32}$$

We shall show below that this expression (20.32) is independent of the quantum number ϵ which is not summed over, and make one use of it. However, we shall not discuss the properties of the Racah coefficients in any detail. Since the coefficients relate different j quantum numbers, most of their properties cannot be derived purely from the effects of rotational transformations for much the same reasons as were discussed in connection with the Wigner coefficients. Suffice it to mention that the Racah coefficients have been used to simplify formulae in many branches of physics including the following: the angular dependence of scattering and reaction cross sections, the angular correlation of successive decays, nuclear structure, the theory of hyperfine structure, atomic and molecular spectra of complicated configurations.

The calculation of matrix elements

We shall now use Wigner and Racah coefficients to express the values of some general types of matrix element. Let $T_m^{(k)}$ be a set of operators transforming into one another in the standard way (8.18) under rotations according to the representation $D^{(k)}$. These are called irreducible tensor operators. For example the operators

$$L_x + iL_y, \; -\sqrt{2}L_z, \; -(L_x - iL_y)$$

transform according to $D^{(1)}$ in the standard way. The components of a second rank symmetric tensor with zero sum of diagonal elements transform according to $D^{(2)}$ (see text of § 9 and problem 9.8).

We shall first consider the matrix element

$$\langle j_1 m_1 | T_m^{(k)} | j_2 m_2 \rangle = \int \psi^*(j_1, m_1) T_m^{(k)} \psi(j_2, m_2)\, \mathrm{d}\tau. \quad (20.33)$$

From (20.27) we have

$$\langle j_1 m_1 | T_m^{(k)} | j_2 m_2 \rangle = B_{j_1 j_2 k}(k j_2 m m_2 | j_1 m_1),$$

where from the fundamental theorem of § 13 the constant $B_{j_1 j_2 k}$ does not depend on the m's. This result can be put in a more symmetrical form by using the symmetry property (20.30):

$$\langle j_1 m_1 | T_m^{(k)} | j_2 m_2 \rangle = B_{j_1 j_2 k}(-1)^{j_2+m_2}\left(\frac{2j_1+1}{2k+1}\right)^{1/2}(j_1 j_2 - m_1 m_2 | k - m).$$

It has become conventional to write the constant $B_{j_1 j_2 k}$ as

$$B_{j_1 j_2 k} = (-1)^{j_1+k-j_2}\frac{\langle j_1 \| T^{(k)} \| j_2 \rangle}{(2j_1+1)^{1/2}}.$$

Here $\langle j_1 \| T^{(k)} \| j_2 \rangle$ is another constant called the reduced matrix element which also is independent of m_1, m_2, m. The double bars denote that it is not a proper matrix element but just a quantity associated with the matrix elements (20.33). Thus, we finally obtain

$$\boxed{\begin{aligned}&\langle j_1 m_1 | T_m^{(k)} | j_2 m_2 \rangle \\ &\quad = (-1)^{j_1+k+m_2}\frac{\langle j_1 \| T^{(k)} \| j_2 \rangle}{(2k+1)^{1/2}}(j_1 j_2 - m_1 m_2 | k - m),\end{aligned}} \quad (20.34)$$

where the Wigner coefficient gives the dependence of the matrix element on m_1, m_2, m.

We consider next a matrix element of the type

$$Z = \langle L_1 S_1 J_1 M_1 | \sum_m (-1)^m X_m^{(k)} Y_{-m}^{(k)} | L_2 S_2 J_2 M_2 \rangle. \quad (20.35)$$

Here the wave functions

$$\psi(L_2S_2J_2M_2) = \sum_{M_{L2}M_{S2}} (L_2S_2M_{L2}M_{S2}|J_2M_2)\psi(L_2S_2M_{L2}M_{S2}) \tag{20.36}$$

transform according to $D^{(J_2)}$ under rotations, and are linear combinations of the $\psi(L_2S_2M_{L2}M_{S2})$ corresponding to energy level eigenfunctions in the limit of Russell–Saunders coupling (§§ 11, 12). Similarly

$$\psi^*(L_1S_1J_1M_1) = \sum_{M_{L1}M_{S1}} (L_1S_1M_{L1}M_{S1}|J_1M_1)\psi^*(L_1S_1M_{L1}M_{S1}). \tag{20.37}$$

In (20.35) $X_m^{(k)}$ is an irreducible tensor operator which operates on the orbital variables only, and $Y_{-m}^{(k)}$ on the spin variables only. Thus we have

$$\begin{aligned}\langle L_1S_1M_{L1}M_{S1}|X_m^{(k)}Y_{-m}^{(k)}|L_2S_2M_{L2}M_{S2}\rangle \\ = \langle L_1M_{L1}SM_S|X_m^{(k)}|L_2M_{L2}SM_S\rangle \times \\ \times \langle S_1M_{S1}LM_L|Y_{-m}^{(k)}|S_2M_{S2}LM_L\rangle\end{aligned} \tag{20.38}$$

where the matrix elements are independent of S, M_S and of L, M_L respectively. The value of each matrix element is given by (20.34):

$$\begin{aligned}\langle L_1M_{L1}SM_S|X_m^{(k)}|L_2M_{L2}SM_S\rangle \\ = (-1)^{L_1+k+M_{L2}}\frac{\langle L_1\|X^{(k)}\|L_2\rangle}{(2k+1)^{1/2}}(L_1L_2\,-M_{L1}M_{L2}|k\,-m),\end{aligned} \tag{20.39}$$

and a similar formula holds for the second element. Substituting (20.36) to (20.39) into equation (20.35), we obtain for the matrix element

$$\begin{aligned}Z = \langle L_1\|X^{(k)}\|L_2\rangle\langle S_1\|Y^{(k)}\|S_2\rangle \\ \times \sum_{\substack{m,\,M_{L1},\,M_{L2},\\ M_{S1},\,M_{S2}}} \frac{(-1)^{m+2k+L_1+S_1+M_{L2}+M_{S2}}}{(2k+1)} \\ \times (L_1S_1M_{L1}M_{S1}|J_1M_1)(L_2S_2M_{L2}M_{S2}|J_2M_2) \\ \times (L_1L_2\,-M_{L1}M_{L2}|k\,-m)\,(S_1S_2\,-M_{S1}M_{S2}|km).\end{aligned} \tag{20.40}$$

To simplify this, we first note that as in (20.28), $(-1)^mY_{-m}^{(k)}$ transforms in the same way as $(-1)^kY_m^{(k)*}$, where we assume k to be an integer. Hence

$$\sum_m (-1)^m X_m^{(k)}Y_{-m}^{(k)} \text{ transforms like } (-1)^k\sum_m Y_m^{(k)*}X_m^{(k)},$$

and is therefore invariant under rotations (problems 13.12). Hence from the fundamental theorem of § 13 it follows that the matrix element (20.38) is zero unless $J_1 = J_2$ ($= J$ say) and $M_1 = M_2$ ($= M$ say), and that it is independent of M. We now use the symmetry properties (20.31) to write

$$(L_1L_2 - M_{L1}M_{L2}|k - m) = (-1)^{-L_1-M_{L1}+L_1+k-L_2}\left(\frac{2k+1}{2L_2+1}\right)^{1/2}(L_1kM_{L1} - m|L_2M_{L2})$$

$$(S_1\,S_2 - M_{S1}\,M_{S2}|km) = (-1)^{-S_2-M_{S2}}\left(\frac{2k+1}{2S_1+1}\right)^{1/2}(kS_2 - mM_{S2}|S_1M_{S1}).$$

We also use $M_{L2} - M_{L1} + m = 0$ in the exponent of (-1). The summation in (20.40) then takes on exactly the form (20.32) required for the Racah coefficients and we have

$$Z = \langle L_1\|X^{(k)}\|L_2\rangle\langle S_1\|Y^{(k)}\|S_2\rangle(-1)^{L1+S2-J}W(L_1S_1L_2S_2Jk).$$

The fact that the matrix element is independent of M shows that the summation in (20.32) is independent of ϵ. In the above discussion the use of "orbital variables" and "spin variables" is, of course, unnecessary and any two independent sets of variables will do. We can therefore generalize our result and write

$$\boxed{\begin{aligned}\langle j_1j_2jm|\sum_m(-1)^mX_m{}^{(k)}Y_{-m}{}^{(k)}|j'_1j'_2j'm'\rangle \\ = \delta_{jj'}\delta_{mm'}(-1)^{j_1+j'_2-j}\langle j_1\|X^{(k)}\|j'_1\rangle \times \\ \times\langle j_2\|Y^{(k)}\|j'_2\rangle W(j_1j_2j_1'j_2'jk),\end{aligned}} \qquad (20.41)$$

where $X_m{}^{(k)}$ operates on one set of variables corresponding to the quantum numbers j_1, j'_1, and $Y_{-m}{}^{(k)}$ on a different set corresponding to j_2, j'_2.

References

The coupling of angular momenta and in particular the properties of the Racah coefficients have been discussed by many people including Racah (1942), Biedenharn, Blatt and Rose (1952). In addition the latter authors give nearly twenty references covering the general theory, the various fields of application, and the tabulation of the coefficients. We shall merely add the following: Wigner

(1952 and 1931), Schwinger (1952), Edwards (1957) covering the theory of angular momentum from various points of view; Siegbahn (1955, p. 531), a review on the angular correlation of nuclear radiation; Schwartz (1955) on the application to the theory of hyperfine structure.

Summary

We have first discussed the definition of the Wigner coefficients, particularly as regards their phase, and the unitary property. Using only transformational properties under rotations, we derived the formula (20.20) which relates all Wigner coefficients with different m's but the same j. This calculation illustrated the method of spinor invariants, and led the way to a complete formula for the Wigner coefficients, as well as to a study of their symmetry properties. The Racah coefficients were then defined, and two common types of matrix element were expressed in terms of Wigner and Racah coefficients.

PROBLEMS

20.1 Two normalized sets of functions $u_m = f_1(r) Y_{1m}(\theta, \phi)$ and $v_\mu = f_2(r) Y_{1\mu}(\theta, \phi)$ both transform according to $D^{(1)}$ under rotations, where Y_{lm} are the spherical harmonics (8.20). Show that the products $u_m v_\mu$ are neither normalized nor orthogonal, and that the vectors

$$\sum_{m,\,\mu} (1, 1, m, \mu|JM) u_m v_\mu$$

are orthogonal but not normalized. Calculate their normalization constants.

20.2 By operating on both sides of (20.1) with I_- (§ 8), prove the recurrence relation

$$A_\gamma(ab\alpha\beta|c\gamma - 1) = A_{\alpha+1}(ab\alpha + 1\beta|c\gamma) + A_{\beta+1}(ab\alpha\beta + 1|c\gamma)$$

where $A_m{}^2 = j(j+1) - m(m-1)$.

20.3 Calculate the Wigner coefficients $(j, \frac{1}{2}, m, \pm\frac{1}{2}|j \pm \frac{1}{2}, m \pm \frac{1}{2})$ in each of the following three ways: (i) from the general formula (20.26); (ii) by carrying out the procedure of § 9 in detail as in problem 9.3 starting with the vector with $J = j + \frac{1}{2}$, $M = j + \frac{1}{2}$; and (iii) by using the recurrence relation of problem 20.2 and the other recurrence relation given by Condon and Shortley (1951, p. 74) starting with $(j\frac{1}{2}j\frac{1}{2}|j + \frac{1}{2}, j + \frac{1}{2}) = 1$. Which method is the quickest?

20.4 By squaring (20.29), summing both sides of α, β and γ, and using the unitary property (20.23), show that one obtains directly

$$A_{abc} = \pm\left(\frac{2c+1}{2a+1}\right)^{1/2}.$$

Show how the correct sign can be determined from (20.3) and (20.20) without using the general formula (20.26) for the Wigner coefficients. Note that this would be a more elegant way than the method used in the text for establishing the symmetry property (20.30), if it were not for the difficulty of establishing the correct sign.

20.5 Prove all the symmetry properties (20.31) of the Wigner coefficients. Show that $(ab00|c0) = 0$ if $a + b + c =$ an odd integer.

20.6 Using the symmetry properties (20.31) and the fact that the sum (20.32) is independent of ϵ, show that

$$W(abcdef) = \sum_{\alpha\beta\delta\epsilon\phi} [(2e+1)(2f+1)]^{-1/2} \times (ab\alpha\beta|e\epsilon)(bd\beta\delta|f\phi)(af\alpha\phi|c\gamma)(ed\epsilon\delta|c\gamma).$$

This is the more usual definition of the Racah coefficients.

20.7 By comparing problem 20.6 with equation (20.32), show that

$$W(abcdef) = (-1)^{e+f-b-c}W(afedcb).$$

Similarly prove that

$$W(abcdef) = W(badcef) = W(cdabef) = W(acbdfe) = (-1)^{e+f-a-d}W(ebcfad) = (-1)^{e+f-b-c}W(aefdbc).$$

20.8 Write down the value of the matrix element

$$\langle LSJM|\mathbf{L}\cdot\mathbf{S}|LSJM\rangle$$

in terms of a Racah coefficient, and by comparison with (13.20), (13.22) calculate a formula for $W(LSLSJ1)$.

20.9 Express the spin-spin coupling

$$\mathbf{s}_i\cdot\mathbf{s}_j/r_{ij}^3 - 3(\mathbf{s}_i\cdot\mathbf{r}_{ij})(\mathbf{s}_j\cdot\mathbf{r}_{ij})/r_{ij}^5$$

in the form $\Sigma(-1)^m X_m^{(k)} Y_{-m}^{(k)}$ as in (20.41), where the X's operate on the orbital variables only and the Y's on the spin variables only. Also express $1/|\mathbf{r}_i - \mathbf{r}_j|$ in a series of terms of this form, where in each term the X's operate on $\mathbf{r}_i$ only and the Y's on $\mathbf{r}_j$ only.

20.10* Discuss the use of Racah coefficients in one of the following fields (see references in the text): (i) the angular dependence of scattering and reaction cross sections, (ii) the angular correlation of successive decays, (iii) nuclear structure, (iv) the theory of hyperfine structure, (v) atomic spectra of complicated configurations.

21. Hyperfine Structure

Introduction

In Chapter II, we found that the energy levels of an atom can be described in terms of configurations, which are split into terms by the electrostatic repulsion between the electrons. These terms are split again into levels by the spin-orbit coupling, where the levels are designated by a J quantum number. We now consider the previously omitted term $\mathscr{H}_{\text{nucl}}$ of (10.1), which takes into account

(i) the motion of the nucleus, its finite size, and the deviation of the potential from a pure Coulomb field near and inside the nucleus, and

(ii) the spin, magnetic moment, quadrupole moment, etc. of the nucleus.

The items in category (i) only introduce small quantitative shifts of the energy levels, known as isotope shifts because they are different for different isotopes (Ramsey 1953). We shall not discuss them further. On the other hand the items in category (ii) above lead to interesting qualitative splittings of the atomic energy levels. These splittings due to the interactions between the electrons and the nucleus are called the hyperfine structure.

It might at first sight be thought that to treat the electron-nuclear interactions we would have to know in detail the motion of the nucleons inside the nucleus, which in turn depends on the nature of nuclear forces. This is to some extent true as regards the items in category (i) above, i.e. if one wants to calculate some of the isotope shifts. However, as regards the major features of the electron-nuclear interactions (category (ii) above), we shall show that the nucleus can be considered as a mass point with charge Ze, with an intrinsic angular momentum† $\mathbf{I}$ of magnitude $\hbar[I(I+1)]^{1/2}$, a magnetic moment $\boldsymbol{\mu}_n$, and with electric quadrupole and higher moments. Here I is the nuclear angular momentum quantum number, called the nuclear spin for short. We shall show that the

† In this section we shall have no occasion to use the infinitesimal rotation operators, so that the present use of $\mathbf{I}$ and I should cause no confusion.

magnetic moment $\boldsymbol{\mu}_n$ is always parallel to the angular momentum $\mathbf{I}$, so that we write

$$\boxed{\boldsymbol{\mu}_n = \mu_n \frac{\mathbf{I}}{\hbar I}.} \tag{21.1}$$

Here μ_n is a measure of the magnetic moment, and is by convention loosely referred to as *the* magnetic moment. Strictly of course the magnitude of $\boldsymbol{\mu}_n$ is

$$\mu_n \frac{\sqrt{[I(I+1)]}}{I}$$

from (21.1).

We shall then show how the interactions between the nucleons and the electrons, and the energy splittings these give rise to, can be expressed in terms of the various nuclear moments. From the point of view of this section, we shall regard the values of these moments as arbitrary parameters to be determined by experiment. Clearly it is one of the aims of nuclear theory to calculate the moments from first principles, but at the present time this can only be done very approximately, and we shall not consider this aspect further.

Nuclear spin and parity

Let us first consider a system of A interacting nucleons not acted on by any external fields. The Hamiltonian is

$$\mathscr{H} = -\hbar^2 \sum_1^A \frac{1}{2m_i} \nabla_i^2 + \tfrac{1}{2} \sum_{i,j=1}^A U_{ij}(\mathbf{r}_i - \mathbf{r}_j). \tag{21.2}$$

Here U_{ij} is the interaction energy between nucleons i and j, which is not necessarily the same for all pairs of particles, and which may also depend on the nucleon spins and even their momenta. This Hamiltonian can be expressed in terms of the co-ordinates $\mathbf{R}(X, Y, Z)$ of the centre of mass, and the co-ordinates $\mathbf{r}'_i$ of the particles relative to the centre of mass.

$$M\mathbf{R} = \sum_1^A m_i \mathbf{r}_i, \qquad M = \sum_1^A m_i, \qquad \mathbf{r}'_i = \mathbf{r}_i - \mathbf{R}.$$

$\mathbf{R}$ and the $\mathbf{r}'_i$ $(i = 2, \ldots, A)$ are independent co-ordinates, and $\mathbf{r}'_1$ is a dependent co-ordinate given by

$$m_1 \mathbf{r}'_1 = -\sum_2^A m_i \mathbf{r}'_i.$$

Then (21.2) becomes

$$\mathscr{H} = \mathscr{H}_{\text{cm}} + \mathscr{H}_{\text{int}}, \tag{21.3}$$

$$\mathscr{H}_{\text{cm}} = -\frac{\hbar^2}{2M}\nabla^2_{\text{cm}} = -\frac{\hbar^2}{2M}\left(\frac{\partial^2}{\partial X^2} + \frac{\partial^2}{\partial Y^2} + \frac{\partial^2}{\partial Z^2}\right),$$

$$\mathscr{H}_{\text{int}} = -\hbar^2 \sum_2^A \frac{1}{2m_i}(\nabla'_i)^2 + \frac{\hbar^2}{2M}\mathbf{G}^2 + \tfrac{1}{2}\sum U_{ij}(\mathbf{r}'_i - \mathbf{r}'_j),$$

$$\nabla'_i = \left(\frac{\partial}{\partial x'_i}, \frac{\partial}{\partial y'_i}, \frac{\partial}{\partial z'_i}\right), \quad \mathbf{G} = \sum_2^A \nabla'_i.$$

Here $\mathscr{H}_{\text{cm}}$ determines the motion of the centre of mass and $\mathscr{H}_{\text{int}}$ the internal motion of the nucleons relative to the centre of mass. There are no "cross terms" involving both $\mathbf{R}$ and the $\mathbf{r}'_i$, so that in the absence of external potentials the eigenfunctions of $\mathscr{H}$ can be written rigorously in the form

$$\Psi = \psi_{\text{cm}}(\mathbf{R})\phi_{\text{int}}(\mathbf{r}'_2, \ldots \mathbf{r}'_A), \tag{21.4}$$

where ϕ_{int} specifies the internal state of the nucleus. Now $\mathscr{H}_{\text{int}}$ is invariant under rotations of the $\mathbf{r}'_i$, so that the eigenfunctions ϕ_{int} can be sorted into degenerate sets transforming according to $D^{(I)}$. This defines the nuclear spin I (nuclear angular momentum quantum number) already introduced. In addition the nuclear interactions U_{ij} are known to be invariant under inversion to a very good approximation (§ 29), so that we can associate a definite parity w with the set of states belonging to each energy level (§ 15).

We next suppose that the nucleons are moving in an external electric potential $V(\mathbf{r})$, and add

$$\sum_1^A e_i V(\mathbf{r}_i) = eZV(\mathbf{R}) + \left[\sum_1^A e_i V(\mathbf{r}_i) - eZV(\mathbf{R})\right] \tag{21.5}$$

to the Hamiltonian, where eZ is the total nuclear charge. The first term gets included in $\mathscr{H}_{\text{cm}}$ and determines the motion of the centre of mass, whereas the bracketed term in (21.5) is a cross term that mixes different states of the form (21.4). However, it is extremely small, of the order of 10^{-2} electron volts, compared with the excitation energy of the levels of $\mathscr{H}_{\text{int}}$ which is of the order of 10^6 electron volts. Thus the amount of mixing of levels with different I and w is negligible, and for almost all practical purposes the wave function for the whole nucleus can be written as a single function of the form (21.4). Expressed in physical terms, the external field does not influence the internal motions of the nucleons appreciably, so that we can regard the nucleus as a rigid thing with fixed angular momentum and fixed magnetic and electric moments. These properties are unaffected by the surroundings, in particular by the motion of the atom's electrons, although the nucleus can of course rotate as a whole.

The absence of an electric dipole moment. Higher moments

Classically the electric dipole moment of the nucleus relative to the centre of mass is given by the operator

$$\mathbf{D}_{\text{class}} = \sum_{1}^{A} e_i \mathbf{r}'_i, \tag{21.6}$$

where e_i ($= e$ or 0) is the charge of the ith nucleon. Let ϕ be any state in the vector space spanned by the ground state wave functions $\phi_{\text{int}}(I, M_I)$. Then from the theorem (13.8c) we have that

$$\boxed{\int \phi^* \mathbf{D}_{\text{class}} \phi \, d\tau = 0,}$$

because $\phi^*\phi$ has even parity $w^2 = +1$, and $\mathbf{D}_{\text{class}}$ has odd parity, i.e. changes sign under $\mathbf{r}_i \Rightarrow -\mathbf{r}_i$. The expectation value of the classical orbital dipole moment $\mathbf{D}_{\text{class}}$ is thus always zero. However, there still remains the possibility that the nucleons have an intrinsic dipole moment, in the same way as electrons and nucleons have intrinsic spin angular momenta and spin magnetic moments. Suppose this is represented by an operator $\mathbf{D}_{\text{spin}}$. For us to interpret this quantity as an electric dipole moment, it must manifest itself through the occurrence in the Hamiltonian of an interaction term

$$-\mathscr{E} \cdot (\mathbf{D}_{\text{class}} + \mathbf{D}_{\text{spin}})$$

with an external electric field $\mathscr{E}$. Now all known electromagnetic interactions are invariant under inversion, so that $\mathbf{D}_{\text{spin}}$ must behave in the same way as $\mathbf{D}_{\text{class}}$ under inversion. Consequently the expectation value of $\mathbf{D}_{\text{spin}}$ is also zero, and nuclei exhibit no permanent electric dipole moments.†

In the same way it follows that nuclei can have no electric multipole moments of order 2^l where l is odd, and no magnetic multipole

† The same result applies of course to any system interacting with parity-conserving forces. In this strict sense the ground state of an asymmetrical molecule does not have a permanent dipole moment, and it corresponds to a spherically symmetric wave function with the molecule pointing in no particular direction in space. However a molecule has a large number of rotational levels separated from the ground state by energies normally small compared with kT (§§ 23 and 24). This gives the system certain properties at ordinary temperatures which are analogous to those of a classical dipole moment.

moments of order 2^l where l is even. The other moments are allowed, e.g. magnetic dipole, electric quadrupole, etc. (For definitions of higher order multipole moments, see Stratton (1941) and Schwartz (1955).) A nuclear 2^l-pole moment must therefore be an electric one if l is even and a magnetic one if l is odd. Hence the words "electric" and "magnetic" are usually omitted without causing any confusion.

It should be noted how the proof that the nuclei have no dipole moments depends on the fact that the internal wave functions ϕ have a definite parity ± 1, which in turn followed from the invariance under inversion of the nuclear interactions U_{ij}. Real nuclear interactions can be divided into three classes; the strong interactions of the order of 10 MeV. which are responsible for nuclear binding and most other properties, the electromagnetic interactions of the order of 1 MeV., and the much weaker interactions of the order of 1 eV. which are responsible for beta decay (§ 29). Of these, the last is not invariant under inversion, so that nuclei are expected to have some electric dipole moment which, however, is too small to detect experimentally (Ramsey 1953, Lee and Yang 1956).

Limitation of multipole moments by I

Let $M^{(l)}$ be an electric or magnetic multipole moment of order 2^l. This can be expressed as a linear combination of $2l+1$ irreducible multipole moments $M_m{}^{(l)}$ which transform according to the irreducible representation $D^{(l)}$ of the rotation group (Stratton 1941, p. 182). The products $M_m{}^{(l)}\phi_{\text{int}}(I,M_I)$ transform according to $\sum D^{(j)}$ where (§ 9)

$$|I-l| \leqslant j \leqslant I+l.$$

Hence by the fundamental theorem of § 13, the ground state matrix elements $\langle I, M'_I|M_m{}^{(l)}|I, M_I\rangle$ are all zero unless

$$|I-l| \leqslant I \leqslant I+l$$

$$\boxed{\text{i.e. unless } l \leqslant 2I.}$$

Thus a single nucleon or any $I=\frac{1}{2}$ nucleus can only have a magnetic (dipole) moment, and no quadrupole or higher moments. Similarly, an $I=0$ nucleus has only a zero order moment, namely its charge Ze.

The magnetic dipole interaction

The operator of the nuclear magnetic moment due to orbital motion can be written down classically,

$$\boldsymbol{\mu}_{n,\text{orb}} = \sum_1^A (e_i/2mc)\mathbf{r}'_i \wedge \mathbf{p}'_i,$$

where $\mathbf{p}'_i$ is the momentum of particle i relative to the centre of mass. Clearly $\boldsymbol{\mu}_{n,\text{orb}}$ transforms under rotations and inversion in the same way as the orbital angular momentum operator $\Sigma \mathbf{r}'_i \wedge \mathbf{p}'_i$, and this is still true if we include spin magnetic moment and spin angular momentum. (This latter fact was proved in detail in connection with the electron spin; see particularly below equation (11.18) and appendix F.) It follows from § 13 that the matrix elements of $\boldsymbol{\mu}_n$ and $\mathbf{I}$ are proportional within the vector space of the internal ground state wave functions $\phi_{\text{int}}(I, M_I)$,

$$\langle I, M_I|\boldsymbol{\mu}_n|I, M'_I\rangle = (\text{const})\langle I, M_I|\mathbf{I}|I, M'_I\rangle.$$

The proportionality constant is conventionally written as $\mu_n/(\hbar I)$, and since we are only concerned with the ground state of the internal Hamiltonian $\mathscr{H}_{\text{int}}$, we can write effectively

$$\boldsymbol{\mu}_n = \mu_n \frac{\mathbf{I}}{\hbar I}. \tag{21.1}$$

This proves that $\boldsymbol{\mu}_n$ and $\mathbf{I}$ are always parallel. Similarly $\mathbf{H}_{\text{elect}}$, the magnetic field at the nucleus due to the orbital motion and spin magnetic moments of the electrons, is proportional to their angular momentum $\mathbf{J}$, provided we restrict ourselves to considering only one particular (say the lowest) electronic energy level:

$$\mathbf{H}_{\text{elect}} = (\text{const})\mathbf{J}. \tag{21.7}$$

Now the magnetic interaction energy between the electrons and the nucleus is

$$\mathscr{H}_M = -\boldsymbol{\mu}_n \cdot \mathbf{H}_{\text{elect}}, \tag{21.8}$$

and after substituting (21.1) and (21.7) this becomes

$$\mathscr{H}_M = a\mathbf{I}\cdot\mathbf{J} \tag{21.9}$$

$$\text{where } a = -\frac{\mu_n}{\hbar I}\left\langle\left|\frac{\mathbf{H}_{\text{elect}}\cdot\mathbf{J}}{\mathbf{J}^2}\right|\right\rangle. \tag{21.10}$$

We now turn to consider the kind of energy levels given by an interaction Hamiltonian of the form (21.9). We defer to the next subsection the problem of calculating $\mathbf{H}_{\text{elect}}$ so that $\boldsymbol{\mu}_n$ can be determined from (21.10) once a has been measured experimentally. We consider only the set of low lying states.

$$\Psi(I, M_I, J, M_J) = \phi_{\text{nucl}}(I, M_I)\psi_{\text{elect}}(J, M_J)$$

in which the nucleus and the electrons are both in their respective ground states $\phi_{\text{nucl}}(I, M_I)$ and $\psi_{\text{elect}}(J, M_J)$. The interaction

$\mathscr{H}_M$ (21.9) is invariant under combined rotations of the electronic and nuclear co-ordinates together, so that the eigenstates of the whole atom can be designated by combined angular momentum quantum numbers F, M_F.

$$\Psi(I,J,F,M_F) = \sum_{M_I,M_J} (IJM_IM_J|FM_F)\Psi(I,M_I,J,M_J),$$
$$|I - J| \leqslant F \leqslant I + J. \qquad (21.11)$$

The relative energies $E(F)$ of these $(2F + 1)$-fold degenerate levels are obtained by applying the perturbation $\mathscr{H}_M$ (21.9) to the first order of approximation

$$E(F) = \langle IJFM_F|a\mathbf{I} \cdot \mathbf{J}|IJFM_F\rangle.$$

These matrix elements can be calculated by exactly the same procedure as was used in § 13 to calculate the fine structure splitting due to the spin-orbit coupling $\zeta\mathbf{L} \cdot \mathbf{S}$. If $\mathbf{F} = \mathbf{I} + \mathbf{J}$ is the total angular momentum of the atom, we have

$$2\mathbf{I} \cdot \mathbf{J} = \mathbf{F}^2 - \mathbf{I}^2 - \mathbf{J}^2 \qquad (21.12)$$

and hence

$$\boxed{E(F) = \tfrac{1}{2}a\hbar^2[F(F + 1) - I(I + 1) - J(J + 1)].} \qquad (21.13)$$

Usually it is the difference between successive levels that is measured by nuclear resonance techniques.

$$E(F) - E(F - 1) = a\hbar^2F.$$

Thus the intervals should bear simple numerical ratios to one another, and any deviation from these simple ratios indicates some omission in the theory, e.g. the existence of a quadrupole moment and second order perturbation effects from $\mathscr{H}_M$.

If the atom is in an externally applied magnetic field $\mathbf{H}_0$, the Hamiltonian must contain also interaction terms $-\boldsymbol{\mu} \cdot \mathbf{H}_0$ for both the electronic and the nuclear magnetic moments. For the same reason that $\boldsymbol{\mu}_n$ and $\mathbf{I}$ are parallel, the electronic magnetic moment $\boldsymbol{\mu}_e$ is parallel to $\mathbf{J}$. From problem 13.6, it is given by

$$\boldsymbol{\mu}_e = -g_L\beta\frac{\mathbf{J}}{\hbar}$$

where g_L is the Landé g-factor

$$g_L = 1 + \frac{J(J + 1) + S(S + 1) - L(L + 1)}{2J(J + 1)}$$

and β the electronic Bohr magneton. The Hamiltonian then becomes

$$\mathscr{H} = a\mathbf{I} \cdot \mathbf{J} + (g_L\beta/\hbar)\mathbf{J} \cdot \mathbf{H}_0 - \boldsymbol{\mu}_n \cdot \mathbf{H}_0. \tag{21.14}$$

Here $\boldsymbol{\mu}_n$ is of the order of the nuclear Bohr magneton $e\hbar/2m_nc$ which is about 2000 times smaller than the electronic Bohr magneton because of the larger mass m_n of a nucleon. We shall therefore neglect the last term in (21.14). We first consider the case of a very weak external field (< 100 gauss) so that $\beta H_0 \ll a\hbar^2$. To a first order of approximation the nuclear and electronic angular momenta are still coupled as described by the total quantum number F, and from (21.13) the energy levels are

$$E(F, M_F) = E(F) + (g_L\beta/\hbar)\langle IJFM_F|J \cdot \mathrm{H}_0|IJFM_F\rangle. \tag{21.15}$$

Within a manifold of states with the same F, the matrix elements of $\mathbf{J}$ must be proportional to those of $\mathbf{F}$ because they are both operators transforming according to $D^{(1)}$. We can therefore write effectively [see (13.20)]

$$\mathbf{J} = k\mathbf{F}, \tag{21.16}$$

where the constant k is evaluated from the identity

$$\begin{aligned} \mathbf{I}^2 = (\mathbf{F} - \mathbf{J})^2 &= \mathbf{F}^2 + \mathbf{J}^2 - 2\mathbf{F} \cdot \mathbf{J} \\ &= \mathbf{F}^2 + \mathbf{J}^2 - 2k\mathbf{F}^2. \end{aligned}$$

If we choose the z-axis along the direction of $\mathbf{H}_0$, the energy levels (21.15) finally become

$$\boxed{\begin{aligned} E(F, M_F) = {}& \tfrac{1}{2}a\hbar^2[F(F+1) - I(I+1) - J(J+1)] + \\ & + g_L\beta H_0 \frac{F(F+1) + J(J+1) - I(I+1)}{2F(F+1)} M_F. \end{aligned}} \tag{21.17}$$

Now if the field is rather larger (of the order of 10^4 gauss), we have that $\beta H_0 \gg a\hbar^2$, so that the electronic and nuclear angular momenta are decoupled and the states are designated to a first approximation by the quantum numbers I, M_I, J, M_J. Also βH_0 is less than the fine structure separation, so that electronic energy levels with different J do not get mixed and the approximation (21.9) is still valid. We have

$$\mathbf{I} \cdot \mathbf{J} = \tfrac{1}{2}(I_+J_- + I_-J_+) + I_zJ_z,$$

and only the last term gives any diagonal matrix elements in the M_I, M_J representation. The energy levels of (21.14) therefore are

$$E(M_I, M_J) = a\hbar^2 M_I M_J + g_L \beta H_0 M_J. \tag{21.18}$$

Fig. 12 shows schematically the variation of the energy levels with H_0 as they change from the low field region in which the approximation (21.17) is valid, to the approximation of (21.18).

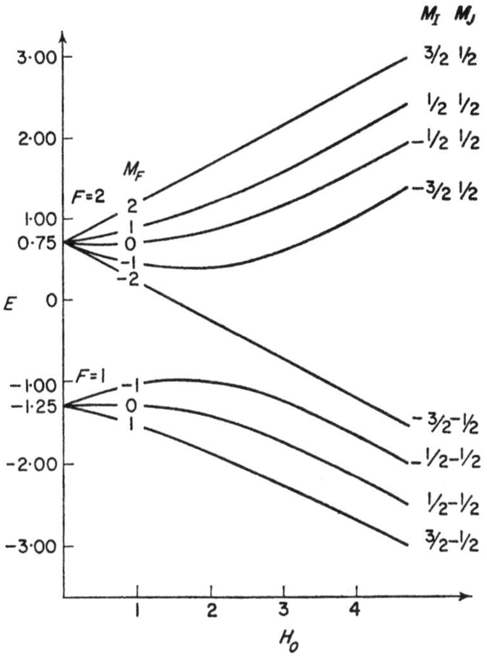

FIG. 12. Variation of the energy levels in a magnetic field for the case $I = 3/2$, $J = 1/2$, $a > 0$. The energy is plotted in units of $a\hbar^2$ and the field in units of $a\hbar^2/g_L\beta$.

Calculation of the interaction constant a

We shall now calculate the constant a in the magnetic interaction $\mathscr{H}_M$ (21.9). We regard the nucleus as a magnetized sphere with total magnetic moment $\boldsymbol{\mu}_n$, and we have to calculate separately

(i) the interaction with the main part of the electron distribution which lies outside the nucleus, and

(ii) the interaction with that part of the electron distribution which penetrates inside the nucleus.

We consider part (i) first. The orbital motion of the electrons produces a magnetic field $\mathbf{H}_{\text{orb}}$ at the nucleus, given by

$$\mathbf{H}_{\text{orb}} = \frac{-e}{c}\sum_i \frac{\mathbf{v}_i \wedge (-\mathbf{r}_i)}{r_i^3} = \frac{e}{mc}\sum \frac{\mathbf{p}_i \wedge \mathbf{r}_i}{r_i^3} = -2\beta \sum \frac{\mathbf{l}_i}{\hbar r_i^3}.$$

The interaction energy of the nucleus with the electronic spins is

$$-\mathbf{H}_{\text{spin}}\cdot\boldsymbol{\mu}_n = \sum_i \frac{-2\beta}{\hbar}\left[\frac{\mathbf{s}_i\cdot\boldsymbol{\mu}_n}{r_i^3} - \frac{3(\mathbf{s}_i\cdot\mathbf{r}_i)(\boldsymbol{\mu}_n\cdot\mathbf{r}_i)}{r_i^5}\right].$$

Putting $\mathbf{H}_{\text{elect}} = \mathbf{H}_{\text{orb}} + \mathbf{H}_{\text{spin}}$, we obtain from (21.10)

$$\boxed{a\hbar^2 = \frac{2\mu_n\beta}{IJ(J+1)}\left\langle\left|\frac{\mathbf{J}}{\hbar^2}\cdot\sum_i\left\{\frac{\mathbf{l}_i - \mathbf{s}_i}{r_i^3} + \frac{3(\mathbf{s}_i\cdot\mathbf{r}_i)\mathbf{r}_i}{r_i^5}\right\}\right|\right\rangle.} \tag{21.19}$$

The matrix element must be evaluated for the particular electronic level considered. Closed shells make no contribution because the $\mathbf{l}_i$ and $\mathbf{s}_i$ add up to zero. For a single electron outside closed shells (21.19) reduces to

$$\boxed{a\hbar^2 = \frac{2\mu_n\beta}{I}\frac{L(L+1)}{J(J+1)}\left\langle\left|\frac{1}{r^3}\right|\right\rangle,} \tag{21.20}$$

as the reader should have no difficulty in showing by following the hints in problem 2.15.

We now turn to the contribution (ii) above to the interaction constant a. Near the origin, a single electron wave function behaves like (const)r^l, so that only an s-electron has any appreciable probability of being inside the nucleus. Furthermore its orbital motion is spherically symmetrical so that we need only consider the interaction of its spin magnetic moment $-(2\beta/\hbar)\mathbf{s}$ with the nucleus. The nucleus is idealized as a small sphere of volume V with a uniform magnetization

$$\mathbf{M} = \boldsymbol{\mu}_n/V.$$

The magnetic flux density $\mathbf{B}_n$ inside the nucleus is then

$$\mathbf{B}_n = \frac{8\pi}{3}\mathbf{M} = \frac{8\pi\boldsymbol{\mu}_n}{3V}. \tag{21.21}$$

It does not matter for the purposes of the present calculation whether the magnetic moment is thought of as being due to circulating currents corresponding to the orbital motion of the nucleons, or due to a permanent magnet corresponding to the intrinsic moments of the nucleons, or indeed as due to a mixture of both. The flux density $\mathbf{B}_n$ is given by (21.21) in all cases.† The potential energy of a magnetic moment $\boldsymbol{\mu}_e$ introduced into this field is‡

$$-\boldsymbol{\mu}_e\cdot\mathbf{B}_n. \tag{21.22}$$

The interaction energy is therefore

$$\begin{aligned}\mathscr{H}_M &= -\mathbf{B}_n\cdot\sum_i(-2\beta\mathbf{s}_i/\hbar)\ \text{(probability of electron being inside the nucleus)}\\ &= -\left(\frac{8\pi}{3V}\boldsymbol{\mu}_n\right)\cdot\sum_i(-2\beta\mathbf{s}_i/\hbar)(|\psi_i(0)|^2V)\end{aligned}$$

i.e.

$$\boxed{\mathscr{H}_M = \frac{16\pi}{3}\frac{\mu_n\beta}{I}\frac{\mathbf{I}\cdot\sum\mathbf{s}_i|\psi_i(0)|^2}{\hbar^2}.} \tag{21.23}$$

This can easily be expressed in the form

$$\mathscr{H}_M = a\mathbf{I}\cdot\mathbf{J}$$

using the trick of (21.10). A pair of electrons differing only in having opposite spins give no resultant contribution to (21.23). However, for a single s-electron outside closed shells, we have $\mathbf{s}_i = \mathbf{J}$ and obtain

$$\boxed{a\hbar^2 = \frac{16\pi}{3}\frac{\mu_n\beta}{I}|\psi(0)|^2.} \tag{21.24}$$

This is the most important case for practical purposes.

Quadrupole interaction

We start by writing down in full the Coulomb interaction between the protons in the nucleus and the electrons. Let $\mathbf{r}_{n\alpha}$ and $\mathbf{r}_{e\beta}$ be the positions of the αth proton in the nucleus and the βth electron.

† It is the magnetic field intensity $\mathbf{H}$ that is not the same for the different cases. We have $\mathbf{H} = (8\pi/3)\mathbf{M}$ if the magnetic moment is produced by a current distribution in vacuum, and $\mathbf{H} = -(4\pi/3)\mathbf{M}$ if the sphere is a permanent magnet. However, the magnitude of $\mathbf{H}$ is irrelevant to the calculation.

‡ Formula (21.22) is completely equivalent to (21.9). Firstly the interaction energy between two magnetic moments $\boldsymbol{\mu}_e$ and $\boldsymbol{\mu}_n$ can always be written in the two ways $W = -\boldsymbol{\mu}_e\cdot\mathbf{B}_n = -\boldsymbol{\mu}_n\cdot\mathbf{B}_e$ depending on which one is thought of as being brought from infinity up to the other one. Secondly in (21.9) and throughout this book we should use $\mathbf{B}$ instead of $\mathbf{H}$ (Stratton 1941, p. 130). However, since it only makes a difference in the present paragraph, convention has been followed and we have everywhere else written $\mathbf{H}$ although we mean the magnetic flux density $\mathbf{B}$ measured in gauss.

If we neglect the electron density actually inside the nucleus, we have $\mathbf{r}_e$ always greater than $\mathbf{r}_n$, so that the Coulomb interaction energy can be expanded in terms of Legendre polynomials $P_n(\cos\omega)$ (Stratton 1941, p. 172).

$$\begin{aligned}\mathscr{H}_{\text{Coulomb}} &= \sum_{\alpha\beta} \frac{-e^2}{|\mathbf{r}_{n\alpha} - \mathbf{r}_{e\beta}|} \\ &= -e^2 \sum_{\alpha,\beta} \left[\frac{1}{r_{e\beta}} + \frac{r_{n\alpha}}{r_{e\beta}^2} P_1(\cos\omega_{\alpha\beta}) + \right. \\ &\qquad \left. + \frac{r_{n\alpha}^2}{r_{e\beta}^3} P_2(\cos\omega_{\alpha\beta}) + \ldots\right] \qquad (21.25)\end{aligned}$$

where $\omega_{\alpha\beta}$ is the angle between $\mathbf{r}_{n\alpha}$ and $\mathbf{r}_{e\beta}$,

$$r_{n\alpha}r_{e\beta}\cos\omega_{\alpha\beta} = x_{n\alpha}x_{e\beta} + y_{n\alpha}y_{e\beta} + z_{n\alpha}z_{e\beta}, \qquad (21.26)$$
$$P_1(\cos\omega) = \cos\omega, \qquad P_2(\cos\omega) = \tfrac{1}{2}(3\cos^2\omega - 1).$$

In this expansion, the first term gives the potential $-e^2Z/r_e$ of the nuclear charge, and this has already been included in the Hamiltonian $\mathscr{H}_{\text{orb}}$ of Chapter II. The next term can be put equal to zero, because it leads to expressions involving the components $e\Sigma z_{n\alpha}$, etc., of the nuclear electric dipole moment which we have shown to be zero. The third term in the expansion (21.25) is called the electric quadrupole interaction $\mathscr{H}_Q$. Using (21.26), we may write it as

$$\mathscr{H}_Q = -\tfrac{1}{6}\sum_{ij} Q_{ij}(\nabla\mathscr{E})_{ij}, \qquad (21.27)$$
$$Q_{ij} = e\sum_{\alpha}[\tfrac{3}{2}(r_{n\alpha i}r_{n\alpha j} + r_{n\alpha j}r_{n\alpha i}) - \delta_{ij}r_{n\alpha}^2],$$
$$(\nabla\mathscr{E})_{ij} = e\sum_{\beta}\frac{1}{r_{e\beta}^5}[\tfrac{3}{2}(r_{e\beta i}r_{e\beta j} + r_{e\beta j}r_{e\beta i}) - \delta_{ij}r_{e\beta}^2],$$

where the r_i are the components x, y, z of $\mathbf{r}$. Here the Q_{ij} are the quadrupole moment operators of the nuclear charge distribution, and the $(\nabla\mathscr{E})_{ij}$ are the operator components of the electric field gradient at the nucleus due to the external electrons.

The expression (21.27) can be written in a much more convenient form. The $r_{n\alpha i}$ transform under rotations in the same way as the component I_i of the angular momentum $\mathbf{I}$, and hence the Q_{ij} transform in the same way as

$$\tfrac{3}{2}(I_iI_j + I_jI_i) - \delta_{ij}\mathbf{I}^2. \qquad (21.28)$$

Furthermore, these expressions form a symmetric second rank tensor with zero sum of diagonal terms, so that they transform according to the irreducible representation $D^{(2)}$ of the rotation group (problem 9.8). Thus the fundamental theorem (13.8) can be applied, and inside a set of states with the same I the matrix elements of Q_{ij} and (21.28) are proportional, so that we write effectively

$$Q_{ij} = C[\tfrac{3}{2}(I_iI_j + I_jI_i) - \delta_{ij}\mathbf{I}^2], \tag{21.29}$$

where the constant C does not depend on i, j. C is customarily expressed in terms of another constant Q called "the" quadrupole moment of the nucleus and defined by

$$\begin{aligned} eQ &= \langle I, M_I = I|Q_{zz}|I, M_I = I\rangle \\ &= \int \rho_n(M_I = I)\,[3z_n{}^2 - r_n{}^2]\,\mathrm{d}\tau_n \\ &= C\langle I, M_I = I|3I_z{}^2 - \mathbf{I}^2|I, M_I = I\rangle \\ &= C\hbar^2[3I^2 - I(I+1)] = \hbar^2CI(2I-1), \end{aligned} \tag{21.30}$$

where $\rho_n(M_I = I)$ is the nuclear charge distribution for the state $M_I = I$. Similarly $(\nabla\mathscr{E})_{ij}$ can be written

$$(\nabla\mathscr{E})_{ij} = \frac{-eq_J}{\hbar^2J(2J-1)}[\tfrac{3}{2}(J_iJ_j + J_jJ_i) - \delta_{ij}\mathbf{J}^2] \tag{21.31}$$

in terms of the constant

$$q_J = -\frac{1}{e}\int \rho_e(M_J = J)\left[\frac{3z_e{}^2 - r_e{}^2}{r_e{}^5}\right]\mathrm{d}\tau_e. \tag{21.32}$$

Therefore

$$\begin{aligned} \mathscr{H}_Q = \frac{e^2q_JQ}{6I(2I-1)J(2J-1)\hbar^4}\sum_{ij}&[\tfrac{3}{2}(I_iI_j + I_jI_i) - \delta_{ij}\mathbf{I}^2] \\ &\times[\tfrac{3}{2}(J_iJ_j + J_jJ_i) - \delta_{ij}\mathbf{J}^2]. \end{aligned}$$

This can be simplified further. Since $\mathscr{H}_Q$ is invariant under simultaneous rotation of nuclear and electron co-ordinates, it must be expressible in terms of the only available invariant combinations $\mathbf{I}\cdot\mathbf{J}$ and $\mathbf{I}^2\mathbf{J}^2$:

$$\mathscr{H}_Q = \alpha(\mathbf{I}\cdot\mathbf{J})^2 + \beta(\mathbf{I}\cdot\mathbf{J}) + \gamma\mathbf{I}^2\mathbf{J}^2.$$

The coefficients α, β, γ can be evaluated by putting $i\hbar J_z = J_xJ_y - J_yJ_x$, etc., in the $\beta\mathbf{I}\cdot\mathbf{J}$ term, imagining everything multiplied out, and comparing coefficients of $I_x{}^2J_x{}^2$, $I_x{}^2J_y{}^2$ and $I_xI_yJ_yJ_x$. We obtain (Ramsey 1953)

$$\boxed{\mathscr{H}_Q = \frac{3e^2q_JQ}{8I(2I-1)J(2J-1)}\left[\frac{4(\mathbf{I}\cdot\mathbf{J})^2}{\hbar^4} + \frac{2\mathbf{I}\cdot\mathbf{J}}{\hbar^2} - \frac{4\mathbf{I}^2\mathbf{J}^2}{3\hbar^4}\right].} \tag{21.33}$$

The contribution of $\mathscr{H}_Q$ to the energy levels can be evaluated easily in the limit of very small or zero magnetic field where F and M_F are good quantum numbers. We have from (21.12)

$$\langle IJFM_F|2\mathbf{I}\cdot\mathbf{J}|IJFM_F\rangle$$
$$= \hbar^2K = \hbar^2[F(F+1) - I(I+1) - J(J+1)], \quad (21.34)$$

and hence

$$E_Q = \langle FM_F|\mathscr{H}_Q|FM_F\rangle$$
$$= \frac{3e^2q_JQ}{8I(2I-1)J(2J-1)}[K(K+1) - \tfrac{4}{3}I(I+1)J(J+1)]. \quad (21.35)$$

Alternative treatment of quadrupole interaction

The following, more elegant, derivation of (21.35) illustrates the method developed by Schwartz (1955) for handling multipole interactions of any order using Racah coefficients. As shown below, $\mathscr{H}_Q$ can be expressed in the form

$$\mathscr{H}_Q = -\tfrac{1}{4}\sum_m (-1)^m Q_m(\nabla\mathscr{E})_{-m}, \quad (21.36)$$

where

$$Q_0 = Q_{zz}, \qquad Q_{\pm 1} = \sqrt{\tfrac{2}{3}}(Q_{xz} \pm iQ_{yz}),$$
$$Q_{\pm 2} = \sqrt{\tfrac{1}{6}}(Q_{xx} - Q_{yy} \pm 2iQ_{xy}),$$

and the $(\nabla\mathscr{E})_m$ are defined similarly with $(\nabla\mathscr{E})_0 = (\nabla\mathscr{E})_{zz}$. By referring back to the definition (21.27) of the Q_{zz}, etc., it is seen that the five quantities Q_m or $(\nabla\mathscr{E})_m$ are related to one another just like the spherical harmonics $Y_{2,m}$ (see below equation (8.21)). The Q_m and the $(\nabla\mathscr{E})_m$ therefore transform under rotations as standard base vectors according to the representation $D^{(2)}$. The result (21.36) is now proved as follows. From (21.25), $\mathscr{H}_Q$ transforms irreducibly according to $D^{(2)}$ under rotations of the nuclear co-ordinates alone, keeping $\mathbf{r}_{e\beta}$ fixed, say along the z-axis. Thus $\mathscr{H}_Q$ is a linear function of the Q_m, and similarly of the $(\nabla\mathscr{E})_m$. Also $\mathscr{H}_Q$ is invariant under rotations of nuclear and electronic co-ordinates together, and so is (21.36) (see the discussion below 20.40)). The products $Q_m(\nabla\mathscr{E})_{m'}$ transform according to $D^{(2)} \times D^{(2)}$, and since this includes the representation $D^{(0)}$ only once, $\mathscr{H}_Q$ must be proportional to (21.36). Finally the numerical constant can be written down by inspection from the coefficient of $z_n{}^2z_e{}^2$.

In the absence of a magnetic field, the contribution of $\mathscr{H}_Q$ (21.36) to the energy levels can be evaluated using formula (20.41) for the value of the matrix element.

$$\begin{aligned} E_Q(F) &= \langle IJFM_F| -\tfrac{1}{4} \sum (-1)^m Q_m (\nabla \mathscr{E})_{-m} | IJFM_F \rangle \\ &= -\tfrac{1}{4}(-1)^{I+J-F} \langle I \| Q_m \| I \rangle \langle J \| (\nabla \mathscr{E})_m \| J \rangle W(IJIJF2). \end{aligned} \tag{21.37}$$

The reduced matrix element $\langle I\|Q_m\|I\rangle$ can be expressed in terms of the quadrupole constant Q by using formula (20.34). From (21.30) we have

$$\begin{aligned} eQ &= \langle I, M_I = I | Q_0 | I, M_I = I \rangle \\ &= (-1)^{2I+2} \frac{\langle I\|Q_m\|I\rangle}{\sqrt{5}} (I, I, -I, I|2, 0). \end{aligned}$$

Here the last bracket is a Wigner coefficient which is evaluated from formula (20.26).

$$(I, I, -I, I|2, 0) = (-1)^{2I} \frac{\sqrt{5}(2I)!}{[(2I-2)!(2I+3)!]^{1/2}}.$$

Therefore

$$\langle I\|Q_m\|I\rangle = eQ \frac{[(2I-2)!(2I+3)!]^{1/2}}{(2I)!}, \tag{21.38}$$

and similarly

$$\langle J\|(\nabla \mathscr{E})_m\|J\rangle = -eq_J \frac{[(2J-2)!(2J+3)!]^{1/2}}{(2J)!}. \tag{21.39}$$

Making use of the symmetry properties of the Racah coefficients (problem 20.7), we have

$$W(IJIJF2) = W(IIJJ2F),$$

and in the latter form the coefficient can be looked up in the tables given by Biedenharn *et al.* (1952).

$$\begin{aligned} W(IIJJ2F) = (-1)^{F-I-J} \left[\frac{(2I-2)!(2J-2)!}{(2I+3)!(2J+3)!}\right]^{1/2} \times \\ \times 6[K(K+1) - \tfrac{4}{3}I(I+1)J(J+1)], \end{aligned} \tag{21.40}$$

where K is defined by (21.34). Now substituting (21.38), (21.39) and (21.40) into (21.37), we just reobtain the result (21.35) for E_Q.

References

The subject of hyperfine interactions is covered in the books of Kopfermann (1940), Casimir (1936) and Ramsey (1953), where further references may be found. Schwartz (1955) has developed the general theory in a form applicable to multipole moments of all orders. The theory of hyperfine structure of paramagnetic resonance spectra in crystals has been given by Abragam and Pryce (1951a), and the subject reviewed by Bleaney and Stevens (1953) and Nierenberg (1958).

Summary

It has been shown that the nucleus may be considered as a rigid system with a fixed amount of angular momentum and various magnetic and electric multipole moments. The magnetic dipole moment is always parallel to the angular momentum. Not all orders of multipole moments exist, the number of non-zero ones being limited by parity considerations and by the nuclear spin I. In particular there is no electric dipole moment. The interactions of the magnetic dipole moment and the electric quadrupole moment with the atomic electrons have been discussed in detail.

Problems

21.1 Show from (21.5) that a strong external electric field can produce an *induced* electric dipole moment in the nucleus. This polarization of the nucleus occurs for instance when the atom's electrons penetrate the nucleus, and it can lead to an appreciable shift of the energy levels in heavy nuclei, the magnitude of the shift depending on the particular isotope. Make an order of magnitude estimate of this effect (Kopfermann 1940).

21.2 Draw an energy level diagram analogous to Fig. 12 for the cases $J = 1, I = \frac{1}{2}$, and $J = \frac{3}{2}, I = \frac{1}{2}$.

21.3 Calculate $\langle IJFM_F|-\mu_n \cdot \mathbf{H}_0|IJFM_F\rangle$ assuming $\mathbf{H}_0$ to be along the z-axis.

21.4 Calculate the selection rules for magnetic dipole transitions between the levels of Fig. 12, (i) for low magnetic field and (ii) for high magnetic fields. Neglect those transitions which are induced by the coupling with the nuclear moment because their intensity is only about $(1/2000)^2$ times that of the electronically induced transitions.

21.5 Deduce the value (21.20) of the coupling constant a for a single electron from equation (21.19). Hints: put $\mathbf{J} = \mathbf{l} + \mathbf{s}$; $\mathbf{l} \cdot \mathbf{r} = \mathbf{r} \wedge \mathbf{p} \cdot \mathbf{r} = 0$; use the relations of problem 11.1 valid for a single spin only.

21.6 With the notation of (21.27), (21.28), why is it *not* correct to write

$$\langle I, M_I|r_ir_j|I, M'_I\rangle = C\langle I, M_I|I_iI_j|I, M'_I\rangle$$

where C is independent of i and j?

21.7 Someone tried to calculate the constant k of (21.16) by the following procedure. Why is it wrong? He put $\mathbf{I} = \mathbf{F} - \mathbf{J} = (1 - k)\mathbf{F}$, and then $\mathbf{F}^2 - \mathbf{I}^2 - \mathbf{J}^2 = 2\mathbf{I} \cdot \mathbf{J} = 2k(1 - k)\mathbf{F}^2$,

$$\text{i.e.} \qquad 2k(1 - k) = \frac{F(F+1) - I(I+1) - J(J+1)}{F(F+1)},$$

which he solved as a quadratic equation for k.

21.8 In deriving (21.35) from (21.34), one has to use the fact that

$$\langle F, M_F|(\mathbf{I} \cdot \mathbf{J})^2|F, M_F\rangle = \{\langle F, M_F|\mathbf{I} \cdot \mathbf{J}|F, M_F\rangle\}^2.$$

Prove this.

21.9 Calculate $\langle IJM_IM_J|\mathscr{H}_Q|IJM_IM_J\rangle$ from either (21.33) or (21.36).

21.10 Use the method of equations (21.36) to (21.40) to calculate the contributions of the magnetic dipole moment and the octupole moment to the energy levels in zero external magnetic field (Schwartz 1955).

21.11 A paramagnetic ion with a single $3d$ electron is in a crystalline electric field, such that the wave function of the lowest (orbitally non-degenerate) level has the symmetry of $x^2 - y^2$. Calculate the hyperfine interaction, and show that it can be expressed in the spin Hamiltonian form (§ 18) $I.A.S$, where A is the tensor

$$A = \begin{bmatrix} -b & . & . \\ . & -b & . \\ . & . & 2b \end{bmatrix}.$$

Calculate the value of b in terms of the expectation value $\langle|1/r^3|\rangle$ for the $3d$ orbit. Note that the orbital momentum is completely quenched.

21.12 The Mn^{++} ion has the configuration . . . $(3s)^2(3p)^6(3d)^5$, the lowest term being ^6S. Is there any s-electron contribution (21.23) to the hyperfine structure (i) in the independent-electron-orbitals, Hartree–Fock approximation, and (ii) with a correct many-electron wave function? How does the situation differ in a term with $L \neq 0$? (Pratt, 1956; Abragam *et al.*, 1955; Heine, 1957).

Chapter V

THE STRUCTURE AND VIBRATIONS OF MOLECULES

22. Valence Bond Orbitals and Molecular Orbitals

Electron wave functions

The purpose of this section is to discuss the electronic wave functions of molecules. We shall assume for the present that the nuclei of the atoms in a molecule are stationary, fixed to their equilibrium positions. The Hamiltonian for the electrons moving around the molecule is

$$\mathscr{H} = -\frac{\hbar^2}{2m}\sum_i \nabla_i^2 - \sum_i V_n(\mathbf{r}_i) + \sum_{i<j}\sum \frac{e^2}{r_{ij}}, \qquad (22.1)$$

where the first term is the kinetic energy, $-V_n(\mathbf{r})$ is the attractive potential of the stationary nuclei, and the last term is the mutual Coulomb repulsion between the electrons. As in § 12, the wave function Ψ has to be antisymmetric under the interchange of electron co-ordinates $\mathbf{r}_i$, and this is achieved automatically by writing Ψ as a determinant or as a sum of determinants. As in the case of an atom (§ 10), an exact wave function has to be represented by an infinite series of determinants because of "configurational interaction", but in practice the first one or two determinants serve as a reasonable approximation for many purposes. The elements of these determinants are one-electron wave functions $\psi(\mathbf{r})u_\pm$ called *orbitals*. As we shall see, these are usually chosen to belong to one of three types, atomic orbitals, valence bond orbitals and molecular orbitals, depending on the circumstances. In each case one wants orbitals with definite transformation properties, so that one can use these properties in calculating matrix elements and energy values, etc.

The hydrogen molecule

In order to introduce the different types of orbital, we start by considering the hydrogen molecule. We take the two protons to be at the positions

$$\mathbf{R}_1 = (0, 0, -\tfrac{1}{2}a), \qquad \mathbf{R}_2 = (0, 0, \tfrac{1}{2}a).$$

The potential energy $V(\mathbf{r})$ of an electron is shown in Fig. 13a, and includes the nuclear potential $V_n(\mathbf{r})$ as well as the self-consistent average potential of the second electron. The potential is symmetrical about the origin and the electron can penetrate easily from the one hydrogen atom to the other one. The eigenfunction ψ_b with lowest energy is the orbital shown qualitatively in Fig. 13b. We can have

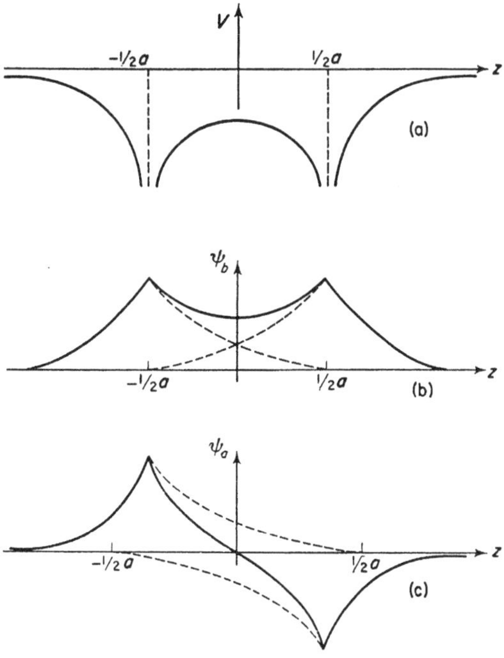

FIG. 13. The hydrogen molecule: (a) the potential energy of one electron in the molecule; (b) the bonding orbital ψ_b; (c) the anti-bonding orbital ψ_a. In each case only the value along the line of centres (z-axis) is plotted.

two electrons in this orbital, one with up spin and one with down spin, and the wave function for the two electrons in the H_2 molecule becomes

$$\Psi_{MO} = (2!)^{-1/2} \begin{vmatrix} \psi_b(\mathbf{r}_1)u_{+1} & \psi_b(\mathbf{r}_1)u_{-1} \\ \psi_b(\mathbf{r}_2)u_{+2} & \psi_b(\mathbf{r}_2)u_{-2} \end{vmatrix}. \tag{22.2}$$

Since the orbital ψ_b has a low energy, it gives the whole system of electrons plus nuclei a low energy, which means physically that the two atoms are bound together in a stable molecule. The orbital is therefore called a *bonding orbital*. The orbital ψ_a with next lowest energy has the form of Fig. 13c. It is called an *anti-bonding orbital* because it has a rather higher energy than ψ_b and hence would detract from the binding of a molecule. For example in H_2 only the bonding orbital is occupied, which leads to a well-bound molecule. On the other hand in the molecule He_2, the bonding and anti-bonding orbitals are both occupied by two electrons each, so that the molecule is unstable. The origin of the energy difference between a bonding and an anti-bonding orbital can easily be seen from Fig. 13. The charge density $\psi^*\psi$ is approximately the same for ψ_a and ψ_b, so that the potential energy $\int \psi^* V(\mathbf{r})\psi \, dv$ does not give the major contribution to the energy difference between them. However, the kinetic energy is

$$\frac{\int \psi^*[-(\hbar^2/2m)\nabla^2]\psi \, dv}{\int \psi^*\psi \, dv} = \frac{(\hbar^2/2m)\int |\nabla\psi|^2 \, dv}{\int |\psi|^2 \, dv}. \tag{22.3}$$

From Fig. 13, $\nabla\psi$ is clearly much smaller for ψ_b than for ψ_a in the central region between the two nuclei, so that ψ_b has a considerably lower kinetic energy than ψ_a.

Near the nucleus $\mathbf{R}_1$, the attraction of this nucleus is much stronger than that of the $\mathbf{R}_2$ one and vice versa, so that ψ_b and ψ_a are approximately hydrogen $1s$ wave functions near the nuclei. This gives us a rough analytical representation of ψ_b and ψ_a (see the dotted lines in Fig. 13b and 13c);

$$\psi_b = \psi_{1s}(\mathbf{r} - \mathbf{R}_1) + \psi_{1s}(\mathbf{r} - \mathbf{R}_2) \tag{22.4a}$$

$$\psi_a = \psi_{1s}(\mathbf{r} - \mathbf{R}_1) - \psi_{1s}(\mathbf{r} - \mathbf{R}_2). \tag{22.4b}$$

Clearly the more overlap there is between the functions centred on the two atoms, the lower the kinetic energy (22.3), and the stronger the bond formed by the electrons in bonding orbitals, other things being equal.

The covalent bond

It has been found that the electrons in polyatomic molecules can also be described often by *bond orbitals*, such as ψ_b and ψ_a of Fig. 13, which straddle two neighbouring atoms. If $\partial\psi/\partial z = 0$ midway between the atoms (or nearly midway between them if they are different elements), then the orbital is a bonding one ψ_b and if $\psi = 0$ halfway or nearly halfway between the atoms then it is an antibonding one ψ_a. Each type can be occupied by two electrons, one with up and one with down spin. As in (22.4), ψ_b and ψ_a can be written

$$\boxed{\begin{aligned} \psi_b &= \psi_1(\mathbf{r} - \mathbf{R}_1) + \psi_2(\mathbf{r} - \mathbf{R}_2) \quad (22.5a) \\ \psi_a &= \psi_1(\mathbf{r} - \mathbf{R}_1) - \psi_2(\mathbf{r} - \mathbf{R}_2) \quad (22.5b) \end{aligned}}$$

in terms of some atomic-like orbitals ψ_1 and ψ_2. These need not be ordinary $1s$, $2s$, $2p$, etc., atomic orbitals which are symmetrical about the nucleus $\mathbf{R}_1$ or $\mathbf{R}_2$. As already mentioned, the more overlap there is between $\psi_1(\mathbf{r} - \mathbf{R}_1)$ and $\psi_2(\mathbf{r} - \mathbf{R}_2)$, the stronger the bond that is formed. The strength of the bond therefore tends to be increased by having lopsided orbitals, stretching out towards one another from their respective nuclei $\mathbf{R}_1$ and $\mathbf{R}_2$ as shown for the carbon-carbon bond in Fig. 14. These lopsided atomic orbitals are called *directed valence orbitals.*

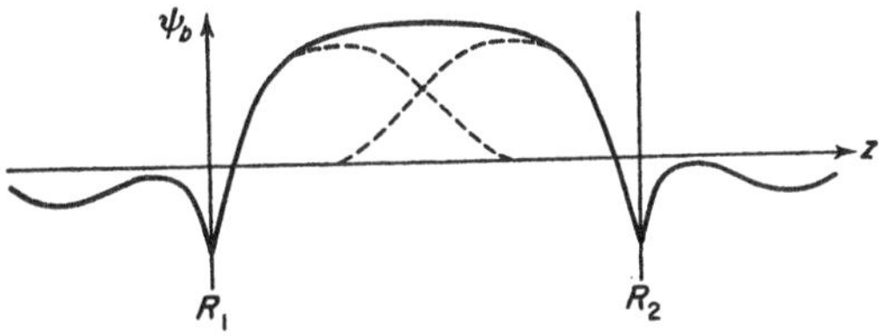

FIG. 14. Carbon-carbon bonding orbital. The bonding orbital is the sum of the two directed valence orbitals shown dotted.

A single directed valence orbital for a carbon atom is shown in Fig. 15a. Near the nucleus, the wave function must approximate to a $2s$ or $2p$ orbital, since the $1s$ ones are already filled and higher ones like $3s$, $3p$, . . . have too high an energy. Fig. 15 shows how

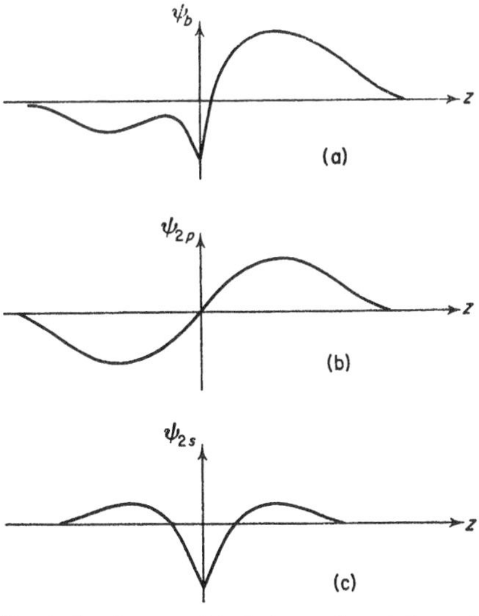

FIG. 15. A directed valence orbital for carbon (a) can be seen to be expressible as the sum of the $2p$ and $2s$ atomic orbitals (b) and (c).

the directed valence orbital can actually be formed as a linear combination of a $2s$ and a $2p$ atomic orbital. This is very satisfactory as regards one directed orbital by itself. However, as is well known, the carbon atom tends to form four valence bonds with its neighbours, directed as if towards the corners of a regular tetrahedron (i.e. alternate corners of a cube) as indicated by A, B, C, D in Fig. 16. The main problem is to show now that *four* orbitals

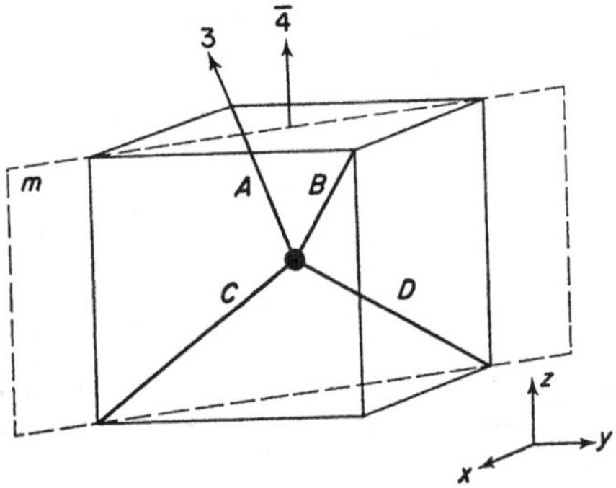

FIG. 16. Symmetry of the directed valence orbitals of carbon. A, B, C, D are the directions of the four orbitals, shown with reference to a cube. The four orbitals transform into one another under the point-group $\bar{4}3m$, a few representative symmetry elements being shown.

ψ_A, ψ_B, ψ_C, ψ_D, directed as described and all equivalent to one another, can indeed be formed from atomic $2s$, $2p_x$, $2p_y$, $2p_z$ functions. This can be proved as follows. The directed orbitals are transformed into one another by the point-group $\bar{4}3m$, and form a representation D of this group. The 3-fold axis shown in Fig. 16 induces the transformation

$$A \to A, \quad B \to C, \quad C \to D, \quad D \to B,$$

the transformation matrix being

$$\begin{bmatrix} 1 & \cdot & \cdot & \cdot \\ \cdot & \cdot & \cdot & 1 \\ \cdot & 1 & \cdot & \cdot \\ \cdot & \cdot & 1 & \cdot \end{bmatrix}.$$

We have therefore for the three-fold rotation the character

$$\chi(3) = 1.$$

Since the character is the sum of *diagonal* matrix elements, we only get a contribution to it when an orbital is transformed into itself, such as A above. Thus we can write down by inspection $\chi(\bar{4}) = 0$, $\chi(\mathrm{m}) = 2$, $\chi(2_z) = 0$, $\chi(\mathrm{E}) = 4$. From the character table of the group $\bar{4}3\mathrm{m}$ (Table 16) and from the decomposition law (14.2), we can reduce the representation D into

$$D = A_1 + T_2. \tag{22.6}$$

Also $\psi(2s)$ is spherically symmetrical and transforms according to A_1. $\psi(2p_x)$, $\psi(2p_y)$, $\psi(2p_z)$ transform like x, y, z, according to the representation T_2. This can be seen by inspection, or from (14.4), or it can be looked up in appendix K or Table 16. Thus $\psi(2s)$,

TABLE 16

Character Table of Point-group $\bar{4}3\mathrm{m}$

Representation	Characters					
	E	$\bar{4}_z$	2_z	3	m	
A_1	1	1	1	1	1	r^2
A_2	1	-1	1	1	-1	
E	2	0	2	-1	0	
T_1	3	1	-1	0	-1	
T_2	3	-1	-1	0	1	x, y, z
$D = A_1 + T_2$ (see (22.6))	4	0	0	1	2	

$\psi(2p_x)$, $\psi(2p_y)$, $\psi(2p_z)$ also transform according to D (22.6), whence it follows that we can form linear combinations of $\psi(2s)$, $\psi(2p_x)$, $\psi(2p_y)$, $\psi(2p_z)$ which have the required symmetry properties of ψ_A, ψ_B, ψ_C, ψ_D. The correct linear combinations, which can be found by using the projection operators (14.11) or by inspection, are (not in normalized form)

$$\begin{aligned}
\psi_\mathrm{A} &= \psi(2s) - \psi(2p_x) - \psi(2p_y) + \psi(2p_z)\\
\psi_\mathrm{B} &= \psi(2s) + \psi(2p_x) + \psi(2p_y) + \psi(2p_z)\\
\psi_\mathrm{C} &= \psi(2s) + \psi(2p_x) - \psi(2p_y) - \psi(2p_z)\\
\psi_\mathrm{D} &= \psi(2s) - \psi(2p_x) + \psi(2p_y) - \psi(2p_z).
\end{aligned} \tag{22.7}$$

Each of these directed orbitals forms a bonding orbital of the type Fig. 14 with the neighbouring atom in the appropriate direction. Combined with the two spin functions u_+ and u_-, they make eight *spin-orbitals*. These are all occupied, four of the electrons coming from the central carbon atom and one each from the neighbours. Thus near the carbon atom the wave function is like a full shell

$(2s)^2(2p)^6$ atomic wave function. However, the spin orbitals have to be normalized to unity, and since they straddle two atoms, each electron has only a 50% probability of being on the central carbon atom. The total charge on the carbon atom is therefore correctly given as $8 \times \frac{1}{2} = 4$ electrons.

We can now see why the directed valence bonds of carbon must point in the tetrahedral directions of Fig. 16. From the fact that only one type of methyl chloride (CH_3Cl) molecule is known to exist, we infer that all the four valencies of carbon are completely equivalent to one another. They must therefore make equal angles with one another, which limits their arrangement to two possibilities, namely the tetrahedral one of Fig. 16 and the coplanar one Fig. 17a. We can now repeat the above analysis for the planar configuration Fig. 17a, and soon find that it is impossible to make such a set of directed valence orbitals out of the atomic orbitals $\psi(2s)$, $\psi(2p_x)$, $\psi(2p_y)$, $\psi(2p_z)$. The directed valencies of carbon must therefore point in the tetrahedral directions.

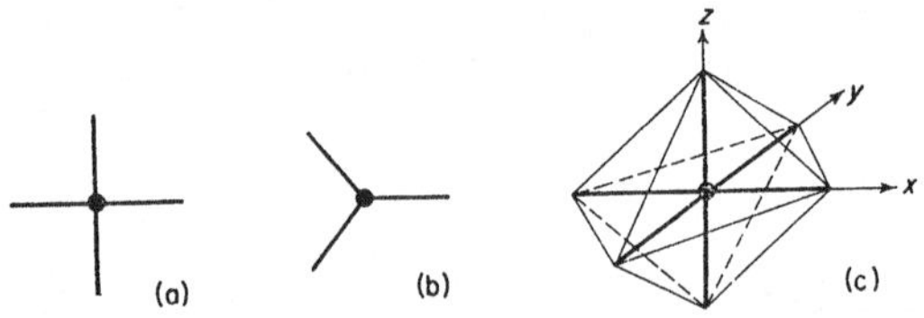

FIG. 17. (a) Symmetry of four directed valence orbitals in a plane; (b) three directions at 120° in a plane; (c) six directions at 90° to one another. These are called octahedral directions because they point to the vertices of a regular octahedron as shown.

From the above example, the properties of the type of binding between atoms which we have been discussing can be summarized as follows:

(i) *The electrons are in localized bonding orbitals straddling two atoms, so that each electron is "shared" between the two atoms.*

(ii) *There are two electrons with opposite spin in each bonding orbital, and by means of the sharing each atom tends to build up a full shell of electrons.* Usually the atoms contribute one electron each to the bond, but one atom may contribute both.

(iii) *The bonds are directed along certain directions in space.*

Such a bond is known as a *covalent* one. Table 17 shows what directed valence orbitals can be formed from various sets of atomic orbitals. Each of these orbitals is axially symmetric about the line joining the atoms and gives what is called a σ bond, which is the

TABLE 17

Directed Valence Orbitals

Co-ordination number	Configuration used	Directions of valence orbitals
2	sp	linear
3	sp^2	trigonal, plane
	ds^2	trigonal, plane
	p^3	trigonal pyramid
4	sp^3	tetrahedral
	dsp^2	tetragonal, plane
5	d^4s	tetragonal pyramid
6	d^2sp^3	octahedral
	d^4sp	trigonal prism
8	sp^3d^3f	cube vertices

usual type of saturated covalent bond. Unsaturated bonds can be described by an extension of the theory (see below).

When an atom has two incompletely filled electronic shells, there clearly tends to be a considerable degree of choice in which atomic orbitals to use in making up directed valence orbitals. Not only can we vary which orbitals to use out of each shell but also how many electrons are to be allocated to each shell. For instance in the transition elements Ti to Ni, the $3d$, $4s$ and $4p$ orbitals have comparable energy and are only partly occupied, so that they are all available for bond formation. This explains why these metals form such a variety of covalently bonded complexes, such as the ferricyanides, ferrocyanides, and the $[Co(NH_3)_5]^{+++}$ complex (see problem 22.4 and Table 17).

Molecular orbitals and the structure of benzene

If an electron can penetrate from one atom to the neighbouring one and be described by a bond orbital as in Figs. 13 and 14, then there would appear to be no reason why it could not go on to the next atom and so on. This can indeed happen, and electrons are therefore often described by *molecular orbitals* which extend throughout the molecule.† As an extreme example, a macroscopic crystal of a solid can be considered as one gigantic molecule, and Fig. 18 shows an orbital in metallic sodium. It is an infinite chain made up out of $3s$ atomic orbitals.

By way of a more detailed example, we shall consider the structure of benzene. The six carbon atoms are known from X-ray structure

† In the case of a diatomic molecule, molecular orbitals and valence bond orbitals are clearly the same thing.

work to form a regular hexagon in a plane, which we shall take as the x, y-plane. The C—C and C—H bonds make 120° angles with one another, and this suggests (Table 17) that we can describe them

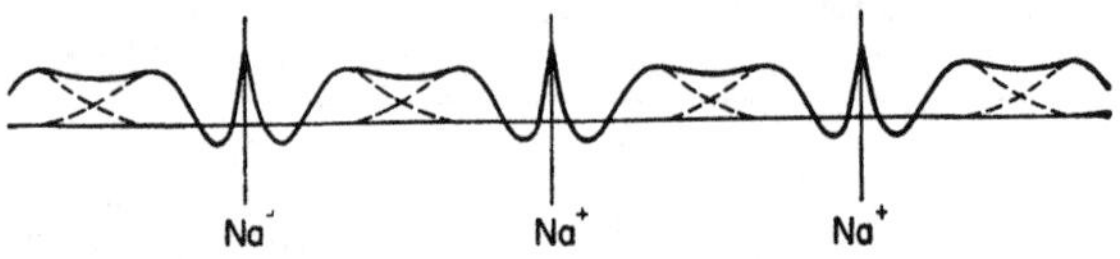

FIG. 18. Lowest energy one-electron orbital in sodium metal.

by bonding orbitals formed from $\psi(2s)$, $\psi(2p_x)$, $\psi(2p_y)$ atomic orbitals on the carbon atoms and $\psi(1s)$ on the hydrogen. This gives the bonds shown as lines in Fig. 19, and leaves six electrons to be

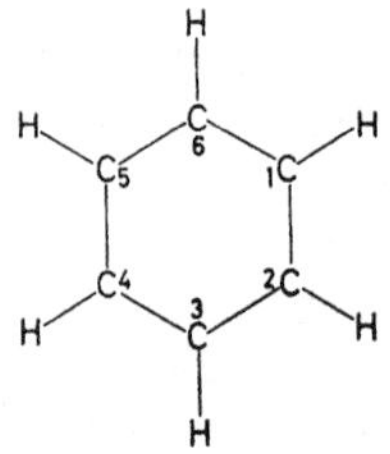

FIG. 19. Bond structure of benzene. The bonds corresponding to (22.9) are not indicated.

fitted into orbitals made up out of the $\psi(2p_z)$ functions on the six carbon atoms. Let us write ψ_z as short for

$$\psi_z \equiv \psi(2p_z) = zf(r), \tag{22.8}$$

and let $\mathbf{R}_n$ ($n = 1$ to 6) be the positions of the six carbon nuclei in Fig. 19. The six orbitals have symmetry 6mm but it is simplest to consider just the sixfold rotation (point-group 6). A symmetrically linked orbital forming a chain round the ring similar to Fig. 18 is

$$\psi_0 = \sum_n \psi_z(\mathbf{r} - \mathbf{R}_n). \tag{22.9a}$$

This transforms according to the identity representation A of the point-group 6 (see appendix K). Analogously the five orbitals

$$\psi_{\pm 1} = \sum_n \exp(\pm in\frac{\pi}{3})\psi_z(\mathbf{r} - \mathbf{R}_n), \tag{22.9b}$$

$$\psi_{\pm 2} = \sum_n \exp(\pm 2in\frac{\pi}{3})\psi_z(\mathbf{r} - \mathbf{R}_n), \tag{22.9c}$$

$$\psi_3 = \sum_n (-1)^n \psi_z(\mathbf{r} - \mathbf{R}_n),$$

transform respectively according to the representations E_1, E_2 and B. These orbitals are not σ ones like those we discussed previously because the ψ_z orbitals are directed up and down perpendicular to the plane of the hexagon as shown by equation (22.8). They form what are called π orbitals (see below). Nevertheless ψ_0 is still a bonding orbital because of the way neighbouring atomic orbitals join up to reduce the kinetic energy. Likewise ψ_3 has a nodal plane half-way between each pair of carbon atoms, and is an antibonding orbital. The other four $\psi_{\pm 1}$, $\psi_{\pm 2}$ take up intermediate positions. In the case of benzene we take only the three orbitals ψ_0, ψ_{+1}, ψ_{-1} with lowest energy, putting two electrons in each (up and down spin). The 1s inner core electrons on the carbon atoms are held tightly to their own parent atoms, and are therefore described by atomic orbitals. The total wave function is then a 42×42 determinant of 12 atomic spin-orbitals (carbon 1s electrons), 24 valence bond spin-orbitals (C—C and C—H σ-type covalent bonds), 6 molecular spin-orbitals (C—C π-type covalent bonds).

There is a point of great chemical importance attached to the use of the molecular orbitals (22.9). If we apply a perturbation at one carbon atom by replacing the hydrogen atom by some radical such as $-NO_2$ or $-Cl$ or $-OH$, then the sixfold symmetry is destroyed and the molecular orbitals become linear combinations of the simple ones ψ_0, $\psi_{\pm 1}$, $\psi_{\pm 2}$, ψ_3. Since the energy difference between $\psi_{\pm 1}$ and $\psi_{\pm 2}$ is not very great, there will be a considerable amount of the latter mixed into the occupied orbitals. Thus since the orbitals extend round the whole ring, the charge density on every atom is affected, whereas in a valence bond picture only the two nearest neighbours would be affected to a first approximation. For instance in the trimethylphenylammonium ion $[N(CH_3)_3(C_6H_5)]^+$ analogous to $(NH_4)^+$, the positive charge resides mainly on the nitrogen atom and thus attracts electrons from the benzene ring towards the nearest carbon atom, say number 6 in Fig. 19, resulting in a general deactivation of the *whole* molecule. Moreover it can easily be shown (problem 22.6) that the removal of electrons takes place preferentially from atoms 1, 3 and 5. These atoms are therefore completely deactivated, so that any reaction of the molecule takes place at the meta position (atoms 2 and 4). On the other hand most radicals (—Cl, —Br, —I, —OH, —NH_2) tend to lose electrons to the benzene ring for reasons which we shall not go into here (Pauling 1939, p. 142; Coulson and Longuet-Higgins 1947, p. 56). If, however, we accept this as a fact, then the above argument shows that the ortho and para positions (atoms 1, 3 and 5) will be preferentially activated.

In solids a molecular orbital such as shown in Fig. 18 which traverses the whole of a macroscopic crystal, is called a *Bloch orbital*. In metals the fact that the electrons have such extended wave functions is intimately connected with their ability to move easily through the crystal and carry an electric current.

Energy considerations

So far we have put electrons into atomic orbitals, valence bond orbitals or molecular orbitals in a rather *ad hoc* fashion, and by using different orbitals, we can make up different types of wave function for the molecule. For example, we could write down an atomic orbital type of trial wave function for the hydrogen molecule

$$\Psi_{AO} = (2!)^{-1/2} \begin{vmatrix} \psi_{1s}(\mathbf{r}_1 - \mathbf{R}_1)u_{+1} & \psi_{1s}(\mathbf{r}_1 - \mathbf{R}_2)u_{-1} \\ \psi_{1s}(\mathbf{r}_2 - \mathbf{R}_1)u_{+2} & \psi_{1s}(\mathbf{r}_2 - \mathbf{R}_2)u_{-2} \end{vmatrix}. \qquad (22.10)$$

This contrasts with (22.2) in that here each electron is permanently on one atom. The deciding factor in the matter is, of course, how can the electrons arrange themselves into orbitals so as to achieve the lowest energy for the molecule as a whole. In general the wave function giving the lowest energy in this way is closest to the true wave function. It is therefore one of the tasks of theoretical chemists and of solid state physicists to find out what type of wave function is the appropriate one for any given situation, and why. We shall therefore consider in more detail the effect on the total energy of using different types of orbital. Our remarks will apply to all the types of orbital, Bloch, molecular, valence bond, and atomic, considered as a sequence ranging from the very extended Bloch orbitals to the highly localized atomic ones.

The first principle is that the more extended an orbital, the lower is the kinetic energy (22.3), *provided it is a bonding type orbital.* We have already seen that this is so for a valence bond orbital compared with atomic orbitals (Figs. 13 and 14), and from the sodium wave function of Fig. 18 it is clear that we tend to achieve a continually lower energy in this way as we string more and more atomic orbitals together. The kinetic energy therefore favours extended orbitals over localized ones.

The second principle is that the Coulomb repulsive energy in (22.1) *has exactly the opposite tendency: it favours localized orbitals.* An orbital always has to be normalized to one electron, so that the more extended an orbital the less the probability of the electron being on a given atom. The atom therefore has to be covered by more electron orbitals to achieve the correct charge on it. Although only a few of these electrons will be found on the atom on the average, there is still a non-zero chance of them all being there together. In this case the electrons are very close together, giving a high Coulomb energy $\Sigma e^2/r_{ij}$. For example, we have seen in a saturated organic molecule that the use of bond orbitals (22.5a) and (22.7) means that a given carbon atom has in a sense acquired a complete shell of electrons $(2s)^2(2p)^6$. On the average only half of these will be on the carbon atom at one time, but there is a probability of $(\frac{1}{2})^8$ of them all being there together. Although this probability is small, it corresponds to an enormous Coulomb energy and may therefore be expected to give an appreciable contribution to the total energy. Similarly, we can consider the effect of 5, 6 and 7 electrons being simultaneously on the one atom.

The third principle is that small interatomic distances (relative to the size of the atomic-like orbitals) favour extended orbitals and large distances the localized ones. With extended orbitals, the coming together of many electrons on one atom gives a high Coulomb repulsive energy which does not vary greatly with interatomic distance since it is concerned with the state of affairs inside one atom. On the other hand, the lowering of the kinetic energy of an extended orbital depends on the amount of overlap between atomic-like orbitals on neighbouring atoms, and this is greatest for small interatomic separations. Thus the kinetic energy tends to win over the potential energy at small separations and vice versa at large ones. This is in accordance with common sense, as we shall see by considering the example of the hydrogen molecule. Suppose that we can alter the internuclear distance $R = |\mathbf{R}_1 - \mathbf{R}_2|$ arbitrarily at will. For very large R, we have two separate atoms, each with one electron. We do not expect to find both electrons ever on one atom, and the localized atomic orbitals are the right ones to use, giving the atomic orbital type of total wave function Ψ_{AO} (22.10). At the opposite extreme for very small R, the two protons may be considered almost like a doubly charged helium nucleus, and the electrons revolve in molecular orbitals round the pair as a whole without being identified with the one proton or the other. This gives the molecular orbital wave function Ψ_{MO} (22.2), in accordance with the third principle. At the equilibrium internuclear distance the molecular orbital function is still the better, corresponding to a covalent chemical bond.

This example of the hydrogen molecule illustrates another point. The preceding discussion suggests that at an intermediate internuclear distance, neither Ψ_{MO} nor Ψ_{AO} is a satisfactory wave function and that we should take a linear combination between the two. Only in this way can we go over continuously from the small R to the large R limit (see problem 22.7). However, even this would only give an approximate representation of the correct wave function. The true wave function of a molecule or atom cannot be written as a single determinant or as a simple sum of two or three; it requires an infinite series of determinants. The use of a single determinant, as we are mostly doing in this section, is therefore just an attempt to obtain a simple semi-quantitative description of molecules. Only in this context is the discussion of when to use which type of orbital meaningful. Also the use of valence bond or molecular orbitals with special symmetry properties is just a part of this approximation. Nevertheless, as in the case of atoms (§§ 10–12), the total Hamiltonian of the molecule (22.1) has a definite symmetry group. Thus the wave function Ψ of the whole molecule can be assigned rigorously to an irreducible representation of this group, no matter how complicated Ψ may be when expressed in terms of one-electron orbitals. Any symmetry arguments about Ψ are therefore precise.

Diatomic molecules

We shall now discuss diatomic molecules in more detail, continuing for the present to use the approximation of writing Ψ as a single determinant of one-electron orbitals. We shall further assume that with the atoms at their ordinary equilibrium distance apart, the molecular orbital description like (22.2) is a better approximation to the true wave function than the Heitler–London atomic orbital one like (22.10). In the case of diatomic molecules, valence bond orbitals and molecular orbitals are of course the same thing. If the two atoms are different, e.g. CO, the symmetry group of the molecule is ∞ m. We choose each electron orbital to transform according to one of the irreducible representations of this group (Table 18). The notation A_1, A_2, E_1, . . . is analogous to that for the finite groups, the alternative one Σ^+, Σ^-, Π, . . . being the usual chemical one. Capital letters, $\Sigma, \Pi, \Delta, \ldots$ are used when the representations refer to states of the whole molecule, and small letters are often used for the same representations when these refer to one-electron orbitals. This is analogous to the use of S, P, D, . . . and s, p, d, . . . states in atomic theory. The bonding and anti-bonding orbitals of Figs. 13 and 14 are seen to belong to the representation σ^+. It is impossible to devise a one-electron orbital of the σ^- type, because if it is cylindrically symmetrical under the rotations then it must be invariant under the reflection m_x. The symmetry Σ^- can, however, apply to a whole molecular wave function (problem 22.11).

TABLE 18

Character Table for the Group ∞ m

Irred. rep.	Characters E	$R(\phi, z)$	m_x
A_1 or Σ^+	1	1	1
A_2 or Σ^-	1	1	−1
E_1 or Π	2	$2\cos\phi$	0
E_2 or Δ	2	$2\cos 2\phi$	0
.....	...	.....	...
E_l or Λ	2	$2\cos l\phi$	0

The irreducible representations of the group ∞/mm are the same with an additional suffix g or u according to whether they are even (gerade) or odd (ungerade) under inversion.

Orbitals of the other symmetry types can be constructed as follows. If the axis of the molecule is chosen as the z-axis, then in the notation of (22.8),

$$\boxed{\begin{aligned}\psi(\pi_x, \text{bond}) &= \psi_x(\mathbf{r}-\mathbf{R}_1) + \psi'_x(\mathbf{r}-\mathbf{R}_2)\\ \psi(\pi_y, \text{bond}) &= \psi_y(\mathbf{r}-\mathbf{R}_1) + \psi'_y(\mathbf{r}-\mathbf{R}_2)\end{aligned}} \tag{22.11}$$

form a pair of bonding π orbitals. If the two atoms are unlike, ψ and ψ' may be different p functions, e.g. $2p$ and $3p$ ones. Together with the σ orbital, these orbitals can be used to describe double and treble bonds as in acetylene H—C$\equiv$C—H. The antibonding orbital can be formed similarly. When forming δ and higher orbitals, it is helpful to note that a pair of functions transforming according to the general representation Λ (or E_l) can always be chosen so as to belong to the representations $\exp(\pm il\phi)$ of the axial rotation group. For example instead of the pair (22.11), we could use the linear combinations

$$\begin{aligned}\psi(\pi_+, \text{bond}) &= \psi(\pi_x, \text{bond}) + i\psi(\pi_y, \text{bond}),\\ \psi(\pi_-, \text{bond}) &= \psi(\pi_x, \text{bond}) - i\psi(\pi_y, \text{bond}),\end{aligned} \tag{22.12}$$

which have this property.

If the molecule consists of two like atoms, the symmetry group is ∞/mm which contains the inversion as an extra symmetry element. The irreducible representations are the same as those in Table 18, except that each one has an additional suffix g (gerade) or u (ungerade) according to whether it is even (g) or odd (u) under the inversion. (See the preamble to appendix K, and appendix L for further details.)

The oxygen molecule

Table 19 gives the different possible orbitals that can be formed out of the $2s$ and $2p$ functions on the two oxygen atoms, and which of these are occupied. They have been arranged in what is the

TABLE 19

Orbitals in the Lowest Energy Configuration of O_2

Orbital	Number of orbitals occupied	Number of independent spin orbitals	Irreducible representation	Atomic functions	Bonding or anti-bonding
$\sigma_g 2s$	2	2	σ_g^+	$2s$	bond.
$\sigma_g 2p$	2	2	σ_g^+	$2p_z$	bond.
$\pi_u 2p$	4	4	π_u	$2p_x, 2p_y$	bond.
$\sigma_u 2s$	2	2	σ_u^+	$2s$	anti-b.
$\pi_g 2p$	2	4	π_g	$2p_x, 2p_y$	anti-b.
$\sigma_u 2p$	0	2	σ_u^+	$2p_z$	anti-b.

probable order of increasing energy. The $2s$ atomic function has a lower energy than the $2p$ one, so that the $\sigma_g 2s$ bonding orbital has the lowest energy. Of the $2p$ orbitals, the one made out of two $2p_z$ functions ($\sigma_g 2p$) will have the maximum amount of overlap and the lowest energy. Likewise the corresponding antibonding orbital $\sigma_u 2p$ has the highest energy. The $2p_x$ and $2p_y$ atomic functions have less overlap, so that the orbitals $\pi_u 2p$ and $\pi_g 2p$ made up from them have an intermediate energy. Note that for σ orbitals the even one (g) is the bonding one, whereas for π orbitals the bonding one is odd (u). The precise order of the first four rows of Table 19 is actually somewhat uncertain, particularly as regards the position of the $2s$ antibonding orbital. This uncertainty is, however, quite immaterial, because if we put the twelve available valence electrons into the orbitals in order, only the last two types of orbital are incompletely filled and there is no doubt about their order. Thus the configuration of the molecule is

$$(\sigma_g 1s)^2(\sigma_u 1s)^2(\sigma_g 2s)^2(\sigma_g 2p)^2(\pi_u 2p)^4(\pi_u 2s)^2(\pi_g 2p)^2 \ldots \quad (22.13)$$

In this configuration, the $\pi_g 2p$ shell is incompletely filled, only two of the available four spin-orbitals being taken up. Thus we can write down ${}^4C_2 = 6$ different wave functions all belonging to this configuration. As in the case of atoms, these wave functions group themselves into terms. All the terms can easily be found by extending Slater's procedure for atoms, and this is left as problem 22.13. As in the atomic case (§ 12), each full shell like $(\sigma_g 2s)^2$ or $(\pi_u 2p)^4$ is completely symmetrical and need not be considered in deriving the symmetry of the total wave function. Here we shall just write down the ground term by inspection, using arguments analogous to those relating to Hund's rule in § 12. The four $\pi_g 2p$ spin-orbitals, of which we have to choose two, are

$$\psi(\pi_x)u_+, \qquad \psi(\pi_x)u_-, \qquad \psi(\pi_y)u_+, \qquad \psi(\pi_y)u_-$$

in the notation of (22.11), all being anti-bonding. The kinetic energy and the average self-consistent potential energy are the same for all of these, so what we need to minimize is the Coulomb energy e^2/r_{12} between the two electrons. This is achieved by keeping the electrons as far apart as possible, i.e. by having one electron in a $\psi(\pi_x)$ orbital and the other one in a $\psi(\pi_y)$ orbital since each orbital is zero where the other is at its maximum. Also as explained in § 12, the antisymmetry of the wave function keeps electrons of parallel spin apart, and in this way the energy can be lowered still further by choosing the spins parallel. The total wave function is therefore the determinant based on

$$\psi(\pi_x;\, r_1)u_{+1} \qquad \psi(\pi_y;\, r_2)u_{+2}. \tag{22.14a}$$

Since inside a determinant we can take linear combinations of the columns without altering the value of the determinant, we can equally well use

$$\psi(\pi_+;\, r_1)u_{+1} \qquad \psi(\pi_-;\, r_2)u_{+2}, \tag{22.14b}$$

in the notation of (22.12). From the latter form, the total wave function Ψ is seen to be invariant under rotations so that we have a Σ term. Since both orbitals are even under inversion, Ψ is also even (g). In (22.14a), $\psi(\pi_x)$ and $\psi(\pi_y)$, respectively, do and do not change sign under m_x, so that the product changes sign. We therefore have a Σ_g^- term. Since the two electron spins are parallel, the term has a threefold spin degeneracy corresponding to $S = 1$. We therefore write it as ${}^3\Sigma_g^-$ with $2S+1$ as a pre-superscript as in the case of atomic terms. This resultant spin gives the oxygen molecule a magnetic moment and thus accounts for the paramagnetism of oxygen.

Separating the atoms

We consider a molecule in some definite state, and imagine the nuclei of the two atoms to be pulled further and further apart while retaining the same symmetry throughout. This process can be followed spectroscopically, because as the molecule is excited into higher vibrational states, the average interatomic distance increases due to the asymmetrical form of the interatomic potential (Fig. 20).

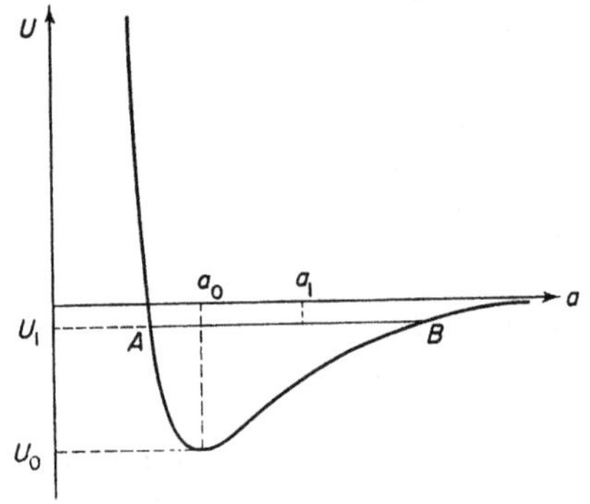

FIG. 20. Potential energy U of two atoms as a function of their separation a. U_0 and a_0 are the equilibrium values. For an energy U_1, the atoms oscillate between the points A and B on the curve, the mean separation a_1 being larger than a_0.

As the separation tends to infinity, the wave function tends to a combination of atomic functions like (22.10) and the energy to a sum of the corresponding atomic energy levels. The question is, which atomic orbitals and atomic energy levels? This can be answered group theoretically.

Let the molecular wave function transform according to the representation $D^{(\alpha)}$, and let a be a parameter proportional to the interatomic distances. Then the wave function Ψ must vary continuously with a and hence its irreducible representation cannot suddenly change. As we shall prove presently, its energy $E^{(\alpha)}(a)$ may cross the energy $E^{(\beta)}(a)$ of another level for some value of a only if $D^{(\alpha)}$ and $D^{(\beta)}$ are not equivalent. Thus we write down all the irreducible representations of the energy levels of the molecule in order of increasing energy in one column, and the same for the set of separated atoms in a parallel column. The nth state belonging to $D^{(\alpha)}$ in one column must then change continuously into the nth state belonging to $D^{(\alpha)}$ in the other column (Fig. 21).

Before going on to discuss a specific example, it is important to prove the theorem used above, namely that two energy levels $E_1(a)$ and $E_2(a)$ do not cross one another if they belong to the same irreducible representation $D^{(\alpha)}$. Let the corresponding wave functions be Ψ_1 and Ψ_2, and let us suppose that E_1 and E_2 are close

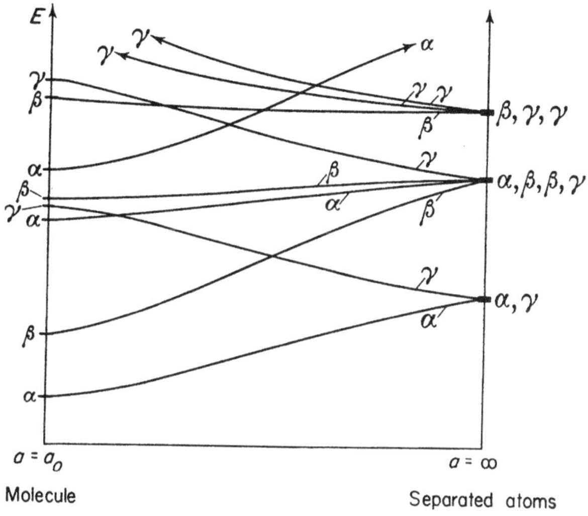

FIG. 21. Variation of energy levels E with a. The labels α, β, γ, . . . refer to the irreducible representations $D^{(\alpha)}$, etc., of the symmetry group of the molecule.

together but not equal for some value say a' of a. Small variations of a and a' can be treated by perturbation theory and we shall investigate whether we can expect the energy $E_1(a) - E_2(a)$ to vanish for some value of a near a'. Let the variation of the Hamiltonian be $\Delta\mathscr{H}(a)$. The interaction between the levels Ψ_1 and Ψ_2 cannot be treated by ordinary non-degenerate perturbation theory, and we need to consider the eigenvalues of the 2×2 matrix†

$$\begin{bmatrix} A & C \\ C^* & B \end{bmatrix}. \tag{22.15}$$

However, the interaction with all the other levels can be included in this matrix in the spirit of the spin-Hamiltonian formalism of § 18 by putting

$$A = \langle 1|\Delta\mathscr{H}|1\rangle + \sum_n \frac{\langle 1|\Delta\mathscr{H}|n\rangle\langle n|\Delta\mathscr{H}|1\rangle}{E_1 - E_n} + \ldots$$

$$C = \langle 1|\Delta\mathscr{H}|2\rangle + \sum_n \frac{\langle 1|\Delta\mathscr{H}|n\rangle\langle n|\Delta\mathscr{H}|2\rangle}{E_1 - E_n} + \ldots$$

† For an exposition of the matrix formulation of quantum mechanics, see Schiff 1955.

with an analogous expression for B (Pryce 1950). Since this in principle includes up to infinite order of perturbation, we expect (22.15) to give a good picture of the energy levels for some finite range of a. We only neglect $E_2 - E_1$ compared with $E_n - E_1$, and this error can be made arbitrarily small by taking a' arbitrarily close to the assumed point of degeneracy. The difference in energy between the two levels is given by

$$(E_1 - E_2)^2 = (A - B)^2 + 4CC^*,$$

and this can only vanish if

$$f(a) \equiv A(a) - B(a) = 0, \tag{22.16a}$$

and

$$g(a) \equiv C(a) = 0. \tag{22.16b}$$

If Ψ_1 and Ψ_2 belong to the same representation, then (22.16a) and (22.16b) are two independent equations (actually three if we note that C is in general complex), and it is vanishingly improbable that both should be satisfied for a single value of a. We conclude therefore that the energy levels $E_1(a)$ and $E_2(a)$ will *not* cross. If, however, Ψ_1 and Ψ_2 belong to different irreducible representations, then (22.16b) is automatically satisfied by the fundamental theorem of § 13. Thus only the one equation (22.16a) need be considered, and it is therefore quite possible that this gives a zero energy separation for some values of a. This proves the theorem. Note that the theorem may not hold if $\mathscr{H}$ varies as a function of two or more arbitrary parameters $a, b, c, \ldots$, an example of which is given in problem 26.8.

By way of a detailed example, we shall consider the ground state of HCl. The first step is to decide what the symmetry of the ground state is. The configuration of a neutral chlorine atom is $(1s)^2(2s)^2(2p)^6(3s)^2(3p)^5$. Of the $3s$ and $3p$ atomic orbitals, the $3p_z$ one or some hybrid of it with $3s$ will have maximum overlap with the hydrogen $1s$ function. Thus we shall put four $3p$ electrons in the spin-orbitals $\psi(3p_x)u_\pm$, $\psi(3p_y)u_\pm$, and the remaining two electrons in the covalent bonding spin-orbitals

$$[\psi(3p_z : \mathbf{r} - \mathbf{R}_{\text{Cl}}) + \psi(1s : \mathbf{r} - \mathbf{R}_{\text{H}})]u_\pm. \tag{22.17}$$

This gives a total symmetry of ${}^1\Sigma^+$. So far we have assumed that HCl is covalently bonded. We might alternatively assume it is ionic H^+Cl^-, analogously to Na^+Cl^-. In this case the proton has no electron on it, and the chlorine atom has a closed shell $(3p)^6$. The symmetry is therefore again ${}^1\Sigma^+$, and the ground state may be considered to be a linear combination of the two configurations.

Measurements of the dipole moment of the molecule show that it can be regarded as about 83% covalent and 17% ionic. This illustrates how one can have intermediate types of bonds by writing the wave function of the molecule as a linear combination of several wave functions corresponding to several different simple structures. This of course can only be done when the separate wave functions all have identical symmetry, e.g. $^1\Sigma^+$ in the present case, since the whole wave function must belong to some definite symmetry type.

We now consider the states of the separated atoms. From Table 6, the atomic states are hydrogen ^2S and chlorine ^2P. There are therefore $2 \times 6 = 12$ independent wave functions for the pair. We write these as

$$\sum (\pm 1) P \psi_{\text{Cl}}(M_L, M_S = \pm)\psi_{\text{H}}(M_L = 0, M_S = \pm),$$

abbreviated to $$(M_L, \pm)(0, \pm),$$

where $\Sigma(\pm 1)P$ is the antisymmetrizing operator. We can group these into terms by following Slater's procedure (see Table 5), and obtain the terms $^3\Pi$, $^3\Sigma^+$, $^1\Pi$, $^1\Sigma^+$ as shown in Table 20. These

TABLE 20

Lowest Terms for Two Separated Chlorine and Hydrogen Atoms

Atomic states Cl	H	Total M_L	Total M_S	Term†
(1 +)	(0 +)	1	1	a
(1 −)	(0 +)	1	0	a
(1 +)	(0 −)	1	0	b
(0 +)	(0 +)	0	1	c
(0 +)	(0 −)	0	0	c
(0 −)	(0 +)	0	0	d

†(a) $^3\Pi$; (b) $^1\Pi$; (c) $^3\Sigma^+$; (d) $^1\Sigma^+$.

The M_L refers to the representation for rotations about the line joining the atoms. All the $M_L = 0$ functions are even under the reflection m_x.

all have the same energy in the limit of infinite separation. One of these is a $^1\Sigma^+$ state, the same as the state of the molecule, and since we are dealing with the lowest energy states these join up in the sense of Fig. 21. That is, the molecular state goes over into a state of the separated atoms which leaves each atom in its *lowest* state. This therefore gives the limiting energy of $E(a)$ as $a \to \infty$. It should be noted that if $^1\Sigma^+$ had not occurred among the terms found from the ground states of the atoms, then as $a \to \infty$ one or more of the atoms would have been left in an excited state. An example of the latter situation is the methane molecule (problem 22.14).

Clearly similar arguments can be applied to the way the one-electron orbitals change with interatomic distance, and we can also investigate what happens in the limit $a \to 0$ as the nuclei of the two atoms coalesce into a new single nucleus.

Spin-orbit coupling

So far we have based our discussion on the Hamiltonian (22.1) without spin-orbit coupling, and all symmetry considerations have referred to the orbital part of the wave function only. However, if the spin-orbit coupling is important, the Hamiltonian is not invariant under orbital transformations alone, and we have to apply the symmetry transformations to the orbital and spin variables simultaneously. As with the full rotation group, the spin functions u_+, u_- transform according to double-valued representations, and thus so do all spin-orbitals. It has already been shown in § 16 how to work out these double-valued representations of the point-groups, and we shall not consider the matter further. As mentioned there, they have been tabulated for all crystallographic point-groups by Koster (1957).

Further references

The whole subject of the quantum mechanical basis of molecular structure is covered in rather greater detail than here by Eyring, Walter and Kimball (1944). Pauling (1939) and Coulson (1952) apply the quantum mechanical ideas to give a description of the chemical properties of a wide range of compounds. Van der Waerden (1932) discusses the spectra and symmetries of diatomic molecules in great detail. Coulson and Longuet-Higgins (1947–48) treat the chemical properties of conjugated hydrocarbons in terms of molecular orbitals. Roothan (1951) reviews some aspects of molecular orbital theory. Boys, Cook, Reeves and Shavitt (1956) show how one can calculate the properties of simple molecules quantitatively from first principles by systematically approximating to the exact wave function by a linear combination of many determinants. Mott and Stevens (1957) discuss the use of localized versus extended orbitals in solids.

Summary

We have discussed the different types of orbitals, atomic, directed valence, valence bond, molecular and Bloch orbitals, and in what physical situations they are used to describe the electrons in molecules and solids. Their symmetry properties have been used to describe, classify, and enumerate them. We have also considered the symmetry of the whole molecular wave function, and how this affects the way the energy changes as the molecule is pulled apart into single atoms.

Problems

22.1 Calculate the kinetic energies of the orbitals ψ_{b} and ψ_{a} (22.4), and hence show that ψ_{b} has the lower kinetic energy.

22.2 Show that a set of three directed valence orbitals making 120° angles with one another in the x, y-plane (Fig. 17b), can be constructed out of $\psi(2s)$, $\psi(2p_x)$, $\psi(2p_y)$ atomic orbitals.

22.3 Show that the following are four directed valence orbitals making 90° angles with one another in a plane (Fig. 17a):

$$\psi_{\mathrm{A}} = \psi(2s) + \psi(2p_x), \qquad \psi_{\mathrm{B}} = \psi(2s) - \psi(2p_x),$$
$$\psi_{\mathrm{C}} = \psi(2s) + \psi(2p_y), \qquad \psi_{\mathrm{D}} = \psi(2s) - \psi(2p_y).$$

Why cannot these be used to describe the four valences of carbon?

22.4 Show that six octahedrally directed valence orbitals can be formed from the set of atomic orbitals $(3d)^2 4s (4p)^3$. Hence discuss the electronic structure of the octahedral, covalently bonded, complex ion $[Co(NH_3)_6]^{+++}$, and show that it must be diamagnetic, in contrast to the paramagnetic moment that one would expect a Co^{+++} ion to have (Table 15). The nickelous ion Ni^{++} forms complexes with four covalent bonds, sometimes at 90° to one another in a plane (Fig. 17a) and sometimes tetrahedrally directed like the carbon valences. What would you expect the magnetic moments of the two types of complex to be (Pauling 1939, p. 111)?

22.5 Calculate the kinetic energies of the molecular orbitals (22.9) compared with that of a simple two-atom π bonding orbital.

22.6 Referring to Fig. 19 and the discussion below (22.9) in the text, suppose that the hydrogen atom on carbon atom number 6 has been replaced by a radical which attracts the electrons of the benzene ring towards this atom. Treat this effect as a perturbation and show that the occupied orbitals have the form

$$\psi_\lambda = \psi_0 + \text{other terms}, \qquad \psi_\mu = \psi_{+1} - \psi_{-1},$$
$$\psi_\nu = \cos\alpha(\psi_{+1} + \psi_{-1}) + \sin\alpha(\psi_{+2} + \psi_{-2}) + \text{other terms},$$

where for the roughest picture we may neglect the "other terms". Hence calculate the charge density on all the carbon atoms and show that the electrons are removed preferentially from the ortho and para positions (carbon atoms 1, 3 and 5). Check this conclusion by taking into account all the first order perturbation terms in the orbitals.

22.7 Let the wave function of the hydrogen molecule for all values of the interatomic distance R be written

$$\Psi = \cos\alpha\, \Psi_{\mathrm{MO}} + \sin\alpha \frac{1}{\sqrt{2}} (\Psi_{\mathrm{AO},1} - \Psi_{\mathrm{AO},2}).$$

Here Ψ_{MO} is the molecular orbital function (22.2), $\Psi_{\mathrm{AO},1}$ the atomic orbital function (22.10), and $\Psi_{\mathrm{AO},2}$ the same with u_+ and u_- interchanged, each function being suitably normalized. Choose α to minimize the expectation value of the energy (22.1), and hence show that $\alpha \to 0$ for small R and $\alpha \to 90°$ for large R. (Coulson and Fisher 1949.) Note: the linear combination $\Psi_{\mathrm{AO},1} - \Psi_{\mathrm{AO},2}$ is required to give an $S = 0$ function: the corresponding $S = 1$ functions refer to excited states of the molecule.

22.8 The radii of the $4f$ and $6s$ shells of the gadolinium atom are about 0.3 and 2 Angstrom units. In gadolinium metal, the distance between nearest neighbour atoms is 3.6 A.U. What types of orbital would you expect to use to describe the magnetic $4f$ electrons and the conduction $6s$, $6p$ electrons in the metal respectively?

22.9* What types of wave function are commonly used for the electrons in pure solids? Illustrate your answer by particular reference to sodium metal, sodium chloride, diamond, solid hydrogen, iron, nickel. References: Seitz 1940, Hall 1952, Mott and Stevens 1957, Heine 1958.

22.10 Prove the following theorem: if a single determinant $\Psi = |\det \psi_j(\mathbf{r}_i, \sigma_{zi})|$ of orbitals ψ_j describes a particular *non-degenerate* state of a molecule, then the ψ_j form a complete basis for a representation of the symmetry group $\mathfrak{G}$ of the molecule. Hint: (a) show that $T\Psi = c_T\Psi$ where T is any element of $\mathfrak{G}$ and c_T a constant; (b) suppose that the ψ_j have to be supplemented by a few more functions ϕ_r to make up an invariant vector space. Show that this would imply that $T\psi_j$ must include some ϕ_r for some T and some ψ_j, and hence that it leads to a contradiction to (a).

22.11 Construct a simple function of the variables $\mathbf{r}_1$ and $\mathbf{r}_2$ that transforms according to the irreducible representation Σ^- of the group ∞ m.

22.12 Valence bond orbitals are formed out of the following pairs of atomic orbitals on two neighbouring atoms: (a) $2s$–$2p_z$

(b) $2s$–$2p_x$ (c) $3d$–$3d$. What are the irreducible representations of these valence bond orbitals? Consider both cases of like and unlike atoms.

22.13 Using Slater's procedure (see Tables 5 and 20), show that the lowest configuration (22.13) of the oxygen molecule gives rise to the three terms ${}^3\Sigma_g{}^-$, ${}^1\Sigma_g{}^+$, ${}^1\Delta_g$. In what order do you expect their energies to lie?

22.14 Consider the ground state of the methane molecule, and suppose that the internuclear distances are being gradually increased to infinity while retaining the symmetry of the molecule, until the wave function describes a set of five separate atoms. What atomic states (terms) will the atoms be left in? Are these the ground states for the atoms?

22.15 A bonding and an anti-bonding molecular orbital are formed from the $2s$ functions on two identical nuclei a long distance a apart. What functions do the two molecular orbitals change into as a is decreased continuously to zero (Coulson 1952, p. 93)?

23. Molecular Vibrations

In the last section, we used a model in which the nuclei of all the atoms in a molecule are fixed at positions $\mathbf{R}_\alpha = (X_\alpha, Y_\alpha, Z_\alpha)$. We showed how one could then discuss the motion of the electrons round the nuclei, and in principle one could obtain the electronic energy of the molecule to any desired degree of approximation. In this section we shall assume that this has been done, so that we start with a total energy $E(\mathbf{R}_\alpha)$ which clearly depends on what positions $\mathbf{R}_\alpha$ we assume for the nuclei. Then

$$\boxed{V(\mathbf{R}_\alpha) = E(\mathbf{R}_\alpha) - E(\mathbf{R}_{\alpha 0})} \tag{23.1}$$

is the potential energy of the system when the nuclei are displaced from their normal equilibrium positions $\mathbf{R}_{\alpha 0}$ to an arbitrary position $\mathbf{R}_\alpha$, though strictly this potential energy function $V(\mathbf{R}_\alpha)$ applies only to static displacements. However, molecular vibrations are in general so slow compared with the electronic motion, because the mass of the nuclei is so much greater than that of the electrons, that we may regard the vibrations as slowly varying semi-static displacements. In this section, therefore, we shall base our discussion of molecular vibrations on the assumption that the nuclei move subject only to the potential (23.1). For a quantum mechanical justification of this approach, we refer the reader to Eyring, Walter and Kimball (1944, p. 190).

Separation of co-ordinates

The Hamiltonian for our system of vibrating nuclei is

$$\mathscr{H} = T + V(\mathbf{R}_\alpha), \tag{23.2}$$

where T is the kinetic energy

$$T = \sum \tfrac{1}{2} m_\alpha \mathbf{V}_\alpha{}^2 = \sum (1/2m_\alpha)\mathbf{P}_\alpha{}^2, \tag{23.3}$$

and m_α, $\mathbf{V}_\alpha$, $\mathbf{P}_\alpha$ the mass, velocity, and momentum of the αth nucleus. $\mathbf{R}_\alpha$, $\mathbf{V}_\alpha$, $\mathbf{P}_\alpha$ all refer to a system of co-ordinates OX, OY, OZ which is fixed in the laboratory. However this is rather an inconvenient system of co-ordinates to use. If we picture the movement of a molecule, we naturally think of this movement as made up out of three components; a sideways motion of the whole molecule (translation), a rotation of the molecule about itself as a rigid whole, and an internal vibration of the nuclei relative to one another. It is therefore convenient to define a new system of co-ordinates. We use three co-ordinates (X_{cm}, Y_{cm}, Z_{cm}) for the centre of mass

$$\mathbf{R}_{\text{cm}} = (1/M)\Sigma m_\alpha \mathbf{R}_\alpha, \qquad M = \Sigma m_\alpha,$$

and any variation of $\mathbf{R}_{\text{cm}}$ means the molecule is moving sideways as a whole. We next define the positions of the nuclei by their co-ordinates

$$\mathbf{R}'_\alpha = \mathbf{R}_\alpha - \mathbf{R}_{\text{cm}} \tag{23.4}$$

relative to the centre of mass. Then we redescribe the $\mathbf{R}'_\alpha$ further in terms of a frame of reference Ox, Oy, Oz which is rigidly fixed to the equilibrium configuration of the molecule. The orientation of this frame of reference, and of the molecule as a whole, is specified by three angles θ_i $(i = 1, 2, 3)$ with respect to the laboratory axes OX, OY, OZ. The vibrations of the molecule are then described in terms of the displacements

$$\boldsymbol{\delta}_\alpha{}^{\text{vib}} = \mathbf{r}_\alpha - \mathbf{r}_{\alpha 0} \tag{23.5}$$

of the nuclei from their equilibrium positions $\mathbf{r}_{\alpha 0}$ in the rotating Ox, Oy, Oz frame of reference. The $\boldsymbol{\delta}_\alpha{}^{\text{vib}}$ are still not our final co-ordinates because there are too many of them so that they are not linearly independent. To specify the positions of all the N nuclei, we need $3N$ co-ordinates. Six of these are X_{cm}, Y_{cm}, Z_{cm}, and the three angles θ_i. We therefore choose any $3N$–6 independent linear combinations of the displacements $\boldsymbol{\delta}_\alpha{}^{\text{vib}}$, which we shall refer to as the vibrational co-ordinates Q_β ($\beta = 1$ to $3N$–6).

We can also divide up the total Hamiltonian (23.2) corresponding to the new set of co-ordinates. We first expand the potential energy as

$$V(Q_\beta) = d_{\beta\gamma}Q_\beta Q_\gamma + \text{higher terms}, \quad (\beta, \gamma \text{ summed}) \qquad (23.6)$$

where we may neglect all higher terms than the second because for the present we shall discuss only small, simple harmonic vibrations.† The vibrational velocities of the nuclei can be specified by the co-ordinates $\dot{Q}_\beta = dQ_\beta/dt$. We can now write the total Hamiltonian for the nuclei as

$$\mathscr{H} = \mathscr{H}_{\text{cm}} + \mathscr{H}_{\text{rot}} + \mathscr{H}_{\text{vib}}, \qquad (23.7)$$

$$\mathscr{H}_{\text{cm}} = (1/2M)\mathbf{P}_{\text{cm}}^2,$$

$$\mathscr{H}_{\text{rot}} = \frac{L_x^2}{2\Omega_x} + \frac{L_y^2}{2\Omega_y} + \frac{L_z^2}{2\Omega_z}, \qquad (23.8)$$

$$\mathscr{H}_{\text{vib}} = c_{\beta\gamma}\dot{Q}_\beta\dot{Q}_\gamma + d_{\beta\gamma}Q_\beta Q_\gamma. \qquad (23.9)$$

Here $\mathscr{H}_{\text{cm}}$ describes the sideways motions of the whole molecule. $\mathscr{H}_{\text{rot}}$ is the kinetic energy of rotation, where Ox, Oy, Oz have been chosen along the principal axes of inertia of the molecule. The L_i are the corresponding components of the angular momentum, and Ω_x, Ω_y, Ω_z the principal moments of inertia. $\mathscr{H}_{\text{vib}}$ is the remaining vibrational energy, consisting of a kinetic term in the vibrational velocities and the potential term (23.6).

Actually this division (23.7) of the Hamiltonian is not quite correct. If we contemplate a gross rearrangement of our molecule such that the nuclei are moved far from their original positions in arbitrary directions, we come to the conclusion that there is no really unique way of dividing this displacement of the nuclei into so much pure rotation and so much vibrational displacement. The division is only unique in a first order of approximation for small distortions. Thus the Hamiltonian (23.7) contains terms that depend in fact on both the θ_i and the Q_β: for example the mean moments of inertia Ω_x, Ω_y, Ω_z are not really constants but are effected by large vibrations, and the vibrations can contribute to

† The linear and constant terms in (23.6) are of course absent because by definition the position $Q_\beta = 0$, for all β, gives a potential energy minimum $V(0) = 0$. Also V is independent of $\mathbf{R}_{\text{cm}}$ and θ_i since physically the potential energy of the molecule depends on its internal configuration only and not on its position or orientation relative to the laboratory OX, OY, OZ axes.

the angular momentum **L** if we have two degenerate modes as in a conical pendulum. References to further discussions of these difficulties are given at the end of this section.

Normal co-ordinates

The vibrational Hamiltonian (23.9) can be brought into a more convenient form by using a new set of co-ordinates

$$q_\beta = N_{\beta\gamma} Q_\gamma, \qquad \dot{q}_\beta = N_{\beta\gamma} \dot{Q}_\gamma, \tag{23.10}$$

where for the moment N is an unspecified matrix. With these co-ordinates the Hamiltonian becomes

$$\mathscr{H}_{\text{vib}} = (\tilde{N}^{-1}CN^{-1})_{\beta\gamma}\dot{q}_\beta\dot{q}_\gamma + (\tilde{N}^{-1}DN^{-1})_{\beta\gamma}q_\beta q_\gamma. \tag{23.11}$$

Now there is an important theorem in matrix algebra (Margenau and Murphy 1943, p. 311) which states that a matrix N can be found which satisfies the two conditions

$$\tilde{N}^{-1}CN^{-1} = \tfrac{1}{2}\begin{bmatrix} 1 & \cdot & \cdot & & \cdot \\ \cdot & 1 & \cdot & & \cdot \\ \cdot & \cdot & 1 & & \cdot \\ & & & & \\ \cdot & \cdot & \cdot & & 1 \end{bmatrix},$$

$$\tilde{N}^{-1}DN^{-1} = \tfrac{1}{2}\begin{bmatrix} \omega_1^{\,2} & \cdot & \cdot & & \cdot \\ \cdot & \omega_2^{\,2} & \cdot & & \cdot \\ \cdot & \cdot & \omega_3^{\,2} & & \cdot \\ & & & & \\ \cdot & \cdot & \cdot & & \omega_{3N-6}^{2} \end{bmatrix}.$$

Indeed there exist systematic methods for finding N as we shall see later. The corresponding co-ordinates q_β defined by (23.10) are called *normal co-ordinates*, and we have from (23.11)

$$\mathscr{H}_{\text{vib}} = \tfrac{1}{2}\sum_\beta (\dot{q}_\beta^{\,2} + \omega_\beta^{\,2} q_\beta^{\,2}).$$

The conjugate normal momenta are

$$p_\beta = \frac{\partial T}{\partial \dot{q}_\beta} = \dot{q}_\beta, \tag{23.12}$$

so that we finally obtain $\mathscr{H}_{\text{vib}}$ in the form required in quantum mechanics, namely in terms of co-ordinates and conjugate momenta only

$$\boxed{\begin{aligned}\mathscr{H}_{\text{vib}} &= \tfrac{1}{2}\sum_{\beta}(p_\beta{}^2 + \omega_\beta{}^2 q_\beta{}^2) \\ &= T_{\text{vib}} + V.\end{aligned}} \tag{23.13}$$

Incidentally we notice that the ω_β are all real ($\omega_\beta{}^2$ positive) since the potential energy for any distortion is positive. Each bracket

$$\tfrac{1}{2}(p^2 + \omega^2 q^2) \tag{23.14}$$

is the Hamiltonian for a simple harmonic oscillator, and the quantum mechanical energy levels are†

$$E_n = (n + \tfrac{1}{2})\hbar\omega.$$

Thus the energy levels and eigenstates of $\mathscr{H}_{\text{vib}}$ can be designated by a set of $3N$–6 numbers n_β (positive integers or zero).

$$E(n_1, n_2, \ldots n_{3N-6}) = \sum_{\beta}(n_\beta + \tfrac{1}{2})\hbar\omega_\beta. \tag{23.15}$$

Symmetry transformations

We now consider the group $\mathfrak{G}$ of symmetry transformations of the molecule when all the nuclei are in their equilibrium positions. By this we do not mean the rotation of the molecule in X, Y, Z space, which has already been described by the rotational co-ordinates θ_i. We mean the symmetry transformations of the molecule (rotations, reflections, inversion) in the x, y, z space, for instance the point-group of transformations $\bar{6}$m2 which was discussed for the ozone molecule in § 4. Let the equilibrium positions of the three oxygen atoms in the ozone molecule shown in Fig. 22a be given by some fixed co-ordinates in the x, y, z space. Then the symmetry operation consisting of a 180° rotation about the y-axis interchanges the nuclei 2 and 3, producing a new position of the molecule (Fig. 22b) which is indistinguishable from the original position. Consider now a distorted molecule, e.g. that shown in Fig. 22c where one atom has been displaced. After rotation by 180° about Oy, the molecule looks like Fig. 22d, and we can consider this as a new distortion of the original molecule. We say, therefore, that the distortion of Fig. 22c has been transformed

† The energy eigenvalues and eigenfunctions for a simple harmonic oscillator are derived in appendix G.

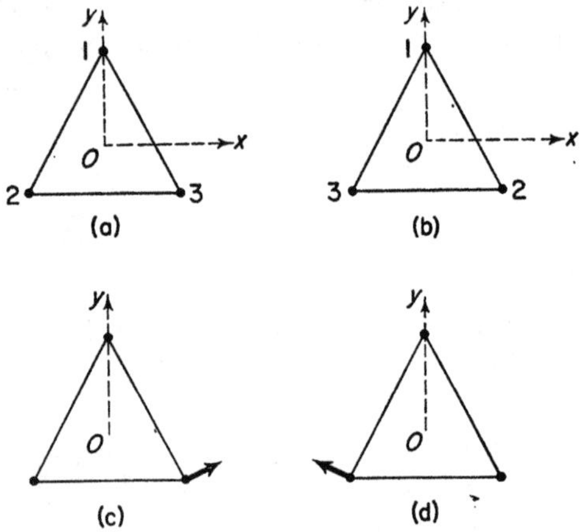

FIG. 22. Rotation of an ozone molecule by 180° about the y-axis: (a) and (b), two equilibrium positions of the molecule, related by the rotations and physically identical: (c) and (d), two distortions of the molecule related by the rotation.

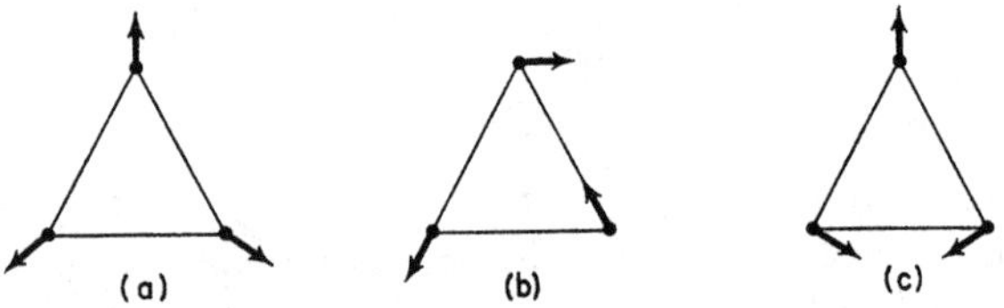

FIG. 23. Vibrational normal co-ordinates of the ozone molecule, (a) transforming according to the irreducible representation A'_1, and (b) and (c) according to E'.

by the rotation into the distortion of Fig. 22d. In this sense distortions can be transformed just like functions $f(x, y, z)$ can, and we can have distortions that transform according to definite representations. For example Fig. 23c shows a distortion which is invariant under a 180° rotation about Oy, whereas the distortion of Fig. 23b changes sign. The distortion of Fig. 23a is invariant under any transformation of the point-group $\bar{6}$m2.

From (23.13), the potential energy (23.1) is given in terms of normal co-ordinates by

$$V = \frac{1}{2} \sum_{\beta} \omega_\beta{}^2 q_\beta{}^2. \tag{23.16}$$

This must have the same value for two distortions such as Figs. 22c and 22d which are related by a symmetry transformation. That is, $V(q_\beta)$ is invariant under the group $\mathfrak{G}$. If a particular ω_β is not equal to any other ω_γ, then from (23.16) the invariance of V implies the invariance of q_β^2. Thus q_β has to transform according to a one-dimensional representation in which every element is represented by $+1$ or -1, or what is the same thing, by a *real one-dimensional representation* (problem 23.3). If two or three ω_β's are equal, say $\omega_\beta = \omega_{\beta+1} = \omega_{\beta+2}$, then we require the invariance of

$$\omega_\beta^2(q_\beta^2 + q_{\beta+1}^2 + q_{\beta+2}^2). \tag{23.17}$$

It is clearly necessary that q_β, $q_{\beta+1}$, $q_{\beta+2}$ transform only among themselves under $\mathfrak{G}$, i.e. form the basis of a representation. This representation must in general be irreducible for the following reason. Suppose the representation were reducible, say into a one-dimensional representation (base vector $\bar{q}_\beta$) and a two-dimensional component (base vectors $\bar{q}_{\beta+1}$, $\bar{q}_{\beta+2}$). Then from problem C.4 (appendix C), we have that

$$A\bar{q}_\beta^2 + B(\bar{q}_{\beta+1}^2 + \bar{q}_{\beta+2}^2)$$

is invariant, where $A = \bar{\omega}_\beta^2$ and $B = \bar{\omega}_{\beta+1}^2 = \bar{\omega}_{\beta+2}^2$ are arbitrary constants. It is not impossible that A and B are accidentally equal, but the probability of this happening exactly is zero. We therefore conclude that q_β, $q_{\beta+1}$, $q_{\beta+2}$ in (23.17) transform according to an irreducible representation, accidental degeneracies excepted as usual. The frequency ω_β and the co-ordinates q_β, $q_{\beta+1}$, $q_{\beta+2}$ are said to be three-fold degenerate. More generally, *a frequency is said to be n-fold degenerate if the corresponding co-ordinates transform according to an n-dimensional irreducible representation.*

The case of complex representations is somewhat anomalous. Let D and D^* be two complex conjugate representations, e.g. the two representations of the point-groups 4 which together are labelled E in appendix K. Such *complex conjugate representations always have to occur in pairs with equal values of* ω_β. We can see this in two ways. If q transforms according to a complex representation D under real rotations, etc., then q must itself be complex. To describe the molecule by real co-ordinates, we must use

$$q_1 = q + q^*, \qquad q_2 = q - q^*.$$

Here q^* transforms according to the complex conjugate representation D^*, so that q_1 and q_2 are not invariant but transform into one another under $\mathfrak{G}$. Thus they must occur together in (23.16) with equal values of ω_β. Another way of looking at this is from the point of view of time-reversal. It was shown in § 19 that states transforming according to D and D^* are degenerate. The complex co-ordinate q describes a travelling wave, rather than a standing wave oscillation as a real co-ordinate does. For instance q may represent a compression travelling round a benzene ring. The co-ordinate q^* then refers to the same compression travelling in the reverse direction around the ring. Clearly if one is a possible motion of the molecule, then so is the other, and they have the same frequency ω_β. It is for this reason that a pair of complex conjugate representations in appendix K have always been bracketed together and regarded as a single two-dimensional representation.

From (23.12), p_β transforms in the same way as q_β, so that in all of the above cases the kinetic energy $T = \sum p_\beta{}^2$ remains invariant. Thus the whole vibrational Hamiltonian is invariant under the group $\mathfrak{G}$ of symmetry transformations.

Enumeration of normal modes

We shall now show how the normal co-ordinates of vibration can be enumerated, by discussing the specific example of ozone. An arbitrary distortion or displacement of the molecule can be specified by the nine quantities $\delta_{x\alpha}$, $\delta_{y\alpha}$, $\delta_{z\alpha}$ ($\alpha = 1, 2, 3$) where $\boldsymbol{\delta}_\alpha$ is the displacement of nucleus α from its equilibrium position. These quantities form the basis of a representation D_{tot} of the symmetry group $\bar{6}$m2 of the molecule, whose characters can easily be written down. From Fig. 22 a rotation of 180° about Oy turns the initial distortion $\delta_{i\alpha}$ into a new distortion $\delta'_{i\alpha}$ given by

$$\begin{aligned} \delta'_{x1} &= -\delta_{x1}, & \delta'_{y1} &= \delta_{y1}, & \delta'_{z1} &= -\delta_{z1}, \\ \delta'_{x2} &= -\delta_{x3}, & \delta'_{y2} &= \delta_{y3}, & \delta'_{z2} &= -\delta_{z3}, \\ \delta'_{x3} &= -\delta_{x2}, & \delta'_{y3} &= \delta_{y2}, & \delta'_{z3} &= -\delta_{z2}. \end{aligned}$$

This rotation is represented therefore by the matrix shown in Table 21, and the character (sum of diagonal terms) is -1. Actually

TABLE 21

The Matrix Representing the Rotation by 180° about the y-axis

	δ_{x1}	δ_{y1}	δ_{z1}	δ_{x2}	δ_{y2}	δ_{z2}	δ_{x3}	δ_{y3}	δ_{z3}
δ'_{x1}	-1	.	.	.	.	.	.	.	.
δ'_{y1}	.	1	.	.	.	.	.	.	.
δ'_{z1}	.	.	-1	.	.	.	.	.	.
δ'_{x2}	.	.	.	.	.	.	-1	.	.
δ'_{y2}	.	.	.	.	.	.	.	1	.
δ'_{z2}	.	.	.	.	.	.	.	.	-1
δ'_{x3}	.	.	.	-1	.	.	.	.	.
δ'_{y3}	.	.	.	.	1	.	.	.	.
δ'_{z3}	.	.	.	.	.	-1	.	.	.

there is no need to go through all this business of writing down the matrices. For any transformations, *only those nuclei can contribute to the character which are transformed into themselves*, i.e. not interchanged with some other nucleus. Thus the 120° rotation about Oz has character zero because every nucleus is shifted round to a new position in the molecule. Moreover from (2.2), *each nucleus which is transformed into itself contributes* $\pm(1 + 2 \cos \theta)$ *to the character*, where the plus and minus refer to proper and improper rotations respectively. The different classes of symmetry elements of the group $\bar{6}$m2 are shown in Table 22, and applying the above two rules we have

$$\chi(E) = 9, \qquad \chi(m_z) = 3, \qquad \chi(3_z) = 0,$$
$$\chi(\bar{6}_z) = 0, \qquad \chi(2_y) = -1, \qquad \chi(m_x) = 1.$$

TABLE 22

Character Table of $\bar{6}$m2

6m2	E	m_z	3_z	$\bar{6}_z$	2_y	m_x		
A'_1	1	1	1	1	1	1		$x^2 + y^2; z^2$
A'_2	1	1	1	1	−1	−1	I_z	
A''_1	1	−1	1	−1	1	−1		
A''_2	1	−1	1	−1	−1	1	z	
E'	2	2	−1	−1	0	0	x, y	$x^2 - y^2, xy$
E''	2	−2	−1	1	0	0	I_x, I_y	yz, zx

Using Table 22 and equation (14.2) or (14.10), we can reduce this representation, and find

$$D_{\text{tot}} = A'_1 + A'_2 + A''_2 + 2E' + E'' \tag{23.18}$$

Since we started by considering an arbitrary displacement of the molecule, D_{tot} includes besides the rotational co-ordinates also the representations of the translational and rotational co-ordinates. The latter we now want to subtract out. The translational displacements X_{cm}, Y_{cm}, Z_{cm} transform into one another like the functions x, y, z. This can be seen pictorially from Fig. 24a for instance, which shows δX_{cm}. Thus from Table 22 δZ_{cm} belongs to the representation A''_2, and δX_{cm}, δY_{cm} to E': i.e.

$$D_{\text{trans}} = A''_2 + E'.$$

Turning to the rotations, we have from (8.2) that a small rotation $R(\theta, z)$ induces the displacements

$$\boldsymbol{\delta}_\alpha^{\mathrm{rot},\,z} = i\theta I_z \mathbf{r}_{\alpha 0}$$

of the nuclei from the original positions. The set of equilibrium positions $\mathbf{r}_{\alpha 0}$ of all the atoms is invariant under $\mathfrak{G}$. Therefore if we write R_z for the set of displacements $\boldsymbol{\delta}_\alpha^{\mathrm{rot},\,z}$, we have that R_x, R_y, R_z transform in the same way as I_x, I_y, I_z. The latter are the components of a pseudo-vector, i.e. they transform like x, y, z under

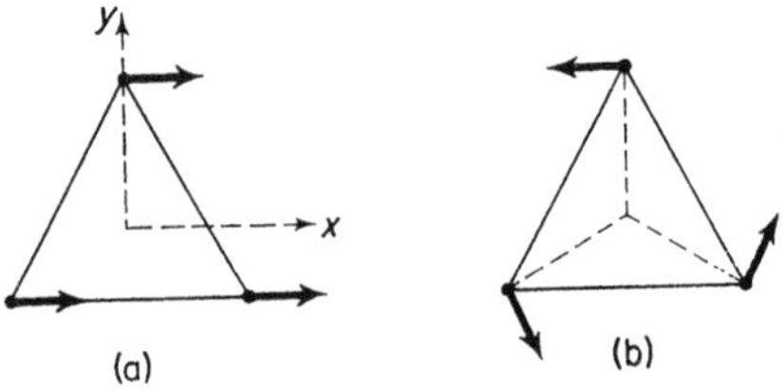

Fig. 24. Displacements of the ozone molecule; (a) the translational displacement δX_{cm}, (b) the rotational displacement R_z.

proper rotations but remain invariant under inversion. This was proved in appendix F and § 11. Thus the characters of the representation formed by R_x, R_y, R_z or I_x, I_y, I_z are given by

$$\begin{aligned} \chi(\text{of } I_x, I_y, I_z) &= \chi(\text{of } x, y, z) \textit{ for proper rotations,} \\ &= -\chi(\text{of } x, y, z) \textit{ for improper rotations.} \end{aligned} \tag{23.19}$$

This is illustrated for instance by the representations A'_2 and A''_2 (Table 22) appropriate to I_z and z. The rotational displacement R_z is shown in Fig. 24b, and this also makes it clear that R_z transforms according to A'_2. Thus from Table 22 we have for the group $\bar{6}$m2

$$D_{\mathrm{rot}} = A'_2 + E''.$$

Now subtracting D_{trans} and D_{rot} from (23.18), we obtain finally

$$D_{\mathrm{vib}} = A'_1 + E'. \tag{23.20}$$

The ozone molecule therefore has one non-degenerate vibrational frequency (A'_1) and one doubly degenerate frequency (E'). The modes of vibration are shown in Fig. 23.

The normal mode A'_1 can easily be written down by inspection (Fig. 23a), and so can the other two with a little practice. But the projection operators (14.11) enable one to obtain any mode systematically. We recall that if δ represents symbolically an arbitrary displacement, then

$$q = \sum_T \chi^{(\lambda)*}(T)T\delta \tag{23.21}$$

transforms according to the irreducible representation $D^{(\lambda)}$, and in particular

$$q_i = \sum_T D_{ij}^{(\lambda)*}(T)T\delta$$

according to the ith row of that representation. In order to use these formulae, we need to represent a displacement mathematically. There are two ways, both of which we have already used to some extent. The first method is to represent a displacement by the nine co-ordinates (in the case of ozone) $\delta_{x\alpha}$, $\delta_{y\alpha}$, $\delta_{z\alpha}$ or by some linear combination of them. For example

$$\delta X_{\text{cm}} = \tfrac{1}{3}(\delta_{x1} + \delta_{x2} + \delta_{x3}) \tag{23.22}$$

clearly measures the X displacement of the centre of mass. In (23.21) we can put δ equal to any linear combination of the $\delta_{j\alpha}$, $T\delta$ then being obtained from Table 21 plus the corresponding matrices for the whole group. Choosing $D^{(\lambda)}$ as the representation A'_1, it can easily be found in this way that the corresponding co-ordinate is

$$q^{(A'_1)} = \delta_{y1} - \tfrac{1}{2}\delta_{y2} - \tfrac{1}{2}\sqrt{3}\delta_{x2} - \tfrac{1}{2}\delta_{y3} + \tfrac{1}{2}\sqrt{3}\delta_{x3}. \tag{23.23}$$

What does this mean? It means that if the arbitrary displacement described by the $\delta_{j\alpha}$ is decomposed into a translation, a rotation, and a sum of normal vibrational distortions, then the amplitude of the A'_1 mode (Fig. 23a) is given by (23.23). The second method of representing a normal co-ordinate is a more pictorial one, and is obtained if we associate with each nucleus a displacement vector such as in Figs. 22 to 24. Starting with an arbitrary set of displace, ment vectors and applying (23.21), we easily obtain Fig. 23a. The relationship between the two ways of representing a normal co-ordinate is the following. Fig. 23a represents such a displacement of the molecule, that the value of $q(A'_1)$ (23.23) is non-zero while the values of all the other q's are zero. Fig. 24a and (23.22) are similarly related. A little difficulty arises when a given representation occurs more than once in D_{tot}. For instance when we apply the construction (23.21) to the representation E', we will in general obtain a linear combination of the vibrational mode, Fig. 23b or c, and the translation X_{cm} or Y_{cm}, Fig. 24a, since each transforms according to E'. We then have to subtract out some δX_{cm}, δY_{cm} until there is no displacement of the centre of mass, in order to obtain a purely vibrational mode.

Infra-red and Raman spectra

The main discussion of infra-red and Raman spectra will be deferred till the next section, but if we neglect all overtones and combination frequencies, the situation is briefly as follows. *A normal vibration q_β gives an absorption line in the infra-red spectrum if q_β transforms in the same way as one of the dipole moment components μ_x, μ_y, μ_z. Further, q_β gives a line in the Raman spectrum if it transforms in the same way as one of the polarizability tensor components α_{ij}* (or linear combination thereof). The dipole moment is an ordinary vector, and the *components μ_x, μ_y, μ_z transform like x, y, z.* In the case of ozone this is according to $E' + A''_2$. Hence the doubly-degenerate vibrational mode E' of (23.20) is infra-red active, whereas the A'_1 mode is not. *The polarizability α_{ij} is a symmetric tensor and hence the components transform in the same way as x^2, y^2, z^2, xy, yz, zx.* We have already obtained the transformation properties of these quantities in (5.20) for the point-group 32, and from this they can be assigned by inspection to the representations of $\bar{6}m2$ as shown in Table 22. Thus any vibrational mode transforming according to A'_1, E' or E'' is Raman active, and this includes all of the vibrational modes (23.20) for ozone.

The transformation properties of these quantities are summarized in Table 23. From this table together with the group

TABLE 23

Transformation Properties of Some Often Used Quantities

Quantity	Transforms like
translation of centre of mass, X_{cm}, Y_{cm}, Z_{cm}	x, y, z (vector)
Rotational displacements R_x, R_y, R_z	I_x, I_y, I_z (pseudo-vector)
dipole moment, μ_x, μ_y, μ_z	x, y, z
polarizability tensor, α_{ij}	x^2, y^2, z^2, xy, yz, zx

character tables (appendix K), the symmetry species of the vibrational normal co-ordinates of any molecule can easily be derived by the method described above, and the infra-red and Raman active ones among them written down.

Linear molecules

The procedure for obtaining the normal vibrations of a linear molecule is the same as that given for ozone above, with one minor difference. If we choose the axis of the molecule as the z-axis and consider the molecule in its equilibrium state, then a rotation about Oz does not change the position of the molecule at all. *We have therefore only two rotational co-ordinates* θ_i or R_x, R_y. This has to be remembered when subtracting R_{rot} out from D_{tot}.

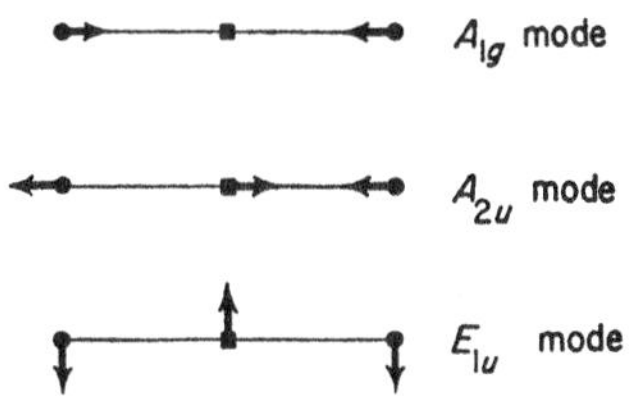

FIG. 25. Vibrational normal modes of the CO_2 molecule. The second E_{1u} mode is the same as the one shown above, but in the plane perpendicular to the page.

The symmetry group of a linear molecule is either ∞m or ∞/mm, depending on whether it has a centre of inversion symmetry or not. Consider for example carbon dioxide, which has symmetry ∞/mm. Applying our rules for the characters, we obtain for D_{tot}

$$\chi(E) = 9, \qquad \chi[R(\phi, z)] = 6 \cos \phi + 3, \qquad \chi(2_x) = -1,$$
$$\chi(\bar{1}) = -3, \qquad \chi[\bar{1}R(\phi, z)] = -2 \cos \phi - 1, \qquad \chi(m_x) = 3,$$

whence from the character table for ∞/mm (appendix L)

$$D_{\text{tot}} = A_{1g} + 2A_{2u} + E_{1g} + 2E_{1u}.$$

We subtract out the three translational components in accordance with Table 23 and the character table

$$D_{\text{trans}} = A_{2u} + E_{1u},$$

and the *two* rotational co-ordinates

$$D_{\text{rot}} = E_{1g},$$

and obtain

$$D_{\text{vib}} = A_{1g} + A_{2u} + E_{1u}. \tag{23.24}$$

There are therefore two non-degenerate modes (A_{1g} and A_{2u}) and a pair of degenerate ones (E_{1u}). Of these A_{1g} is Raman but not infra-red active, and the others are infra-red but not Raman active. The normal modes are pictured in Fig. 25.

The ammonia inversion line

At the beginning of this section we adopted the model that the electrons in a molecule move very rapidly around the nuclei, whose vibrations in comparison are slow semi-static displacements. The wave function corresponding to this model has the form

$$\boxed{\Psi = \psi_{\text{elect}}(\mathbf{r}_{i\alpha}; \mathbf{R}_\alpha)\, \psi_{\text{nucl}}(\mathbf{R}_\alpha),} \tag{23.25a}$$

and it can indeed be shown that this is in general a very good approximation to the wave function. Here ψ_{elect} is a wave function in the electronic co-ordinates $\mathbf{r}_i$, with the nuclear co-ordinates $\mathbf{R}_\alpha$ entering only as fixed parameters. The approximate separation (23.7) of the nuclear Hamiltonian further implies that ψ_{nucl} can be written approximately as

$$\boxed{\psi_{\text{nucl}}(\mathbf{R}_\alpha) = \psi_{\text{trans}}(\mathbf{R}_{\text{cm}})\, \psi_{\text{rot}}(\theta_i)\, \psi_{\text{vib}}(q_\beta).} \tag{23.25b}$$

However, our argument so far implicitly assumes that the nuclei never penetrate far from a definite equilibrium configuration, as witnessed for instance by the fact that our discussion has been couched in terms of small displacements. If each nucleus departs arbitrarily from its position by something of the order of the internuclear distance, then it is difficult to regard such a displacement as the sum of a rotation of the whole molecule and an internal distortion, nor are the approximations (23.25) valid any more.

For this reason, our model breaks down when a molecule can take up more than one distinct equilibrium configuration. An example of this is the ammonia molecule (Fig. 26) in which the nitrogen nucleus can either be above or below the place of the three hydrogen nuclei (position N_1 or N_2). We would expect that the nitrogen nucleus has quite a large potential barrier to overcome in passing from one position to the other, so that this transition will be slow compared with the vibrational frequencies. We can therefore write down two approximate wave function $\psi_{\text{nucl},1}$ and $\psi_{\text{nucl},2}$ to describe the molecule vibrating about the two equilibrium

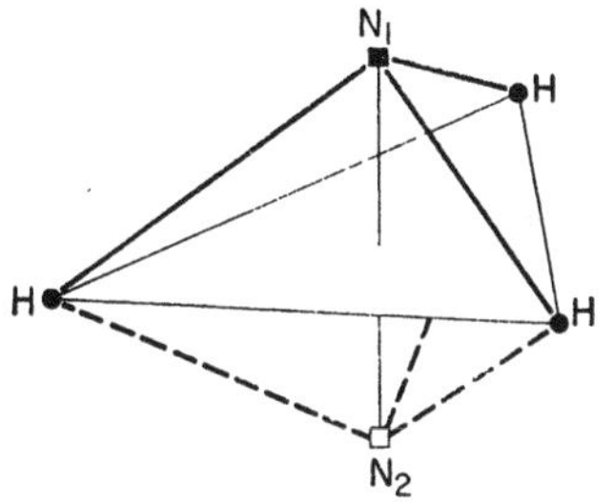

FIG. 26. The two possible positions N_1 and N_2 of the nitrogen nucleus in an ammonia molecule.

positions respectively. However, the transition from N_1 to N_2 is a symmetry transformation for the system of nuclei (though it is not a symmetry of each equilibrium configuration individually), and the whole molecule wave function must be either the symmetric or antisymmetric combination

$$\psi_{\text{elect}}(\psi_{\text{nucl},1} \pm \psi_{\text{nucl},2}). \tag{23.26}$$

Now the time t for a transition from N_1 to N_2 is very long, and the energy difference $\Delta E \sim \hbar/t$ between the two states (23.26) is rather small. This has two consequences. Firstly, the transition N_1 to N_2 gives an extra spectral line, namely the so-called inversion line of ammonia which is in the microwave (1.25 cm) region of the spectrum, compared with the vibration frequencies which lie in the infra-red. This is the line used in the ammonia maser. Secondly, it means the rest of the vibrational spectrum is just that of one configuration alone, with possibly some fine structure of magnitude ΔE.

Other examples of molecules with several equilibrium configurations occur among organic molecules with hindered rotation. For instance in C_2Cl_6, the two $C\,Cl_3$ groups are not free to rotate about the central C—C bond because the big chlorine atoms get in one another's way. However, the two groups can take up three different relative positions, in which the two sets of chlorine atoms interlock.

References

For the separation (23.25a) between the electronic and nuclear motion, the reader is referred to Eyring, Walter and Kimball (1944, p. 190) and to the original paper by Born and Oppenheimer 1927. The separation of the nuclear Hamiltonian into the centre of mass, the rotational and vibrational motions is discussed by Wilson, Decius and Cross (1955, Chapter 11) and by Bodi and Curtis (1956). The energy levels of the rotational Hamiltonian are conveniently summarized by Wilson *et al.* (1955, appendix 16), the results for a linear molecule being derived by Eyring *et al.* (1944, p. 72). The whole subject of molecular vibrations is very fully and clearly discussed in the book by Wilson, Decius and Cross (1955). Two shorter reviews having a scope similar to the present section are Rosenthal and Murphy (1936) and Sponer and Teller (1941).

Summary

The motion of the nuclei in a molecule can be represented approximately as the sum of three parts; a sideways translation of the whole molecule, a rotation of the molecule as a whole, and internal vibrations or distortions of the molecule. The latter are expressed in terms of the vibrational normal modes which have different symmetries. This symmetry determines whether or not a particular vibration gives a line in the infra-red spectrum and the Raman spectrum of the substance.

PROBLEMS

23.1 What are the symmetry groups for the following molecules: C_2H_2, H_2, H_2O, DHO (heavy water), NH_3, CH_4, benzene, cyclohexane? Note: H_2O is not a linear molecule because of the lone pair electrons on the oxygen; benzene is planar.

23.2 Enumerate the symmetries of the normal vibrations of the above molecules and illustrate some of them by figures. Determine which ones contribute as fundamentals to the infra-red and Raman spectra.

23.3 Prove that every element of a real one-dimensional representation is ± 1. Hint: use the unitary property of problem C.2.

23.4 By looking at the character of m_z, sort out the normal modes of benzene into those that are vibrations in the plane of the molecule and those perpendicular to it.

23.5* In order to interpret the vibration spectra of a complicated molecule, one usually makes a theoretical calculation of the frequencies from a theoretical model, in which some force constants are assumed for the various inter-atomic bonds to represent their resistance to stretching or bending. Give an account of how the normal modes and frequencies are calculated from the model, and in particular of how group theory can help in the calculation. Reference: Wilson, Decius and Cross (1955).

23.6* Outline how the energy levels of a rigidly rotating molecule are derived (rigid rotor), distinguishing between the linear, spherical, symmetric and asymmetric cases. Obtain the degeneracies of the levels, and selection rules for transitions between them. References: Casimir 1931; Wilson *et al.* 1955, appendix 16; Herzberg 1945.

23.7* Show how to incorporate the nuclear spins (§ 21) into the wave function (23.25) of a whole molecule, and discuss their effect on the spectra of diatomic molecules with particular reference to H_2, HD, D_2, and O_2^{16}. The nuclei of H, D, O^{16} have spins of $\frac{1}{2}$, 1, 0 respectively, and D stands for a deuterium atom (Eyring *et al.* 1944, p. 265).

24. Infra-red and Raman Spectra

In the last section the wave function for a whole molecule was written in the approximate form

$$\boxed{\Psi = \psi_{\text{elect}}(\mathbf{r}_i;\, \mathbf{R}_\alpha)\, \psi_{\text{trans}}(\mathbf{R}_{\text{cm}})\, \psi_{\text{rot}}(\theta_i)\, \psi_{\text{vib}}(q_\beta).} \tag{24.1}$$

Here the vibrational wave function

$$\boxed{\psi_{\text{vib}} = \psi_{\text{vib}}(n_1,\, n_2,\, \ldots\, n_{3N-6})} \tag{24.2a}$$

corresponds to an excited state of the molecule with n_β quanta of energy in the βth normal mode of vibration, the energy of the state being

$$\boxed{E(n_\beta) = \sum_\beta (n_\beta + \tfrac{1}{2})\hbar\omega_\beta.} \tag{24.2b}$$

The existence of these vibrational excited states and the transitions between them give rise to various types of line spectra, of which we shall discuss the infra-red absorption spectrum and the Raman spectrum.

Infra-red absorption spectrum

If two energy levels E_1 and E_2 of the molecule are separated by an amount

$$\boxed{\hbar\omega = E_2 - E_1,} \tag{24.3}$$

then the molecule can absorb light of frequency ω and make a transition from the state E_1 to E_2. For light polarized in the X-direction, the transition probability, and hence the absorption coefficient of the bulk liquid or gas, is proportional to the modulus squared of the matrix element

$$\langle 2|D_X|1\rangle = \int \Psi_2{}^* D_X \Psi_1\, dv. \tag{24.4}$$

Analogously to the atomic case (13.7), D_X is the X component of the dipole moment operator

$$\mathbf{D} = -e\sum \mathbf{r}_i + e\sum Z_\alpha \mathbf{R}_\alpha, \tag{24.5a}$$

where the summation is over all the electrons (i) and nuclei (α) Z_α being the atomic number.

The formula (24.4) can be reduced using the wave functions (24.1). In almost all cases, the *electronic* energy levels of a molecule are separated by about 1 electron volt (10^4 cm^{-1}) or more, corresponding to visible and ultra-violet frequencies. Infra-red light has insufficient energy to excite the electrons to above the ground state ψ_{eo}, so that in an infra-red transition we must have

$$\psi_{\text{elect},\,1} = \psi_{\text{elect},\,2} = \psi_{\text{eo}}.$$

Turning now to the translational wave function, we have

$$\psi_{\text{trans}} = \exp\,(i\mathbf{k}\cdot\mathbf{R}_{\text{cm}})$$

for a freely moving molecule, so that

$$\psi_{\text{trans}}(\mathbf{R}_\alpha + \mathbf{\Delta}) = \exp(i\mathbf{k}\cdot\mathbf{\Delta})\psi_{\text{trans}}(\mathbf{R}_\alpha).$$

Since $\mathbf{D}$ is invariant under a translation $\mathbf{\Delta}$ of the molecule, we have from the fundamental theorem (13.8b) that $\psi_{\text{trans},1}$ and $\psi_{\text{trans},2}$ must belong to the same translational representations for (24.4) to be non-zero, i.e. we must have $\mathbf{k}_1 = \mathbf{k}_2$. We consider finally the rotational and vibrational co-ordinates, and refer $\mathbf{D}$ to the x, y, z co-ordinate axes fixed to the molecule as defined in the last section;

$$D_X = l_{Xx}D_x + l_{Xy}D_y + l_{Xz}D_z, \tag{24.5b}$$

Here the l_{Ij} are the direction cosines of the x, y, z set of axes relative to the laboratory set of X, Y, Z, and are functions of the rotational co-ordinates θi only. With these substitutions (24.4) becomes

$$\langle 2|D_X|1\rangle = \sum_j \int \psi^*_{\text{rot},2} l_{Xj}\psi_{\text{rot},1}\,dv \int \psi^*_{\text{vib},2}\,\mu_j\,\psi_{\text{vib},1}\,dv, \tag{24.6}$$

where

$$\boxed{\mu_j(q_\beta) = \int \psi_{\text{eo}}{}^*(\mathbf{r}_i;\mathbf{R}_\alpha)\;D_j(\mathbf{r}_i,\mathbf{R}_\alpha)\;\psi_{\text{eo}}(\mathbf{r}_i;\mathbf{R}_\alpha)\;dv_{\text{e}}.} \tag{24.7}$$

In (24.7) only the electron co-ordinates are integrated over. Physically $\mu_j(j = x, y, z)$ is the value of the dipole moment of the molecule when the nuclei are in the positions specified by the q_β or $\mathbf{R}_\alpha$.

The form (24.6) for the transition matrix element is particularly convenient for discussing selection rules. The first integral depends on the θ_i only, and because of the factor $l_{Xj}(\theta_i)$ the transition Ψ_1 to Ψ_2 can involve a change in the rotational as well as the vibrational state of the molecule simultaneously. However, the *rotational* quanta of energy are of the order of only 1 cm^{-1}, whereas the *vibrational* frequencies are in the infra-red with $\hbar\omega \approx 10^2$ to 10^4 cm^{-1}. Consequently the purely rotational transitions are seen more in the microwave region of the spectrum. In the infra-red, a transition involving a vibrational quantum can be accompanied by a simultaneous rotational transition, so that the vibrational lines have a fine structure. This structure is determined purely by the first integral in (24.6) and we shall not consider it further. However, we obtain a selection rule for whether the whole vibrational line is absent or present from whether the second integral of (24.6), namely the matrix element

$$\begin{aligned}\langle 2|\mu_j|1\rangle &= \int \psi^*_{\text{vib},2}\,\mu_j\,\psi_{\text{vib},1}\,dv_n\\ &= \int (\psi_{\text{vib},2}\,\psi^*_{\text{vib},1})^*\,\mu_j\,dv_n,\end{aligned} \tag{24.8}$$

is zero or non-zero. From the matrix element theorem (13.8b), (24.8) can be non-zero only if the representation of μ_j is contained in the reduction of the representation $D^{(\text{vib},2)} \times D^{(\text{vib},1)*}$. Moreover, as usual in such symmetry arguments, we can assume that the converse is also true; i.e. if the representation of μ_j is contained in $D^{(\text{vib},2)} \times D^{(\text{vib},1)*}$, then we expect (24.8) to be non-zero (apart from accidental cases) as long as we have included *all* the symmetry properties of the system in the group $\mathfrak{G}$ to which the representations refer. From (24.5a), the μ_j transform like a vector in the same way as x, y, z (Table 23), according to the representation $D^{(\text{vect})}$ say. Also in (24.6) there is a summation over j, so that the transition probability is finite if at least *one* of the matrix elements $\langle 2|\mu_j|1\rangle$ is non-zero. We have therefore derived the general rule:

there is an infra-red absorption line corresponding to the transition $\psi_{\text{vib},1}$ to $\psi_{\text{vib},2}$ if and only if the representation $D^{(\text{vect})}$ of one or more of the dipole moment components μ_x, μ_y, μ_z is contained in the reduction of the representation $D^{(\text{vib},2)} \times D^{(\text{vib},1)*}$.

(24.9)

We shall not discuss till later how to derive the transformation properties of the vibrational wave functions, but suppose for the sake of example that we are interested in the absorption spectrum of ozone, and that $\psi_{\text{vib},1}$ and $\psi_{\text{vib},2}$ transform according to the representations A'_2 and E' of the symmetry group $\bar{6}$m2 (Table 22). The μ_j transform as a vector like x, y, z, i.e. in the present case μ_x, μ_y according to the representation E' and μ_z according to A''_2. Also the product $\psi_{\text{vib},2}\psi^*_{\text{vib},1}$ transforms according to $E' \times A'_2$, which from (14.3) and the character Table 22 turns out to be equal to E'. This contains the representation of μ_x, μ_y but not that of μ_z. Hence the matrix element $\langle 2|\mu_z|1\rangle$ is zero whereas $\langle 2|\mu_x|1\rangle$ and $\langle 2|\mu_y|1\rangle$ are non-zero, and from (24.6) the latter two are sufficient to guarantee an absorption line in the spectrum corresponding to the energy $E_2 - E_1$.

Raman spectrum

In the Raman effect, one shines light of a fixed frequency ω_1 usually from the mercury spectrum on to a substance, and one observes light of a different frequency ω_2 emerging scattered by the substance. The molecules of the substance meanwhile make a vibrational and/or rotational transition Ψ_1 to Ψ_2, the total energy of course being conserved

$$\boxed{\hbar\omega_1 + E_1 = \hbar\omega_2 + E_2.}$$

Thus it is the *frequency shift* $(\omega_1 - \omega_2)$ that corresponds to the energy difference $E_2 - E_1$. The intensity I_2 of the scattered light is given by

$$\boxed{I_2 = (\omega_2{}^4/c^4)|\langle\alpha_{21}\rangle|^2 I_1.} \tag{24.10}$$

Here α_{21} is a component of the polarizability tensor of the molecule when the nuclei are in fixed positions $\mathbf{R}_\alpha$. The suffices 1 and 2 refer to the directions X, Y or Z of polarization of the incident and emergent beams. As well as being derivable quantum mechanically as in problem 24.11, (24.10) is suggested by a semi-classical argument. The electric field $\mathscr{E}_1$ of the incident beam induces a dipole moment in the molecule of strength

$$\mu = \alpha\mathscr{E}_1.$$

This oscillating dipole radiates energy at a rate proportional to μ^2, and $|\mathscr{E}_1|^2$ is proportional to the indicent intensity. Thus

$$I_2 \propto \mu^2 = \alpha^2\mathscr{E}_1{}^2 \propto \alpha^2 I_1.$$

Since the molecule is itself vibrating with frequencies ω_β, the various combination frequencies

$$\omega_2 = \omega_1 \pm \omega_\beta, \text{ etc.} \tag{24.11}$$

occur among the radiated light. Just as in the direct optical absorption theory the classical μ_X gets replaced the quantum mechanical matrix element (24.6), so analogously α_{21} gets replaced by

$$\langle 2|\alpha_{IJ}|1\rangle = \int \psi^*_{\text{rot},2} l_{Ii} l_{Jj} \psi_{\text{rot},1}\, dv \int \psi^*_{\text{vib},2}\, \alpha_{ij}\, \psi_{\text{vib},1}\, dv. \quad (i, j \text{ summed}) \tag{24.12}$$

As in the case of infra-red absorption, the first integral determines the purely rotational Raman effect and the rotational fine structure of each vibrational Raman line, though these effects are only rarely observable in practice. The last integral in (24.12) determines the selection rules for the possible vibrational transitions. The α_{ij} are the components of a tensor and transform respectively like x^2, y^2, z^2, xy, yz, zx. Analogously to (24.9), we obtain the rule:

> *a possible vibrational transition $\psi_{\text{vib},1}$ to $\psi_{\text{vib},2}$ can give rise to a line in the Raman spectrum if and only if the representation $D^{(\text{tensor})}$ of one or more of the polarizability component α_{ij} occurs in the reduction of $D^{(\text{vib},2)} \times D^{(\text{vib},1)*}$*

(24.13)

For example consider again the transition $\psi_{\text{vib},1}(A'_2)$ to $\psi_{\text{vib},2}(E')$ for ozone already discussed above. The representation $D^{(\text{vib},2)} \times D^{(\text{vib},1)*}$ is

$$E' \times A'_2 = E'$$

as before, and from Table 22 this is seen to be the same as the representation formed by the components $\alpha_{xx} - \alpha_{yy}$, α_{xy}. The Raman spectrum therefore contains a line due to this transition.

Fundamental frequencies

The selection rules (24.9) and (24.13) are completely general. To apply them, we would first have to work out the irreducible representations and energies of all the vibrational states of the molecule, and then start testing all possible pairs of states to see whether the transition between them is an allowed one. While this may be a useful thing to do in the later stages of analysing a spectrum to understand all its detail, it is clearly unsuited to giving a quick overall idea of what kind of spectrum to expect from any given molecule. It is too tedious, and does not distinguish between strong lines and ones too weak to observe. Moreover the selection rules (24.9) and (24.13) are concerned with the initial and final *states* and not at all with the *energy difference* between them. For instance a transition from a state with $n_\beta = 6$ to one with $n_\beta = 7$ gives the same spectral line as a transition from $n_\beta = 2$ to $n_\beta = 3$, or indeed as any transition involving an energy difference of one quantum $\hbar\omega_\beta$. The above selection rules can tell us individually which of these transitions are allowed, but the argument we shall now develop will tell us whether the line as such is allowed, i.e. whether or not there are *any* allowed transitions with a given energy difference. Furthermore the allowed lines will be grouped into sets in approximately decreasing order of intensity.

The dipole moment μ_x, μ_y, μ_z (24.7) is a function of the internal co-ordinates q_β only, and we can expand it (using the summation convention) as

$$\mu_j = \mu_{j0} + q_\beta \left(\frac{\partial \mu_j}{\partial q_\beta}\right)_0 + q_\beta q_\gamma \left(\frac{\partial^2 \mu_j}{\partial q_\beta \partial q_\gamma}\right)_0 + \text{higher terms.} \tag{24.14}$$

Here the suffix zero refers to the equilibrium position of the nuclei. The first term μ_{j0} is the permanent dipole moment of the molecule, and from (24.6) we see that it determines the intensity of the purely rotational transitions. We can tell whether μ_{j0} is zero or not as follows. On the one hand it has already been noted from (24.5a) and (24.7) that μ_z and hence μ_{z0} transforms like the function z, which in the case of the ozone molecule (point-group $\bar{6}$m2) means the representation A''_2. On the other hand μ_{z0} is a property of the equilibrium configuration of the molecule, and must therefore be left invariant under the group $\bar{6}$m2 (since each operation of the group simply moves the nuclei around into a new set of positions which is physically equivalent to the original set: this is what is meant by the symmetry group of the molecule!) That is, μ_{z0} transforms according to the identity representation A'_1. Now μ_{z0} cannot at the same time transform according to both A''_2 and A'_1, and the only way out of this self-contradiction is to have μ_{z0} zero. Similarly μ_{x0}, μ_{y0} are zero and we conclude that the ozone molecule has no permanent dipole moment (as is also intuitively obvious).

A similar argument can be applied to the second terms of (24.14). The suffix zero on the quantity $(\partial\mu_j/\partial q_\beta)_0$ denotes that it is evaluated at the equilibrium position. It is a numerical quantity which is a property of the molecule in its equilibrium configuration. It is therefore invariant under the symmetry group, i.e. transforms according to the identity representation which in the present case is A'_1. Thus the term

$$q_\beta \left(\frac{\partial \mu_j}{\partial q_\beta}\right)_0 \tag{24.15}$$

transforms in the same way as q_β. But it also has to transform in the same way as μ_j, and hence (24.15) *can be non-zero only if* q_β *transforms in the same way as* μ_j. In the case of ozone the vibrational co-ordinates are

$$\begin{aligned} &q_1 \text{ transforming according to } A'_1, \\ &q_2, q_3 \text{ transforming according to } E', \end{aligned} \tag{24.16}$$

and μ_x, μ_y, μ_z transform according to E' and A''_2 as already noted. Hence the q_1 term is completely absent, but the q_2, q_3 terms occur in μ_y, μ_z.

The question now is, what kinds of spectral line does the term (24.15) give rise to? It is shown in most elementary books on quantum mechanics and also later in this section (see equation (24.31)), that for simple harmonic oscillator wave functions the matrix elements of q_β are zero, except between pairs of states differing in energy by one quantum $\hbar\omega_\beta$. In symbols

$$\int \psi^*(n_{\beta 2})\, q_\beta\, \psi(n_{\beta 1}) dv = 0 \qquad \text{unless } n_{\beta 2} - n_{\beta 1} = \pm 1 \text{ and } n_{\gamma 2} = n_{\gamma 1} \text{ for } \gamma \neq \beta. \tag{24.17}$$

The absorption lines are thus at the *fundamental frequencies* ω_β. Combining this with the italicized result above (24.16), we have the rule:

> *there is an infra-red absorption line at the fundamental frequency* ω_β *if the corresponding normal co-ordinate* q_β *transforms in the same way as one of the dipole moment components* μ_x, μ_y, μ_z.

(24.18)

This is just the rule stated without proof and illustrated in § 23.

The theory for Raman lines is completely analogous. We expand the polarizability components;

$$\alpha_{ji} = \alpha_{ij,0} + q_\beta \left(\frac{\partial q_\beta}{\partial \alpha_{ij}}\right)_0 + q_\beta q_\gamma \left(\frac{\partial^2 \alpha_{ij}}{\partial q_\beta \partial q_\gamma}\right)_0 + \text{higher terms.} \tag{24.19}$$

The first term on the right side determines the intensity of the purely rotational Raman spectrum, which only exists if at least one of the components α_{ij} (or a linear combination thereof) transforms according to the identity representation. The next term gives rise to the fundamental Raman lines and we have the rule:

the Raman spectrum contains the fundamental line of frequency shift ω_β if q_β transforms in the same way as one of the polarizability components α_{ij} (or a linear combination of them).

(24.20)

Overtones and combination tones

The same type of argument can be applied to determine which quadratic and higher terms are present in the expansions (24.14) and (24.19). All the derivatives such as $(\partial^2\mu_j/\partial q_\beta \partial q_\gamma)$ in (24.14) remain invariant under symmetry transformations of the molecule, and they are non-zero only if the appropriate product like $q_\beta q_\gamma$ transforms according to the same irreducible representations as μ_j does. For instance from (24.16), we have for ozone

$$\begin{aligned} q_1^2 &\text{ transforms according to } A'_1 \times A'_1 = A'_1, \\ q_1q_2,\, q_1q_3 &\text{ transforms according to } A'_1 \times E' = E', \qquad (24.21\text{a}) \\ q_2^2 + q_3^2 &\text{ transforms according to } A'_1, \\ q_2^2 - q_3^2,\, q_2q_3 &\text{ transforms according to } E'. \qquad (24.21\text{b}) \end{aligned}$$

Thus all of the above terms can appear in one or another of the α_{ij}, but only the terms (24.21a) and (24.21b) can occur in the μ_j expansions.

Before going on to derive what lines these quadratic and higher terms give rise to, it is necessary to recall some results about simple harmonic oscillator wave functions which are derived fully in appendix G. If we put

$$a = (2\hbar\omega)^{-1/2}(p + i\omega q), \qquad a^* = (2\hbar\omega)^{-1/2}(p - i\omega q), \qquad (24.22)$$

then the Hamiltonian for each mode of vibration becomes

$$\mathscr{H} = \tfrac{1}{2}(p^2 + \omega^2 q^2) = (aa^* + \tfrac{1}{2})\hbar\omega. \qquad (24.23)$$

The eigenvalues of the operator aa^* are the positive integers n (including zero). The energy levels are therefore

$$E_n = (n + \tfrac{1}{2})\hbar\omega$$

and n is the number of energy quanta in the oscillation above the "zero-point energy" $\frac{1}{2}\hbar\omega$. The corresponding wave functions can be written

$$\psi_n = a^n \psi_0. \tag{24.24}$$

The wave functions in this form are *not* normalized, but this is irrelevant in the present context since we are not interested here in calculating absolute intensities of lines. Similarly the vibrational Hamiltonian (23.14)

$$\mathscr{H}_{\text{vib}} = \sum_\beta \tfrac{1}{2}(p_\beta^2 + \omega_\beta^2 q_\beta^2) \tag{24.25}$$

has eigenvalues

$$E(n_\beta) = \sum_\beta (n_\beta + \tfrac{1}{2})\hbar\omega_\beta \tag{24.26a}$$

and eigenfunctions

$$\boxed{\psi_{\text{vib}}(n_\beta) = \prod_\beta (a_\beta)^{n_\beta}\, \psi_0,} \tag{24.26b}$$

where

$$\psi_0 = \exp\left[-\frac{1}{2\hbar}\sum_\beta \omega_\beta q_\beta^2\right]. \tag{24.27}$$

Let us first consider the case of only one fundamental frequency ω. Suppose symmetry allows a term Aq^N to be present in the expansion of μ_j or α_{ij}. From (24.12) we have

$$q^N = B(a - a^*)^N = B[a^N \pm (a^*)^N + \text{other terms}],$$

where

$$B = (-i)^N (\hbar/2\omega)^{\frac{1}{2}N},$$

so that the transition matrix element (24.8) is

$$\langle 2|\mu_j|1\rangle = AB\int \psi_{n_2}{}^*[a^N + (-a^*)^N + \text{other terms}]\psi_{n_1}\, dv, \tag{24.28}$$

where A is some constant. This expression can be evaluated with the help of the symmetry properties of the Hamiltonian. (24.23) is invariant under the group of transformations T_ϕ,

$$\boxed{T_\phi\, a = a\, e^{i\phi}, \qquad T_\phi\, a^* = a^*\, e^{-i\phi},} \tag{24.29}$$

where ϕ is any angle, so that the wave functions ψ_n belong to definite representations of this group. Actually (24.29) is a rather unphysical transformation which cannot be applied easily to ψ_0. However, when we use the wave functions (24.24), ψ_0 always turns up in matrix elements in the combination $\psi_0^*\psi_0$ which is invariant.† It is therefore irrelevant how ψ_0 itself transforms, and we arbitrarily suppose that ψ_0 is invariant under T_ϕ. Hence from (24.24), ψ_n transforms according to

$$\chi^{(n)}(T_\phi) = \exp(in\phi). \tag{24.30}$$

If we take only the first term in the square bracket of (24.28), we have that the matrix element transforms according to

$$\chi(T_\phi) = \exp i(-n_2 + N + n_1)\phi.$$

From the result (13.8c) of the matrix element theorem of § 13, this part of the matrix element is non-zero only if $\chi(T_\phi)$ is the identity representation, i.e. if

$$n_2 - n_1 = N. \tag{24.31a}$$

Similarly the second term in (24.28) gives a contribution to the matrix element if

$$n_2 - n_1 = -N \tag{24.31b}$$

Physically therefore, (24.28) always allows transitions in which N quanta are absorbed or emitted. These N-quanta transitions are referred to as *overtones* of the fundamental line of frequency ω_β. If $N > 1$, (24.28) contains other terms leading to further lines which we shall mention again below. For $N = 1$ we just obtain the selection rule (24.17). More generally, *if*

$$q_\beta{}^b q_\gamma{}^c q_\delta{}^d \ldots \text{ with } N = b + c + d + \ldots \tag{24.32}$$

transforms in the same way as one of the μ_j *or* α_{ij}, *then this term occurs with non-zero coefficient in the expansion of* μ_j *or* α_{ij}, *and all the* N*-quantum transitions*

$$\frac{E_2 - E_1}{\hbar} = \pm b\omega_\beta \pm c\omega_\gamma \pm d\omega_\delta \pm \ldots \tag{24.33}$$

have a non-zero intensity. Such transitions are called *combination tones*. The choice of $\pm$ signs throughout (24.33) derives from the fact that (24.32) always contains equal powers of a_β and a_β^*, etc., like in (24.28). This has the consequence that a summation tone $\omega_\beta + \omega_\gamma$ and the difference tone $\omega_\beta - \omega_\gamma$ are either both present or both absent from a spectrum. It also means that in a Raman spectrum the lines are symmetrically placed about the exciting

† This follows from the fact that ψ_0 is non-degenerate and therefore transforms according to a one-dimensional representation, $\chi(T) = \exp(i\alpha)$ say. For a one-dimensional representation we have $\chi^*(T)\,\chi(T) = \chi(T^{-1})\,\chi(T) = \chi(E) = 1$.

frequency ω, the absorption and emission of a given energy difference $|E_1 - E_2|$ being either both possible or both forbidden processes.

We now mention two complications. Firstly the expansion (24.28) contains other terms besides a^N and $(a^*)^N$, for instance the term $a^{N-1}a^*$ which by the same argument is seen to give an $(N-2)$-quantum transition. If the corresponding frequency ω_β is non-degenerate, it can easily be shown† that this $(N-2)$-quantum transition would already have been found in analysing the term q_β^{N-2}. In the case of degenerate frequencies however, it is possible to obtain new lines this way. The second complication arises only with doubly or triply degenerate frequencies. Consider the term $q_\beta^2 q_\gamma$. From (28.33) this gives the transitions

$$E_2 - E_1 = (2\omega_\beta + \omega_\gamma)\hbar, \qquad (2\omega_\beta - \omega_\gamma)\hbar,$$
$$(-2\omega_\beta + \omega_\gamma)\hbar, \qquad (-2\omega_\beta - \omega_\gamma)\hbar,$$

in general all obviously 3-quantum transitions. However, if $\omega_\beta = \omega_\gamma$, these transitions become

$$E_2 - E_1 = 3\hbar\omega_\beta, \qquad \hbar\omega_\beta, \qquad -\hbar\omega_\beta, \qquad -3\hbar\omega_\beta,$$

and the middle two appear to be one-quantum transition. However, they have only the intensity of a 3-quantum process (see below), if the genuine fundamental line is not allowed.

Relative intensities

We shall now calculate the order of magnitude of the relative intensity of an N-quantum transition (28.33). The usual Hamiltonian for a simple harmonic oscillator is

$$\frac{1}{2m}p^2 + \frac{1}{2}m\omega^2 q^2,$$

and comparison with (23.14) shows that our q_β are more or less

$$q_\beta = m^{1/2} \times (\text{displacement of atoms})$$

where m is the mass of the vibrating atoms. Thus the matrix elements of the term (28.32) are of order

$$(m^{1/2}\delta)^N$$

† If q_β transforms according to a *real* one-dimensional representation, then q_β^2 transforms according to the identity representation. Thus q_β^N and q_β^{N-2} transform in the same way, and are either both present or both absent from the expansions (24.14) and (24.19).

where δ is the distance through which the atoms vibrate. Also

$$\left|\frac{\partial^N \mu_j}{\partial q_\beta{}^b\, \partial q_\gamma{}^c\, \partial q_\delta{}^d \ldots}\right| \approx \frac{|\mu_j|}{(m^{1/2}R)^N}$$

where R is the internuclear distance. Thus the intensity of the lines is proportional to

$$\left|\langle 2|\mu_j|1\rangle\right|^2 \approx |\mu_j|^2\left(\frac{\delta}{R}\right)^{2N}.$$

We can estimate δ as follows. Classically, when the atoms are at the end of their swing, all the energy is in the form of potential energy and

$$E = (n + \tfrac{1}{2})\hbar\omega = \tfrac{1}{2}m\omega^2\delta^2.$$

Taking typical values $n = 5$, $\hbar\omega = 200\ \text{cm}^{-1}$, $m = 15$ proton masses, $R = 10^{-8}$cm, we find $\delta/R \approx 0{\cdot}1$, so that successive sets of N-quantum transitions differ in intensity by a factor of about $(\delta/R)^2 \approx 1/100$. We shall see below that the presence of anharmonicity increases the intensities of some lines rather, but it nevertheless remains true that transitions with high values of N become increasingly weaker. For example, in the observed spectrum of CO_2 as listed by Herzberg (1945, p. 274), the line with highest N is a 7-quantum transition.

All these intensities have to be multiplied, of course, by the fraction f_1 of molecules actually in the state $\psi_{\text{vib},1}$ initially, and hence able to make the particular transition. If the sample is in equilibrium at an absolute temperature T, then

$$f_1 = (\text{const})\, \exp(-E_1/kT). \tag{24.34}$$

Molecular vibrations have an energy of about $2000\ \text{cm}^{-1}$, whereas at room temperature $kT = 200\ \text{cm}^{-1}$. Thus transitions *from* excited states will tend to be very weak. This is seen very clearly in the Raman effect, where the lines on the high frequency side of the incident radiation (anti-Stokes lines) are much weaker than those on the low frequency side (Stokes lines), the intensity ratio of corresponding lines being

$$\exp - \frac{E_1 - E_2}{kT}.$$

Anharmonicity

We shall now examine one of the simplifying assumptions that has been made throughout § 23 and the present section so far. In § 23, we started with an arbitrary potential $V(\mathbf{R}_\alpha)$ governing the nuclear motions, and expanded this about the equilibrium configuration as a power series (23.6) in the Q_β. Assuming that the amplitude δ of a nuclear vibration is very small compared with the internuclear distance R, we retain in $\mathscr{H}_{\text{vib}}$ (23.9) only the quadratic terms, and this then led to the simple harmonic oscillators Hamiltonian (23.13) or (24.25). However, the assumption is not strictly valid because we showed above that $\delta/R \approx 0{\cdot}1$. The true situation is that the vibrations are sufficiently small for (24.25) to be a very good first approximation to the Hamiltonian, but that the real Hamiltonian contains anharmonic perturbing terms such as

$$g_\beta q_\beta^3 + g_{\beta\gamma} q_\beta^2 q_\gamma + f_\beta q_\beta^4 + \ldots, \tag{24.35}$$

which have small but observable effects. These effects are of two kinds, (i) an increase in the intensity of some lines, and (ii) a splitting of some degenerate levels and lines.

As regards intensities, the presence of perturbing anharmonic terms in the Hamiltonian means that the eigenstates will not have the simple harmonic oscillator form (24.26), and those arguments which depend on this specific form have to be re-examined. For instance the selection rule (24.17) no longer holds, and the term q_β in the expansions of μ_j (24.14) and α_{ij} (24.19) can give rise to various lines in addition to the fundamental frequency ω_β. These other lines will have an intensity proportional to the amount of anharmonicity in the potential. This can be seen explicitly by considering the $g_\beta q_\beta^3$ term in (24.35). We restrict ourselves to the case of a single frequency ω and drop the suffix β. From ordinary perturbation theory and the selection rules (24.31) for the matrix elements $\langle 2|gq^3|1\rangle$ of the perturbation, we have that the perturbation mixes wave functions ψ_n with $\Delta_n = \pm 3$ (Schiff 1949, p. 151). The perturbed wave functions are

$$\psi_{\text{pert}}(n) = c_n\psi_n + c_{n+3}\psi_{n+3} + \text{other terms}$$
$$\psi_{\text{pert}}(n+4) = d_{n+4}\psi_{n+4} + d_{n+1}\psi_{n+1} + \text{other terms},$$

where $c_n \approx d_{n+4} \approx 1$, $c_{n+3} \approx d_{n+1} \approx g$, and where the wave functions on the right-hand side are the unperturbed ones (24.24). If the expansion of μ_j contains the term Aq, then we have

$$\int \psi^*_{\text{pert}}(n+4)\mu_j\psi_{\text{pert}}(n)\,\mathrm{d}v$$
$$\approx Ac_{n+3}d_{n+4}\int \psi^*_{n+4}q\psi_{n+3}\,\mathrm{d}v + Ac_n d_{n+1}\int \psi^*_{n+1}q\psi_n\,\mathrm{d}v \propto g,$$

so that the linear term Aq in the expansion of μ_j leads to a 4-quantum transition with intensity proportional to $g^2(\delta/R)^2$. In this kind of way the allowed N-quantum transitions with high N usually have their intensities increased considerably above that expected from the unperturbed simple harmonic oscillator calculation.

Some of the energy levels (24.26a) of the simple harmonic oscillator Hamiltonian are highly degenerate, and the anharmonic perturbation (24.35) produces a splitting in it. Consider a pair of degenerate frequencies $\omega_\beta = \omega_\gamma = \omega$ say. The energy

$$E(n) = (n_\beta + \tfrac{1}{2})\hbar\omega_\beta + (n_\gamma + \tfrac{1}{2})\hbar\omega_\gamma = (n_\beta + n_\gamma + 1)\hbar w \tag{24.36a}$$

depends only on the total number of quanta $n = n_\beta + n_\gamma$. The level is therefore $(n + 1)$-fold degenerate,† corresponding to the pairs of values of n_β and n_γ,

$$\begin{aligned} n_\beta &= n,\ n-1,\ n-2,\ \ldots,\ 1,\ 0; \\ n_\gamma &= 0,\ 1,\ 2,\ \ldots,\ n-1,\ n. \end{aligned} \tag{24.36b}$$

For the point-groups, the irreducible representations are at most 3-dimensional, so that for n greater than 2 or 3, the wave functions of the level must form the basis of a *reducible* representation. This will be split by the perturbation (24.35) into irreducible components in accordance with the general theory of § 6. The splittings are usually up to about 50 cm^{-1}. For example in ozone the $n = 2$ level of the degenerate frequencies $\omega_2 = \omega_3$ (24.16) is three-fold degenerate in the simple harmonic oscillator approximation, the wave functions being

$$a_2{}^2\psi_0, \qquad a_2a_3\psi_0, \qquad a_3{}^2\psi_0. \tag{24.37}$$

These transform according to the representation $A'_1 + E'$ of the symmetry group $\bar{6}$m2 (see below), and the anharmonic perturbation therefore splits the level into a non-degenerate one (A'_1) and a doubly degenerate one (E'). The spectral lines are split correspondingly.

Symmetry of the wave functions

In order to study the splittings of the energy levels by anharmonicity in detail, it is necessary to know the transformation properties of the wave functions (24.26b). From (24.27), ψ_0 remains invariant under the symmetry group of the molecule just like the

† Although this degeneracy is obvious by inspection, it is an interesting exercise to derive it group theoretically. (See problems 24.9 and 24.10.)

potential energy $\frac{1}{2} \sum \omega_\beta^2 q_\beta^2$ does. We also have† $p_\beta = \dot{q}_\beta$ (23.12), so that p_β transforms in the same way as q_β, and hence so do a_β and a_β^* (24.22). If we write the representation of q_β and a_β as $D^{(\beta)}$, we have that the state $\psi(n_\beta)$ (24.26b) with energy $\sum (n_\beta + \frac{1}{2})\hbar\omega_\beta$ transforms according to the product representation

$$D = D(n_\beta) \times D(n_\gamma) \times \ldots, \tag{24.38}$$

where $D(n_\beta) = D^{(\beta)} \times D^{(\beta)} \times D^{(\beta)} \times \ldots$ to n_β factors.

For a real one-dimensional representation, we have that $D^{(\beta)} \times D^{(\beta)}$ is the identity representation,‡ so that if ω_β is a non-degenerate frequency

$$\begin{aligned} D(n_\beta) &= D^{(\beta)} \text{ if } n_\beta = \text{odd}, \\ &= \text{identity representation if } n_\beta = \text{even}. \end{aligned}$$

The reduction of $D(n_\beta)$ is rather more complicated for degenerate frequencies, $\omega_\beta = \omega_\gamma = \omega$ say, $D^{(\beta)} = D^{(\gamma)}$. The degenerate states (24.36) now transform according to

$$D(n) = D^{(\beta)} \times D^{(\beta)} \times D^{(\beta)} \times \ldots \text{ to } n = n_\beta + n_\gamma \text{ factors.} \quad \text{(symmetrical product)} \tag{24.39}$$

Here the words "symmetrical product" denote that this representation is not quite the same as what is usually meant by a product representation. We first illustrate this by taking the case $n = 2$ as an example. Suppose that we have four quantities a_β, a_γ and b_β, b_γ, each pair transforming according to the two-dimensional representation $D^{(\beta)}$. There are in general $2 \times 2 = 4$ second degree products between them,

$$a_\beta b_\beta,\ a_\beta b_\gamma,\ a_\gamma b_\beta,\ a_\gamma b_\gamma,$$

the important point being that these reduce to only three independent ones when we put $a_\beta = b_\beta$, $a_\gamma = b_\gamma$ because then $a_\beta b_\gamma$ and $a_\gamma b_\beta$ become identical. Specializing to the case of the degenerate frequency (24.16) of ozone, we have the four linear combinations

$$(a_2b_2 + a_3b_3); \qquad (a_2b_2 - a_3b_3,\ a_2b_3 + a_3b_2); \qquad (a_2b_3 - a_3b_2); \tag{24.40a}$$

† This result is directly connected with the form (23.14) of the Hamiltonian, and thus assumes implicitly that the q_β are all real. It is therefore essential in this connection to regard a pair of one-dimensional complex conjugate representations as a single real two-dimensional one.

‡ There are three simple ways of seeing this: from the footnote below equation (24.29), from problem 23.3, and from (14.13) or problem 13.13.

transforming according to

$$E' \times E' = A'_1 + E' + A''_2. \tag{24.40b}$$

This is the reduction of the *ordinary product representation* $E' \times E'$ However, to obtain our wave functions (24.37) we want to put $a_2 = b_2$, $a_3 = b_3$, in which case the last term in (24.40a) becomes identically zero and the representation A''_2 drops out of (24.40b). We have what is called the *symmetrical product*

$$E' \times E' \text{ (sym)} = A'_1 + E'. \tag{24.40c}$$

More generally if a set of base vectors ϕ_i transform according to D, the products $\phi_i\phi_j$ are said to transform according to the symmetrical product $D \times D$ (sym). This is equal to the ordinary product $D \times D$ with some of the irreducible components dropped out, because $\phi_i\phi_j$ and $\phi_j\phi_i$ are not independent quantities (see problem 9.4). The same applies to repeated products $D \times D \times D \times \ldots$ (sym). Clearly in the case of degenerate frequencies it is the symmetrical product (24.39) that we have to use in (24.38) to obtain the total representation.

From (14.3) the characters of the ordinary n^{th} product are $[\chi(T)]^n$, where the $\chi(T)$ are the characters of $D^{(\beta)}$, but from what has been said above this cannot be applied to the symmetric product $D(n)$ and we have to develop a new formula for the characters of $D(n)$ which we denote by $\chi^{(n)}(T)$. For a doubly-degenerate frequency, the matrix $D_{ij}^{(\beta)}(T)$ can always be reduced by an equivalence transformation (5.15) to the diagonal form

$$\begin{bmatrix} b & \cdot \\ \cdot & c \end{bmatrix}.$$

Then $\chi(T) = b + c$. Let a_β and a_γ be the base vectors corresponding to this form of $D_{ij}^{(\beta)}(T)$. With these base vectors the other matrices of the representation will not be diagonal, but it is sufficient that a pair of such base vectors exists for each transformation T. We have

$$Ta_\beta = ba_\beta, \qquad Ta_\gamma = ca_\gamma,$$
$$T^n a_\beta = b^n a_\beta. \qquad T^n a_\gamma = c^n a_\gamma,$$
$$T(a_\beta{}^{n-r} a_\gamma{}^r) = b^{n-r} c^r (a_\beta{}^{n-r} a_\gamma{}^r).$$

The base vectors of the representation $D(n)$ are the eigenfunctions

$$a_\beta{}^n \psi_0, \; a_\beta{}^{n-1} a_\gamma \psi_0, \; a_\beta{}^{n-2} a_\gamma{}^2 \psi_0, \ldots a_\gamma{}^n \psi_0$$

of the degenerate energy level (24.36). The characters of $D(n)$ are therefore

$$\begin{aligned}\chi^{(n)}(T) &= b^n + b^{n-1}c + b^{n-2}c^2 + \ldots c^n \\ &= \tfrac{1}{2}[(b+c)(b^{n-1} + b^{n-2}c + \ldots + c^{n-1}) + (b^n + c^n)] \\ &= \tfrac{1}{2}[\chi(T)\chi^{(n-1)}(T) + \chi(T^n)]. \end{aligned} \tag{24.42}$$

This recurrence relation serves to determine the characters $\chi^{(n)}(T)$ of the symmetrical product representation $D(n)$. The total representation D (24.38) can then be built up. For the sake of completeness we quote an alternative formula for $\chi^{(n)}(T)$ (Wilson *et al.* 1955, p. 355):

if $\chi(T) = 0$ and $\chi(T^2) = 2$, then

$$\begin{aligned}\chi^{(n)}(T) &= 0 \text{ for } n \text{ odd}, \\ &= 1 \text{ for } n \text{ even};\end{aligned}$$

otherwise $$\chi^{(n)}(T) = \frac{\sin(n+1)\theta_T}{\sin\theta_T} \tag{24.43}$$

where $$\cos\theta_T = \tfrac{1}{2}\chi(T);$$

in the limiting cases of $\theta_T = 0$ or π,

$$\begin{aligned}\chi^{(n)}(T) &= n+1 \text{ for } \theta_T = 0, \\ &= (-1)^{n+1}(n+1) \text{ for } \theta_T = \pi.\end{aligned}$$

Analogous formulae exist for three-fold degenerate frequencies (problems 24.5 and 24.6).

Summary

The intensities of the lines in the infra-red absorption spectrum and the Raman spectrum are largely determined by the matrix elements

$$\int \psi^*_{\text{vib},2}\,\mu_i\,\psi_{\text{vib},1}\,dv \quad \text{and} \quad \int \psi^*_{\text{vib},2}\,\alpha_{ij}\,\psi_{\text{vib},1}\,dv \tag{24.44}$$

of the dipole moment μ_j and the polarizability tensor α_{ij} of the molecule. The μ_j and α_{ij} are expanded as power series (24.14) and (24.19), only certain terms being allowed in the expansions on account of symmetry. The allowed terms give one immediately the corresponding fundamental frequencies, overtones and combination tones (24.33) which will be present in the spectra. One also obtains a rough lower limit estimate of the intensity. Such an analysis is very quick and helpful in arriving at a first tentative interpretation of a newly observed spectrum. If, however, one tries to derive the complete and detailed form of the spectrum in this way, one runs into the following difficulties: (i) the "other terms" in (24.28) give rise to lines in addition to those we have considered; (ii) with degenerate frequencies some of the N-quanta transitions degenerate into ones of apparent lower order; (iii) the anharmonicity of the potential splits the lines quite appreciably and upsets the calculation of the expected intensities.

Having made a preliminary identification of the spectrum, it is therefore better to draw up a complete table of the allowed energy levels with their associated irreducible representations, taking into account the splittings of the levels due to anharmonicity. One can then determine systematically all the allowed transitions between these levels from the rigorous selection rules (24.9) and (24.13), which are derived directly from (24.44). This is not too arduous a task because the temperature dependent factor (24.34) in the intensity usually allows one to restrict one's attention to just the few lowest lying states for the initial state $\psi_{\text{vib},1}$. It should be emphasized that this procedure is rigorously exact. Each energy level has a definite irreducible representation of the symmetry group of the molecule associated with it. Although these representations are calculated with the help of the simple harmonic oscillator wave functions, they remain unaffected by the anharmonic part of the potential which only affects their energies and thus produces splittings (see § 6). On the other hand the transformation T_ϕ (24.29) is a purely mathematical device, unrelated to the symmetry of the molecule, and applies only to the simple harmonic oscillator Hamiltonian. Thus the simple harmonic oscillator selection rules and degeneracies derived from T_ϕ do not apply rigorously any more when the anharmonic part (24.35) of the potential is taken into account.

References

Wilson, Decius and Cross (1955) discuss the theory of vibrational spectra in detail, and Herzberg (1945) tabulates and interprets the actual spectra of a large number of molecules.

Epilogue

In the last two sections and earlier in this book, we have often used the ozone molecule for illustrative purpose, assuming that the oxygen atoms form an equilateral triangle. An analysis such as described above then predicts a vibrational spectrum which is different from the observed one, so that the assumption cannot be correct (Herzberg 1945, p. 286). Electron diffraction studies and an analysis of the binding energy of the molecule also support this conclusion. The correct configuration appears to be an obtuse-angled isosceles triangle.

PROBLEMS

24.1 What strong two-quanta transitions do you expect to find in the infra-red and Raman spectra of some of the molecules of problems 23.1 and 23.2?

24.2 A certain molecule with point-group symmetry ∞m has three low lying vibrational energy levels, the associated irreducible representations being A_1, E_1 and E_2. What infra-red and Raman transitions are allowed between these levels, and between them and the ground state?

24.3 The methane molecule has a triply degenerate frequency corresponding to the normal co-ordinates q_1, q_2, q_3 which transform according to the representation T_2 of the point-group $\bar{4}3m$. Determine the degeneracy of the $n = 2$ level and how it is split by the anharmonic part of the potential. What infra-red and Raman transitions are allowed between these states and the ground state of the molecule?

24.4 The vibrational frequencies of the carbon dioxide molecule are about 1350 cm^{-1} (A_{1g}), 667 cm^{-1} (E_{1u}) and 2349 cm^{-1} (A_{2u}) (see equation (23.24)). Taking the ground state as the zero of energy, make a list of the approximate energies of the first few excited states. List also all the transitions $E_2 - E_1$ that are allowed in the infra-red and Raman spectra with $E_2 < 10{,}000$ cm^{-1} and $E_1 < 1000$ cm^{-1}, and are less than 4-quanta transitions. Check this against the list of observed frequencies (Herzberg 1945, p. 274).

24.5 Derive for the three-fold degenerate case the expression analogous to (24.42):

$$\chi^{(n)}(T) = \tfrac{1}{3}\big(2\chi(T)\chi^{(n-1)}(T) + \\ + \tfrac{1}{2}\{\chi(T^2) - [\chi(T)]^2\}\chi^{(n-2)}(T) + \chi(T^n)\big).$$

24.6* Derive (24.43) and the analogous result for the three-fold degenerate case (Wilson *et al.* 1955, p. 352).

24.7 D is a real representation of some group. Show that the symmetrical product $D \times D$ (sym) contains the identity representation once.

24.8 Show using the method of § 9, that

$$D^{(j)} \times D^{(j)}\ (\text{sym}) = \sum D^{(J)}$$

with $$J = 2j,\ 2j-2,\ 2j-4,\ \ldots,\ 2,\ 0,$$

where $D^{(j)}$ is a representation of the full rotation group, and j is an integer. What is the corresponding result when j is half an odd integer?

24.9 The Hamiltonian $\mathscr{H}$ for a doubly-degenerate simple harmonic oscillator is invariant under (24.29) and under $a_\beta \to a_\beta \cos\theta - a_\gamma \sin\theta$, $a_\gamma \to a_\beta \sin\theta + a_\gamma \cos\theta$. Hence show that the symmetry group of $\mathscr{H}$ is isomorphic with the full rotation group (§ 8), and derive the degeneracy of the higher energy levels. Compare with problem 24.10 and with the method of Wilson *et al.* (1955, p. 352).

24.10 Deduce from the symmetry of the Hamiltonian that the energy levels of a two-dimensional isotropic harmonic oscillator are $(n + 1)$-fold degenerate, where n may be any positive integer or zero (Baker 1956). Use the fact that the Hamiltonian in the form $\hbar\omega(a_1{}^*a_1 + a_2{}^*a_2 + 1)$ is invariant under the group of two-dimensional unitary matrices operating on a_1, a_2, and this group is almost isomorphic with the full rotation group (problems 8.13, 8.15 and 8.17).

24.11* Extend the results of the previous problem to an n-dimensional isotropic harmonic oscillator (Baker 1956). Can this approach be used to calculate the splitting of the levels by anharmonic terms in the potential?

24.12* *Raman scattering intensity.* Give a quantum mechanical derivation of the scattered Raman intensity formula (24.10).

Notes: the author has been unable to find an elementary modern derivation of this, but had no difficulty in obtaining one along the following lines. Consider the whole system of electrons and nuclei and electromagnetic radiation interacting with one another. The transition probability to a new state in which one photon ω_1 is absorbed and ω_2 is emitted is given by second order perturbation theory (equation (29.20) of Schiff 1955). It is necessary to consider two types of intermediate state, one in which ω_2 is emitted after ω_1 is absorbed and one in which ω_2 is emitted first. The one-photon matrix elements are given by equations (50.9) and (50.13) of Schiff 1955. To reduce the resultant expression to the desired form, it is necessary to use some identities of Dirac (1927) and Schiff (1955, equation (35.20)). The assumptions that the ground state is electronically non-degenerate and that the vibrational quantum is small, plus the use of time-reversal symmetry, allow one to cast the result into the form of (24.10) (Placzek 1934). To obtain the connection with the polarizability α_{21}, it is necessary to obtain α_{21} by calculating the dipole moment induced by a time-dependent electric field $\mathscr{E} \cos \omega_1 t$.

Chapter VI

SOLID STATE PHYSICS

25. Brillouin Zone Theory of Simple Structures

The translation group and the Brillouin zone

In this and the next section we shall develop the irreducible representations of the symmetry groups of crystalline solids, with applications to the motion of an electron in a perfect crystal. Such a group is called a *space group* $\mathfrak{G}$ and includes both translational and rotational symmetries. The distinguishing feature of a crystal is, of course, that it consists of a regular periodic array of identical cells. Thus fundamental to any space group is the *translational group* $\mathfrak{T}$ consisting of all translations $\mathbf{t}_n$ of the form

$$\boxed{\mathbf{t}_n = n_1\mathbf{a}_1 + n_2\mathbf{a}_2 + n_3\mathbf{a}_3.} \tag{25.1}$$

Here $\mathbf{a}_1$, $\mathbf{a}_2$, and $\mathbf{a}_3$ are three *primitive translation vectors* (not coplanar) along the crystal axes, and n_1, n_2, n_3 are any set of integers. For instance the face centred cubic lattice (Fig. 27) is obtained by taking

$$\mathbf{a}_1 = \tfrac{1}{2}a(\mathbf{i}_y + \mathbf{i}_z), \qquad \mathbf{a}_2 = \tfrac{1}{2}a(\mathbf{i}_z + \mathbf{i}_x), \qquad \mathbf{a}_3 = \tfrac{1}{2}a(\mathbf{i}_x + \mathbf{i}_y), \tag{25.2}$$

where a is the cube length and $\mathbf{i}_x$, $\mathbf{i}_y$, $\mathbf{i}_z$ unit vectors along the edges.

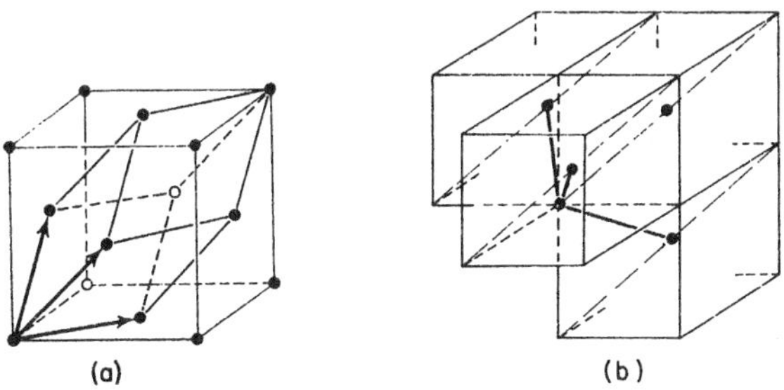

FIG. 27. (a) Primitive translation vectors and primitive unit cell of the face centred cubic structure. (b) Primitive translation vectors of the body centred cubic structure. The primitive cell is obtained by completing the rhombohedron.

Now a finite crystal is not really invariant under any of the $\mathbf{t}_n$ (25.1). Whereas $\mathbf{t}_n$ leaves the middle of the crystal invariant where the structure is really periodic, it displaces all the sides of the crystal to new positions. To make use of the translational symmetry therefore, we have to modify our problem so that it has $\mathfrak{T}$ as an exact symmetry group. We do this by one of two artifacts. Take as a sample a crystal of size $N_1\mathbf{a}_1$ by $N_2\mathbf{a}_2$ by $N_3\mathbf{a}_3$ where N_1, N_2, N_3 are large numbers. We either consider an infinite number of identical such samples stacked together or we imagine opposite faces of the crystal to be made mathematically to join up in the same sense as we can join up the two ends of a straight line by bending it into a circle. In either case we regard the three translations $N_1\mathbf{a}_1$, $N_2\mathbf{a}_2$, $N_3\mathbf{a}_3$ as affecting no change at all and we put them equal to the identity element, i.e. to zero translation

$$\mathbf{t}_{N_100} = \mathbf{t}_{0N_20} = \mathbf{t}_{00N_3} = \mathbf{t}_{000}. \tag{25.3}$$

Expressed in terms of the wave function ψ, this procedure gives the usual *periodic boundary conditions*

$$\psi(\mathbf{r}) = \psi(\mathbf{r} + N_1\mathbf{a}_1) = \psi(\mathbf{r} + N_2\mathbf{a}_2) = \psi(\mathbf{r} + N_3\mathbf{a}_3). \tag{25.4}$$

The artifact does not, clearly, correspond to anything physical and occasionally it has to be used with care, e.g. in discussing any surface effects.

The translations (25.1) with the restriction (25.3) now form a finite group of $N_1N_2N_3$ elements. The group is Abelian, so that all the irreducible representations are one-dimensional (appendix E) and can be written down by the method of § 7. They are labelled by a wave vector $\mathbf{k}$ such that†

$$\mathbf{t}_n \text{ is represented by } \exp(i\mathbf{k}\cdot\mathbf{t}_n), \tag{25.5a}$$

$$\text{or}\quad \mathbf{t}_n\psi_{\mathbf{k}}(\mathbf{r}) \equiv \psi_{\mathbf{k}}(\mathbf{r} + \mathbf{t}_n) = \exp(i\mathbf{k}\cdot\mathbf{t}_n)\psi_{\mathbf{k}}(\mathbf{r}), \tag{25.5b}$$

where $\psi_{\mathbf{k}}(\mathbf{r})$ transforms according to the irreducible representation $\mathbf{k}$. A function which has the property (25.5b) is called a *Bloch function*, and (25.5b) is taken to define its $\mathbf{k}$ vector.

From (14.14), the group $\mathfrak{T}$ has only $N_1N_2N_3$ distinct representations, and we obtain these by putting two types of restriction on $\mathbf{k}$. The

† No confusion should arise from using $\mathbf{t}_n$ both as a translation operator as in $\mathbf{t}_n\psi_{\mathbf{k}}(\mathbf{r})$ and as an ordinary vector in $\exp(i\mathbf{k}\cdot\mathbf{t}_n)$.

first is obtained by noting that there exists in **k**-space a whole lattice of vectors

$$\boxed{\mathbf{K}_m = m_1\mathbf{K}_1 + m_2\mathbf{K}_2 + m_3\mathbf{K}_3,} \tag{25.6}$$

called the *reciprocal lattice*, each satisfying the condition

$$\exp(i\mathbf{K}\cdot\mathbf{t}_n) = 1 \text{ for all } \mathbf{t}_n. \tag{25.7}$$

This is easily shown along the lines of problem 25.2. It follows that **k** is not uniquely defined by the property (25.5), since **k** and $\mathbf{k} + \mathbf{K}_m$ give identical representations

$$\exp i\mathbf{k}\cdot\mathbf{t}_n = \exp i\,(\mathbf{k} + \mathbf{K}_m)\cdot\mathbf{t}_n \text{ for all } \mathbf{t}_n. \tag{25.8}$$

To label the representations uniquely, it is therefore necessary to restrict **k** to some region which is a unit cell of the reciprocal lattice. The particular unit cell which has been adopted by convention is obtained by bisecting the lines joining $\mathbf{K} = 0$ to the nearest reciprocal lattice points. It is called the *Brillouin zone* and from now on we shall assume without further mention that the **k**'s which we use

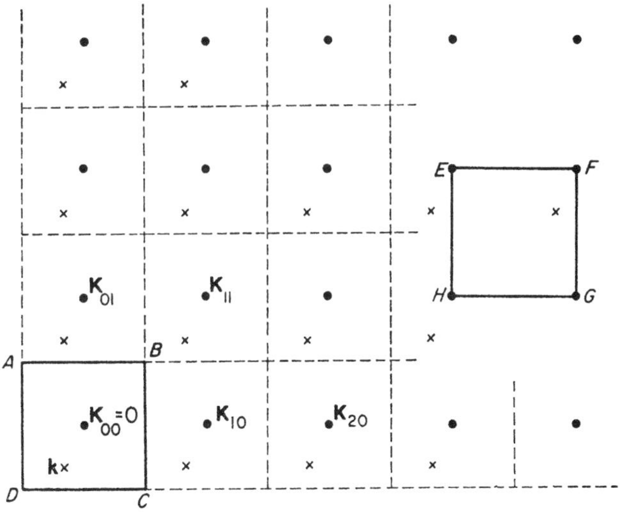

FIG. 28. Reciprocal lattice of the square lattice. The dots are the reciprocal lattice points $\mathbf{K}_m$. A B C D is the conventional Brillouin zone, but any other unit cell of the reciprocal lattice such as E F G H could in principle be used instead. All the points marked x denote the same irreducible representation in the sense of (25.8) as the point **k** inside the Brillouin zone.

to denote representations always lie in the Brillouin zone. With this convention $\mathbf{k}$ is often called the *reduced* wave vector and the Brillouin zone the reduced zone. Fig. 28 illustrates the situation for the two-dimensional square lattice, which we shall use for

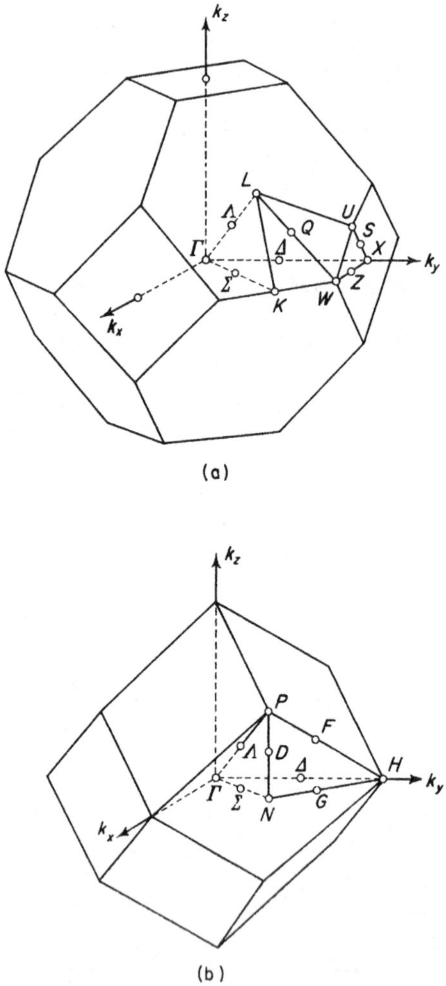

FIG. 29. Brillouin zones of (a) face centred cubic structure, and (b) body centred cubic structure, with special points shown.

illustrative purposes throughout this section although all our arguments apply equally to one, two and three dimensions. Fig. 29 shows the Brillouin zones of the body centred and face centred structures. The second restriction on $\mathbf{k}$ comes from applying (25.3) or (25.4), and we obtain

$$\mathbf{k} = \frac{r_1}{N_1}\mathbf{K}_1 + \frac{r_2}{N_2}\mathbf{K}_2 + \frac{r_3}{N_3}\mathbf{K}_3 \tag{25.9}$$

where r_1, r_2, r_3 are integers. This gives correctly the total number of $N_1\,N_2\,N_3$ different irreducible representations in the Brillouin zone see problem 25.4).

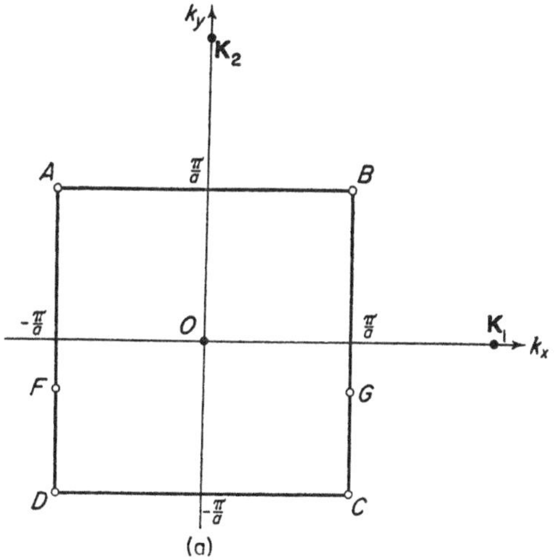

FIG. 30. Brillouin zone of the square lattice, showing some points related by reciprocal lattice vectors.

In this scheme, there is still some ambiguity about $\mathbf{k}$'s on the surface of the zone. Since the Brillouin zone is a unit cell of the reciprocal lattice, opposite faces are always separated by a reciprocal lattice vector. For example in Fig. 30

$$\mathbf{k}_F = \mathbf{k}_G + \mathbf{K}_1,$$

so that $\mathbf{k}_F$ and $\mathbf{k}_G$ denote the same representation and we shall use either one of them indiscriminately. Likewise there is one representation which can be referred to by either $\mathbf{k}_A$, $\mathbf{k}_B$, $\mathbf{k}_C$ or $\mathbf{k}_D$ (Fig. 30).

Other symmetry elements

As well as the translation group $\mathfrak{T}$, a space group $\mathfrak{G}$ in general contains purely rotational symmetry elements R and combined rotational-translational elements $\{R|\mathbf{t}\}$. Among the latter, $\mathbf{t}$ is not necessarily restricted to the lattice translations (25.1): if the crystal symmetry involves screw axes or glide planes, then $\mathbf{t}$ may be a fractional lattice translation (§ 26). In any case a translation $\mathbf{t}$ does not alter the $\mathbf{k}$ vector of a function,†

$$\mathbf{t}_n\psi_{\mathbf{k}}(\mathbf{r}+\mathbf{t}) = \psi_{\mathbf{k}}(\mathbf{r}+\mathbf{t}+\mathbf{t}_n) = \exp(i\mathbf{k}\cdot\mathbf{t}_n)\psi_{\mathbf{k}}(\mathbf{r}+\mathbf{t}).$$

However, *the rotation R transforms a function $\psi_{\mathbf{k}}(r)$ with wave vector $\mathbf{k}$ into a new function $\psi_{\mathbf{k}'}(r)$ with wave vector $\mathbf{k}'$, where $\mathbf{k}'$ is derived from $\mathbf{k}$ by the rotation R applied in $\mathbf{k}$-space.* This result which is intuitively obvious, follows from the fact that in (25.5) the scalar product $\mathbf{k}\cdot\mathbf{t}_n$ remains invariant if we apply the same rotation of axes in both $\mathbf{k}$-space and real space.

The question now arises, is the concept of a $\mathbf{k}$ vector any use in sorting out the irreducible representations of $\mathfrak{G}$, in view of all the rotational symmetry elements? Consider some vector space $\mathfrak{R}$ transforming according to an irreducible representation of $\mathfrak{G}$. Since $\mathfrak{G}$ contains the translation group $\mathfrak{T}$ as a subgroup, $\mathfrak{R}$ is invariant under $\mathfrak{T}$ and we can reduce it with respect to $\mathfrak{T}$ to give a set of base vectors‡ $\psi(\mathbf{k}_1;\, \mathbf{r})$, $\psi(\mathbf{k}_2;\, \mathbf{r})$, . . . with definite $\mathbf{k}$ vectors. Thus

in setting up the irreducible representations of any space group $\mathfrak{G}$, we can restrict ourselves to base vectors which are Bloch functions,

(25.10)

i.e. functions each of which has the property (25.5b) with some definite $\mathbf{k}$ vector.

Consider now the energy of an electron moving in a crystal. The Hamiltonian is

$$-(\hbar^2/2m)\nabla^2 + V(\mathbf{r}) \tag{25.11}$$

where the crystal potential $V(\mathbf{r})$ has the symmetry of the crystal. From the result (25.10) we may use the Bloch functions $\psi_{\mathbf{k}}(\mathbf{r})$ as our basic energy eigenstates, the energy E being given by some function $E(\mathbf{k})$ varying continuously with $\mathbf{k}$ through the Brillouin zone. $E(\mathbf{k})$ is normally a multivalued function of $\mathbf{k}$, the branches being numbered one, two, three, . . . in order of increasing energy, starting arbitrarily with the lowest one of interest, e.g. in metallic sodium

† Note that in general $\psi_{\mathbf{k}}(\mathbf{r}+\mathbf{t}) = \exp(i\mathbf{k}\cdot\mathbf{t})\psi_{\mathbf{k}}(\mathbf{r})$ is *only* true if $\mathbf{t}$ is a lattice translation $\mathbf{t}_n$.

‡ To avoid multiple suffices, we shall on occasions write $\psi_{\mathbf{k}}(\mathbf{r})$ as $\psi(\mathbf{k};\, \mathbf{r})$, or by way of abbreviation as $\psi(\mathbf{k})$.

with the one relating to the valence electron. Because of the discrete nature of the allowed **k** vectors (25.9), a branch of $E(\mathbf{k})$ really represents a *band* of closely spaced energy levels, and the form of $E(\mathbf{k})$ throughout all bands is called the *band structure* of the crystal as illustrated in Fig. 31.†

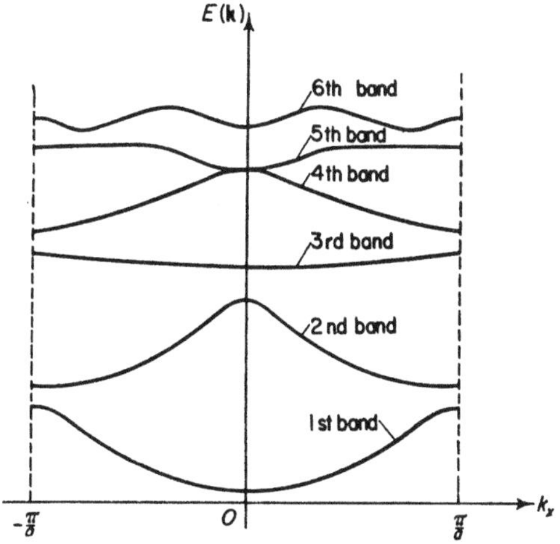

FIG. 31. A typical band structure $E(\mathbf{k})$ with several bands, plotted along the k_x axis.

Let $\psi(\mathbf{k}_1; \mathbf{r})$ be an energy eigenstate with $\mathbf{k}_1$ at a general point in the Brillouin zone (Fig. 32a). Then the various rotations will generate from this other functions $\psi(\mathbf{k}_2; \mathbf{r})$, $\psi(\mathbf{k}_3; \mathbf{r})$, . . . , and from the fundamental theory of § 6 we have that these are also energy eigenfunctions with

$$E(\mathbf{k}_1) = E(\mathbf{k}_2) = E(\mathbf{k}_3) = \ldots \tag{25.12}$$

Thus $E(\mathbf{k})$ *has the full point-group symmetry of the crystal*, and this conclusion is independent of whether the axes are pure rotation axes or screw axes. This theorem allows us to investigate the form of $E(\mathbf{k})$ near the surface of the Brillouin zone. Let us consider one such band, which we shall assume has no degeneracy with any other band. In Fig. 32a, we have $E(\mathbf{k}_1) = E(\mathbf{k}_4)$ by rotational

† The branches of $E(\mathbf{k})$ are sometimes referred to as the "first zone", "second zone", etc., because of the way they can be remapped in the extended zone scheme (problem 25.11). For our purposes the region of **k**-space, namely the Brillouin zone, is the same for all of them: it is just that there are different states with different energies having the same irreducible representation **k**.

symmetry. Now $\mathbf{k}'_4$ and $\mathbf{k}''_4$ are removed from $\mathbf{k}_4$ by reciprocal lattice vectors $\mathbf{K}_1$ and $-\mathbf{K}_2$, and they may be used instead of $\mathbf{k}_4$ to designate the states. In fact the choice of the Brillouin zone as our fundamental cell in **k**-space was rather arbitrary, and we could

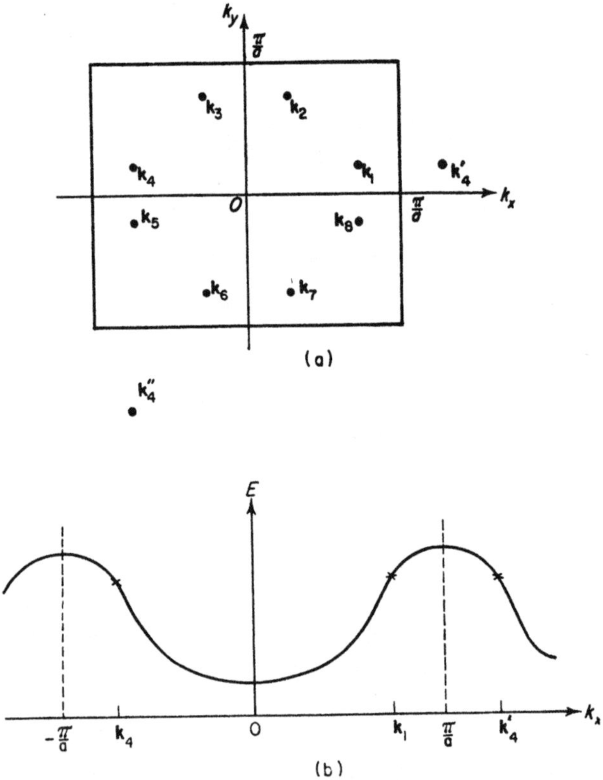

FIG. 32. Symmetry of the band structure. (a) Brillouin zone of the square lattice with points related by rotational symmetry. (b) $E(\mathbf{k})$ plotted along the line $\mathbf{k}_4\ \mathbf{k}_1\ \mathbf{k}'_4$ of Fig. 32a, with a maximum at the zone boundary $k_x = \pm\pi/a$.

equally well have chosen a cell displaced sideways somewhat to include $\mathbf{k}'_4$ and to exclude $\mathbf{k}_4$. Thus we can conveniently extend $E(\mathbf{k})$ as a continuous periodic function in the whole of **k**-space with

$$E(\mathbf{k}_4) = E(\mathbf{k}'_4) = E(\mathbf{k}''_4) = \ldots \text{ etc.}$$

as shown in Fig. 32b. By choosing $\mathbf{k}_1$ near enough to the Brillouin zone boundary, we see from the relation $E(\mathbf{k}_1) = E(\mathbf{k}'_4)$ that $E(\mathbf{k})$

has a minimum or maximum as it crosses the zone surface. I.e. in the case of the square lattice, the normal component of $\mathrm{grad}_{\mathbf{k}}E$ is zero on the zone boundary.

General representations of a space group $\mathfrak{G}$

At this stage we shall restrict ourselves to space groups involving ordinary rotations but without screw axes and glide planes, and return to the general case in the next section. In symbols, all the symmetry elements have the form $\{R|\mathbf{t}_n\}$. We have already shown (25.10) that in setting up the irreducible representations of $\mathfrak{G}$ we can choose as our base vectors Bloch functions $\psi_{\mathbf{k}}(\mathbf{r})$. These are all invariant under $\mathbf{t}_n$ (apart from the factor $\exp i\mathbf{k}\cdot\mathbf{t}_n$) so that we need only consider the effect of the rotations R which by themselves form a point-group $\mathfrak{P}$.

Let $\psi(\mathbf{k}_1)$ be a base vector of some irreducible representation of $\mathfrak{G}$, where $\mathbf{k}_1$ is at some general point in the Brillouin zone (Fig. 32a). We shall construct from this the rest of the representation. By operating on $\psi(\mathbf{k}_1)$ with all the rotations R_i of $\mathfrak{P}$ (where R_1 is the identity element E), we generate other functions $\psi(\mathbf{k}_i)$. These $\mathbf{k}_i$ are all different for, by assumption, $\mathbf{k}_1$ is at a general point in the zone without any rotational symmetry properties. They form a set of vectors having the symmetry of the whole point-group like the stereograms of Fig. 9 or the set $\mathbf{k}_1$ to $\mathbf{k}_8$ in Fig. 32a. The first thing to prove is that these $\psi(\mathbf{k}_i)$'s really form a representation. Let R_j rotate the vector $\mathbf{k}_i$ into $\mathbf{k}_l$. Then R_j can be written

$$R_j = R_l R_i^{-1},$$

and hence

$$R_j\psi(\mathbf{k}_i) = R_l R_i^{-1}\psi(\mathbf{k}_i) = R_l\psi(\mathbf{k}_1) = \psi(\mathbf{k}_l). \tag{25.13}$$

Thus each R_j transforms any $\psi(\mathbf{k}_i)$ into some other $\psi(\mathbf{k}_l)$, so that the functions

$$\psi(\mathbf{k}_1),\ \psi(\mathbf{k}_2),\ \ldots\ \psi(\mathbf{k}_i),\ \ldots \tag{25.14}$$

definitely give a representation of $\mathfrak{G}$. Moreover the representation is irreducible. For since each $\mathbf{k}_i$ occurs only once in the set (25.14), any way we might try to reduce the representation would involve taking linear combinations of functions with different $\mathbf{k}$'s. This we have already proved in (25.10) that we need never do, and we conclude the representation cannot be reduced.

Special points in the zone

The above argument fails for special points in the Brillouin zone because two or more of the $\mathbf{k}_i$ become identical. The $\mathbf{k}_i$ may become identical either (i) by virtue of lying on a rotation axis or mirror plane, or (ii) by being separated by a reciprocal lattice vector from one another (see Fig. 30), or (iii) by (i) and (ii) combined. The six kinds of special point for the square lattice are shown in Fig. 33a, Γ, Δ and Σ belonging to type (i), Z belonging to type (ii), and X and M to type (iii).

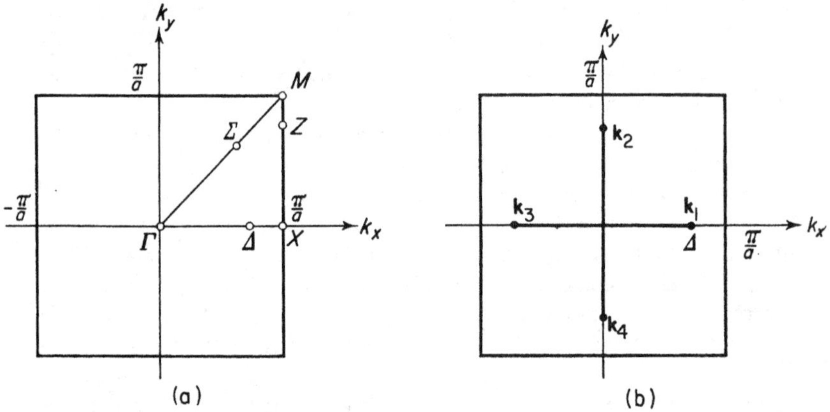

FIG. 33. Brillouin zone of the square lattice, showing special points. (a) The six types of special point. (b) The star of Δ.

We first consider Δ as a representative of the class (i) type of point. The point-group symmetry $\mathfrak{P}$ for the square lattice is† 4mm (see § 16). Let $\psi(\mathbf{k}_1)$ be a Bloch function with $\mathbf{k}_1$ at Δ (Fig. 33b), belonging to some irreducible representation of $\mathfrak{G}$. The elements of $\mathfrak{P}$ can be divided into two classes which leave $\mathbf{k}_1$ unaltered or change the $\mathbf{k}$ of the function respectively, as follows:

class I, $\mathbf{k}_1$ unaltered:—E, m_y;

class II, $\mathbf{k}$ changed:—4_z, 2_z, $4_z{}^3$, m_x, m_d, $m_{d'}$.

The transformations of class II produce functions with $\mathbf{k} = \mathbf{k}_2$, $\mathbf{k}_3$, $\mathbf{k}_4$, and these together with $\mathbf{k}_1$ make up what is called *the star of* Δ as shown in Fig. 33b. Now the rotations of class I satisfy each of the group properties of § 4 because they all leave $\mathbf{k}_1$ invariant. This can be verified in detail. Alternatively it follows from the general result of § 4 that symmetry properties always form groups. This

† In two dimensions, 4mm is the same as 422. It should be noted in what follows that m_x denotes reflection in a line (or plane) *perpendicular to* the x-axis. Suffixes d and d' denote the diagonals of the square. Rotations of 90°, 180°, 270° about the z-axis are denoted by 4_z, 2_z, $4_z{}^3$.

small point-group which leaves $\mathbf{k}_1$ invariant is called the group of $\mathbf{k}_1$ or *the group of* Δ, and is denoted by $\mathfrak{K}$. In the present case it is the point-group m with two representations, the symmetric and the antisymmetric one which we shall call Δ_1 and Δ_2.

We can now take one of these irreducible representations Δ_1 or Δ_2 of $\mathfrak{K}$, and build it up into an irreducible representation of $\mathfrak{G}$. We first choose a $\psi(\mathbf{k}_1)$ that is symmetric or antisymmetric under m_y (Δ_1 or Δ_2). Then using the elements of class II (*not* leaving $\mathbf{k}_1$ invariant), we generate functions with $\mathbf{k} = \mathbf{k}_2$, $\mathbf{k}_3$ and $\mathbf{k}_4$. Since $m_d = 4_z m_y$ and since m_y is an element of $\mathfrak{K}$, the two operations m_d and 4_z produce the same function: likewise m_x and 2_z do, and also $m_{d'}$ and $4_z{}^3$. Thus we obtain four functions $\psi(\mathbf{k}_1)$ to $\psi(\mathbf{k}_4)$, and in the same way as for the general representation (25.14), it follows that they always transform into one another and give an irreducible representation of $\mathfrak{G}$. Moreover the representation is completely specified by the representation Δ_1 or Δ_2 of $\mathfrak{K}$ that we started with. This procedure is quite general for obtaining an irreducible representation of the whole space group $\mathfrak{G}$ from an irreducible representation of $\mathfrak{K}$. The matrices look like

$$\begin{bmatrix} D & \cdot & \cdot & \cdot & \cdots \\ \cdot & \cdot & D' & \cdot & \\ \cdot & \cdot & \cdot & D'' & \\ \cdot & D''' & \cdot & \cdot & \\ \cdot & & & & \\ \cdot & & & & \\ \cdot & & & & \end{bmatrix}$$

where D, D', D'', etc., are matrices of the irreducible representation of $\mathfrak{K}$. However, for most practical purposes it is sufficient to discuss the irreducible representations of $\mathfrak{K}$ plus the overall full symmetry of $E(\mathbf{k})$ without ever writing out the full representation.

In the same way we obtain the irreducible representations of $\mathfrak{G}$ associated with the point Γ, $\mathbf{k} = 0$ (Fig. 33a). In this case the star of Γ consists of the single point $\mathbf{k} = 0$. $\mathfrak{K}$, the group of Γ, is the point-group 4mm, whose irreducible representations can be looked up in appendix K. They are shown in Table 24 numbered arbitrarily Γ_1 to Γ_5 as is the fashion in solid state physics, and these symbols are also used for the corresponding representations of $\mathfrak{G}$. The situation for special points on the zone boundary (types (ii) and (iii) above) is exactly the same. One only has to remember that corresponding points on opposite sides of the zone have to be regarded as the same $\mathbf{k}$ vector. For example the group of X (Fig. 33a) includes the transformations 2_z and m_x, even though these turn the point

$(\pi/a,\ 0)$ into $(-\pi/a,\ 0)$. Thus the group of X is the point-group mm with the irreducible representations shown in Table 24. Similarly with M, all the zone corners correspond to the same **k** vector which may be taken as any one of the four vectors $(\pm\pi/a,\ \pm\pi/a)$. The group of M is 4mm, the same as that of Γ, so that the irreducible representations for the two points are shown together on the same table (Table 24). Those of Σ and Z are likewise shown together with those of Δ.

TABLE 24

Irreducible Representations at Special Points

Γ, M	E	2_z	$4_z, 4_z^3$	m_y, m_x	$m_d, m_{d'}$,
Γ_1, M_1	1	1	1	1	1
Γ_2, M_2	1	1	1	-1	-1
Γ_3, M_3	1	1	-1	1	-1
Γ_4, M_4	1	1	-1	-1	1
Γ_5, M_5	2	-2	0	0	0

X	E	2_z	m_x	m_y
X_1	1	1	1	1
X_2	1	1	-1	-1
X_3	1	-1	1	-1
X_4	1	-1	-1	1

Δ	E	m_y
Σ	E	m_d
Z	E	m_x
Δ_1, Σ_1, Z_1	1	1
Δ_2, Σ_2, Z_2	1	-1

Energy level splittings in a crystal

Our precise specification of the irreducible representations of $\mathfrak{G}$ allow us to make applications to the splittings of energy levels and degeneracies between bands, to which we now proceed. Let us approximate to the eigenfunctions of the Hamiltonian (25.11) by plane waves,† and consider in particular the four

$$\exp i(\pi/a)(x+y), \qquad \exp i(\pi/a)(x-y),$$
$$\exp i(\pi/a)(-x+y), \qquad \exp i(\pi/a)(-x-y). \qquad (25.15)$$

† Or better by orthogonalized plane waves (Herring 1940).

These all have the same reduced **k** vector, which may be taken arbitrarily as any one of the four $(\pm\pi/a, \pm\pi/a)$, say $\mathbf{k} = (\pi/a, \pi/a)$. That is, they belong to the point M in the Brillouin zone (Fig. 33a). The four plane waves (25.15) are equivalent to one another by rotational symmetry and have the same energy. Thus in this approximation, we have four consecutive bands degenerate with one another at M. However, the four functions (25.15) form a representation which is reducible into the components M_1, M_4 and M_5 as shown in Table 25, so that the presence of the potential

TABLE 25

Representation Formed by (25.15)

		E	2_z	4_z	m_x	m_d
Complete representation		4	0	0	0	2
Irreducible components	M_1	1	1	1	1	1
	M_4	1	1	-1	-1	1
	M_5	2	-2	0	0	0

$V(\mathbf{r})$ in (25.11) will split the level into two singlet levels and a doubly degenerate one. The particular order of the levels depends on $V(\mathbf{r})$ and can only be established by detailed calculation (problem 25.13) but the M_5 two-fold degeneracy will remain in any approximation.

Compatibility relations

In any one band certain compatibility relations must be satisfied between the irreducible representations of states along the lines specified by Δ, Σ and Z on the one hand, and the irreducible representations of Γ, X and M at the corners on the other hand (Fig. 33a). This is because X for instance is both a special point in its own right, and yet also a point on the lines Δ and Z. Consider for example the wave function $\psi(\Delta_2; \mathbf{r})$ belonging to the irreducible representation Δ_2: it changes sign under the reflection m_y. Now in any one band $\psi(\mathbf{k}; \mathbf{r})$ varies continuously† with **k**, so that as $\mathbf{k} = \Delta$ approaches X, the function $\psi(\Delta_2; \mathbf{r})$ must turn into a function which is still odd under m_y, i.e. into a function belonging to either the representation X_2 or X_3 since these are the only ones odd under m_y as can be seen from the character table (Table 24). All such compatibility relations for the square lattice are given in Table 26.

† Except possibly at points of accidental degeneracy (see below).

The case of the two-dimensional representations Γ_5 and M_5 is slightly more complicated. Taking Γ_5 for instance, we have two degenerate wave functions $\psi_1(\Gamma_5;\ \mathbf{r})$ and $\psi_2(\Gamma_5;\ \mathbf{r})$ with $\mathbf{k} = 0$.

TABLE 26

Compatibility Relations

Representation	Compatible with
Δ_1	$\Gamma_1, \Gamma_3, \Gamma_5; X_1, X_4$.
Δ_2	$\Gamma_2, \Gamma_4, \Gamma_5; X_2, X_3$.
Σ_1	$\Gamma_1, \Gamma_4, \Gamma_5; M_1, M_4, M_5$.
Σ_2	$\Gamma_2, \Gamma_3, \Gamma_5; M_2, M_3, M_5$.
Z_1	$X_1, X_3; M_1, M_3, M_5$.
Z_2	$X_2, X_4; M_2, M_4, M_5$.

Γ_5 reduces into $\Delta_1 + \Delta_2$ or $\Sigma_1 + \Sigma_2$
M_5 reduces into $\Sigma_1 + \Sigma_2$ or $Z_1 + Z_2$

As we go along the line Δ (Fig. 33a), the only symmetry elements remaining are E and m_y, and under this more restricted group the representation Γ_5 becomes reducible into $\Delta_1 + \Delta_2$. Thus according to the general principles of § 6, the doubly-degenerate level at Γ_5 splits into two singlets as we go in the direction Δ. The same applies to the line Σ, so that we have two bands degenerate at the single point $\mathbf{k} = 0$.

Degeneracy

There are three different basic types of degeneracy between bands. The first type, which we have already encountered, is exemplified by the representations Γ_5 and M_5. These are two-dimensional, so that we have a degeneracy between two bands at a definite point in the Brillouin zone. This type is called an *essential* degeneracy because it is due to symmetry properties and cannot be altered or split by changing the crystal potential $V(\mathbf{r})$ slightly. Both of the other two types of degeneracy depend sensitively on $V(\mathbf{r})$ as regards the exact value of $\mathbf{k}$ at which they occur, and are called *accidental degeneracies.*

The second type of degeneracy is between two *inequivalent* representations and is illustrated by Fig. 34. Suppose we find by direct calculation from the Hamiltonian (25.11) that the lowest two levels at Γ and X have the representations Γ_1, Γ_4 and X_2, X_1

as shown. From the compatibility relations (Table 26), we obtain the representations at Δ between Γ and X. We note that Γ_1 turns into Δ_1 which cannot change into X_2 but must join up with X_1. Similarly Γ_4 turns into Δ_2 which joins on to X_2. Thus the curve $\Gamma_1\Delta_1X_1$ must cross $\Gamma_4\Delta_2X_2$ at some point P, and we obtain a degeneracy between the two bands at P. Incidentally it is important that the bands should be numbered strictly in accordance with

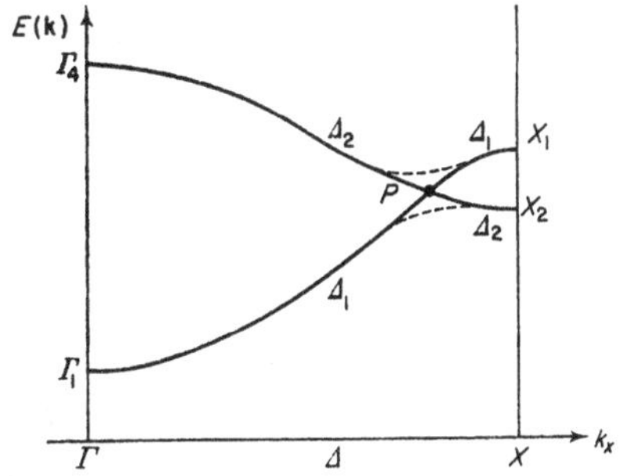

FIG. 34. Representations and $E(\mathbf{k})$ along the line ΓX, showing degeneracy between two inequivalent representations Δ_1 and Δ_2 at P. The dotted curves show how $E(\mathbf{k})$ would vary if we took a curved path from Γ to X that by-passes P slightly.

increasing energy. Thus in Fig. 34, $\Gamma_1\Delta_1P\Delta_2X_2$ belongs to the first band and $\Gamma_4\Delta_2P\Delta_1X_1$ to the second. The necessity for this can be seen by considering a path from Γ to X which is slightly deformed from the straight line $\Gamma\Delta X$ so as to avoid the point P: in this case the energy varies as shown by the dotted line with $\psi_{\mathbf{k}}$ varying very rapidly with $\mathbf{k}$ near the point P, and we see that Γ_1 and X_2 must clearly belong to the same band.

The third type of degeneracy is between equivalent representations, in particular at general points in the Brillouin zone where there is only the one type of representation. This sort of degeneracy can occur along a line of degeneracy in three-dimensional crystals with a centre of symmetry, but not over a whole surface in $\mathbf{k}$-space. For an outline of the proof of this result, see problem 26.8, from which the general method of tackling such a problem in any number of dimensions or in any special case should be clear. In two dimensions with a centre of symmetry, only points and not lines of degeneracy are allowed. The whole subject of accidental degeneracy has been fully discussed by Herring (1937b). The third type seems to be rarely met with in practice, though the first and second types are common.

Summary and references

Various effects of crystal symmetry have been described, and in particular the irreducible representations of space groups not containing screw axes or glide planes. The translational symmetry of wave functions gives a **k** vector, which can be used to label the base vectors of a representation under all conditions. The wave vector **k** conventionally varies over a restricted region of **k**-space called the Brillouin zone. Most irreducible representations, i.e. those involving general **k** vectors, are of dimension n, where n is the number of elements in the point-group symmetry of the structure. The situation is more complicated for **k** vectors lying on symmetry axes, mirror planes or the surface of the Brillouin zone. The irreducible representations have been used to discuss energy level splittings and degeneracies (Herring 1937b) between different bands of the energy function $E(\mathbf{k})$ of an electron travelling in the solid. The case of the two-dimensional square lattice has been treated in detail. Analogous results for the simple cubic, face centred cubic, and body centred cubic structures have been given by Bouckaert, Smoluchowski and Wigner (1936). Further references are given at the end of § 26.

Problems

25.1 Verify that the 1 by 1 matrices (25.5a) really give a representation of $\mathfrak{C}$.

25.2 *The reciprocal lattice.* Prove that the condition (25.7) defines a lattice of vectors (25.6). Hint: in the notation of (25.1), show that

$$\mathbf{K}_1 = \frac{2\pi\mathbf{a}_2 \wedge \mathbf{a}_3}{\mathbf{a}_1 \wedge \mathbf{a}_2 \cdot \mathbf{a}_3}, \qquad \mathbf{K}_2 = \frac{2\pi\mathbf{a}_3 \wedge \mathbf{a}_1}{\mathbf{a}_1 \wedge \mathbf{a}_2 \cdot \mathbf{a}_3}, \qquad \mathbf{K}_3 = \frac{2\pi\mathbf{a}_1 \wedge \mathbf{a}_2}{\mathbf{a}_1 \wedge \mathbf{a}_2 \cdot \mathbf{a}_3}$$

satisfy the relation

$$\mathbf{K}_i \cdot \mathbf{a}_j = 2\pi\delta_{ij}.$$

Put $\mathbf{K} = m_1\mathbf{K}_1 + m_2\mathbf{K}_2 + m_3\mathbf{K}_3$ and show that m_1, m_2, m_3 are integers.

25.3 Verify that the vectors (25.2) form a set of primitive vectors for the face centred cubic structure, and find a set of primitive vectors for the body centred cubic and the simple cubic structures (Fig. 27). Derive and describe their reciprocal lattices and Brillouin zones.

25.4 Show from (25.3) that **k** is restricted to the points (25.9). Calculate the density of points in **k**-space from (25.9) and the volume of the Brillouin zone, and verify that there are $N_1N_2N_3$ distinct representations, i.e. one per primitive unit cell in the crystal. Thus, counting spin degeneracy, the Brillouin zone contains *two electron states per primitive unit cell* of the crystal. It should be noted that in this statement the cell referred to must be the primitive one, i.e. not the cube of Fig. 27a but the cell defined by (25.2). Furthermore it is irrelevant how many atoms there are in each primitive unit cell: while this number is one for body centred and face centred cubic metals, it is two for the hexagonal close packed metals, for the diamond, tin and bismuth structures, and is higher still for more complicated structures such as most salts and molecular crystals.

25.5 Use the idea of translational invariance to discuss the eigenstates of a particle moving freely under no forces.

25.6 Test whether the following functions transform irreducibly under the *translation* group of the square lattice (length of square $= a$), and if so what are their reduced $\mathbf{k}$ vectors: $\exp(\frac{1}{2}i\pi x/a)$, $\exp\, i(\pi/a)(-\frac{3}{2}x - \frac{5}{2}x)$, $\cos(\frac{1}{2}x\pi/a)$, $\sin(y\pi/a)$, $\exp\, i(\pi/a)(x+y)$, $\exp\, i(\pi/a)(x-y)$, $\cos(x\pi/a)\exp(iy\pi/a)$, $\cos(8x\pi/a)$, $\cos\, \pi/a$.

25.7 Show that a Bloch function (25.5b) can be written in the form

$$\psi_{\mathbf{k}}(\mathbf{r}) = \exp(i\mathbf{k}\cdot\mathbf{r})\, u_{\mathbf{k}}(\mathbf{r})$$

where $u_{\mathbf{k}}(\mathbf{r})$ is periodic in the lattice. Hence show that in general (25.5b) holds only for lattice translations $\mathbf{t}_n$ (25.1) and not for arbitrary translations $\mathbf{t}$.

25.8 The energy $E(\mathbf{k})$ in a particular band is a single-valued function of $\mathbf{k}$ inside and on the surface of the Brillouin zone of the face centred cubic structure (Fig. 29a) i.e. there is no degeneracy with other bands. Show that the normal component of $\text{grad}_{\mathbf{k}}E$ is zero over the whole of the square faces of the zone, but that in general it is zero on the hexagonal faces only on the lines joining opposite corners of the hexagon.

25.9 Show that

$$\sum_n \exp i\mathbf{k}\cdot\mathbf{t}_n = N_1 N_2 N_3 \delta_{\mathbf{k}\mathbf{0}},$$

$$\sum_{\mathbf{k}} \exp i\mathbf{k}\cdot\mathbf{t}_n = N_1 N_2 N_3 \delta_{n0},$$

where the summation over $\mathbf{k}$ is over all allowed $\mathbf{k}$'s (25.9) in the Brillouin zone, and where $\delta_{\mathbf{k}\mathbf{0}}$ is the Kroenecker delta symbol (A.15) applied to each component $\mathbf{k}_x$, $\mathbf{k}_y$, $\mathbf{k}_z$ of $\mathbf{k}$. Note that these relations are just the orthogonality relations (14.8) and (14.18).

25.10 A function $\phi = a\psi(\mathbf{k}_1) + b\psi(\mathbf{k}_2)$ belongs to an irreducible representation of a space group $\mathfrak{G}$. Show that $\psi(\mathbf{k}_1)$ and $\psi(\mathbf{k}_2)$ belong to this same irreducible representation and hence may be used as base vectors for the representation. This is an alternative proof of (25.10). Hint: let $\mathbf{t}_n$ be a lattice translation such that $\exp i\mathbf{k}_1 \cdot \mathbf{t}_n \neq \exp i\mathbf{k}_2 \cdot \mathbf{t}_n$. Consider ϕ and $\mathbf{t}_n\phi$, and show

$$a\psi(\mathbf{k}_1) = \frac{[\exp(i\mathbf{k}_2 \cdot \mathbf{t}_n) - \mathbf{t}_n]\phi}{\exp(i\mathbf{k}_2 \cdot \mathbf{t}_n) - \exp(i\mathbf{k}_1 \cdot \mathbf{t}_n)}.$$

25.11 *The free electron model and the extended zone scheme.* Consider a square lattice but with the potential $V(\mathbf{r})$ in (25.11) set equal to zero. The energy eigenfunctions are then $\exp(i\mathbf{k}' \cdot \mathbf{r})$ with energy $E(\mathbf{k}') = \hbar^2(\mathbf{k}')^2/2m$. With the aid of Fig. 35, redescribe $E(\mathbf{k}')$ by a multivalued function $E(\mathbf{k})$ with $\mathbf{k}$ in the Brillouin zone,

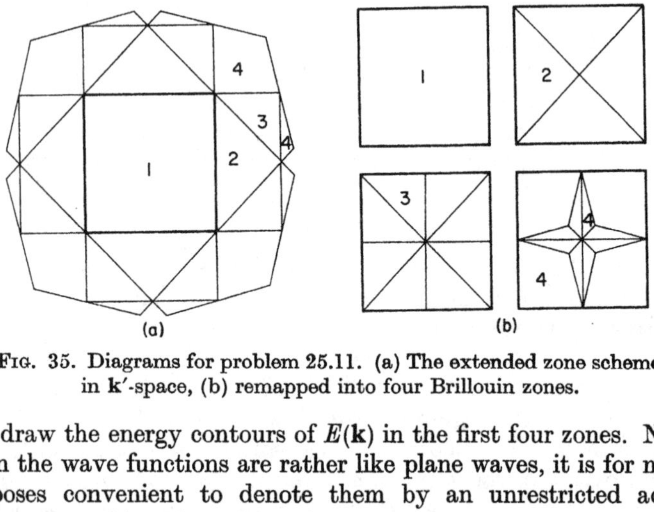

FIG. 35. Diagrams for problem 25.11. (a) The extended zone scheme in $\mathbf{k}'$-space, (b) remapped into four Brillouin zones.

and draw the energy contours of $E(\mathbf{k})$ in the first four zones. Note: when the wave functions are rather like plane waves, it is for many purposes convenient to denote them by an unrestricted actual wave vector $\mathbf{k}'$ rather than the reduced wave vector $\mathbf{k}$. Then $E(\mathbf{k}')$ is a single-valued function of $\mathbf{k}'$, with energy discontinuities at the lines of Fig. 35a. This is called the extended zone scheme.

25.12 Write down the group of $\mathbf{k}$ and the irreducible representations of the face centred cubic structure associated with the point X (Fig. 29a).

25.13 By inspection or using (14.11), write down the linear combinations of the plane waves (25.15) which transform respectively according to M_1, M_4, and M_5. Calculate their energies in the *nearly free electron approximation*, i.e. treating $V(\mathbf{r})$ in (25.11) by first order perturbation theory. Note: in actual metals the potential $V(\mathbf{r})$ is so strong as to make nodes in the wave functions near the atomic nuclei, and this may alter the order of levels from that found in the nearly free electron approximation.

25.14 Consider the functions

$$\psi_m = \sum_n \phi_m(\mathbf{r} - \mathbf{R}_n)$$

where the ϕ_m, $m = 2, 1, 0, -1, -2$ are atomic $3d$ functions and the $\mathbf{R}_n$ are the atomic sites in a face centred cubic metal, e.g. copper. What representation of the space group do the ψ_m form, and to what extent will they be degenerate in the solid? (In an isolated atom all five ϕ_m are, of course, degenerate.)

25.15* Derive the energy levels and irreducible representations at all special points for the first two bands of the body centred cubic structure, using the nearly free electron approximation (see problem 25.13; Mott and Jones 1936, p. 63; Bouchert *et al.* 1936). Compare with the results of a detailed calculation for potassium by Callaway (1956).

25.16* Discuss the energy $E(\mathbf{k})$ of Bloch functions made up out of overlapping p functions in a simple cubic structure using the tight binding method in the nearest neighbour approximation (Mott and Jones 1936, p. 70). Derive the irreducible representations at all special points in the Brillouin zone, discuss all degeneracies and state which degeneracies (if any) would disappear on taking a better approximation (Bouckaert *et al.* 1936).

25.17 Verify the compatibility relations of Table 26.

25.18 *Perturbation theory for* $E(\mathbf{k})$. If the Bloch function $\psi_{\mathbf{k}}(\mathbf{r})$ is written in the form of problem 25.7, show that $u_{\mathbf{k}}(\mathbf{r})$ satisfies

$$[-(\hbar^2/2m)\nabla^2 - (i\hbar^2/m)\mathbf{k}\cdot\nabla + V(\mathbf{r}) + (\hbar^2\mathbf{k}^2/2m)]u_{\mathbf{k}}(\mathbf{r}) = E(\mathbf{k})u_{\mathbf{k}}(\mathbf{r}).$$

Hence, discuss how $E(\mathbf{k})$ at a point $\mathbf{k} = \mathbf{k}_1 + \delta\mathbf{k}$ near $\mathbf{k}_1$ can be investigated using the perturbation operator P where

$$P = (-i\hbar^2/m)\nabla\cdot\delta\mathbf{k} + (\hbar^2/m)\mathbf{k}_1\cdot\delta\mathbf{k} + (\hbar^2/2m)(\delta\mathbf{k})^2.$$

Verify that

$$\int u^*(\mathbf{k}_1, n; \mathbf{r})Pu(\mathbf{k}_1, n'; \mathbf{r})\, dv$$
$$= (\hbar^2/2m)(\delta\mathbf{k})^2\delta_{nn'} + (\hbar/m)\delta\mathbf{k}\cdot\int \psi^*(\mathbf{k}_1, n)\,(-i\hbar\nabla)\,\psi(\mathbf{k}_1, n')\, dv,$$

where n and n' refer to the different bands $E_n(\mathbf{k})$. Using this technique, prove that $\mathrm{grad}_{\mathbf{k}}E = 0$ at the point X in the square lattice (Fig. 33a), and show that one of the effective mass parameters

$$m_{xx} = \frac{\hbar^2}{\partial^2E/\partial k_x^2} \quad \text{or} \quad m_{yy} = \frac{\hbar^2}{\partial^2E/\partial k_y^2}$$

is likely to be small if the energy gap to the next zone is small.

25.19* Use the irreducible representations of space groups to describe the elastic lattice vibrations of some simple solids such as the face centred cubic structure (Peierls 1955, Houston 1948, see also § 23 and Herring 1954). Derive selection rules for the Raman spectrum and show that in first order it should consist of lines rather than bands (Peierls 1955, Stephen 1958). Discuss in this light the ideas of Raman on this subject (see Bhagavantam *et al.* 1948, Menzies 1953).

25.20 Show that the space group of the square lattice is *not* a "direct product" of the translation group $\mathfrak{T}$ and the point-group 4mm in the sense of § 15, and verify that the representation theory of § 15 does not apply.

25.21 In aluminium (face centred cubic), a detailed calculation shows that the irreducible representations in the first and second bands are probably as follows: first band Γ_1, X'_4, W_3; second band Γ'_{25}, X_1, W_3 (Heine 1957b). The order of levels at K is uncertain but assume that it is K_1 (band 1) and K_3 (band 2). Derive the compatibility relations at these points and show that there must be a line of degeneracy between the second and third bands in the plane $k_x = 0$. (For the character tables and notation, see Bouckaert *et al.* 1936. See also problem 26.8.)

26. Further Aspects of Brillouin Zone Theory

In the latter half of the last section we limited ourselves to space groups which do not contain screw axes or glide planes. In this section we shall remove this restriction, and then go on to consider the effects of time-reversal and of spin-orbit coupling. Finally the theory will be applied to a qualitative description of $E(\mathbf{k})$ near the top of the valence band in germanium and indium antimonide.

Screw axes and glide planes: bands sticking together

In § 25 we wrote an arbitrary element of a space-group in the form

$$\{R|\mathbf{t}\} \tag{26.1}$$

where $\mathbf{t}$ is a translation, applied first, and R is a proper or improper rotation following. All the possible rotations that are consistent with crystal symmetry have already been derived in § 16. The translations $\mathbf{t}$ include all the lattice translations $\mathbf{t}_n$ (25.1), and in § 25 when we started discussing the irreducible representations in detail we limited ourselves to space-groups in which $\mathbf{t}$ is always one of these lattice translations. However, there are many space-groups for which this is not so, e.g. that of the diamond structure and the hexagonal-close-packed structure. These contain *screw axes* and *glide planes*, i.e. symmetry elements which are rotations or reflections respectively combined with a translation by a fraction of a lattice displacement. For example a two-fold screw axis is a translation by half a lattice displacement, say $\frac{1}{2}\mathbf{a}_1$, with a 180° rotation about the axis $\mathbf{a}_1$. If we apply this screw axis twice, we obtain a total rotation by $360° = 0°$ and a displacement of $2 \times \frac{1}{2}\mathbf{a}_1$. This is a pure translation, and therefore it has to be a lattice translation $\mathbf{t}_n$ (as of course it is in our example). This condition limits the kind of screw axes and glide planes that one can have. It can also be shown that the displacement in the case of a screw is parallel to the axis of rotation, and in the case of a glide plane is parallel to the reflection plane (problem 26.2).

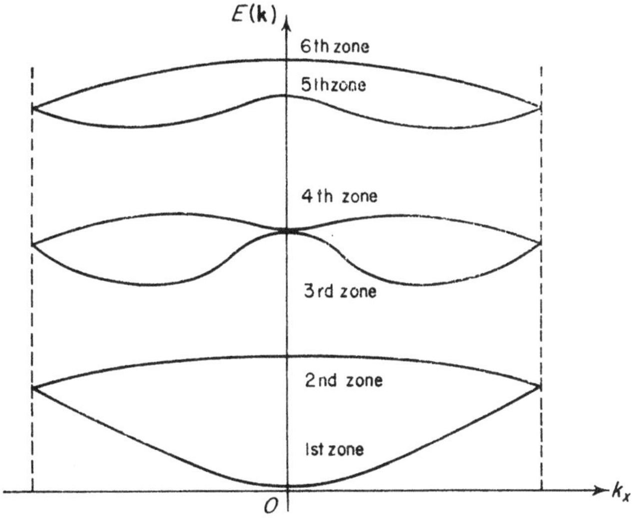

FIG. 36. Energy band structure showing bands "sticking together" in pairs at the zone surface.

The irreducible representations of the more general space-groups with screws and glides are derived by methods analogous to those of § 25. The discussion there is completely general as regards the definition of Brillouin zones and the reciprocal lattice, and with very little extension the arguments there give all the irreducible representations of **k** vectors *not* lying on the surface of the Brillouin zone (problems 26.4 and 26.5). However, when it comes to the surface points of the Brillouin zone a new phenomenon can arise, described as "*bands sticking together*". It can happen that some special points, usually along a whole line on the surface of the Brillouin zone or sometimes over a whole zone face, have *only* two-dimensional irreducible representations. Along such a line or face we always have two bands degenerate, and $E(\mathbf{k})$ looks something like Fig. 36 with the bands going in pairs. It might at first sight appear convenient just to choose a new Brillouin zone of double the volume of the usual one, and to describe each pair of bands in Fig. 36 as a single band in the new zone scheme: but Hund (1936) has shown that this leads to more ambiguities and inconveniences that it removes, and we shall not consider the idea further.

Following Hund (1936), we can set up a necessary condition for the sticking together of bands to occur. Consider the 1*s* states in an array of atoms with total symmetry equal to some space-group $\mathfrak{G}$ all atoms being of the same element. These 1*s* states will be

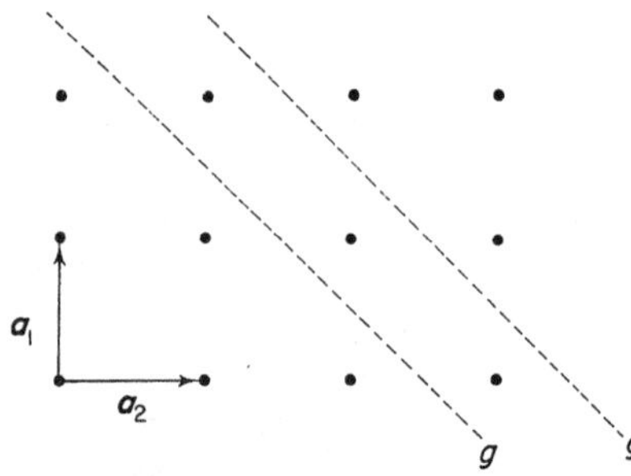

FIG. 37. Glide planes gg in a simple square array with glide $\frac{1}{2}\mathbf{a}_1 - \frac{1}{2}\mathbf{a}_2$. This type of glide plane does not produce sticking together of bands because it is an inherent property of the translational lattice.

separated by several electron volts from other states like 2*s* or 2*p*, but they will be broadened very slightly into a narrow band of energies due to overlap between neighbouring atoms (Mott and Jones 1936, p. 68). There must be two 1*s* states per atom including spin degeneracy, and the Brillouin zone contains two states per unit cell of the crystal (problem 25.4). Thus to have a pair of bands stuck together, we must have at least two 1*s* bands of nearly the same energy and that means at least two atoms per unit cell. Moreover, we saw in § 25 that a space-group with elements only of the form $\{R|\mathbf{t}_n\}$ produces occasional degeneracy between bands but not a systematic sticking together in pairs, since each character table such as Table 24 always contains the *one*-dimensional identity representation. We therefore have to fall back on screws and glides to produce the two atoms per unit cell. Furthermore the glide plane must not be an inherent property of the primitive translation lattice such as in Fig. 37, for this does not give two atoms per unit cell either. Thus *a necessary condition for the sticking together of bands is that the space-group contains screw axes and/or glide planes which are not inherent in the primitive translational lattice.* It is not known whether the condition is also a sufficient one, though the rather general considerations of (26.12) and problem 26.12 suggest that it may very well be so.

We now consider some structure with the space group illustrated in Fig. 38a, and before going into any detailed discussion of the irreducible representations we shall show in the most direct way

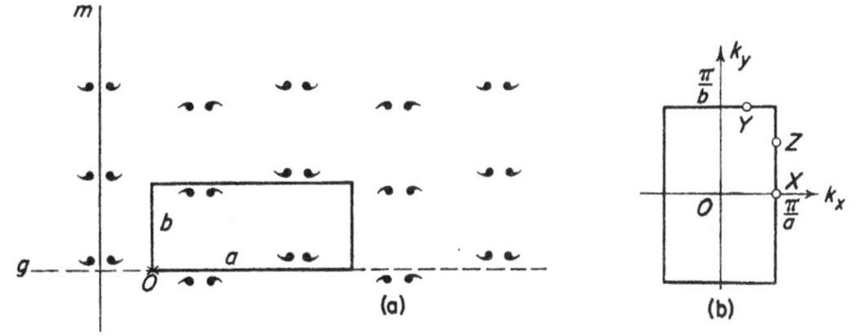

FIG. 38. (a) Pattern with the two-dimensional symmetry *Pbl* (Buerger 1942), showing a glide plane g, a mirror plane m, a centre of inversion O, and a unit cell of size $a \times b$. Note that the commas are not supposed to represent atoms, but just to form a pattern which exhibits the total symmetry as clearly as possible. It is irrelevant to the discussion where and how many atoms there are as long as they exhibit this symmetry. In two dimensions a glide plane and a two-fold screw axis become identical. (b) The Brillouin zone for the above space group.

possible that the presence of the glide plane produces a sticking together of bands at the point $X = (\pi/a, 0)$ in the Brillouin zone (Fig. 38b). Actually the sticking together occurs along the whole side Z, $k_x = \pi/a$, of the Brillouin zone, but this will only become apparent when we consider time-reversal symmetry. Consider now a function

$$\psi(x, y) \qquad \text{with} \qquad k = (\pi/a, 0). \tag{26.2a}$$

Then the following functions must also have the same **k** and must be degenerate with ψ:

$$g\psi = \psi(x + \tfrac{1}{2}a, -y), \tag{26.2b}$$

$$\Pi\psi = \psi(-x, -y), \tag{26.2c}$$

$$m\psi = \psi(-x - \tfrac{1}{2}a, y), \tag{26.2d}$$

where g, Π, m are the glide, inversion and mirror reflection of Fig. 38. We have

$$g\psi = \psi(x + \tfrac{1}{2}a, -y) = m\psi(-x, -y) = m\Pi\psi \qquad (26.3)$$

It is now easy to prove that all the four functions (26.2) cannot be proportional to one another. For let us suppose on the contrary that they are: then since $\Pi^2\psi = \psi = m^2\psi$, we must have

$$\Pi\psi = \alpha\psi, \quad \alpha = \pm 1, \quad \text{and} \quad m\psi = \beta\psi, \quad \beta = \pm 1. \qquad (26.4)$$

Now from (26.3) and (26.4) we have

$$g^2\psi = m\Pi m\Pi\psi = \alpha m\Pi m\psi, \text{ etc. } = \alpha^2\beta^2\psi,$$

and from (26.2b)

$$g^2\psi = \psi(x + a, y) = \exp(i\mathbf{k}_x a)\psi(x, y) = -\psi. \qquad (26.5)$$

Since $\alpha^2\beta^2$ cannot equal -1 we have a contradiction, and it follows that the functions (26.2) cannot be proportional to one another. Every irreducible representation associated with the point X in the Brillouin zone is therefore at least two-dimensional, so that we have sticking together of bands there.

Irreducible representations at surface points

To show how one sets up the irreducible representations of a space-group with screws and glides for points on the Brillouin zone surface, we will now work these out in detail for the point X (Fig. 38b) already considered. As in § 25, we first pick out $\mathfrak{K}$, the group of $\mathbf{k}$, consisting of all elements of the space group $\mathfrak{G}$ which leave $\mathbf{k}$ invariant. In the case of our point X this includes the whole of $\mathfrak{G}$. Now in § 25 we dropped from $\mathfrak{K}$ all the translations $\mathbf{t}_n$ because their effect on any function $\psi_\mathbf{k}$ was completely defined by $\mathbf{k}$, and this left $\mathfrak{K}$ as a simple point-group of rotations. However, this device will not work here because it does not leave us with a group; for instance the element g would be left, but g^2 which from (26.5) is the translation $\mathbf{t}_1 = (a, 0)$ would be excluded. Thus we consider instead the matrices E', g', m', $\mathbf{t}'_n$ etc. which *represent* $\mathfrak{K}$ in some arbitrary representation. Now all translations of the form $\mathbf{t}_{\text{even}} = (2na, mb)$ are represented by the unit matrix E', since $\mathbf{k} = (\pi/a, 0)$ and $\exp(i\mathbf{k} \cdot \mathbf{t}_{\text{even}}) = 1$. Likewise the translations $[(2n + 1)a, mb]$

are all represented by $-E'$. In this way we see that there are only the following eight different matrices in any representation of $\mathcal{K}$

$$\begin{array}{lcl} E' & \text{representing} & \{E|\mathbf{t}_{\text{even}}\} \\ g' & ,, & \{m_y|\boldsymbol{\tau} + \mathbf{t}_{\text{even}}\} \\ \Pi' & ,, & \{\Pi|\mathbf{t}_{\text{even}}\} \\ m' & ,, & \{m_x|\mathbf{t}_{\text{even}}\} \\ -E' & ,, & \{E|\mathbf{t}_1 + \mathbf{t}_{\text{even}}\} \\ -g' & ,, & \{m_y|\boldsymbol{\tau} + \mathbf{t}_1 + \mathbf{t}_{\text{even}}\} \\ -\Pi' & ,, & \{\Pi|\mathbf{t}_1 + \mathbf{t}_{\text{even}}\} \\ -m' & ,, & \{m_x|\mathbf{t}_1 + \mathbf{t}_{\text{even}}\} \end{array} \tag{26.6}$$

$\mathbf{t}_1$ = translation $(a, 0)$: $\boldsymbol{\tau}$ = translation $(\frac{1}{2}a, 0)$.
The reflection m_x is in the line shown in Fig. 38a.

These matrices in themselves clearly form a group, but since in our group postulates (§ 4) we only have provision for multiplying elements and not adding or subtracting them at the same time, we must for the moment consider $-E'$ as quite an independent element form E' etc., and to show this we shall write $\bar{E}'$, $\bar{g}'$, $\bar{\Pi}'$, $\bar{m}'$ instead of $-E'$, $-g'$, $-\Pi'$, $-m'$. The group multiplication table is easily obtained from (26.3) and similar relations, for instance we have

$$\begin{aligned} &(g')^2 = \bar{E}', \qquad (g')^3 = \bar{g}', \qquad (g')^4 = E' = (\bar{E}')^2, \\ &(m')^2 = (\bar{m}')^2 = E' = (\Pi')^2 = (\bar{\Pi}')^2, \\ &m'\Pi' = g', \text{ etc.} \end{aligned} \tag{26.7}$$

In fact inspection shows that the group is isomorphic with the point-group 4mm and we obtain the character Table 27. However,

TABLE 27

Character Table for (26.7)

E'	$\bar{E}'$	g', $\bar{g}$	Π', $\bar{\Pi}'$	m', $\bar{m}'$
1	1	1	1	1
1	1	1	−1	−1
1	1	−1	1	−1
1	1	−1	−1	1
2	−2	0	0	0

this character table cannot take account of the relation $\bar{E} = -E'$ for the reasons already mentioned. In fact some of the irreducible representations of Table 27 are definitely incompatible with this relation, so that for our purposes we retain only the ones satisfying $\chi(\bar{E}') = -\chi(E')$, namely the two-dimensional one, and reject all the rest. It should be noted that this additional requirement stems originally from the particular **k** vector which we are considering. Thus, finally, we obtain the character Table 28 for the group of X, showing as predicted the sticking together of bands in pairs.

TABLE 28

Character Table for the Group of X

Irred. rep.	Characters $\{E\|\mathbf{t}_e\}$	$\{E\|\mathbf{t}_1+\mathbf{t}_e\}$	$\{m_y\|\boldsymbol{\tau}+\mathbf{t}_e\}$ $\{m_y\|\boldsymbol{\tau}+\mathbf{t}_1+\mathbf{t}_e\}$	$\{\Pi\|\mathbf{t}_e\}$ $\{\Pi\|\mathbf{t}_1+\mathbf{t}_e\}$	$\{m_x\|\mathbf{t}_e\}$ $\{m_x\|\mathbf{t}_1+\mathbf{t}_e\}$
X_1	2	-2	0	0	0

The notation for the elements follows (26.6).

Time-reversal symmetry without spin

In the preceding discussion we have only considered the orbital part $\psi(\mathbf{r})$ of the electron wave functions, and then just added the spin functions u_+ or u_- at the end (see for instance problem 25.4). This two-fold spin degeneracy corresponds to neglecting spin-orbit coupling and other spin-dependent effects. In this approximation the Schrödinger equation for $\psi(\mathbf{r})$ is a purely real differential equation, and if $\psi(\mathbf{r})$ is an eigenfunction with energy E then, by taking the complex conjugate of the whole equation, we see that $\psi^*(\mathbf{r})$ must also satisfy the equation with the same energy E. The transformation $\psi(\mathbf{r})$ to $\psi^*(\mathbf{r})$ is just the time-reversal operation T already discussed in § 19, omitting the spin part, and for the present we define

$$T\psi(x, y, z) = \psi^*(x, y, z). \tag{26.8}$$

Since T is not a linear change of co-ordinates in the manner of § 2, we cannot incorporate it into our ordinary group-theoretical scheme. Instead we first have to set up all the irreducible representations of our group and then test whether or not time-reversal symmetry leads to any additional degeneracy.† If $\psi(\mathbf{r})$ belongs to the irreducible representation D, then $T\psi$ transforms according to D^* which consists of the complex conjugates of all the matrices of D. From §19 we have the following three cases.

Effect of time-reversal symmetry *without* spin: (26.9)

(a) D can be transformed to real form: there is no additional degeneracy.

(b) D and D^* are inequivalent: there is an additional doubling of the degeneracy, and D and D^* always occur together as a pair.

(c) D and D^* are equivalent but cannot be transformed to real form: there is an additional doubling of the degeneracy, and D always occurs twice.

Wigner has given a test (19.22) that shows which case a given representation belongs to, but it is not very convenient to apply to space-group representations as it stands, because it involves a summation over all elements of the group. However, Herring (1937a) has simplified the test for a space-group $\mathfrak{G}$ to the following form.

$$\begin{aligned} \sum_{Q_0} \chi(Q_0{}^2) &= n \qquad \text{case (a)}, \\ &= 0 \qquad \text{case (b)}, \\ &= -n \qquad \text{case (c)}, \end{aligned} \tag{26.10}$$

Here Q_0 is an element of $\mathfrak{G}$ which turns $\mathbf{k}$ into $-\mathbf{k}$, and n is the number of such elements. Thus $Q_0{}^2$ leaves $\mathbf{k}$ invariant and is an element of $\mathfrak{K}$, the group of $\mathbf{k}$. The χ in (26.10) refers to the character of $Q_0{}^2$ in an irreducible representation of $\mathfrak{K}$ (*not* of $\mathfrak{G}$). Furthermore *for space-groups containing the inversion Π, if Π is an element of $\mathfrak{K}$ then the Q_0 are just the elements of $\mathfrak{K}$; if Π is an element of $\mathfrak{G}$ but not of $\mathfrak{K}$, then the Q_0 are the elements $\Pi \times \mathfrak{K}$.* The test is therefore very easy to apply once we have the character table for $\mathfrak{K}$. There is obviously no need to sum explicitly over all elements represented by the same matrices, as long as we take one of each type. For

† Clearly from the general considerations of § 6, an additional symmetry element can never split an existing degeneracy and at most only produce new ones.

example for the representation X_1 (Table 29), we choose one value for $\mathbf{t}_{\text{even}}$, say $\mathbf{t}_{\text{even}} = 0$, and use the following eight Q_0:

$$\begin{array}{lllll} Q_0 = & \{E|0\} & \{E|\mathbf{t}_1\} & \{m_y|\tau\} & \{m_y|\tau + \mathbf{t}_1\} \\ Q_0{}^2 = & \{E|0\} & \{E|2\mathbf{t}_1\} & \{E|3\mathbf{t}_1\} & \{E|0\} \\ \chi(Q_0{}^2) = & 2 & 2 & -2 & -2 \end{array}$$

$$\begin{array}{lllll} Q_0 = & \{\Pi|0\} & \{\Pi|\mathbf{t}_1\} & \{m_x|0\} & \{m_x|\mathbf{t}_1\} \\ Q_0{}^2 = & \{E|0\} & \{E|0\} & \{E|0\} & \{E|0\} \\ \chi(Q_0{}^2) = & 2 & 2 & 2 & 2 \end{array}$$

$\sum \chi(Q_0{}^2) = 8$, the representation belongs to case (a).

The test shows we have case (a), and there is no additional degeneracy due to time-reversal symmetry.

By way of a further example, we shall consider the more general point Z, $\mathbf{k} = (\pi/a, k_y)$, of Fig. 38. The character table for the group of Z is derived as for the point X and is shown in Table 29. Herring's

TABLE 29

Character Table for the Group of Z

Irred. rep.	Characters $\{E\|\mathbf{0}\}$	$\{E\|\mathbf{t}_1\}$	$\{m_x\|\mathbf{0}\}$	$\{m_x\|\mathbf{t}_1\}$	$\{E\|\mathbf{t}_n\}$
Z_1	1	−1	1	−1	$\exp i\mathbf{k}\cdot\mathbf{t}_n$
Z_2	1	−1	−1	1	$\exp i\mathbf{k}\cdot\mathbf{t}_n$

Only typical elements are given, the others being easily derivable. The notation follows (26.6).

Time-reversal test

$$\begin{array}{lllll} Q_0 = \Pi \times K & \{\Pi|\mathbf{0}\} & \{\Pi|\mathbf{t}_1\} & \{m_y|\tau\} & \{m_y|\tau + \mathbf{t}_1\} \\ Q_0{}^2 & \{E|\mathbf{0}\} & \{E|\mathbf{0}\} & \{E|\mathbf{t}_1\} & \{E|\mathbf{t}_1\} \end{array}$$

Note that in forming $Q_0{}^2$, we have $\{\Pi|\mathbf{t}_n\}^2 = \{E|\mathbf{0}\}$ for any t_n, so that it is sufficient to consider $\mathbf{t}_n = \mathbf{0}$.

$\Sigma\chi(Q_0{}^2) = 0$ for both Z_1 and Z_2, i.e. each belongs to case (b) and there is an additional degeneracy due to time-reversal.

test is also applied in the table and shows that each representation belongs to case (b), so that time-reversal symmetry gives an extra degeneracy and we have pairs of bands sticking together along the whole side $k_x = \pi/a$ of the Brillouin zone. This sticking together of bands can also be proved by a simple *ad hoc* argument. Let $\psi(x, y)$ have wave vector $k = (\pi/a, k_y)$. Taking the complex conjugate of the defining relation (25.5b), we see that

$$\boxed{T\psi_{\mathbf{k}} = \psi^*(x, y) \text{ has wave vector } -\mathbf{k}} \tag{26.11}$$

i.e. in the present case $(\pi/a, -k_y)$. Thus $gT\psi$ has wave vector $\mathbf{k}$ again, and let us suppose that

$$gT\psi = \alpha\psi, \qquad (\alpha = \text{some constant}). \tag{26.12}$$

We obtain the two relations

$$\begin{aligned} gTgT\psi &= \psi(x + a, y) = -\psi(x, y), \\ gTgT\psi &= gT(\alpha\psi) = \alpha^* gT\psi = \alpha^*\alpha\psi. \end{aligned}$$

Now $\alpha^*\alpha$ cannot possibly equal -1 so that we have a contradiction, (26.12) must be false, and $gT\psi$ and ψ are two independent functions. However, since g and T are both symmetry elements, the two functions must have the same energy and we have a sticking together of bands at Z.

The effect of spin-orbit coupling

So far we have neglected spin-dependent terms in the Hamiltonian, written our wave functions in the form

$$\psi = \phi(\mathbf{r})u_+ \text{ and } \phi(\mathbf{r})u_-, \tag{26.13}$$

and considered the symmetry operations as transforming the orbital variables and orbital wave function $\phi(\mathbf{r})$ only. Now when the Hamiltonian $\mathscr{H}$ contains spin-orbit coupling (§ 11), $\mathscr{H}$ is no longer invariant under orbital transformations alone, but only under the crystal symmetry transformations applied to both the orbital and the spin variables simultaneously. The symmetry group $\mathfrak{G}$ therefore remains the same as before. However, when we apply it to wave functions like (26.13), transforming the spin as well as the orbital part, we need some new irreducible representations to describe the effect on the wave functions. As discussed in detail for the point-groups in § 16, this is because u_+ and u_- behave peculiarly: instead of remaining invariant under a 360° rotation like an ordinary function $f(\mathbf{r})$, they change sign. Let us consider the point $\Gamma(\mathbf{k} = 0)$ in the Brillouin zone of the square lattice (Fig. 33a). The new irreducible spin representations of the group of Γ

are shown in Table 30, as given by Koster (1957) who derived them by the methods of § 16. In the table the element $\bar{2}_z$, etc., is a rotation by 180° about the z-axis followed by a change of sign corresponding to an additional 360° rotation.

TABLE 30

Spin Representations of the Group of Γ for the Simple Square Lattice (Fig. 33a)

Irred. rep.	Characters						
	E	$\bar{E}$	2_z, $\bar{2}_z$	4_z, 4_z^3	$\bar{4}_z$, $\bar{4}_z^3$	$m_x, \bar{m}_x$, $m_y, \bar{m}_y$	$m_d, \bar{m}_d$, $m_{d'}, \bar{m}_{d'}$
Γ_6	2	-2	0	$\sqrt{2}$	$-\sqrt{2}$	0	0
Γ_7	2	-2	0	$-\sqrt{2}$	$\sqrt{2}$	0	0

Time-reversal: each representation belongs to case (c) of (26.15) so that time-reversal never gives any additional degeneracy.

Relationship to the ordinary representations (Table 24):

$$\Gamma_1 \times D^{(1/2)} = \Gamma_6 \qquad \Gamma_2 \times D^{(1/2)} = \Gamma_6$$
$$\Gamma_1 \times D^{(1/2)} = \Gamma_7 \qquad \Gamma_4 \times D^{(1/2)} = \Gamma_7$$
$$\Gamma_5 \times D^{(1/2)} = \Gamma_6 + \Gamma_7$$

Here $D^{(1/2)}$ is the representation formed by the spin functions u_+, u_-, i.e. Γ_6 in the present case.

We can now determine what additional splittings the spin-orbit coupling will produce. Suppose that in the absence of spin-orbit coupling, we have an energy level with a set of orbital wave functions $\phi(\mathbf{r})$ transforming according to one of the representations Γ_i, $i = 1$ to 5 of Table 24. The total wave functions (26.13) then transform according to

$$\Gamma_i \times D^{(1/2)} \tag{26.14}$$

where we have used $D^{(1/2)}$ in a symbolic sense to stand for the representation formed by u_+, u_- in whatever symmetry group we happen to be considering. The usual use of group characters (§ 16) shows that in the present case $D^{(1/2)}$ becomes Γ_6, and that the products (26.14) reduce as in Table 30. We see that only in the case of Γ_5 is the original four-fold degeneracy (double orbital and double spin degeneracy) split by the spin-orbit coupling into two-doublet levels.

With spin-dependent wave functions time-reversal symmetry takes the more complicated form (19.5a and b). We again have the three cases (a), (b), and (c), but the physical consequences are different. Let a set of spin-dependent wave functions ψ transform according to an irreducible representation D. Then we have (§ 19):

> Effect of time reversal symmetry *with* spin:
>
> (a) D can be transformed to real form: there is an additional degeneracy and D always occurs twice.
>
> (b) D and D^* are inequivalent: there is an additional doubling of the degeneracy, and D and D^* always occur together as a pair.
>
> (c) D and D^* are equivalent but cannot be transformed to real form: there is no additional degeneracy.

(26.15)

Note the reversal of the effect in cases (a) and (c) compared with the "without spin" situation (26.9). Herring's test (26.10) applies exactly as before since it is a mathematical property of group representations, independent of any applications to degeneracy in quantum mechanics.

With the use of these techniques, it is in principle quite straightforward to determine what splitting effect spin-orbit coupling will have on a given band structure $E(\mathbf{k})$. The situation for a general point $\mathbf{k}$ in the Brillouin zone is summarized in Fig. 39.

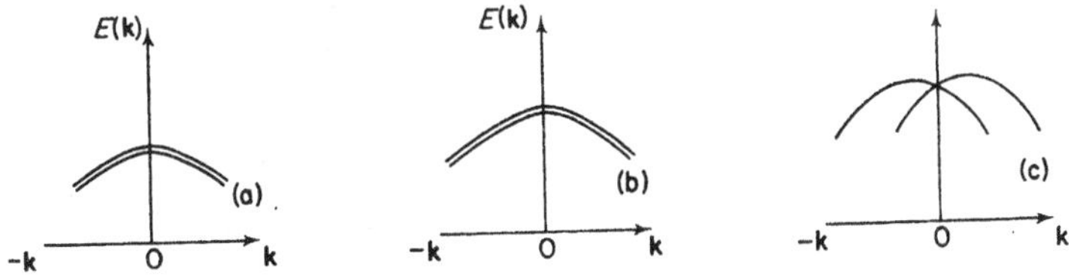

FIG. 39. Symmetry of $E(\mathbf{k})$ in an arbitrary direction about the origin. (a) In the absence of spin-orbit coupling whether or not there is a centre of inversion. The double line denotes the two-fold spin degeneracy and $E(-\mathbf{k}) = E(\mathbf{k})$. (b) With spin-orbit coupling and an inversion centre. There is a two-fold degeneracy at each $\mathbf{k}$, and $E(-\mathbf{k}) = E(\mathbf{k})$. (c) With spin-orbit coupling in a structure without an inversion centre. There is no degeneracy at an arbitrary $\mathbf{k}$, but $E(-\mathbf{k}) = E(\mathbf{k})$. The latter gives the two-fold Kramers degeneracy (§ 19). These results may be derived from (26.9), (26.10) and (26.15), or by simple *ad hoc* arguments (see problem 26.6).

The band structures of indium antimonide and germanium

The crystal structure of indium antimonide is shown in Fig. 40. Each antimony atom tends to loose one electron to the indium, so that we effectively tend to have In^- and Sb^+ ions. Each of these has four valence electrons like carbon, so that like carbon they form four directed covalent bonds in tetrahedral directions as shown in Fig. 40. (For a discussion of the covalence of carbon see § 22 and Fig. 16.) We may therefore expect to form the electron wave

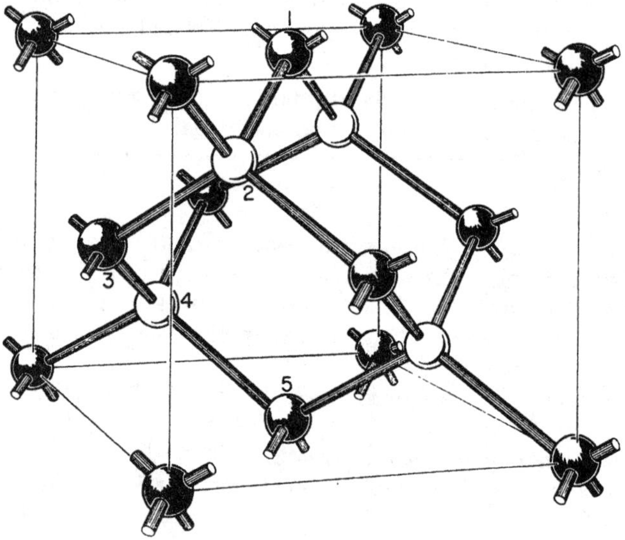

FIG. 40. Structure of indium antimonide and germanium. The double lines show the covalent bonds. The black and white circles correspond to the two types of atoms in indium antimonide. Both black and white circles correspond to germanium atoms in a germanium crystal. Notice how the atoms 1 to 5 form a spiral, indicating a four-fold screw axis in germanium. (From W. SHOCKLEY, *Electrons and Holes in Semiconductors*. D. Van Nostrand Company, Inc., Princeton, New Jersey. Copyright 1950.)

functions $\psi_{\mathbf{k}}$ out of directed valence orbitals of the type (22.7), and indeed detailed calculations of the band structure $E(\mathbf{k})$ can be made on this basis (Hall 1952). However, we shall limit ourselves to a qualitative discussion of the band structure $\mathbf{k} = 0$. It is then more convenient for our purposes to recognize that the directed valence orbitals are just linear combinations of the atomic orbitals

$$\phi_s = \text{an atomic } 5s \text{ function,} \tag{26.16}$$
$$\phi_x,\ \phi_y,\ \phi_z = \text{atomic } 5p_x,\ 5p_y,\ 5p_z \text{ functions,}$$

and to use these atomic orbitals instead. From these we can form the Bloch functions with $\mathbf{k} = 0$

$$\begin{aligned} \psi_{s,k=0}(\mathbf{r}) &= \sum_m \phi_s(\mathbf{r} - \mathbf{r}_m), \\ \psi_{x,k=0}(\mathbf{r}) &= \sum_m \phi_x(\mathbf{r} - \mathbf{r}_m), \\ \psi_{y,k=0}(\mathbf{r}) &= \sum_m \phi_y(\mathbf{r} - \mathbf{r}_m), \\ \psi_{z,k=0}(\mathbf{r}) &= \sum_m \phi_z(\mathbf{r} - \mathbf{r}_m), \end{aligned} \tag{26.17}$$

where the summation is over all atoms $\mathbf{r}_m$. So as not to confuse the notation, we have not indicated explicitly that the atomic orbitals (26.16) are really slightly different on the two types of atom.

The crystal structure is seen from Fig. 40 to have the face centred cubic translational symmetry, and the Brillouin zone is that shown in Fig. 29a. Because we have two types of atom arranged in a particular way, the rotational symmetry is not the full "cubic" point-group but the "tetrahedral" point-group $\bar{4}3m$. There are no screw-axes or glides when there are the two types of atom. The point $\mathbf{k} = 0$ in the Brillouin zone we call Γ, and the group of Γ clearly consists of the whole space group. Its character table is given in Table 31 (Dresselhaus 1955), every translation being represented by the unit matrix because $\mathbf{k} = 0$ (25.5a). It is now

TABLE 31

Group of Γ for the Indium Antimonide Structure

Irred. rep.	Characters: Ordinary Irreducible Representations E	2_x	3	$\bar{4}_x$	m_d
Γ_1	1	1	1	1	1
Γ_2	1	1	1	−1	−1
Γ_3	2	2	−1	0	0
Γ_4	3	−1	0	−1	1
Γ_5	3	−1	0	1	−1

	Irreducible spin representations E	$\bar{E}$	$2_x, \bar{2}_x$	3	$\bar{3}$	$\bar{4}_x$	$\bar{\bar{4}}_x$	$m_d, \bar{m}_d$
Γ_6	2	−2	0	1	−1	$\sqrt{2}$	$-\sqrt{2}$	0
Γ_7	2	−2	0	1	−1	$-\sqrt{2}$	$\sqrt{2}$	0
Γ_8	4	−4	0	−1	1	0	0	0

relationship between ordinary and spin representations

Γ_i :	Γ_1	Γ_2	Γ_3	Γ_4	Γ_5
$\Gamma_i \times D^{(1/2)}$:	Γ_6	Γ_7	Γ_8	$\Gamma_7 + \Gamma_8$	$\Gamma_6 + \Gamma_8$

easy to write down the transformation properties of ψ_s, ψ_x, ψ_y, ψ_z (26.17). Since each atom is turned into another atom of the same type by a rotation and since all atoms in (26.17) have the same coefficient ($\mathbf{k} = 0$), we need only consider the transformations on one atom. By inspection we see that ψ_s transforms according to Γ_1 and ψ_x, ψ_y, ψ_z according to Γ_4. This gives a singly-degenerate and a triply-degenerate level at $\mathbf{k} = 0$ (Fig. 41a).

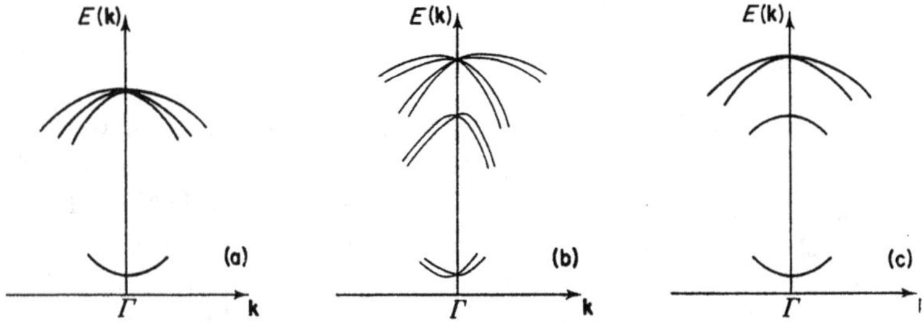

FIG. 41. Energy near Γ in In Sb and Ge. (a) Without spin orbit coupling in both In Sb and Ge, each band has a two-fold spin degeneracy. (b) Splitting of top of the valence band by spin-orbit coupling in In Sb. (c) Splitting by spin-orbit coupling in Ge, with each band being doubly degenerate. The magnitude of the splitting has been exaggerated.

Experimental evidence and detailed calculations show that the Γ_1 and Γ_4 levels are respectively the lowest and highest of all the occupied valence band, as is understandable for the following reason. We have already inferred from the observed crystal structure that the Bloch functions $\psi_{\mathbf{k}}$ will be made up as far as possible from directed valence orbitals like (22.7). In this way the lowest energy will be achieved on the average. However, we have seen that at $\mathbf{k} = 0$ symmetry forbids such mixing of ψ_s with ψ_x, ψ_y, ψ_z, and the levels approximate to atomic ones, i.e. the s-level lowest and the triply-degenerate p-level highest.

We now include the effect of spin-orbit coupling. Including spin degeneracy, the Γ_4 level is six-fold degenerate, and it splits into a doublet (Γ_7) and a quadruplet (Γ_8) (see Table 31) as is easily shown by the method described in connection with Table 30. This is exactly analogous to the splitting of an atomic p level into a $j = \frac{1}{2}$ doublet and $j = \frac{3}{2}$ quadruplet. The magnitude of the splitting is also of the same order of magnitude as in an atom, because the spin-orbit energy (11.8) is proportional to the electron's velocity which is large only when its potential energy is low near the nucleus. The potential here differs only very little in the atom and the solid. In indium antimonide the splitting is thought to be about 0.9 electron volts. Since the crystal structure does not have an inversion centre, $E(\mathbf{k})$ at a general $\mathbf{k}$ has to be in accordance with Fig. 39, so that the variation of $E(\mathbf{k})$ in an arbitrary direction from Γ is as shown in Fig. 41b.

Finally we consider the case of germanium. Here the two types of atom of Fig. 40 become identical, so that the crystal structure has some additional symmetry elements which include four-fold screw axes and a centre of inversion. The type of splitting at $\mathbf{k} = 0$ can be obtained simply in two ways. The group of Γ consists of the whole space-group again, but since $\mathbf{k} = 0$ the translational component of all elements including the screws has no effect and the group of Γ reduces to the "cubic" point-group m3m. The character table can be obtained (appendix L and Koster 1957) and the splitting at $\mathbf{k} = 0$ is found to be qualitatively the same as in indium antimonide. Alternatively this can be seen by noting that the splitting in the less symmetrical indium antimonide is the same as that in the more symmetrical free atom with complete rotational symmetry. Now additional symmetry can only produce extra degeneracy, so that in the present case any intermediate symmetry such as the germanium structure must have the same splitting as the two more extreme cases. In germanium the spin-orbit splitting is 0.29 eV. However, at a general point in the Brillouin zone the inversion symmetry of germanium produces a degeneracy between some of the spin bands in accordance with Fig. 39, so that $E(\mathbf{k})$ looks like Fig. 41c. More detailed calculations of the shape of $E(\mathbf{k})$ can be made using the perturbation approach of problem 25.18, and the three branches of $E(\mathbf{k})$ at the top of the valence band correspond to the possibility of "holes" with three different masses. For details of such calculations and a comparison with experiment the reader is referred to Kane (1956) and the references given there.

Summary

In this section we considered first the irreducible representations of complicated space-groups $\mathfrak{G}$ containing screw axes and/or glide planes. At any interior point of the Brillouin zone the procedure is essentially the same as that of § 25 for simple space-groups, as can be shown in greater detail in the manner of problem 26.5. For points on the surface of the zone, we pick out one $\mathbf{k}$ vector from the star of $\mathbf{k}$ (all $\mathbf{k}_i$ transformed into one another by $\mathfrak{G}$), and set up $\mathfrak{K}$ (the group of $\mathbf{k}$) consisting of all elements of $\mathfrak{G}$ which leave

$\mathbf{k}$ invariant. In any representation, many translations will be represented by the unit matrix because $\mathbf{k}$ lies on the Brillouin zone surface. We lump these all together in $\mathfrak{K}$ and similarly all other sets of elements that are represented by the same matrix. This leaves us with a much simpler group, usually a point-group if $\mathbf{k}$ has a high symmetry or a one-dimensional space-group for more general points, whose character table can be looked up or easily found. A single irreducible representation of $\mathfrak{G}$ is then obtained by taking several sets of functions, as many sets as there are distinct $\mathbf{k}_i$'s in the star of $\mathbf{k}$, each set having one of the $\mathbf{k}_i$ and transforming irreducibly under $\mathfrak{K}_i$.

It is found that screw axes and glide planes generally lead to necessary degeneracy between bands at the surface of the Brillouin zone, this being called "bands sticking together".

In the presence of spin-orbit coupling, the same procedure applies for finding the irreducible representations, except that we now have to use the double-valued spin representations. In either case we always have an additional time-reversal symmetry which may lead to extra degeneracy, as can be determined by a simple test (26.9), (26.10) and (26.15). The theory has been applied to the band structure of indium antimonide and germanium at the top of the valence band.

References

The irreducible representations of simple space-groups were obtained by Bouckaert, Smoluchowski and Wigner (1936) following the mathematical theory of Seitz (1936). The method was extended by Herring (1942) to space-groups with screws and glides, and by Elliott (1954) to include spin-representations and spin-orbit splitting. Herring (1937a and b) discussed the effect of time-reversal symmetry and the occurrence of accidental degeneracy. A systematic presentation of the whole subject, together with a general review and many references, has been given by Koster (1957). Bell (1954) and von der Lage and Bethe (1947) have applied symmetry properties to the types of functions used in detailed calculations of $E(\mathbf{k})$. Herman (1958) has given a compilation of space-groups whose irreducible representations have been tabulated, complete up to 1957.

PROBLEMS

26.1 A space group contains a purely translational symmetry element $\mathbf{t}$. Show that $\mathbf{t}$ must have the form of a lattice displacement $\mathbf{t}_n$ (25.1). Hint: investigate the consequences of having $\mathbf{t} = \frac{1}{2}\mathbf{a}_1$ or $\frac{1}{3}\mathbf{a}_1$.

26.2* Derive the different types of screw axes and glide planes that are possible in a space group. In particular, why do we always take the translational part of a screw parallel to its rotation axis? (Zachariasen 1945, Seitz 1935).

26.3 Make a model of a diamond or germanium crystal, and describe all the symmetry elements in its space group. (See Fig. 40, also Kittel 1956, p. 36, and Herring 1942.)

26.4 *The general point in the zone.* Verify in detail that the argument in § 25 for the irreducible representation connected with a general $\mathbf{k}$ vector in the Brillouin zone, can be applied with very little modification to space-groups containing screw axes and glide planes. (See also problem 26.5.)

26.5 *Irreducible representations at interior points.* $\mathfrak{G}$ is a space-group containing screws and glides, and $\mathbf{k}$ an *interior* point of the Brillouin zone. $\mathfrak{K}$, the group of $\mathbf{k}$, has elements $\{R_{\mathbf{k}}|\mathbf{t}\}$ where the translation $\mathbf{t}$ may not be lattice translations in the case of screws and glides. Show that the rotations $\{R_{\mathbf{k}}|0\}$ form a point-group $\mathfrak{P}_{\mathbf{k}}$. Let the matrices $D(R_{\mathbf{k}})$ form an irreducible representation of $\mathfrak{P}_{\mathbf{k}}$, and show that the matrices $\exp(i\mathbf{k}\cdot\mathbf{t})D(R_{\mathbf{k}})$ form an irreducible representation of $\mathfrak{K}$. Describe how this is then built up into an irreducible representation of $\mathfrak{G}$.

26.6 For a space-group containing (not containing) a centre of inversion Π, time-reversal symmetry does not give (does give) additional degeneracy at a general point in the Brillouin zone under the "without spin" conditions. Show this using the test (26.10) and also by an *ad hoc* argument based on the functions ψ and $\Pi T\psi$. Hence show that with *both* kinds of space-group $E(\mathbf{k}) = E(-\mathbf{k})$.

26.7 Show that with a space-group containing the inversion Π and in the "without spin" situation, the orbital wave functions $\psi_x(\mathbf{r})$ can always be so chosen that $\Pi T\psi_x = \psi_x$.

26.8 *Accidental degeneracy.* A given crystal contains a centre of inversion. Show that it is impossible for two zones to be accidentally degenerate at a whole *surface* of general points in the Brillouin zone, but that it is possible for accidental degeneracy to occur along a *curve* running through the Brillouin zone. (Note that we are excluding all special $\mathbf{k}$-vectors from consideration here.) Hints: (i) Suppose degeneracy occurs at one point $\mathbf{k}$. Using problem 25.18, set up the secular equation

$$\begin{vmatrix} A-E & C \\ C^* & B-E \end{vmatrix} = 0$$

for the energy E at a neighbouring point $\mathbf{k}+\delta\mathbf{k}$. (ii) Using problem 26.7 show that C can be made real. (iii) In the manner of (22.16), obtain the conditions $A - B = 0$ and $C = 0$ for degeneracy to occur at $\mathbf{k} + \delta\mathbf{k}$, and show that these two conditions define a *direction* for $\delta\mathbf{k}$.

26.9 Write down the representations of the whole space-group *Pbl* derived from Z_1 and Z_2 of Table 29, and verify by inspection that they belong to case (b) in agreement with the results of Herrings's test.

26.10 Write down the irreducible representation of the two-dimensional space-group *Pb* (Fig. 42) associated with $\mathbf{k} = (\pi/a, k_y)$, Show directly from first principles that it belongs to case (c) and verify with the test (26.10).

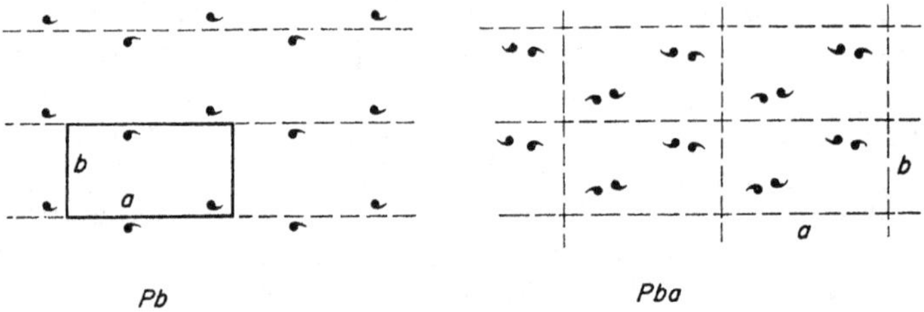

FIG. 42. Patterns exhibiting the two-dimensional space-groups *Pb* and *Pba*, with cell-size $a \times b$. The broken lines denote the glide planes.

26.11 Discuss completely the sticking together of bands in the case of the two-dimensional space-group *Pba* (Fig. 42).

26.12 Show that a two-fold screw axis perpendicular to a face of a Brillouin zone (in three dimensions) gives a double degeneracy over the whole of that face (i) by a direct *ad hoc* argument and (ii) using Herring's test (26.10). This is the only situation in which sticking together of bands can occur over a whole face of the Brillouin zone: in all other cases it is along lines on the surface of the zone.

26.13 Derive all the irreducible spin-representations of the space-group of Fig. 38. Will spin-orbit coupling destroy the sticking together of bands?

26.14* Derive from a simple model a picture of the band structure of graphite, and discuss qualitatively whether any additional splittings would be expected on going to a better approximation and on including spin-orbit coupling (Nozieres 1958 and references given there).

26.15* Write a review on the use of the perturbation method of problem 25.18 ($\mathbf{k}\cdot\mathbf{p}$ method) in theoretical solid state physics for calculating band structures, effective masses, degeneracies, and general theory such as $\langle \mathbf{v} \rangle = \hbar^{-1}$ $\text{grad}_{\mathbf{k}} E$. (Kane 1956, Cohen 1959, Herring 1937b).

26.16* Discuss the effect of crystal symmetry on the thermal conductivity at low temperatures (Herring 1954).

26.17* Discuss the application of group theory to determine the number of exciton lines in the absorption spectrum of an insulator. How does this theory apply to cuprous oxide? (Kittel 1956, Overhauser 1956, Gross 1956).

26.18 The bismuth structure has point-group symmetry $\bar{3}$m and rhombohedral translational symmetry with two atoms per unit cell at (0, 0, 0) and (u, u, u) where u does not bear any relation to the cell size. Describe the Brillouin zone (see Mott and Jones 1936, Fig. 69 and *not* 70). It is found experimentally that the bottom of the conduction band can be fitted by three degenerate ellipsoids

$$E(\mathbf{k}) = (\hbar^2/2m_0)(\alpha_1 k_x{}^2 + \alpha_2 k_y{}^2 + \alpha_3 k_z{}^2 + 2\alpha_4 k_y k_z)$$

with all other coefficients zero. Here x and z are chosen along a diad and the triad axes, and k_x, k_y, k_z are the co-ordinates of $\mathbf{k}$ *relative* to the unknown centre of the ellipsoid. From the form of $E(\mathbf{k})$, show where in the Brillouin zone the centres may lie, and discuss whether symmetry requires the existence of three or a multiple of three such ellipsoids.

26.19 An X-ray of wave vector $\mathbf{k}_1$ is incident on a crystal, and the direction of the diffracted ray is given by $\mathbf{k}_2$ where $|\mathbf{k}_2| = |\mathbf{k}_1|$. Show that the diffracted amplitude is proportional to

$$\int \rho(\mathbf{r}) \exp i(\mathbf{k}_1 - \mathbf{k}_2)\cdot\mathbf{r}\, dv$$

where $\rho(\mathbf{r})$ is the electron density in the crystal and where the integral extends over the whole crystal. Hence, show that the diffracted intensity is zero unless $\mathbf{k}_1 - \mathbf{k}_2 = \mathbf{K}_m$, a reciprocal lattice vector. For this reason an X-ray diffraction photograph consists of a number of spots which can be labelled by the integers (m_1, m_2, m_3) of (25.6). If a crystal has a two-fold screw axis parallel to the vector $\mathbf{K}_1$, show that the intensity of the spot $(m_1, 0, 0)$ is zero if m_1 is an odd integer. This is called an extinction. What extinctions would you expect in a cubic crystal containing a glide plane perpendicular to $\mathbf{K}_1$ with glide direction parallel to $\mathbf{K}_2$? Deduce also the Bragg relation $2d \sin \theta = n\lambda$ from this formalism (Kittel 1956).

27. Tensor Properties of Crystals

Introduction

In anistropic media like crystals, various properties of matter have to be represented as tensors. For instance Ohm's law takes the form

$$J = \boldsymbol{\sigma} \cdot \mathscr{E}$$

where the conductivity of the material is represented by the second order tensor $\boldsymbol{\sigma}$ with components

$$\begin{bmatrix} \sigma_{xx} & \sigma_{xy} & \sigma_{xz} \\ \sigma_{yx} & \sigma_{yy} & \sigma_{yz} \\ \sigma_{zx} & \sigma_{zy} & \sigma_{zz} \end{bmatrix}.$$

Now in a material with no symmetry properties such as a triclinic crystal, these components are all arbitrary (apart from the physical requirement $\sigma_{ij} = \sigma_{ji}$). But when we go to the opposite extreme of a cubic crystal, the two-fold and four-fold symmetries about the x, y, and z-axes impose the requirements†

$$\sigma_{xx} = \sigma_{yy} = \sigma_{zz} = \sigma \text{ say};$$
$$\sigma_{ij} = 0, \ i \neq j.$$

Thus the conductivity tensor reduces to the form

$$\begin{bmatrix} \sigma & 0 & 0 \\ 0 & \sigma & 0 \\ 0 & 0 & \sigma \end{bmatrix}$$

with three non-zero components of which only one is linearly independent.

Similarly many properties of matter form various types of tensor of different orders, for instance the elastic modulus, third order elastic constants, magnetoresistances, piezoelectric constants, magnetostriction constants, etc. The problem is to determine what forms these tensors must have for crystals belonging to the 32 different crystal classes described in § 16. In particular we wish to know which components are zero, and what are the relations between the non-zero ones. A variety of different group theoretical approaches

† This can be proved for instance as follows. Suppose we apply a field $\mathscr{E} = (\mathscr{E}, 0, 0)$ along the x-axis. The component of the current along the y-axis is $\sigma_{yx}\mathscr{E}$. Because of the two fold symmetry about the y-axis, this component must remain the same if we apply the same field along the negative x-direction, i.e. $\mathscr{E} = (-\mathscr{E}, 0, 0)$. Hence $\sigma_{yx}\mathscr{E} = -\sigma_{yx}\mathscr{E}$ and $\sigma_{yx} = 0$.

have been used to discuss this problem. In practice it is probably quickest to use a combination of these depending on the circumstances, e.g. whether one wants to discuss a particular point group or several related ones, whether one wants to discuss a high order tensor *ab initio* or one has available for use the results for lower order tensors, and whether one wants the full form of the tensor or just the numer of independent constants in it. It is therefore not possible to give an exhaustive treatment, but we shall illustrate as many of the ideas as possible by an extensive discussion of a particular example, namely the elastic constants for a crystal of symmetry 422.

Strain, stress and the elastic constants

The strain in a body is the amount by which it is locally deformed. Let **i**, **j**, **k** be unit vectors along the x-, y- and z-axes, and

$$\boldsymbol{\rho} = u\mathbf{i} + v\mathbf{j} + w\mathbf{k} \tag{27.1}$$

be the displacement of an arbitrary point **r** from its undeformed position. The strain at that point is then specified by the six strain components

$$e_{xx} = \frac{\partial u}{\partial x}, \qquad e_{yy} = \frac{\partial v}{\partial y}, \qquad e_{zz} = \frac{\partial w}{\partial z},$$

$$e_{xy} = e_{yx} = \frac{1}{2}\left(\frac{\partial v}{\partial x} + \frac{\partial u}{\partial y}\right), \qquad e_{yz} = e_{zy} = \frac{1}{2}\left(\frac{\partial w}{\partial y} + \frac{\partial v}{\partial z}\right),$$

$$e_{xz} = e_{zx} = \frac{1}{2}\left(\frac{\partial u}{\partial z} + \frac{\partial w}{\partial x}\right). \tag{27.2}$$

Many authors omit the factor $\frac{1}{2}$ in their definitions of e_{xy}, e_{yz}, e_{zx}. We refer to Kittel (1956, p. 85) for a helpful discussion of the strain components in physical terms that can be visualized rather more easily than the formal definitions (27.2).

The stresses in a body are the forces which tend to deform it. Consider a square surface of unit area perpendicular to the x-axis, and let

$$\mathbf{F}_x = (F_{xx}, F_{xy}, F_{xz})$$

be the force acting across this surface. By this we mean that if we cut away the material on one side of the surface, we would have to apply a force $\pm\mathbf{F}_x$ to the material on the other side to keep it in equilibrium. The $\pm$ depends on which side we cut away, the $+$ sign applying if we cut away the material in the positive x-direction from the surface. Similarly we define the other stress components

F_{yx}, F_{yy}, F_{yz}, F_{zx}, F_{zy}, F_{zz}. We now consider a unit cube of material and use the condition for equilibrium that the sum of the moments of the forces must vanish. This gives (Fig. 43) that

$$F_{xy} = F_{yx}, \qquad F_{yz} = F_{zy}, \qquad F_{zx} = F_{xz}.$$

We therefore have six independent stress components F_{xx}, F_{yy}, F_{zz}, F_{xy}, F_{xz}, F_{yz}.

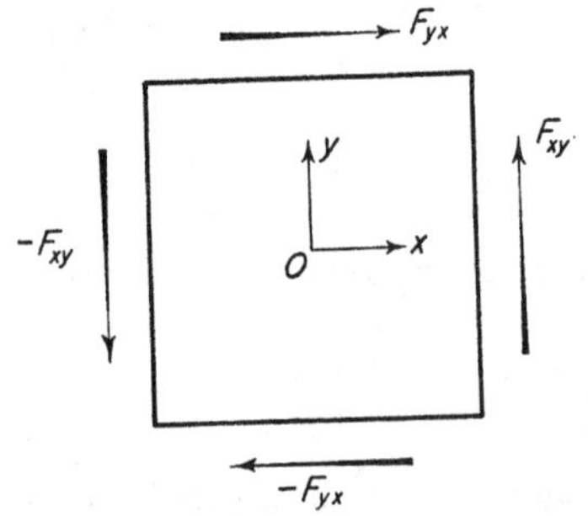

FIG. 43. Demonstration that $F_{xy} = F_{yx}$ for the total torque about the origin to be zero.

In discussing stress and strain components, it is sometimes convenient to employ an alternative notation. This replaces the double sufficies like xy by a single one 1 to 6 as follows:

$$\begin{gathered} 1 \text{ for } xx,\ 2 \text{ for } yy,\ 3 \text{ for } zz, \\ 4 \text{ for } yz \text{ and } zy,\ 5 \text{ for } xz \text{ and } zx,\ 6 \text{ for } xy \text{ and } yx. \end{gathered} \tag{27.3}$$

Thus the stress and strain components will be written as e_{pq}, F_{pq} or e_i, F_i where we shall always use p, q, r, s for suffices taking the values x, y, z (or 1, 2, 3) and i, j for suffices running from 1 to 6.

In a material obeying Hooke's law, the stress and strain components are related by the constants c_{ij} known as the elastic constants or the elastic stiffness constants.

$$F_i = c_{ij}e_j, \quad (j \text{ summed})$$

or

$$F_{pq} = c_{pqrs}e_{rs}. \quad (r,\ s \text{ summed}) \tag{27.4}$$

By considering a small cube and the work done on it by all the forces F_{pq}, we obtain that the total work done per unit volume in an arbitrary small strain δe_{pq} is

$$\delta U = F_{xx}\delta e_{xx} + F_{yy}\delta e_{yy} + F_{zz}\delta e_{zz} + \\ + 2F_{yz}\delta e_{yz} + 2F_{zx}\delta e_{zx} + 2F_{xy}\delta e_{xy}.$$

We have $\partial U/\partial e_1 = F_1$ and $\partial U/\partial e_2 = F_2$, and on further differentiation

$$\frac{\partial F_1}{\partial e_2} = \frac{\partial^2 U}{\partial e_1\,\partial e_2} = \frac{\partial F_2}{\partial e_1}, \text{ etc.}$$

Hence from (27.4) we obtain

$$c_{ij} = c_{ji}. \tag{27.5}$$

This reduces the number of independent elastic constants from thirty-six to twenty-one.

Transformation properties

From (27.4) it follows that the c_{pqrs} transform under rotations as the components of a fourth order tensor, i.e. in the same way as the products $pqrs$. This is an immediate application of a fundamental theorem of tensor analysis (Milne 1948, p. 42), and can also be proved directly as follows. Suppose the co-ordinates x, y, z transform according to

$$Rp' = D_{pp'}p \qquad (p, p' = x, y, z) \quad (p \text{ summed})$$

under a rotation R. D is a unitary matrix and is real, whence $D^{-1} = \tilde{D}$. It can easily be shown that $\partial/\partial x$, $\partial/\partial y$, $\partial/\partial z$ transform like x, y, z, and thus from (27.2) the strain component e_{pq} transforms like the product pq according to $D_{pp'}D_{qq'}$. Similarly F_{pq} transforms in the same way. Each side of (27.4) therefore transforms according to

$$D_{pp'}D_{qq'}\delta_{r'r''}\delta_{s's''} = D_{pqrsp'q'r's'}D_{rr''}D_{ss''},$$

where c_{pqrs} transforms according to $D_{pqrsp'q'r's'}$. Multiplying each side by $D_{r'''r''}D_{s'''s''} = D^{-1}_{r''r'''}D^{-1}_{s''s'''}$ and summing over r'', s'', we obtain

$$D_{pqrsp'q'r's'} = D_{pp'}D_{qq'}D_{rr'}D_{ss'}.$$

Thus

$$c_{pqrs},\ pqrs,\ e_{pq}e_{rs}, \tag{27.6}$$

all transform in the same way.

We now discuss the transformation properties in more detail to take into account the fact that the tensors e_{pq}, F_{pq}, c_{ij} are all symmetric

$$e_{pq} = e_{qp}, \qquad F_{pq} = F_{qp}, \qquad c_{ij} = c_{ji}. \tag{27.7}$$

The components of an arbitrary tensor t_{pq} transform like the nine products x_1x_2, x_1y_2, x_1z_2, . . . according to

$$D^{(1)} \times D^{(1)} = D^{(2)} + D^{(1)} + D^{(0)}.$$

The linear combinations that transform like standard base vectors are given in (9.4), and we notice that the linear combinations transforming according to $D^{(1)}$ are antisymmetric and form the three components of an antisymmetric tensor (which is the same thing as a pseudo-vector (Milne 1948)). They therefore cannot apply to the e_{pq} which form a symmetric tensor, and which transform like the other base vectors, i.e. according to $D^{(2)} + D^{(0)}$. The separation into symmetric and antisymmetric components is not accidental and can be deduced in a more general way as follows. As in § 9, we consider the products $u_m v_s$ transforming according to $D^{(j)} \times D^{(j)}$ and count up the number of times each particular value of $M = m + \mu$ occurs. However, this time we do not consider $u_m v_\mu$ and $u_\mu v_m$ separately, but count the symmetric sum $u_m v_\mu + u_\mu v_m$ once only. Thus the highest values $M = 2j$ and $M = 2j - 1$ occur once only, $M = 2j - 2$ twice, etc., and we have

$$D^{(j)} \times D^{(j)} \text{ (symmetric product)} = D^{(2j)} + D^{(2j-2)} + \ldots + D^{(0)}, \quad (j = \text{integer}) \tag{27.8}$$

in contrast to the usual formula (9.2). For a discussion of symmetric products from a slightly different point of view, see equation (24.39) et seq. We now see that the symmetric tensor e_{pq} transform according to the symmetric product

$$D^{(1)} \times D^{(1)} \text{ (sym)} = D^{(2)} + D^{(0)},$$

and from (27.6) the c_{pqrs} transform like the symmetric products $e_{pq}e_{rs}$ according to

$$\begin{aligned}
&[D^{(2)} + D^{(0)}] \times [D^{(2)} + D^{(0)}] \text{ (sym)} \\
&\quad = [D^{(2)} \times D^{(2)}] \text{ (sym)} + [D^{(2)} \times D^{(0)} + D^{(0)} \times D^{(2)}] \text{ (sym)} + \\
&\qquad\qquad + [D^{(0)} \times D^{(0)}] \text{ (sym)} \\
&\quad = [D^{(4)} + D^{(2)} + D^{(0)}] + D^{(2)} + D^{(0)}.
\end{aligned} \tag{27.9}$$

This gives twenty-one independent base vectors, just equal to the number of elastic constants.

We can now write down from (27.9) the representation that the c_{ij} give under some point-group of rotations (§ 16). Consider for instance the group 422, whose irreducible representations are given in appendix K. In (27.9),

$D^{(4)}$ becomes $2A_1 + A_2 + B_1 + B_2 + 2E$,

$D^{(2)}$ becomes $A_1 + B_1 + B_2 + E$,

$D^{(0)}$ becomes A_1,

where these decompositions follow from (14.2) and (14.4). Thus the c_{ij} transform according to

$$6A_1 + A_2 + 3B_1 + 3B_2 + 4E. \tag{27.10}$$

The consequences of rotational symmetry

Let us assume that the material under discussion has a group $\mathfrak{G}$ of rotational symmetries. In the case of an isotropic material $\mathfrak{G}$ is the full rotation and reflection group, and for crystals $\mathfrak{G}$ is one of the point-groups of proper and improper rotations. By definition, a rotation R of the group moves the co-ordinate axes to a completely equivalent orientation, so that the properties of the material are the same expressed in terms of the two sets of co-ordinates. Thus *if the tensor T represents some property of the material, the value of each component T_{pqr} . . . of the tensor must be invariant under the group of rotations $\mathfrak{G}$.* This we shall refer to as the fundamental theorem of the present section.

We shall now consider the linear combinations

$$L_m^{(\alpha)} = \alpha_{11}c_{11} + \alpha_{12}c_{12} + \dots + \alpha_{66}c_{66} \tag{27.11}$$

which transform as standard base vectors according to the representations $D^{(\alpha)}$ in (27.9), or its equivalent like (27.10). We have two ways of looking at the transformation properties of the $L_m^{(\alpha)}$. On the one hand they transform by definition according to $D^{(\alpha)}$. On the other hand by the fundamental theorem each c_{ij} in (27.11) is invariant so that $L_m^{(\alpha)}$ is invariant. Now a set of non-zero quantities cannot transform according to two different irreducible representations at the same time (Theorem 3, appendix C). We therefore conclude (i) if $D^{(\alpha)}$ is not the identity representation ($D^{(0)}$ in (27.9) or A_1 in (27.10)), then $L_m^{(\alpha)} = 0$; and (ii) if $D^{(\alpha)}$ is the identity representation, then $L_m^{(\alpha)}$ may be some non-zero constant. In the case of complete rotational invariance the identity representation $D^{(0)}$ occurs twice in (27.9), so that (27.11) is a set of 21 equations which determine the c_{ij} in terms of *two* constants. This is the well-known result that an isotropic medium has only two independent elastic constants, namely the bulk modulus and the rigidity modulus.

A similar argument applies if we consider the case of a crystal with point-group symmetry 422. The identity representation A_1 occurs six times in (27.10). The elastic constants c_{ij} are therefore expressible in terms of six constants. There now remains the problem of actually determining the values of the c_{ij} in terms of six suitable constants, in particular determining which c_{ij} are zero. This could in principle be done by following the present line of argument. The correct linear combinations (27.11) could be constructed by using the projection operator (14.11a) on one c_{ij} picked at random, and the equations (27.11) could then be solved for all the c_{ij}. This however would be extremely tedious, and instead we shall start again at the beginning and use the "direct inspection" method.

The direct inspection method

We saw in (27.6) that the elastic constant c_{pqrs} transforms in the same way as the product $pqrs$, e.g. c_{xxyz} like x^2yz. In fact for convenience we shall use the products to represent the c_{pqrs}. The order of the factors $pqrs$ is important, except that because of (27.7) we may interchange the first two, the last two, or the first pair with the last pair, i.e. we put

$$pqrs = qprs = pqsr = rspq.$$

We shall follow the convention of alphabetic order in each pair pq and rs, and not list these equivalent combinations separately. The elastic constants c_{ij} then transform like the following products:

$$\begin{aligned} &x^4,\ y^4,\ z^4,\ yzyz,\ xzxz,\ xyxy,\\ &x^2y^2,\ x^2z^2,\ x^2yz,\ x^3z,\ x^3y,\\ &y^2z^2,\ y^3z,\ y^2xz,\ y^2xy,\ z^2yz,\\ &z^2xz,\ z^2xy,\ yzxz,\ yzxy,\ xzxy. \end{aligned} \tag{27.12}$$

The group 422 contains a 180° rotation about the z-axis, i.e. the transformation

$$x \to -x, \qquad y \to -y, \qquad z \to z.$$

Under this transformation the product x^3z becomes $-x^3z$. However by the fundamental theorem we require each product to be invariant, so that we have

$$x^3z = 0.$$

Similarly any product in (27.12) is zero if it contains z to an odd power. The two-fold rotations about the x- and y-axes eliminate products with odd powers in x or y. Thus all of the products (27.12) are zero except

$$x^4,\ y^4,\ z^4,\ yzyz,\ xzxz,\ xyxy,\ x^2y^2,\ x^2z^2,\ y^2z^2. \tag{27.13}$$

We next consider the effect of a 45° rotation about the z-axis,

$$x \to y, \qquad y \to -x, \qquad z \to z. \tag{27.14}$$

This gives the following relations:

$$x^4 = y^4, \qquad yzyz = xzxz, \qquad x^2z^2 = y^2z^2.$$

We now have three relations among the nine non-zero components (27.13), so that these are expressible in terms of six constants. This is the number already deduced from (27.10), so that there can be no more independent relations among the components, as can easily be verified by applying all the rotations of the group 422 in turn. The matrix c_{ij} of elastic constants therefore has the form

$$\begin{matrix} c_{11} & c_{12} & c_{13} & 0 & 0 & 0 \\ c_{12} & c_{11} & c_{13} & 0 & 0 & 0 \\ c_{13} & c_{13} & c_{33} & 0 & 0 & 0 \\ 0 & 0 & 0 & c_{44} & 0 & 0 \\ 0 & 0 & 0 & 0 & c_{44} & 0 \\ 0 & 0 & 0 & 0 & 0 & c_{66} \end{matrix} \tag{27.15}$$

in terms of the six independent constants c_{11}, c_{12}, c_{13}, c_{33}, c_{44}, c_{66}.

This method of direct inspection is clearly applicable to working out the components of any tensor, provided we can choose a natural set of x, y, z-axes such that each rotation of the group sends a co-ordinate axis into a co-ordinate axis like in (27.14). Thus trigonal groups with 120° rotations about the main axis require special consideration (Fumi 1952b, c).

Summary

We have shown how the rotational properties of tensors are derived. From this we can determine the number of independent constants, in terms of which the tensor components can be expressed. It is also possible to determine the actual scheme of the components, which ones are zero and which non-zero ones are related.

References

General references: Fumi 1952a and Hearmon 1956, and references given there. The direct inspection method: Fumi 1952b, 1952c. Elastic constants, including third order elastic constants, Hearmon 1956.

Problems

27.1 The piezoelectric constants a_{qrs} are defined by the relation

$$P_q = a_{qrs}e_{rs}, \quad (r, s \text{ summed})$$

where P is the electric polarization in the material produced by the strain e_{rs}, and where q, r, s run over the indices x, y, z. Show that $a_{qrs} = 0$ for any crystal having a centre of symmetry. Determine the scheme of piezoelectric constants for a crystal with symmetry 422, and check the number of independent constants by a second method.

27.2* Determine the form of the matrix c_{ij} of elastic constants for a crystal having symmetry 32 (Fumi 1952c).

27.3* Review the different methods that have been used to calculate the scheme of components in tensors subject to some point-group of rotational symmetry. Discuss their usefulness and their limitations. For references see Fumi (1952a).

27.4* Most tensors like c_{ij} that represent properties of matter are either symmetric or antisymmetric. This is called their intrinsic symmetry. Review and relate the different types of argument that are used to establish the intrinsic symmetry of tensors and other quantities, e.g. conductivity, dielectric constants, the potential at A produced by a charge at B and vice versa, the elastic constants; and show how macroscopic reversibility is involved in the case of static properties. Discuss the statement "Time-reversal symmetry is responsible for the intrinsic symmetry of matter tensors" (Fumi 1952a, p. 740).

27.5 In which of the thirty-two crystal classes can a crystal exhibit (i) the piezoelectric effect and (ii) the pyroelectric effect? The piezoelectric effect is defined in problem 27.1. In the pyroelectric effect a crystal exhibits a permanent spontaneous electric dipole moment (Cady, 1947).

Chapter VII

NUCLEAR PHYSICS

28. The Isotopic Spin Formalism

Isotopic spin

Since nuclei contain two quite separate types of particle, protons and neutrons, the wave functions describing them become one stage more complicated than the wave function for the electrons in an atom. The most obvious way of distinguishing which co-ordinates $\mathbf{r}_k$, σ_{zk} in a wave function ψ refer to which type of particle, would be to use always the co-ordinates $k = 1$ to p for describing the p protons, and the remaining $k = p+1$ to $p+n$ for the n neutrons. However, such a system turns out in some ways to be rather clumsy, and a more convenient description is obtained by using an additional *isotopic spin* co-ordinate $\tau_z = \pm\frac{1}{2}$ to distinguish whether a particle with orbital and spin co-ordinates $\mathbf{r}$, σ_z is a proton ($\tau_z = +\frac{1}{2}$) or a neutron ($\tau_z = -\frac{1}{2}$). In a way we are considering protons and neutrons to be two states ($\tau_z = \pm\frac{1}{2}$) of one "particle", the nucleon. The fact that neutrons and protons are rather alike in their masses and interactions makes such a description particularly useful, but the isotopic spin formalism is in itself quite independent of any assumptions about the degree of similarity between them, e.g. as regards nuclear forces.

Following the same argument as in § 11, we introduce two basic functions of the co-ordinate τ_z,

$$\boxed{\begin{aligned} \xi_+(\tau_z) &= 1, \quad \tau_z = \tfrac{1}{2}; & \xi_-(\tau_z) &= 0, \quad \tau_z = \tfrac{1}{2}; \\ &= 0, \quad \tau_z = -\tfrac{1}{2}; & &= 1, \quad \tau_z = -\tfrac{1}{2}, \end{aligned}} \tag{28.1}$$

so that any function of τ_z can be expressed in terms of these two. In particular if we multiply ξ_+ by τ_z, we obtain the function

$$\begin{aligned} \tau_z\xi_+ &= \tfrac{1}{2}, \quad \tau_z = \tfrac{1}{2}, \\ &= 0, \quad \tau_z = -\tfrac{1}{2}. \end{aligned}$$

Thus we can write

$$\tau_z\xi_+ = \tfrac{1}{2}\xi_+, \qquad \tau_z\xi_- = -\tfrac{1}{2}\xi_-, \tag{28.2}$$

and consider τ_z as an operator as well as a co-ordinate, and this can lead to no confusion just as in the case of the orbital variables x, y, z or the spin variable σ_z (problem 11.2). We also consider unitary transformations of ξ_+, ξ_-

$$T\xi_\beta = T_{\alpha\beta}\xi_\alpha, \text{ (summed)},$$

which define 2×2 unitary matrices. Now all 2×2 unitary matrices with determinant $+1$ form the representation $D^{(1/2)}$ of the rotation group (cf. equation (8.24), and problem 8.13). It is therefore convenient to describe these transformations of ξ_+, ξ_- using the language of rotations in a fictitious three-dimensional *isotopic spin space* with axes Ox, Oy, Oz. In these terms the transformations I_x, I_y, I_z defined by the properties

$$\begin{aligned} I_x\xi_+ &= \tfrac{1}{2}\xi_-, & I_y\xi_+ &= \tfrac{1}{2}i\xi_-, & I_z\xi_+ &= \tfrac{1}{2}\xi_+, \\ I_x\xi_- &= \tfrac{1}{2}\xi_+, & I_y\xi_- &= -\tfrac{1}{2}i\xi_+, & I_z\xi_- &= -\tfrac{1}{2}\xi_-, \end{aligned} \tag{28.3a}$$

must be interpreted from (8.18), (8.23) as the infinitesimal rotation operators. In particular *comparison with* (28.2) *gives* $I_z = \tau_z$, *and we shall in future write* τ_x, τ_y, τ_+, τ_- *for* I_x, I_y, $I_x \pm iI_y$. It follows therefore from (11.18) and (8.29) as in § 11 that τ_x, τ_y, τ_z transform like and have the properties of an angular momentum vector, called the isotopic spin, with the difference that we have dropped the $\hbar$. From (28.3a) these operators are represented by the matrices

$$\tau_x = \tfrac{1}{2}\begin{bmatrix} \cdot & 1 \\ 1 & \cdot \end{bmatrix}, \quad \tau_y = \tfrac{1}{2}\begin{bmatrix} \cdot & -i \\ i & \cdot \end{bmatrix}, \quad \tau_z = \tfrac{1}{2}\begin{bmatrix} 1 & \cdot \\ \cdot & -1 \end{bmatrix}$$

$$\tau_+ = \begin{bmatrix} \cdot & 1 \\ \cdot & \cdot \end{bmatrix}, \quad \tau_- = \begin{bmatrix} \cdot & \cdot \\ 1 & \cdot \end{bmatrix}. \tag{28.3b}$$

The interactions of nucleons can now be expressed in terms of these operators. E.g. the electric charge of a nucleon is

$$q = (\tfrac{1}{2} + \tau_z)e.$$

For clearly we have

$$q\xi_+ = e\xi_+, \qquad q\xi_- = 0$$

so that the expectation value of q is e for a proton state and 0 for a neutron state. The Coulomb repulsion between protons can then be written

$$\mathscr{H}_C = \sum_k \sum_{<l} (\tfrac{1}{2} + \tau_{zk})(\tfrac{1}{2} + \tau_{zl})e^2/r_{kl} \tag{28.4}$$

where the summation includes all nucleons. There is therefore no difficulty in distinguishing protons and neutrons in the isotopic spin formalism when required.

For more than one nucleon, we have

$$T_j = T_x,\ T_y,\ T_z \tag{28.5}$$

as the components of the total isotopic spin vector **T** (cf. equation (8.34)). Wave functions for several nucleons can be written as

$$\sum_{\alpha\beta\gamma\ldots} \psi_{\alpha\beta\gamma\ldots}(\mathbf{r}_1, \sigma_{z1}, \mathbf{r}_2, \sigma_{z2}, \ldots)\xi_{\alpha,1}\xi_{\beta,2}\xi_{\gamma,3}\ldots, \tag{28.6}$$

where $\alpha, \beta, \gamma \ldots$ is some arrangement of + and − signs (cf. equation (11.11b)). Using (28.3), (28.5) we can apply rotations in isotopic spin space to these wave functions, and sort them out into $\psi(T, M_T)$ transforming according to the irreducible representation $D^{(T)}$ of the rotation group † as standard base vectors (8.18). Thus the functions ξ_+, ξ_- transforming according to $D^{(1/2)}$ are characterized by the quantum numbers $\tau = \frac{1}{2}$, $m_\tau = \pm\frac{1}{2}$. For any wave function ψ we have by operating with T_z that

$$\boxed{M_T = \sum_k m_{\tau k} = \tfrac{1}{2}(p - n) = p - \tfrac{1}{2}A,} \tag{28.7}$$

where $A = p + n$ is the total number of nucleons described by ψ. Thus for fixed A, M_T is a measure of the nuclear charge pe. In any ordinary situation we know A and the total nuclear charge for a system, so that we can describe it by a wave function ψ with a definite value of M_T, though ψ may contain components with various values of T. In particular we can form pure isotopic functions Ξ_{TM} as linear combinations of the $\xi_{\alpha 1}\xi_{\beta 2}\xi_{\gamma 3}\ldots$. For instance for two nucleons we have

$$\begin{aligned} \Xi_{1,1} &= \xi_{+1}\xi_{+2}, \qquad & \Xi_{1,0} &= 2^{-1/2}(\xi_{+1}\xi_{-2} + \xi_{-1}\xi_{+2}), \\ \Xi_{1,-1} &= \xi_{-1}\xi_{-2}, \qquad & \Xi_{0,0} &= 2^{-1/2}(\xi_{+1}\xi_{-2} - \xi_{-1}\xi_{+2}). \end{aligned} \tag{28.8}$$

The exclusion principle

We shall now discuss how the exclusion principle is to be expressed in terms of our isotopic spin formalism, and start by considering

† The quantum number T should not be confused with the vector **T** (T_x, T_y, T_z). We shall use capital letters to denote total quantum numbers and operators for several nucleons, and lower case letters for individual nucleons, as in chapter II and (28.5).

two nucleons, a proton and a neutron. Let the proton be in the state $\psi(\mathbf{r}, \sigma_z)$ and the neutron in $\phi(\mathbf{r}, \sigma_z)$. The combined state is

$$\psi(\mathbf{r}_p, \sigma_{zp})\phi(\mathbf{r}_n, \sigma_{zn}). \tag{28.9}$$

and in terms of isotopic spin this can be written

$$\psi(\mathbf{r}_1, \sigma_{z1})\phi(\mathbf{r}_2, \sigma_{z2})\xi_{+1}\xi_{-2}$$

or

$$\psi(\mathbf{r}_2, \sigma_{z2})\phi(\mathbf{r}_1, \sigma_{z1})\xi_{+2}\xi_{-1}, \tag{28.10}$$

depending on which particle we call number 1. Now these two wave functions (28.10), which mathematically are linearly independent functions, describe only one and the same physical state, namely (28.9), and not two separate degenerate states. Moreover the same state can also be described by any linear combination of the functions (28.10), for example by the antisymmetric one

$$\Psi = \psi(1)\phi(2)\xi_{+1}\xi_{-2} - \psi(2)\phi(1)\xi_{+2}\xi_{-1}, \tag{28.11}$$

where for short $\psi(1) = \psi(\mathbf{r}_1, \sigma_{z1})$ etc. Clearly this redundancy among the wave functions (28.10), (28.11) is undesirable because it makes for ambiguity in writing down a wave function for a given state. *We shall therefore make the convention that we shall always use the antisymmetric wave function* (28.11), for there is always one, and only one, of these.

This convention is also convenient in another respect. Using the spin functions (28.8), (28.11) can be written

$$\Psi = \Psi(T{=}1, M_T{=}0) + \Psi(T{=}0, M_T{=}0)$$

where

$$\begin{aligned} \Psi(T{=}1, M_T{=}0) &= 2^{-1/2}[\psi(1)\phi(2) - \psi(2)\phi(1)]\Xi_{1,0} \equiv \Psi_1\Xi_{1,0}, \\ \Psi(T{=}M_T{=}0) &= 2^{-1/2}[\psi(1)\phi(2) + \psi(2)\phi(1)]\,\Xi_{0,0} \equiv \Psi_0\Xi_{0,0}. \end{aligned} \tag{28.12}$$

Here we have separated out the $T = 1$ and $T = 0$ components. Since permutations commute with isotopic spin rotations, permutations cannot mix the two components with $T = 1$, $T = 0$, each of which must therefore be antisymmetric as can easily be verified explicitly. Let us consider the function $\Psi_1\Xi_{1,0}$. Operating on it with T_+ we obtain using (8.18)

$$T_+\Psi_1\Xi_{1,0} = 2^{1/2}\Psi_1\Xi_{1,1}, \tag{28.13}$$

which again must be antisymmetric. Since T_+ increase the value of M_T by one, by (28.7) it turns a given wave function $\Psi(M_T)$ into a new function $\Psi(M_T{+}1)$ describing a state in which one neutron has been turned into a proton, as is obvious from (28.7). Moreover the two protons described by (28.13) are seen to be in the *same* orbital and spin state Ψ_1 as the neutron-proton pair in $\Psi_1 \Xi_{1,0}$. It is now clear what advantage there is in *choosing* $\Psi_1 \Xi_{1,0}$ antisymmetric by convention: it means the n-p state $\Psi_1 \Xi_{1,0}$ is directly and simply related by (28.13) to the corresponding p-p state, where the latter *has* to be antisymmetric because of the exclusion principle.

The above argument is quite general. Let

$$\psi_p(\mathbf{r}_1, \sigma_1; \ldots; \mathbf{r}_p, \sigma_p)$$

and

$$\phi_n(\mathbf{r}_{p+1}, \sigma_{p+1}; \ldots; \mathbf{r}_{p+n}, \sigma_{p+n})$$

be space and spin functions for p protons and n neutrons. By permuting co-ordinate numbers, the combined state can be described by up to $(p+n)!/(p!n!)$ different wave functions in the isotopic spin formalism. However, by convention we always choose the one and only antisymmetric function

$$\boxed{\Psi = [(p+n)!]^{-1/2} \sum_P \delta_P P \psi_p \phi_n \xi_{+1} \cdots \xi_{+,p} \xi_{-,p+1} \cdots \xi_{-,p+n}} \tag{28.14}$$

where

$$[(p+n)!]^{-1/2} \sum_P \delta_P P$$

is the antisymmetrizing operator of (12.2). The function clearly satisfies the requirements of the exclusion principle, namely antisymmetry under the permutation of proton co-ordinates alone, or of neutron co-ordinates alone. In addition the function is antisymmetric under permutations of proton and neutron co-ordinates in the sense of the isotopic spin formalism. This additional antisymmetry, besides conferring a certain uniqueness on the function, has the following advantage. By operating on (28.14) with the operators T_+ and T_- we can generate new functions having different values of M_T, and hence describing different numbers of protons and neutrons, say p' and n'. Simultaneously, the detailed requirements of the exclusion principle have changed, since we now want antisymmetry in the first p' co-ordinates instead of the first p, and in the last n' instead of n co-ordinates. However, since permutations commute with isotopic spin rotations, the new functions are still of the totally antisymmetric type (28.14) and thus satisfy the exclusion principle automatically. *Therefore requiring the wave functions to be antisymmetric in everything prevents us from ever violating the exclusion principle in any calculation.* The fact that functions of the form (28.14) transform under isotopic spin rotations into functions of the same antisymmetry type allows us to sort them out according to the irreducible representations $D^{(T)}$ of the rotation group. Then the set of functions

$$\Psi(T, M_T) \quad -T \leqslant M_T \leqslant T \tag{28.15}$$

describes a fixed number of nucleons in a fixed orbital and spin state, with the number of protons differing from state to state as given by (28.7) and with each state satisfying the exclusion principle.

An alternative formulation

We shall now sketch briefly an alternative method of setting up the wave functions. Apart from generally clarifying the isotopic spin formalism and being of practical use on occasions, this formulation is of interest in two connections. Firstly, there is the historical interest. From what has been said in this section, there is evidently a complete analogy between isotopic spin and the ordinary electron spin. In the latter case the exclusion principle demands antisymmetry in the wave function for electrons with parallel spin, but for antiparallel spin the usual antisymmetry requirement is of a more formal nature and does not for instance prevent the two electrons being in the same orbital state. In § 12 we showed, using determinant wave functions following Slater (1929), how to work out the different terms arising from a configuration and write down their wave functions. However, before Slater's method was discovered, the same results were obtained by the much more cumbersome method to be outlined below (cf. problem 28.4). It is described in detail in Wigner (1931) and Weyl (1931).

Secondly, this formulation does shed more light on one interesting question which was already raised in § 12. In the case of atomic energy levels, the energy of a term is calculated purely from $\mathscr{H}_{\text{orb}}$ (10.2) without any spin-dependent forces. Why then does the energy depend on the spin quantum number S? In the nuclear case the analogous situation is as follows. The isotopic spin only distinguishes neutrons from protons, and in the next section we shall assume that nuclear forces are, to a good approximation, charge independent, i.e. are the same for protons and neutrons. Then T becomes a good quantum number, but how can the energy of a state depend on T?

As in (28.14) let ψ_p and ϕ_n be antisymmetric orbital and spin wave functions for p protons and n neutrons. Then any of the $(p+n)!/(p!n!)$ distinct functions $P\psi_p\,\phi_n$ also describe the same combined state, where P is any permutation of the $p + n$ coordinates. All the functions $P\psi_p \phi_n$ form a vector space which is invariant under the permutation group $\mathfrak{P}_A$ where $A = p + n$ is the total number of particles. Let us suppose this space has been reduced into irreducible representations $\Delta^{(\lambda)}$ of $\mathfrak{P}_A$, with base vectors $\Psi_i^{(\lambda)}$. Consider now all the products $\xi_{\alpha 1}\xi_{\beta 2}\xi_{\gamma 3} \ldots$ of (28.6). These transform into one another under $\mathfrak{P}_A$ and under isotopic spin rotations, and from § 15 we can therefore find linear combinations $\Xi_{Mj}^{(T\mu)}$ arranged in rectangles so that they transform according to $\Delta^{(\mu)}$ under $\mathfrak{P}_A$ in each row and according to $D^{(T)}$ in each column. For instance for three nucleons we have the rectangles $\mathscr{S}$, $D^{(3/2)}$ and Γ, $D^{(1/2)}$, where $\mathscr{S}$ and Γ are the representations of Table 3.

$$\begin{aligned}
\Xi^{3/2,\mathscr{S}}_{3/2,1} &= \xi_{+1}\xi_{+2}\xi_{+3} \\
\Xi^{3/2,\mathscr{S}}_{1/2,1} &= 3^{-1/2}(\xi_{+1}\xi_{+2}\xi_{-3} + \xi_{+1}\xi_{-2}\xi_{+3} + \xi_{-1}\xi_{+2}\xi_{+3}) \\
\Xi^{2/3,\mathscr{S}}_{1/2,1} &= 3^{-1/2}(\xi_{-1}\xi_{-2}\xi_{+3} + \xi_{-1}\xi_{+2}\xi_{-3} + \xi_{+1}\xi_{-2}\xi_{-3}) \\
\Xi^{3/2,\mathscr{S}}_{3/2,1} &= \xi_{-1}\xi_{-2}\xi_{-3}
\end{aligned}$$

$$\begin{aligned}
\Xi^{1/2,\Gamma}_{1/2,1} &= 2^{-1/2}(\xi_{+1}\xi_{+2}\xi_{-3} - \xi_{+1}\xi_{-2}\xi_{+3}) \\
\Xi^{1/2,\Gamma}_{-1/2,1} &= 2^{-1/2}(\xi_{-1}\xi_{+2}\xi_{-3} - \xi_{-1}\xi_{-2}\xi_{+3}) \\
\Xi^{1/2,\Gamma}_{1/2,2} &= 6^{-1/2}(\xi_{+1}\xi_{+2}\xi_{-3} + \xi_{+1}\xi_{-2}\xi_{+3} - 2\xi_{-1}\xi_{+2}\xi_{+3}) \\
\Xi^{1/2,\Gamma}_{-1/2,2} &= 6^{-1/2}(-\xi_{-1}\xi_{-2}\xi_{+3} - \xi_{-1}\xi_{+2}\xi_{-3} + 2\xi_{+1}\xi_{-2}\xi_{-3}) \qquad (28.16)
\end{aligned}$$

Corresponding to any representation $\Delta^{(\lambda)}$ of $\mathfrak{P}_A$, we can define another representation $\Delta^{(\lambda\,\mathrm{conj})}$ by the matrices

$$\Delta_{ij}^{(\lambda\,\mathrm{conj})}(P) = \delta_P[\Delta_{ij}^{(\lambda)}(P)]^*, \qquad (28.17)$$

where as before $\delta_p = \pm 1$ as the permutation P is even or odd. Because of the δ_P in 28.17, this is not just the complex conjugate representation $\Delta^{(\lambda)*}$. For instance, for $\mathfrak{P}_3$, $\mathscr{S}$ and $\mathscr{A}$ (Table 3) are a conjugate pair, whereas $\Gamma^{(\mathrm{conj})}$ is equivalent to Γ. If $f_i^{(\lambda)}$ and $g_i^{(\lambda\,\mathrm{conj})}$ transform according to $\Delta^{(\lambda)}$ and $\Delta^{(\lambda\,\mathrm{conj})}$, then from (28.17) and the unitary property (equation (A.20), problems C.2, C.1) we have that the linear combination

$$\sum_i f_i^{(\lambda)} g_i^{(\lambda\,\mathrm{conj})} \qquad (28.18)$$

is antisymmetric. Now the orbital and spin functions $\Psi_i^{(\lambda)}$ represent definite numbers of protons and neutrons, so that to obtain a corresponding wave function in the isotopic spin formalism we have to multiply it by a Ξ with the appropriate M_T (28.7). However, among our rectangles we have a large number of Ξ's with a given

M_T, and there is no compelling reason to choosing any particular one. This situation corresponds to the redundancy in (28.10), (28.11). However we can obtain a convenient unique function

$$\Psi = \sum_i \Psi_i^{(\lambda)} \Xi_{M_T,i}^{T,\lambda\,\mathrm{conj}}, \tag{28.19}$$

which by (28.18) is antisymmetric. By further detailed arguments it can be shown that each T is associated with only one λ conj and vice versa, and that the representations $\Delta^{(\lambda)}$ which turn up among the $\Psi_i^{(\lambda)}$ are in fact such that there is a λ conj among the rectangles of Ξ's. Since we started with an arbitrary state $\psi_p\,\phi_n$, we conclude that there is always in the isotopic spin formalism a totally antisymmetric wave function. Hence stating the exclusion principle in the new form (antisymmetry with respect to *all* co-ordinates) does not eliminate any physically relevant functions, and gives us a unique one where there would otherwise be an ambiguity.

We now return to the question of energies. Let us assume as is approximately true that the forces between all nucleons are the same, irrespective of whether they are protons or neutrons. The energy of a state then depends among other things on how close the nucleons can get to one another. This depends on the representation $\Delta^{(\lambda)}$ of the orbital and spin function $\Psi_i^{(\lambda)}$. For instance it was shown below (12.6) that if $\Delta^{(\lambda)}$ is the completely antisymmetric representation, then there is zero probability of finding two nucleons (protons or neutrons) at the same place. If on the other hand $\Delta^{(\lambda)}$ is completely symmetric, the probability is a maximum. Thus the energy depends on λ. However we do not usually compute λ for our wave functions, but if we always work in the isotopic spin formalism with the totally antisymmetric functions (28.19), then λ uniquely determines λ conj, which in turn uniquely determines T. Thus we can conveniently use the quantum number T to distinguish levels which in general have different orbital symmetry and energy.

Summary

We have developed the isotopic spin formalism for writing a wave function for a system of protons and neutrons. A proton and a neutron are considered as two states of one particle, the nucleon, distinguished by the isotopic spin quantum number $\tau_z = \pm\frac{1}{2}$. This leads to the use of T, M_T as total quantum numbers for a system. The exclusion principle in this formalism takes the form that a wave function must be antisymmetric under any permutation of the nucleon co-ordinates, the orbital, spin and isotopic spin co-ordinates being simultaneously permuted.

Problems

28.1 Consider the operator $P_{12} = -\frac{1}{2}(1+4\tau_1 \cdot \tau_2)$. Show by giving τ_{z1}, τ_{z2} all possible values that

$$P_{12}\psi(r_1, \sigma_{z1}, \tau_{z1}, r_2, \sigma_{z2}, \tau_{z2}) = -\psi(r_1, \sigma_{z1}, \tau_{z2}, r_2, \sigma_{z2}, \tau_{z1}).$$

Also derive this result by showing that $T^2 = 1 - P_{12}$ and operating on the functions (28.8). Hence using the exclusion principle show that P_{12} operating on a wave function interchanges the orbital and spin co-ordinates of nucleons 1 and 2.

28.2 Show that the operator $P = \tau_{+1}\tau_{1-}\tau_{+2}\tau_{-2}$ has the eigenvalue 1 when operating on a two proton state, and 0 otherwise. Hence use it to write down the Coulomb potential energy between nucleons. Reduce this operator analytically to the form (28.4).

28.3 Show that $\Gamma^{(\text{conj})}$ is equivalent to Γ, where Γ is the representation of Table 3 of the group $\mathfrak{P}_3$. In (28.17), show that the conjugate of $\Delta^{(\lambda\,\text{conj})}$ is just $\Delta^{(\lambda)}$. Formulate generally in terms of group characters the condition under which $\Delta^{(\lambda\,\text{conj})}$ is equivalent to $\Delta^{(\lambda)}$.

28.4 Consider the electron configuration $(2p)^3$. Write down all possible *orbital* functions for the three electrons and sort them into rectangles $\psi(L, M_T; \lambda, i)$ whose rows and columns transform according to $D^{(L)}$ and $\Delta^{(\lambda)}$ under rotations and $\mathfrak{P}_3$. Do the same for the spin functions $U(S, M_S; \mu, j)$. By multiplying all orbital by all spin functions, write down what terms one would naïvely expect from this configuration on the basis of § 11 (neglecting the exclusion principle). Using (28.18) determine which are allowed by the exclusion principle. Also prove that the exclusion principle never excludes part of a term, but always allows or excludes the whole of it.

29. Nuclear Forces

Charge independence

There are several kinds of force acting between the nucleons in a nucleus. The most important ones constitute a specifically nuclear force and are referred to as the "strong interactions". Next in importance comes the Coulomb repulsion between the protons. This is considerably smaller: for instance the strength of the nuclear potential is of the order of 25 MeV (Blatt and Weisskopf 1952, p. 54), whereas in an alpha particle the Coulomb energy between the protons at a separation of 2.10^{-13} cm is 0·7 MeV. Then there are effects due to the mass difference between a neutron and a proton, and interactions involving their magnetic moments. These are all thought to be due indirectly to electromagnetic effects and, together with the Coulomb force, constitute the "electromagnetic interactions". Finally there are the "weak interactions" which couple nucleons with electrons and neutrinos, and which are responsible for beta decay (§ 33). In this section we shall only consider the strong interactions, and often refer to them loosely as *the* nuclear forces.

These strong interactions appear to have at least two symmetry properties. The first one is that they are invariant under the space inversion Π (3.11), so that to a very good approximation one can associate a definite parity with each nuclear level (Lee and Yang 1956). The more important second property of the strong interactions is that they are independent of the charges of the particles, i.e. the forces between two protons (*p-p*), two neutrons (*n-n*), and neutron and a proton (*n-p*) are all the same. The evidence for the equality of *p-p* and *n-n* forces comes from the great similarity between "mirror" nuclei like H^3 and He^3, which differ only by the interchange of protons and neutrons (Bethe and Morrison 1956, pp. 9, 116). The equality of *p-p* and *n-p* forces is indicated by *p-p* and *n-p* scattering experiments (Bethe and Morrison 1956, p. 97), by the cross sections for the reactions

$$p + p \rightarrow \pi^+ + D$$
$$p + n \rightarrow \pi^\circ + D$$

(§ 30; also Henley *et al.* 1953), and by the energy levels of nuclei as discussed below. In addition there appears to be no experimental evidence of any other kind contrary to this view.

The charge independence property of the strong interactions can easily be expressed in terms of the isotopic spin formalism of the previous section. Clearly if the Hamiltonian does not involve τ_x, τ_y, τ_z, then it acts equally between any pair of nucleons. However such a form for the Hamiltonian is unnecessarily restrictive. It is sufficient that the Hamiltonian be invariant under rotations in isotopic spin space. Thus it could for instance involve the term $\boldsymbol{\tau}_1 \cdot \boldsymbol{\tau}_2$. For consider a complete set of functions $\psi(\alpha, T, M_T)$ for a system of A nucleons, where α denotes all the other quantum numbers besides T, M_T necessary to specify the function. If $\mathscr{H}_{si}$ (si = strong interactions) is invariant under isotopic spin rotations, then the functions $\mathscr{H}_{si}\psi(\alpha, T, M_T)$ transform just like the ψ's according to the irreducible representation $D^{(T)}$. Thus from the matrix element theorem of § 13, we have that

$$\int \psi^*(\alpha',T',M'_T)\, \mathscr{H}_{si}\, \psi(\alpha,T,M_T) = \delta_{TT'}\, \delta_{MM'}\, a(\alpha,\alpha',T) \tag{29.1}$$

is non-zero only when $T = T'$, $M_T = M_T'$, and then does not depend on M_T. Now the $\psi(\alpha, T, M_T)$ for different values of M_T refer to states in which the nucleons always have a particular orbital and spin wave function, but which differ in the number of protons (28.7). The result (29.1) then indicates that the Hamiltonian $\mathscr{H}_{\mathrm{si}}$ acts the same way on all these states independent of the number of protons, and therefore satisfies our condition of charge independence.

Form of the interaction

We have just proved that if $\mathscr{H}_{\mathrm{si}}$ is invariant under rotations in isotopic spin space, then it merits the description "charge-independent". However, we know experimentally that $\mathscr{H}_{\mathrm{si}}$ is charge independent and we want to express this in the isotopic spin formalism. I.e. we need to prove the converse of the above, namely, that $\mathscr{H}_{\mathrm{si}}$ being charge-independent, is invariant under isotopic spin rotations. To do this we have to consider the form of all possible types of interaction. Let B be any operator and let

$$\begin{aligned} B\xi_+ &= a\xi_+ + b\xi_- , \\ B\xi_- &= c\xi_+ + d\xi_- . \end{aligned} \tag{29.2}$$

The matrix of coefficients can be expressed as a linear combination of the matrices (28.3b) plus the unit matrix, and correspondingly the operator B can be written

$$B = \tfrac{1}{2}(a + d) + c\tau_+ + b\tau_- + \tfrac{1}{2}(a - d)\tau_z. \tag{29.3}$$

At this stage we limit ourselves to interactions involving only pairs of particles. It then follows from (29.3) that we may restrict ourselves to linear combinations of the following 16 terms.

$$\begin{gathered} \tau_{1+} \pm \tau_{2+}, \qquad \tau_{1-} \pm \tau_{2-}, \qquad \tau_{1+}\tau_{2+}, \qquad \tau_{1-}\tau_{2-}, \\ \tau_{1+}\tau_{2z} \pm \tau_{1z}\tau_{2+}, \qquad \tau_{1-}\tau_{2z} \pm \tau_{1z}\tau_{2-} , \end{gathered} \tag{29.4a}$$

$$\tau_{1z} - \tau_{2z}, \qquad i(\tau_{1+}\tau_{2-} - \tau_{1-}\tau_{2+}), \qquad \tau_{1z} + \tau_{2z}, \qquad \tau_{1z}\tau_{2z}, \tag{29.4b}$$

$$1, \quad \tfrac{1}{2}(\tau_{1+}\tau_{2-} + \tau_{1-}\tau_{2+}) + \tau_{1z}\tau_{2z} = \boldsymbol{\tau}_1 \cdot \boldsymbol{\tau}_2 . \tag{29.4c}$$

Now nuclei have a definite charge, corresponding to definite numbers of protons and neutrons. From (28.7) this means that M_T is a good quantum number for designating the eigenstates of $\mathscr{H}_{\mathrm{si}}$. Thus $\mathscr{H}_{\mathrm{si}}$ is invariant under isotopic spin rotations *about the z-axis*, so that it cannot contain any of the terms (29.4a). This argument of course does not apply to the weak interactions leading to beta decay, for these couple states with different charge and M_T. Next, by operating with each of the interactions (29.4b) on wave functions

$$\Psi = \psi(r_1, \sigma_{z1}, r_2, \sigma_{z2})\Xi_{1,M_T} \quad \text{with} \quad M_T = 1, 0, -1,$$

it can be seen that they give quite different results for different values of M_T and hence cannot be considered charge-independent. For instance

$$\begin{aligned}(\tau_{1z} + \tau_{2z})\Psi &= \Psi \quad &\text{if } M_T = 1 \quad &(p\text{-}p),\\ &= 0 \quad &\text{if } M_T = 0 \quad &(p\text{-}n),\\ &= -\Psi \quad &\text{if } M_T = -1 \quad &(n\text{-}n).\end{aligned}$$

It can also be shown that no linear combination of these four interactions can be used in a charge-independent Hamiltonian (problem 29.2). This only leaves the two terms (29.4c), which are invariant under isotopic spin rotations and which give charge independent forces as already shown in the last paragraph. We have therefore proved that *the conditions of charge independence and invariance under isotopic spin rotations are equivalent.*

We can now easily deduce the actual form of the Hamiltonian if we first make the assumption that $\mathscr{H}_{\text{si}}$ does not depend on the momenta (nor the velocities nor angular momenta) of the particles. There is no compelling reason for this assumption; it is made here for the sake of simplicity in the absence of any definite evidence to the contrary (Bethe and Morrison 1956, p. 99). As in (29.4) we need only consider the 16 products up to the second degree of the spin operators s_{i1}, s_{j2} with the identity operator. Using the irreducible representations of the rotation group these can be classified into

two scalars,	$1, \mathbf{s}_1 \cdot \mathbf{s}_2$;	(29.5a)
three vectors,	$\mathbf{s}_1 + \mathbf{s}_2, \mathbf{s}_1 - \mathbf{s}_2, \mathbf{s}_1 \wedge \mathbf{s}_2$;	(29.5b)
one tensor, five components	$t_m, -2 \leqq m \leqq 2$.	(29.5c)

As regards the orbital variables, $\mathscr{H}_{\text{si}}$ can only depend on $\mathbf{r} = \mathbf{r}_1 - \mathbf{r}_2$. The functions of this can be written as

the scalar,	1;	(29.6a)
the vector,	$\mathbf{r}$;	(29.6b)
the tensor,	$t'_m = r^2 Y_{2,m}(\theta, \phi)$;	(29.6c)
etc.		

all multiplied by an arbitrary function of r. The tensors in (29.5c), (29.6c) are not complete nine-component second rank tensors, but are the five parts transforming irreducibly according to $D^{(2)}$. These form a symmetric tensor with zero sum of diagonal terms (see § 9). Now the Hamiltonian has to be invariant under rotations (there is no intrinsic preferred axis among the nucleons), and the only way of obtaining such interactions is by taking the scalar products of corresponding terms in (29.5) and (29.6). We obtain the following six scalars and pseudoscalars:

$$1, \quad \mathbf{s}_1\cdot\mathbf{s}_2, \quad \Sigma t_m{}^* t_m{}' = 3(\mathbf{s}_1\cdot\mathbf{r})(\mathbf{s}_2\cdot\mathbf{r}) - r^2\mathbf{s}_1\cdot\mathbf{s}_2, \qquad (29.7a)$$

$$(\mathbf{s}_1+\mathbf{s}_2)\cdot\mathbf{r}, \quad (\mathbf{s}_1-\mathbf{s}_2)\cdot\mathbf{r}, \quad \mathbf{s}_1\wedge\mathbf{s}_2\cdot\mathbf{r}. \qquad (29.7b)$$

The three terms of (29.7a) can each be multiplied by the two terms of (29.4c) giving the six interactions of Table 32. These are all

TABLE 32

Nuclear Interactions

Wigner	$V_W(r)$
Bartlett (spin exchange)	$V_B(r)\mathbf{s}_1 . \mathbf{s}_2$
Tensor	$V_t(r)(\mathbf{s}_1 . \mathbf{r})(\mathbf{s}_2 . \mathbf{r})$
Heisenberg (space-spin exchange)	$V_H(r)\boldsymbol{\tau}_1 . \boldsymbol{\tau}_2$
Majorana (space exchange)	$V_M(r)(\mathbf{s}_1 . \mathbf{s}_2)(\boldsymbol{\tau}_1 . \boldsymbol{\tau}_2)$
Tensor exchange	$V_{te}(r)(\mathbf{s}_1 . \mathbf{r})(\mathbf{s}_2 . \mathbf{r})(\boldsymbol{\tau}_1 . \boldsymbol{\tau}_2)$

invariant under the space inversion Π, symmetric in the co-ordinates 1 and 2, and charge independent. On the other hand the terms (29.7b) must be rejected because they are not invariant under Π ($\mathbf{s}_1$, $\mathbf{s}_2$ are invariant under Π, appendix F).

Isotopic spin multiplets

Since we have established that $\mathscr{H}_{\mathrm{si}}$ is invariant under isotopic spin rotations, its eigenfunctions can be sorted out to transform according to the irreducible representations $D^{(T)}$. The set of $2T+1$ states have the same energy and from what is called an isotopic spin multiplet or *T-multiplet*. Because of the different values of M_T, these states clearly belong to neighbouring nuclei with the same mass number. Fig. 44 contains two examples of a T-multiplet, one with $T = 1$, $J = 2$ and one with $T = 1$, $J = 0$. The three states of each multiplet are seen to have not exactly the same energy because of the minor charge-dependent electromagnetic interactions mentioned at the beginning of this section. Some of them can easily be corrected for. For instance the mass difference between a neutron and a hydrogen atom amounts to 0·79 MeV.

If we assume the protons in the nucleus to be uniformly spread out, the Coulomb energy is (Blatt and Weisskopf 1952, p. 216)

$$E_C = \tfrac{3}{5}Z(Z-1)e^2/R, \tag{29.8}$$

where R is the nuclear radius. Using (29.8) and $R = 1{\cdot}44\, A^{1/3}$ 10^{-13} cm, we obtain for the corrected relative energy levels of the $J = 0$ T-multiplet 1·97 ($M_T = -1$), 1·77 ($M_T = 0$), and 1·96 ($M_T = 1$). The remaining discrepancy is presumably due to the

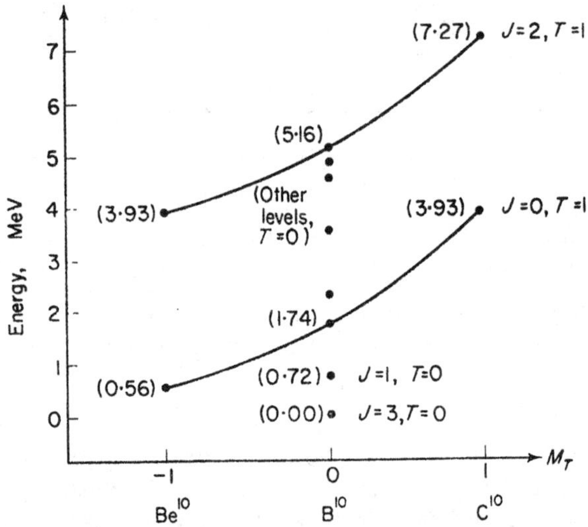

FIG. 44. Relative energies of the ground states and first excited states of the nuclei Be^{10}, B^{10}, C^{10}. The mass energies of the atomic electrons and the nucleons is included. The parity of all states is even.

crudeness of (29.8) and other small charge-dependent effects. Nevertheless the degree of agreement shows that the strong interactions must be charge-independent.

The shell model and nuclear levels

In the theory of atomic energy levels (Chapter II), it was seen to be a useful starting approximation to assume that each electron moves in a fixed net potential, which represents in some average way the effect of all the other electrons and the nucleus. Similarly in nuclei it is fruitful to assume that each nucleon moves approximately independently in a certain potential well, and this leads to 1*s*, 2*s*, 2*p*, 3*s*, etc., orbits as for electrons. As in the case of *jj* coupling in atoms (§ 11), the presence of fairly strong tensor and/or spin-orbit forces couples the orbital and spin angular momenta of each nucleon to give a total angular momentum j (Bethe and Morrison 1956, pp. 99, 165). However the order of the energy levels of the various states is different from that in the electronic case, and this also results in a different grouping into closed shells. The order in which the various individual nucleon levels are filled is shown in Table 33, where levels with nearly the same energy have been

TABLE 33

Levels in the Nuclear Shell Model

Shell	Levels	Number of states	Cumulative total
1	$1s_{1/2}$	2	2
2	$2p_{3/2}$, $2p_{1/2}$	6	8
3	$3d_{5/2}$, $2s_{1/2}$, $3d_{3/2}$	12	20
4	$4f_{7/2}$	8	28
5	$3p_{3/2}$, $4f_{5/2}$, $3p_{1/2}$, $5g_{9/2}$	22	50

The last column gives the total number of states up to and including a given shell. After the fifth shell differences appear between protons and neutrons, and between various authors (Bethe and Morrison 1956, p. 164; Siegbahn 1955, p. 421).

grouped together into shells. The total number of states in each shell is also shown, the level $2p_{3/2}$ being for instance $2j+1 = 4$-fold degenerate. Each of these states can be filled twice in a nucleus, once by a proton and once by a neutron.

As an example, let us calculate J and T for the low lying allowed levels of the $A = 10$ nuclei Be^{10}, B^{10}, C^{10}. From Table 33 the configuration is $(1s_{1/2})^4(2p_{3/2})^6$. By writing down all possible determinant wave functions for this configuration in the manner of § 12, we can tabulate all the allowed values of M_J, M_T and hence group them into J, T multiplets. As in § 12, the closed shell $(1s_{1/2})^4$ always gives zero contribution to M_J, M_T and can therefore be omitted. Also $(2p_{3/2})^6$ gives rise to the same levels as $(2p_{3/2})^2$, because together they form another closed subshell. Abbreviating the wave function $2^{-1/2}\sum\delta_P P\psi(j=3/2, m_{j1}, m_{\tau 1}; \mathbf{r}_1, \sigma_{z1}, \tau_{z1})\psi(j=3/2, m_{j2}, m_{\tau 2}; \mathbf{r}_2, \sigma_{z2}, \tau_{z2})$ to $(m_{j1}, m_{\tau 1} = \pm)(m_{j2}, m_{\tau 2} = \pm)$, we obtain analogously to Table 5 the values of M_J, M_T shown in Table 34. These states can be grouped into levels with

$$J = 2,\ T = 1; \quad J = 0,\ T = 1; \quad J = 3,\ T = 0; \quad J = 1,\ T = 0;$$

their energies being as shown in Fig. 44. Thus the $J = 3$ state can

TABLE 34

States in the Configuration $(2p_{3/2})^2$

Wave function	M_J	M_T	level†
(3/2 +) (1/2 +)	2	1	a
(3/2 +) (−1/2+)	1	1	a
(3/2 +) (−3/2 +)	0	1	a
(1/2 +) (−1/2 +)	0	1	b
(3/2 +) (3/2 −)	3	0	c
(3/2 +) (1/2 −)	2	0	c
(3/2 −) (1/2 +)	2	0	a
(3/2 +) (−1/2 −)	1	0	a
(3/2 −) (−1/2 +)	1	0	c
(3/2 +) (−3/2 −)	0	0	a
(3/2 −) (−3/2 +)	0	0	b
(1/2 +) (1/2 −)	1	0	d
(1/2 +) (−1/2 −)	0	0	c
(1/2 −) (−1/2 +)	0	0	d

and corresponding wave functions with negative M_J, M_T.

† (a) $J = 2$, $T = 1$; (b) $J = 0$, $T = 1$;
(c) $J = 3$, $T = 0$; (d) $J = 1$, $T = 0$.

only apply to B^{10} and not to Be^{10}, C^{10}, because it is an isotopic singlet ($M_T = T = 0$). On the other hand the $J = 0$, $T = 1$ level applies to all three nuclei Be^{10}, B^{10}, C^{10} ($M_T = -1, 0, 1$), though the energy is not exactly the same for all three due to the Coulomb energy (29.8) and the mass difference as already discussed. As in (11.24) the parity is given by

$$\boxed{w = (-1)^{\Sigma l_i},}$$

and is even for all the states mentioned above.

The deuteron

For a system of two free nucleons, the only force acting is the interaction between them and this depends spatially only on the difference $\mathbf{r} = \mathbf{r}_1 - \mathbf{r}_2$ of their co-ordinates. Thus the wave equation becomes "separable" in terms of the centre of mass co-ordinate $\mathbf{R} = \frac{1}{2}(\mathbf{r}_1 + \mathbf{r}_2)$ and $\mathbf{r}$, and the wave function can be written (Schiff 1949, p. 81)

$$\psi_d(\mathbf{R})\phi(\mathbf{r})u_{1\alpha}u_{2\beta}\xi_{1\gamma}\xi_{2\delta}, \tag{29.9}$$

or rather as a linear combination of such wave functions. Here $\psi_d(\mathbf{R})$ gives the motion of the centre of gravity. Instead of the form (29.9), it is convenient to introduce the isotopic spin functions

(28.12) Ξ_{T,M_T} with $T = 0, 1$ and the corresponding spin functions U_{S,M_S} with $S = 0, 1$, and thus to use the set of wave functions

$$\psi = \psi_d(\mathbf{R})d(k) \tag{29.10}$$

where

$$d(k) = \phi_k(r)U_{S,M_S}\Xi_{T,M_T}. \tag{29.11}$$

We have used the symbol d (for deuteron) in (29.11) to denote that $d(k)$ is an internal state of the deuteron. We now assume that in the ground state of the deuteron the orbital function $\phi_k(\mathbf{r})$ is a spherically symmetric $1s$ type of function $\phi_{1s}(r)$ with $L = 0$. Then to make (29.10) antisymmetric, we must either use $S = 1$, $T = 0$ or $S = 0$, $T = 1$. This gives two states for the deuteron, $J = 1$ ($T = 0$) and $J = 0$ ($T = 1$). The former one has the lower energy and is the usual ground state. The $J = 0$, $T = 1$ state is actually completely unstable, as we would expect from the fact that two protons or two neutrons do not form a bound system.

So far we have designated the ground state of the deuteron by the quantum number $J = 1$, but if the wave function were precisely of the form (29.11) then L and S would separately be good quantum numbers. This would be so if the strong interactions did not include any tensor interactions (Table 32) nor spin orbit coupling, since the other interactions are invariant under separate orbital and spin rotations. In particular we would have $L = 0$, and this would imply a zero quadrupole moment $\langle Q\rangle$ for the deuteron with respect to its centre of mass

$$\langle Q\rangle = \int \psi^* Q\psi \, dv$$

where

$$\begin{aligned} Q &= \tfrac{1}{4}\,[3(z_1 - Z)^2 - (\mathbf{r}_1 - \mathbf{R})^2](\tfrac{1}{2} + \tau_{z1}) + \\ &\quad + \tfrac{1}{4}[3(z_2 - Z) - (\mathbf{r}_2 - \mathbf{R})^2](\tfrac{1}{2} + \tau_{z2}) \\ &= \tfrac{1}{4}(3z^2 - r^2)(1 + T_z) \end{aligned} \tag{29.12}$$

(Bethe and Morrison 1956, p. 104). However it is known experimentally that the deuteron does have a quadrupole moment, as we would expect in general for a $J = 1$ nucleus from the considerations of § 21. This can be explained by assuming that $\mathscr{H}_{\text{si}}$ contains some tensor (or tensor exchange) interaction $\mathscr{H}_{\text{tens}}$ (Table 32) which we shall treat as a perturbation. The unperturbed wave function $d(1s)$ then becomes

$$d(1s) + \sum_k \frac{\int d^*(k)\mathscr{H}_{\text{tens}}\, d(1s)dv}{E_{1s} - E_k}\, d_k \tag{29.13}$$

Since $\mathscr{H}_{\text{tens}}$ is invariant under simultaneous orbital and spin rotations and under isotopic spin rotations, from § 13 all $d(k)$ in (29.13) must have $J = 1$, $T = 0$ like $d(1s)$. Since $\mathscr{H}_{\text{tens}}$ transforms according to $D^{(2)}$ under orbital rotations or under spin rotations, the coefficient of $d(k)$ in (29.13) is zero unless $d(k)$ has in addition $L = 2$, $S = 1$, 2 or 3. In the present case of only two nucleons only $S = 1$ applies. Thus the ground state d of the deuteron can be written as the sum of terms ($M_S = \pm 1$, 0) of the form

$$\phi_s(r)U_{1,M_S}\Xi_{0,0} \quad \text{and} \quad \phi_d(r)Y_{2,M}(\theta,\phi)U_{1,M_S}\Xi_{0,0}, \tag{29.14}$$

where $\phi_d Y_{2,M}$ is a d-symmetry ($l = 2$) orbital. In fact it can be shown that this form applies exactly, not only in first order perturbation, if ϕ_s is interpreted as an arbitrary $l = 0$ orbital differing somewhat from the original ϕ_{1s} (problem 29.8). Now from (29.12) the quadrupole moment of the ground state is a linear combination of terms

$$\int \phi_s{}^*Q\phi_d Y_{2,M}\, dv \quad \text{and} \quad \int \phi_d{}^*Q\phi_d Y_{2,M}\, dv,$$

and does not vanish identically.

These conclusions about the wave function of a deuteron can be summarized conveniently as follows. The final wave function can be written in the form

$$\boxed{\psi = \psi_d(\mathbf{R})d_m, \ (m \equiv M_J = 0, \pm 1)} \tag{29.15}$$

where $\psi_d(\mathbf{R})$ describes the motion of the centre of gravity $\mathbf{R}$, and d_m is a function of the form (29.14) describing the internal motion. We also have that d_m transforms according to $D^{(1)}$ under orbital and spin rotations ($J = 1$), according to $D^{(0)}$ under isotopic spin rotations ($T = 0$), and has even parity, which completely describes the symmetry of (29.15). Now given this symmetry, we do not need to consider for most purposes the detailed internal structure of the deuteron. We can regard it as if it were a new type of fundamental particle described by the "spin" functions d_m, where the three states $m = 0, \pm 1$ refer to some internal degree of freedom. This is why we have used small letters d and m in (29.15), in accordance with the usual convention of using small letters for single particles and capitals for composite systems. From this point of view the deuteron is often described as having a "spin" of 1, although the quantum number $J = 1$ is really determined by a combination of internal orbital motion and true spin angular momentum.†

† If there is likely to be confusion with the electronic angular momentum, then I is used instead of J for the total internal angular momentum, i.e. "spin", of a nucleus or other compound particle. This is done in § 21.

The same argument can be applied of course to any nucleus, and we have already made extensive use of this in § 21 where we treated the nucleus as a black box characterized by a certain net "spin", magnetic moment, quadrupole moment, etc. There is thus a complete analogy with electrons described by u_+, or nucleons by $\xi_\pm$, the only difference being that in the case of the deuteron we happen to know explicitly how the internal degree of freedom describes the relative motion of the proton and neutron. This internal motion is described by saying that d_m contains a mixture of orbital states $L = 0$ and $L = 2$, and is an eigenfunction† of S^2 corresponding to $S = 1$.

Other particles

As already mentioned the same type of description (29.15) can also be applied to other particles. For instance the internal wave function α of an alpha particle has $J = T = 0$ as expected from the shell model. Thus we can write for an alpha particle wave function

$$\boxed{\psi = \psi_\alpha(\mathbf{R})\,\alpha} \tag{29.16}$$

where $\psi_\alpha(\mathbf{R})$ corresponds to the motion of the centre of gravity $\mathbf{R}$, and where the function α is invariant under orbital and spin rotations and under isotopic spin rotations, and has even parity.

Similarly for π-mesons we write

$$\boxed{\psi = \psi_m(R)\,\pi^m, \qquad m = 0,\,+,\,-,} \tag{29.17}$$

where π^+, π^0, π^- denote the internal wave functions of the corresponding π-mesons. Experimentally it is known that these particles have zero intrinsic spin angular momentum, so that the functions π^m are all invariant under rotations. Also $\Pi\pi^m = -\pi^m$ where Π is the space inversion, i.e. the π^m all have odd parity.

It is also known experimentally that π-mesons are intimately connected with nuclear forces (Bethe and Morrison 1956, pp. 29, 147). In particular it is found that the forces between π-mesons and nucleons have about the same range and strength as inter-nucleon forces, and that they too are very nearly charge independent. It is natural therefore to look for a unified interaction theory which embraces π-mesons as well as nucleons. We achieve this by regarding the three types of π-meson as three different internal states of

† However the d_m do not transform according to $D^{(1)}$ under spin rotations and M_S is not a good quantum number.

one particle described by the isotopic spin co-ordinate τ_z, the functions π^+, π^0, π^- transforming as standard base vector (8.18) with $M_T = 1, 0, -1$ and $T = 1$ under isotopic spin rotations. The strong interactions can again be expressed in terms of the τ operators and thus again become charge-independent if they are invariant under isotopic spin rotations.

Similarly the isotopic spin formalism can be extended to all particles which interact via the strong interactions and Table 35

TABLE 35

Quantum Numbers of Particles Having Strong Interactions

Particle	Mass	Spin	Parity	T	M_T	N	Strangeness
Nucleons							
Proton	1840	$\frac{1}{2}$	even	$\frac{1}{2}$	$\frac{1}{2}$	1	0
Neutron	1840	$\frac{1}{2}$	even	$\frac{1}{2}$	$-\frac{1}{2}$	1	0
Antiproton	1840	$\frac{1}{2}$	odd	$\frac{1}{2}$	$-\frac{1}{2}$	-1	0
Antineutron	1840	$\frac{1}{2}$	odd	$\frac{1}{2}$	$\frac{1}{2}$	-1	0
Mesons							
π^+, π^0, π^-	270	0	odd	1	$1, 0, -1$	0	0
$K^+, K^0,$	970	0	?	$\frac{1}{2}$	$\pm\frac{1}{2}$	0	1
$\bar{K}^0, K^-$	970	0	?	$\frac{1}{2}$	$\pm\frac{1}{2}$	0	-1
Hyperons							
Λ^0	2180	$\frac{1}{2}$	?	0	0	1	-1
$\Sigma^+, \Sigma^0, \Sigma^-$	2330	$\frac{1}{2}$	?	1	$1, 0, -1$	1	-1
Ξ^0, Ξ^-	2580	$\frac{1}{2}$	?	$\frac{1}{2}$	$\pm\frac{1}{2}$	1	-2

The mass is in units of the electron mass.

shows the values of T and M_T for the wave functions describing some of these particles. In addition it has been found that there are other useful quantum numbers describing the particles, namely the nucleon number N and the strangeness. These are presumably connected with an invariance property of the Hamiltonian under some transformation involving a new co-ordinate not yet understood. In each case the charge of the particle is

$$e[M_T + \tfrac{1}{2}N + \tfrac{1}{2}(\text{strangeness})],$$

this being the generalization of (28.7). The concept of isotopic spin does not appear to apply to electrons, positrons, neutrinos and μ mesons (collectively known as leptons) which do not interact via the strong interactions.

References

Bethe and Morrison 1956, Blatt and Weisskopf 1952, Sachs 1953, and references given there.

Summary

The strong interactions among nucleons and other fundamental particles have been discussed, particularly the consequences of their being charge-independent. The form of the Hamiltonian, the existence of isotopic spin multiplets, and the states of the deuteron and other particles have been described.

Problems

29.1 Suppose that nuclear forces are only charge symmetric but not necessarily charge independent, i.e. the *n-n* and *p-p* forces are equal to one another but not to the *n-p* force. Which of the interactions (29.4) would then be allowable in the Hamiltonian?

29.2 Operate with each of the interactions (29.4b) on the Ξ's (28.8), and show that no charge-independent linear combination can be formed from them. Note that it is necessary to use the conditions (i) that the Hamiltonian must be symmetric as regard interchange of particle co-ordinates, and (ii) that the linear combinations must contain all real coefficients so that the Hamiltonian is Hermitian (Schiff 1949, p. 129). Also list critically all the assumptions that have been used in arriving at Table 32.

29.3 Suppose that the nuclear Hamiltonian is invariant under space inversion, but is not necessarily charge-independent. Which of the following interactions would be allowed? $(\mathbf{s}_1 + \mathbf{s}_2) \wedge \mathbf{p} \cdot \mathbf{r}$, $\mathbf{s}_1 \cdot \boldsymbol{\tau}_1 + \mathbf{s}_2 \cdot \boldsymbol{\tau}_2$, $(\mathbf{s}_1 \wedge \mathbf{s}_2 \cdot \mathbf{r})(\boldsymbol{\tau}_1 \cdot \boldsymbol{\tau}_2)$, $[(\mathbf{s}_1 - \mathbf{s}_2) \wedge \mathbf{p} \cdot \mathbf{r}_0(\tau_{1z} - \tau_{2z})]$ $\mathbf{s}_1 \cdot \mathbf{s}_2 i(\tau_{1+}\tau_{2-} - \tau_{1+}\tau_{2+})$, where $\mathbf{r} = \mathbf{r}_1 - \mathbf{r}_2$, $\mathbf{p} = \mathbf{p}_1 - \mathbf{p}_2$.

29.4 Are the electromagnetic interactions invariant under isotopic spin rotations about the z-axis?

29.5 Derive the J, T and parity quantum numbers for the energy levels of Li^8, Be^8, B^8 arising from the lowest configuration.

29.6 Assuming the Hamiltonian is charge independent and invariant under space inversion, show that for *two* nucleons the energy eigenfunctions are also eigenfunctions of S^2 where S is the total spin angular momentum.

29.7 Assuming that the Hamiltonian contains only the Wigner and Heisenberg interactions (Table 32), what would the sign of $V_H(r)$ have to be to account for the deuteron energy levels.

29.8 Show that (29.14) gives the general form of the deuteron ground state wave function in the presence of tensor interactions.

Using Wigner coefficients, write down the correct linear combinations of the terms (29.14) corresponding to eigenstates with $m \equiv M_J = 1, 0, -1$.

29.9 Suppose that the nuclear Hamiltonian contains the spin-orbit term $V_{LS}(r)(\mathbf{r}_1 - \mathbf{r}_2) \wedge (\mathbf{p}_1 - \mathbf{p}_2) \cdot (\mathbf{s}_1 + \mathbf{s}_2)$ and the interactions of Table 32 *excluding* the tensor ones. Would this account for a quadrupole moment of the deuteron in the ground state?

29.10* Derive all the nuclear interactions that are allowed by symmetry (excluding the assumption of charge independence) which are independent of, or linear in, the momenta of the particles (Eisenbud and Wigner 1941).

29.11 Several nucleons are interacting via charge-independent forces. It is asserted that therefore the charge density in the resultant nucleus must be equal to eZ/A times the particle density. Either prove this assertion, or discuss why it is incorrect.

30. Reactions

In the previous two sections we showed how the symmetry properties of the eigenstates of nuclei are discussed, especially as regards isotopic spin. We shall now apply these concepts to reactions when the nuclei take part in time-dependent processes. In particular we shall illustrate how symmetry properties can be used to determine selection rules for reactions, can be used to discuss the angular distribution of the reaction products, and to calculate the ratios of cross sections if we have several related reactions.

Parity of the π^- meson

As a first example to illustrate the kind of problem that has to be discussed, we consider the reaction

$$\pi^- + d \rightarrow n + n, \tag{30.1}$$

and use it to determine the parity of the π^- meson. What happens in practice is that the π^- meson is stopped in some material containing deuterium. Being negatively charged, it can go into stable electron-like orbits about the deuteron, quickly falling down to the lowest $1s$ level. The radius of this orbit is 273 $(= m_\pi/m_e)$ times smaller than the corresponding electron orbit because of the mass difference. Thus the meson spends quite an appreciable fraction of the time inside the nucleus, during which the reaction (30.1) can occur. The initial wave function before the reaction can be written in the notation of (29.15), (29.17)

$$\Psi_i = \psi_{1s}(|\mathbf{r}_d - \mathbf{r}_\pi|)\pi^- d_m. \tag{30.2}$$

Throughout this section we shall use suffices i and f to stand for "initial" and "final". Since the π-meson has zero spin (Table 35) and the deuteron has spin 1, Ψ_i transforms according to $D^{(1)}$ under rotations. The deuteron has even intrinsic parity (see previous section), so that the initial parity is

$$w_i = w_\pi , \tag{30.3}$$

where w_π is the intrinsic parity of the π-meson which we want to determine. Let us now consider the two neutrons in (30.1) and write down a set of basic wave functions

$$\Psi = \psi_l(\mathbf{r}_1 - \mathbf{r}_2)U_{S,M_s}\xi_{-1}\xi_{-2} \tag{30.4}$$

to describe them, where ψ_l transforms according to $D^{(l)}$ under rotations, and where $S = 0$ or 1. To make Ψ antisymmetric, we must have either $l =$ even, $S = 0$ (i.e. ψ symmetric, U_{SM_s} anti-symmetric), or $l =$ odd and $S = 1$. Using the spectroscopic notation of § 11, we have therefore the states ${}^1S_0$, ${}^2P_{0,1 \text{ or } 2}$, ${}^1D_2$, ${}^3F_{2,3 \text{ or } 4}$, etc., and an arbitrary $J = 2$ state for instance would be a linear combination of the ${}^3P_2$, ${}^1D_2$, ${}^3F_2$ states. We now assert that the final J and parity after the reaction must be the same as the initial ones. Thus the initial value $J = 1$ limits us to the final state ${}^3P_1$. This has odd parity since $l = 1$, so that from (30.3) the π^- meson has odd parity. Clearly the crux of this argument is the assertion that the J and parity do not change during the reaction. To prove this we first have to establish the following theorem.

Fundamental theorem

Consider an atom in a box containing also some electromagnetic radiation, and consider two energy levels of the atom E_1 and E_2 with wave functions ψ_1 and ψ_2. We recall (Schiff 1955, p. 28) that a stationary state of an isolated system depends on the time only through a factor $\exp(-iEt/\hbar)$ and is an eigenfunction of the Hamiltonian. Now ψ_1 is an eigenfunction of the atomic Hamiltonian, but it is not true to say that having the atom in the state ψ_1 gives a stationary state for the whole system because the atom may absorb (emit) a quantum of radiation and make a transition to ψ_2. To obtain a stationary state we would have to take a linear combination of states with the atom in ψ_1 and N photons, and with the atom in ψ_2 and $N - 1$ photons in the box.

Similarly we could write down a Hamiltonian $\mathscr{H}_3$ for the initial three particles in (30.1) (π^-, proton, neutron), using two-particle interactions of the form of Table 32 and the Coulomb force, and we would presumably find (30.2) to be a satisfactory eigenfunction.

However such a Hamiltonian $\mathscr{H}_3$ is incomplete, and the initial state (30.2) is not really a stationary state of the whole pion-nucleon field, as shown by the fact that the reaction (30.1) is observed to occur. $\mathscr{H}_3$ is particularly inadequate in that it refers to a fixed number of particles, namely three, and cannot describe processes involving annihilation or creation of particles. Thus we require a field theoretical Hamiltonian that can. How this is achieved can be seen from the Coulomb repulsion. The coulomb interaction between two particles is usually written e_1e_2/r_{12}, but by using the electric field $\mathscr{E}$ we can express this electrostatic energy for several particles in the form $(1/8\pi)\int \mathscr{E}^2\, dv$ while leaving the number of particles completely unspecified. In the following we shall assume all Hamiltonians to be of this very general field-theoretical type. Furthermore in this picture a reaction is described, not by a sudden change of the wave function from one type to another, but as the continuous development of one wave function under such a general Hamiltonian.

We now prove the following theorem. *If $\mathscr{H}$ is invariant under a group $\mathfrak{P}$ of transformations T (not involving time), and if a set of wave function $\psi_\mu(t)$ transforms according to the representation $D_{\lambda\mu}(T)$ at some time $t = t_0$, then the $\psi_\mu(t)$ transform according to the same representation $D_{\lambda\mu}(T)$ at all times t.* Instead of giving a field theoretical discussion as we should, we shall just use the time-dependent Schrödinger equation

$$\left[\mathscr{H}(q) - i\hbar\frac{\partial}{\partial t}\right]\psi_\mu(q, t) = 0, \tag{30.5}$$

where q represents all the co-ordinates of the system besides t. Now the operator in (30.5) is invariant under $\mathfrak{G}$, and the eigenfunctions $\psi_\mu(q, t)$ with eigenvalue zero can be sorted out to transform according to irreducible representations of $\mathfrak{G}$ or any other convenient way, e.g. according to $D_{\lambda\mu}(T)$. Since the transformations do not affect t, it is just a parameter having any value from $-\infty$ to ∞. The transformation properties thus remain constant in time, which proves the theorem.† Two variations of this proof are indicated in problems 30.1 and 30.2. As an example, in the reaction (30.1) the Hamiltonian is invariant under rotations and space inversion.

† An even more direct way of seeing the theorem is to note that the solution of (30.5) can be written

$$\psi_\mu(q, t) = \exp[i(t - t_0)\mathscr{H}]\psi_\mu(q, t_0).$$

The exponential is invariant under G, and so $\psi_\mu(q, t)$ transforms in exactly the same way as $\psi_\mu(q, t_0)$.

Also the initial state is one of definite J and parity, so that these quantum numbers cannot change during the reaction, which is what we asserted.

This theorem is often expressed in the following somewhat weaker form: *if an operator commutes with the Hamiltonian, it is a constant of the motion.* This follows from equation (17.1)

$$\frac{\mathrm{d}A}{\mathrm{d}t} = (1/i\hbar)(A\mathscr{H} - \mathscr{H}A).$$

It can be related to the earlier form of the fundamental theorem by noting that all symmetry transformations commute with the Hamiltonian and are thus constants of the motion (see equation (17.5)). The weaker form of the theorem is quite adequate for many applications, e.g. the reaction (30.1) already discussed. Here the angular momentum operators J^2 and J_z, and the parity operator Π commute with $\mathscr{H}$ and are therefore constants of the motion. Thus the initial and final wave functions are designated by the same values of the quantum numbers J, M_J and w.

Consider now the reaction

$$N_{\mathrm{i}} + \alpha \rightarrow d + N_{\mathrm{f}},$$

called an $(\alpha,\ d)$ reaction, where N_{i}, N_{f} are the initial and final nuclei. The initial wave function is

$$\Psi_{\mathrm{i}} = \psi_{N_{\mathrm{i}}}\psi_{\alpha}, \tag{30.6}$$

where we really should include an antisymmetrizing operator $[(n+p)!]^{-1/2}\sum \delta_P P$ as in (28.14) but this does not affect the other transformation properties. Since α-particle and deuteron functions are invariant under isotopic spin rotations (see equations (29.15) and (29.16)), we have $T_{\mathrm{i}} = T_{\mathrm{f}}$ where these are the T values of the nuclei, i.e. we have the selection rule

$$\boxed{\Delta T = 0 \text{ for } (\alpha,\ \alpha),\ (\alpha,\ d),\ (d,\ \alpha),\ (d,\ d).} \tag{30.7a}$$

Similarly $(\alpha,\ \alpha)$, $(d,\ \alpha)$, $(d,\ d)$ reactions have the same selection rule as shown. For example consider the reaction

$$\mathrm{B}^{10} + d \rightarrow d + \mathrm{B}^{10*}$$

leaving the final nucleus in an excited state. We suppose that the B^{10} nucleus is initially in its ground state, which from Fig. 44 has $T = 0$. The reaction cannot therefore leave the nucleus in one of its $T = 1$ states, which helps one to determine experimentally what the T values of the various states are (Fig. 44). Similarly we have the selection rule

$$\boxed{\Delta T = 0, \pm 1; \text{ for } (n, n), (n, p), (p, p), (p, n)} \tag{30.7b}$$

π-meson production cross sections

We shall now consider the differential cross sections $\sigma_{\theta\phi}$ (Schiff 1955, p. 96) for the reactions

$$p + p \to \pi^+ + d, \tag{30.8}$$

$$p + n \to \pi^0 + d. \tag{30.9}$$

Let the nucleons initially be in states† ψ and ϕ. For (30.8) the initial wave function is

$$\Psi_{i,pp} = \frac{\psi(1)\phi(2) - \psi(2)\phi(1)}{\sqrt{2}}\Xi_{1,1}. \tag{30.10}$$

Under the influence of the strong interactions this wave function will change with time into

$$\Psi_f = a\Psi(p + p) + b\Psi(\pi^+ + d) \tag{30.11}$$

where the coefficients will depend on the energy and on ψ and ϕ, for instance on whether the nucleons are really heading towards one another! Thus there is a certain probability $|b|^2$ of finding the final state $\pi^+ + d$, and this gives the cross section for reaction (30.8). Similarly if in (30.9) we take the same ψ and ϕ to refer to the proton and neutron respectively, the initial wave function is

$$\begin{aligned}\Psi_{i,pn} &= 2^{-1/2}[\psi(1)\phi(2)\xi_{+1}\xi_{-2} - \psi(2)\phi(1)\xi_{+2}\xi_{-1}] \\ &= \tfrac{1}{2}[\psi(1)\phi(2) + \psi(2)\phi(1)]\Xi_{0,0} \\ &\quad + 2^{-1/2}\frac{\psi(1)\phi(2) - \psi(2)\phi(1)}{\sqrt{2}}\Xi_{1,0}.\end{aligned} \tag{30.12}$$

Now the state $\pi^0 + d$ has $T = 1$, so that the first term in (30.12) with $T = 0$ can never change into a $\pi^0 + d$ state. The second term is related to (30.10) by isotopic spin rotations, and since the strong interactions are invariant under isotopic spin rotations we have,

† The double use of ϕ as the azimuthal angle in $\sigma_{\theta\phi}$ and as a wave function in (30 10) should cause no confusion.

using the fundamental theorem, that this term in the wave function changes with time as follows,

$$\frac{\psi(1)\phi(2) - \psi(2)\phi(1)}{\sqrt{2}} \Xi_{1,0} \rightarrow a\Psi(p + n) + b\Psi(\pi^0 + d)$$

with the same coefficients as in (30.11). Since (30.12) contains an extra factor $2^{-1/2}$, the probability of obtaining the products $\pi^0 + d$ is only half that for (30.11), and we have

$$\sigma_{\theta\phi}(p + n \rightarrow \pi^0 + d) = \tfrac{1}{2}\sigma_{\theta\phi}(p + p \rightarrow \pi^+ + d). \qquad (30.13)$$

This relation holds for any particular initial geometrical arrangement and energy, and for all angles of the reaction products.

The compound nucleus

Suppose a light nucleus is bombarded with particles such as n, p, d, α, etc., with a few MeV energy. The particle may not hit the nucleus at all or it may "bounce off the surface" or just suffer a deflection due to the Coulomb force. But if the particle penetrates the nucleus, it immediately interacts with the nearest nucleons with the strong interactions of the order of 25 MeV, so that it rapidly loses its excess kinetic energy and becomes part of the nucleus. In general the probability of forming the compound nucleus is high only when the amount of energy available corresponds to a definite excited state of this nucleus, and consequently the cross section exhibits sharp resonance peaks around the appropriate energies. Now this new compound nucleus is in general quite unstable because of the extra nucleon(s) and kinetic energy. Nevertheless, it takes relatively quite a long time for the compound nucleus to break up, because statistical fluctuations have to concentrate enough of the extra energy on one particle for this to escape. We therefore have really a double reaction

$$N_{\mathrm{i}} + \text{particle} \rightarrow N_{\mathrm{C}} \rightarrow N_{\mathrm{f}} + \text{particle}, \qquad (30.14)$$

and can apply selection rules to each stage of the reaction. For instance the isotopic spin selection rules (30.7a) and (30.7b) become respectively

$$T_{\mathrm{i}} = T_{\mathrm{C}} = T_{\mathrm{f}} \qquad \text{and} \qquad T_{\mathrm{i}} \pm \tfrac{1}{2} = T_{\mathrm{C}} = T_{\mathrm{f}} \pm \tfrac{1}{2}.$$

Angular distribution of reaction products

Suppose for example the compound nucleus is left in a $J = 0$ state. Then the reaction products must come out with spherical symmetry in the centre of mass co-ordinate system. Conversely if the reaction products emerge completely isotropically, then this indicates strongly that the compound nucleus state has $J = 0$ if it contains an even number of nucleons. The corresponding result for $J_C > 0$ is that *the differential reaction cross section* $\sigma(\theta, \phi)$ *cannot contain terms in* $\cos\theta$ *higher than* $(\cos\theta)^{2J}$ *when J is an integer, and* $(\cos\theta)^{2J-1}$ *when J is half an odd integer.* This we shall now prove, and call the *first angular distribution theorem.*

Let ψ_{Cm} with $m = M_J$ be the initial state of the compound nucleus. The final wave function can be written (Schiff 1949, p. 101)

$$\Psi_{fm} = F_{m\rho}(\theta, \phi)\left(\frac{e^{ikr}}{rv^{1/2}}\right)\Phi_\rho \qquad (\rho \text{ summed})$$

representing an outgoing wave, where v is the speed of the emerging article relative to the final nucleus. Here Φ_ρ is the product of the intrinsic wave functions of the outgoing particle and of the final nucleus, and ρ indicates their spin states. The cross section is

$$\sigma(\theta, \phi) = \sum_m a_m \sigma_m, \tag{30.15}$$

where

$$\sigma_m = \sum_\rho |F_{m\rho}(\theta, \phi)|^2. \tag{30.16}$$

We have summed over ρ in (30.16) because we assume that the spin directions of the final nucleus and particle are not being measured, so that we want the cross section summed over all the different spin states ρ. In (30.15) a_m is the probability of the compound nucleus being formed in the state ψ_{Cm}. If also the initial nuclei and the incident beam are both unpolarized, then everything is symmetrical about the axis of the incident beam which we take as the z-axis. We have $a_m = a_{-m}$, and that a_m is independent of the angle ϕ. Consider now the effect of a rotational co-ordinate transformation T. The ψ_{Cm} transform according to $D^{(J)}_{mm'}$. Suppose the $F_{m\rho}$ transform according to an undetermined representation $D_{m\rho,m'\rho'}$. Also the Φ_ρ transform according to the product representation $D^{(J_f)} \times D^{(J_p)}$ where J_f, J_p are the values of J for the final nucleus and particle. For simplicity we shall write the matrix in this product representation as $t_{\rho\rho'}$. Now by the fundamental theorem above, Ψ_{fm} and ψ_{Cm} transform in the same way, so that we have

$$\begin{aligned} T\Psi_{fm'} &= D^{(J)}_{mm'}\Psi_{fm'} = D^{(J)}_{mm'}\delta_{\rho\rho'}F_{m\rho}\Phi_{\rho'}A \\ &= T[F_{m'\rho''}\Phi_{\rho''}A] \\ &= D_{m\rho m'\rho''}t_{\rho'\rho''}F_{m\rho}\Phi_{\rho'}A, \end{aligned}$$

where A is the spherically symmetrical factor $\exp(ikr)/rv^{1/2}$. Thus we have

$$D_{m\rho m'\rho''}t_{\rho'\rho''} = D^{(J)}_{mm'}\delta_{\rho\rho'}. \qquad (30.17)$$

Since rotations are unitary transformations, t is a unitary matrix (appendix C, lemma 2 and problem C.2) and $\tilde{t}^* = t^{-1}$. Solving (30.17) we have

$$D_{m\rho,m'\rho'} = D^{(J)}_{mm'}t^*_{\rho\rho'}.$$

Thus the functions $F_{m\rho}$ transform in the same way as $\psi_{Cm}\Phi_\rho{}^*$, and from (30.16) σ_m transforms like

$$\psi^*{}_{Cm}\psi_{Cm}(\sum_\rho \Phi_\rho{}^*\Phi_\rho). \qquad (30.18)$$

Now we have (using the summation convention)

$$\begin{aligned} T\Phi_\rho{}^*\Phi_\rho &= t^*_{\rho'\rho}t_{\rho''\rho}\Phi_{\rho'}{}^*\Phi_{\rho''} \\ &= t_{\rho''\rho}t^{-1}{}_{\rho\rho'}\Phi_{\rho'}{}^*\Phi_{\rho''} \\ &= \Phi_{\rho'}{}^*\Phi_{\rho'}, \end{aligned}$$

so that $\sum\Phi_\rho{}^*\Phi_\rho$ is invariant under rotations. Consequently each σ_m and hence also σ transforms like $\psi_{Cm}{}^*\psi_{Cm}$ (m not summed), i.e. according to

$$D^{(J)} \times D^{(J)} = \sum_l D^{(l)}$$

where the highest value of l is $2J$. Thus if $\sigma(\theta, \phi)$ is expressed in spherical harmonics $Y_{lm}(\theta, \phi)$, the expansion cannot contain terms with $l > 2J$. Hence expressed in terms of $\cos\theta$, powers higher than $(\cos\theta)^{2J}$ cannot occur. Also since ψ_{Cm} and σ_ρ have definite parities, from (30.18) σ is invariant under inversion and contains only even powers of $\cos\theta$. Hence when J is half an odd integer, powers of $\cos\theta$ higher than $(\cos\theta)^{2J-1}$ cannot occur, which proves the theorem. These results furnish an important method of determining the J value of excited states of nuclei.

Angular momentum and parity selection rules

From the fundamental theorem, the J value and the parity of a system remain unchanged during a reaction. However, this is not as useful as it might at first appear because in general the initial state does not have a definite angular momentum nor definite parity. Classically a parallel incident beam contains particles having all angular momenta from zero to infinity relative to a given target nucleus, depending on how close the path of the particle is to the nucleus (Fig. 45). Quantum mechanically the incident beam is represented as a plane wave

$$\boxed{\begin{aligned} e^{ikz} &= \sum_{l=0}^{\infty} (2l+1) i^l j_l(kr) P_l(\cos\theta), \\ j_l(x) &= (\pi/2x)^{1/2} J_{l+1/2}(x), \end{aligned}} \tag{30.19}$$

where the components transforming according to different representations $D^{(l)}$ under rotations are exhibited explicitly, and where $j_l(kr)$ is a spherical Bessel function. The $l = 0, 1, 2 \ldots$ components are called the s-, p-, d-wave, etc., and they have parity $(-1)^l$.

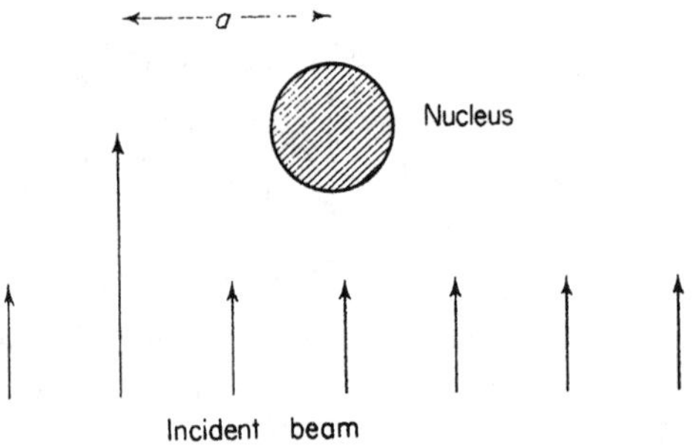

FIG. 45. A beam of particles incident on a nucleus. The angular momentum of a given particle about the nucleus is mva, where a is called the impact parameter. It is the distance of closest approach to the nucleus if the particle followed a straight path.

Although the incident wave contains all these components, not necessarily all of them are involved in a reaction. Classically the larger the angular momentum $l\hbar$, the larger is the distance $a = l\hbar/mv$ (Fig. 45) of closest approach to the nucleus, and if $a \gg R$ where R is the nuclear radius, then the particle will not hit the nucleus at all. Quantum mechanically we have for small r

$$\boxed{j_l(kr) = \frac{(kr)^l}{(2l+1)!!}[1 + \text{higher powers of } kr],}$$

so that from (30.19) the probability of a particle being close enough to the nucleus to interact strongly with it is small for large l and for small velocities (large k). For very low velocities only the s-wave contributes significantly to any reaction, and more generally a reaction is dominated by the wave with the lowest l that can contribute to the reaction due to selection rules.

We shall now show how a reaction has been used to establish some information about a particular excited state of the B^{10} nucleus. Consider the formation

$$p(2{\cdot}56 \text{ MeV}) + \text{Be}^9 \text{ (ground state)} \rightarrow \text{B}^{10*} \qquad (30.20)$$

of the 8·89 MeV excited state of the compound nucleus B^{10}, and the three modes of its decay

$$\begin{aligned} &\text{B}^{10*} \rightarrow \alpha + \text{Li}^{6**}(0, +, 1;\ 3{\cdot}58 \text{ MeV state}), \\ &\text{B}^{10*} \rightarrow \alpha + \text{Li}^{6*}(3, +, 0;\ 2{\cdot}19 \text{ MeV state}), \qquad (30.21) \\ &\text{B}^{10*} \rightarrow \alpha + \text{Li}^{6}(1, +, 0;\ \text{ground state}). \end{aligned}$$

Here the asterisks are used to denote excited states, and the sets of quantum numbers are J, w, T. In the formation (30.20) of the compound nucleus, we do not know *a priori* which l component of the incident wave dominates the reaction, and therefore gain no information about the compound nucleus. The same almost applies to the outgoing α-particle wave in (30.21), but it tells us a little bit. Experimentally only the first of the decays (30.21) is found to occur. Since the first decay occurs and the α-particle has $T = 0$, we conclude that the B^{10*} state has $T = 1$ like the 3·58 MeV Li^{6**} state, and this is consistent with the other two decay modes not occurring. The anisotropy of the angular distribution of the α-particles would then set an upper limit on the value of J. However there is an alternative explanation. In nuclei only the strong interactions are invariant under isotopic spin rotations, whereas the electromagnetic interactions which are about 100 times weaker are not invariant. Thus T is not an exact quantum number for designating a state, and it could be that the B^{10*} nucleus is in a predominantly $T = 0$ state with a little ($\sim 1\%$) of a $T = 1$ state mixed in. This would explain the first reaction (30.21) being observed. Now the isotopic spin selection rule allows the other two reactions (30.21) to occur also, and the fact that they are not observed must be due to an additional selection rule. This restricts the possible values of J and w for the B^{10} state. For suppose $J > 0$. Then the decay to the ground state of lithium is allowed if the α-particle emerges in an l wave with $l = J$ or $l = J - 1$ and such that $(-1)^l = w$, and the decay to the 2·19 MeV state would be allowed also. However no decay into the 2·19 MeV and ground states is possible from a $J = 0$, $w = +1$ state, because this would require an l wave with l odd, which would violate the conservation of parity. We conclude therefore that the B^{10*} state may be a $J = 0$, $w = +1$, $T = 0$ state containing an admixture of $T = 1$. Actually other evidence supports the $T = 1$ alternative suggested first with a little $T = 0$ mixed in with it, and the non-occurrence of the last two reactions (30.21) then indicates that either the B^{10*} state has $J = 0$, $w = +1$ for the same reason as before, or that the admixture of $T = 0$ in it is very small indeed (Malm and Inglis 1954).

The second angular distribution theorem

Consider a reaction which proceeds predominantly through a particular l wave in the incident beam of particles (30.19). The initial wave function can be written

$$\Psi_{\mathrm{i},\rho m} = f(r)\, Y_{lm}(\theta, \phi)\, \phi_\rho$$

where ϕ_ρ describes the particular spin state ρ of the incoming particle and of the initial nucleus. Similarly the final state is

$$\Psi_{\mathrm{f},\rho m} = F_{\rho\lambda m}(\theta, \phi) \left(\frac{e^{ikr}}{rv^{1/2}}\right) \Phi_\lambda$$

where Φ_λ gives the spin state of the final nucleus and the final particle. Suppose that under a rotation the various functions transform with the following unitary matrices:

function:—	Y_{lm},	ϕ_ρ,	$F_{\rho\lambda m}$,	Φ_λ;
matrix:—	$D^{(l)}_{mm'}$,	$t_{\rho\rho'}$,	$D_{\rho\lambda m\rho'\lambda' m'}$,	$T_{\lambda\lambda'}$.

Since by the fundamental theorem $\Psi_{\mathrm{i},\rho m}$ and $\Psi_{\mathrm{f},\rho m}$ transform in the same way, we have analogously to (30.17)

$$D^{(l)}_{mm'}\, t_{\rho\rho'} \delta_{\lambda\lambda'} = D_{\rho\lambda m\rho'\lambda'' m'} T_{\lambda'\lambda''}.$$

Solving this equation we obtain

$$D_{\rho\lambda m\rho'\lambda' m'}, = D^{(l)}_{mm'} t_{\rho\rho'} T_{\lambda\lambda'}{}^*.$$

Thus the functions $\sigma_m(\theta, \phi) = \sum_{\lambda\rho} |F_{\rho\lambda m}|^2$ transform in the same way as

$$Y_{lm}{}^* Y_{lm} (\sum_\rho \phi_\rho{}^* \phi_\rho) \sum_\lambda \Phi_\lambda{}^* \Phi_\lambda).$$

As in (30.18), each bracket is invariant under rotations, so that the σ_m transform like

$$D^{(l)} \times D^{(l)} = \sum D^{(L)}$$

where the highest value of L that can occur is $2l$. If we choose the z-axis parallel to the incident beam, the reaction cross section $\sigma(\theta, \phi)$ is just σ_0, and there is no dependence on ϕ. Hence $\sigma(\theta)$

expressed in powers of cos θ can contain no term higher than $(\cos\theta)^{2l}$, *where l is given by the l-wave in the incident beam through which the reaction proceeds.* This is the *second angular distribution theorem.*

Consider for instance the reaction

$$Li^6(1, +, 0) + \alpha(0, +, 0) \to B^{10}(2, -, 0, \text{excited state}) \to B^{10} + \gamma,$$

where the bracketed figures are the quantum numbers J, parity w, T respectively. If we consider only the lth component in (30.19), the initial wave function transforms according to $D^{(l)} \times D^{(1)} \times D^{(0)}$ under rotations, so that we must have l at least 1 to obtain a $J = 2$ state in the compound nucleus. Also $l = 1$ gives the correct parity, so that the p-wave dominates the reaction. The angular distribution of the final γ-ray will therefore have the form

$$\sigma(\theta) = a + b\cos^2\theta.$$

Of course in practice the argument is used in reverse. The observed angular distribution is used to obtain information about which l-wave dominates the reaction, and hence to limit the possible J value and parity of the compound nucleus. Sometimes they can be determined unambiguously in this way.

Angular correlation of successive decays

Suppose that a nucleus N goes through two successive γ-decays from an excited state a through the state b to c.

$$N_a \to N_b + \gamma(L_1),$$
$$N_b \to N_c + \gamma(L_2). \tag{30.22}$$

Here L_1 and L_2 give the angular momentum $\hbar L$ carried off by the γ-ray in what is called a 2^L-pole transition. If the life time of the state b is very short, then external influences do not have time to affect the nucleus in its b state, and the actual final quantum state of the first reaction becomes the initial state for the second. This leads to a relationship between the two decays. We shall choose as our z-axis the direction in which the first γ quantum has been observed, and shall denote by $\sigma(\theta)$ the angular distribution of the second γ quantum relative to the first one. If $\sigma(\theta)$ is expanded in powers $(\cos\theta)^\nu$ of $\cos\theta$, we shall prove that the highest power ν_{max} is given by

$$\boxed{\nu_{\text{max}} = \text{Min}(2L_1, 2L_2, 2J_b \text{ or } 2J_b - 1).} \tag{30.23}$$

Here $2J_b$ applies when J_b, the angular momentum of the N_b state, is an integer, and $2J_b - 1$ applies when J_b is half an odd integer.

We first note that the forms of the matrix elements governing the absorption and the emission of a quantum are the same, so that all selection rules are the same, and for our purposes we may replace the reaction (30.22) by

$$N_a + \gamma(L_1) \to N_b \to N_c + \gamma(L_2). \tag{30.24a}$$

Similarly the miscroscopic probability of a reaction going one way is the same as for the opposite way, and we may consider

$$N_c + \gamma(L_2) \to N_b \to N_a + \gamma(L_1). \tag{30.24b}$$

Secondly, a photon state $\gamma(L)$ has angular momentum $\hbar L$ and transforms according to $D^{(L)}$ (problem 30.3). We may therefore treat it as a spinless particle in an orbital L-wave. We can now apply the first and second angular distribution theorems to each of the reactions (30.24), and the result (30.23) follows immediately.

Nucleon π-meson scattering

We shall now calculate the relative cross sections of the reactions

$$\pi^+ + p \to \pi^+ + p, \tag{30.25a}$$

$$\pi^- + p \to \pi^0 + n, \tag{30.25b}$$

$$\pi^- + p \to \pi^- + p. \tag{30.25c}$$

Consider first (30.25a). The initial wave function is

$$\Psi_{ia} = \psi_i(\text{orb, spin})\pi^+\xi_+ = \psi_i\Xi_{3/2,3/2}, \tag{30.26}$$

where ψ_i depends on the orbital and spin variables $\mathbf{r}_p$, σ_{zp}, $\mathbf{r}_\pi$ of the initial particles and represents the meson beam incident on the protons. $\Xi_{3/2,3/2}$ is a pure isotopic spin function with $T = \frac{3}{2}$, since the π^m and the ξ_+ transform according to $D^{(1)}$ and $D^{(1/2)}$ under isotopic spin rotations respectively (Table 35) and M_T has to be 3/2 because we have a proton ($M_T = \frac{1}{2}$) and π^+ meson ($M_T = 1$). The initial function (30.26) turns under the effect of the strong interactions into the final state

$$\Psi_{fa} = \psi_f(\text{orb, spin})\Xi_{3/2,3/2} = \psi_f\, \pi^+\xi_+. \tag{30.27}$$

which again can only consist of a π^+ meson and a proton. Here ψ_f represents the outgoing beam (including any unscattered part) and has a certain angular dependence which determines $\sigma(\theta, \phi)$.

If we now assume that all three reactions (30.25) are being carried out at the same energy under identical geometrical conditions, the initial state for reactions (30.25b), (30.25c) is

$$\begin{aligned}\Psi_{ib} &= \psi_i(\text{orb, spin})\pi^-\xi_+ \\ &= \psi_i[\sqrt{(\tfrac{1}{3})}\Xi_{3/2,-1/2} + \sqrt{(\tfrac{2}{3})}\Xi_{1/2,-1/2}],\end{aligned} \tag{30.28}$$

where we have expressed the product $\pi^-\xi_+$ in terms of $T = \frac{3}{2}$ and $T = \frac{1}{2}$ functions using the Wigner coefficients in the manner of (9.8). The numerical values of the coefficients have been taken from appendix I. The reason for writing (30.28) in this way is that the two terms in it change with time under the influence of the strong interactions quite independently. In fact we shall assume that the $T = \frac{1}{2}$ component is affected only so slightly by the interactions that the final wave function is the same as the initial function, so that this component of the meson beam goes straight on without any appreciable scattering. We shall justify this assumption later. However we shall suppose that the $T = \frac{3}{2}$ term in (30.28) does change with time due to the interactions, so that it does give scattering. In fact it is related to (30.26) by isotopic spin rotations and changes finally into $\sqrt{(\frac{1}{3})}\psi_f\Xi_{3/2,-1/2}$ analogously to (30.27). The final wave function for reactions (30.25b), (30.25c) is therefore

$$\begin{aligned}\Psi_{fb} &= \sqrt{(\tfrac{1}{3})}\psi_f\Xi_{3/2,-1/2} + \sqrt{(\tfrac{2}{3})}\psi_i\Xi_{1/2,-1/2} \\ &= \sqrt{(\tfrac{1}{3})}\psi_f[\sqrt{\tfrac{2}{3}}\pi^0\xi_- + \sqrt{(\tfrac{1}{3})}\pi^-\xi_+] \\ &\qquad\qquad + (\text{the unscattered } T = \tfrac{1}{2} \text{ term}),\end{aligned} \tag{30.29}$$

where we have used the Wigner coefficients, this time in the manner of (9.7) to split $\Xi_{3/2,-1/2}$ up again. The important point is that the orbital and spin wave function ψ_f in (30.29) is the same as in (30.27) so that the cross sections for the three reactions are proportional to one another. Moreover the probabilities of forming the states $\pi^- + p$, $\pi^0 + n$ and $\pi^- + p$ are proportional to the squares of the amplitude coefficients in (30.27) and (30.29). Hence the cross sections for the three reactions (30.25) are in the ratio $1 : 2/9 : 1/9$, i.e. $9 : 2 : 1$. This is in agreement with experiment over the range 0 to 300 MeV of meson energies, which justifies the assumption that the $T = \frac{1}{2}$ term leads to only a small amount of scattering in this range. The reason for this can be understood vaguely by thinking of the compound nucleus picture. If the proton and π-meson formed a definite "compound nucleus" in a defined excited state, this would have a particular T value ($T = \frac{3}{2}$ in our case) and so would the final reaction products. Now the proton and π-meson do not form a relatively stable compound nucleus state, as evidenced by the fact that the scattering cross section does not exhibit a sharp resonance peak at some energy. However, we can still think of this tendency towards forming a $T = \frac{3}{2}$ compound nucleus as persisting and dominating the scattering in our energy range. For energies of 800–1200 MeV the $T = \frac{1}{2}$ scattering predominates over the $T = \frac{3}{2}$ and in the range 300–800 MeV the two processes are comparable (Gell-Mann and Watson 1954).

References

Bethe and Morrison 1956, Blatt and Weisskopf 1952, Sachs 1953, and references given there.

Summary

If in a reaction process the interaction Hamiltonian is invariant under some group of transformations, then the initial and final states in the reaction have the same transformation properties with respect to the group of transformations. This has been used for isotopic spin rotations, spatial rotations, and space inversion. Some miscellaneous examples have illustrated its application to selection rules for reactions, the angular distribution of the reaction products, and the ratios of some related cross sections.

Problems

30.1 $\mathscr{H}$ is invariant under some group of transformations, and $\psi_\mu(\mathbf{r}, t_0)$ is a set of wave functions at $t = t_0$ transforming according to the representation D. Using the time-dependent Schrödinger equation, show that $\partial\psi_\mu/\partial t$, $\partial^2\psi_\mu/\partial t^2$ and all higher time derivatives at $t = t_0$ also transform according to D. Hence using a Taylor expansion for $\psi_\mu(\mathbf{r}, t)$, show that the ψ_μ transform according to D at all times t.

30.2 $\mathscr{H}$ is invariant under the space inversion Π. Hence show that $\Pi\mathscr{H} = \mathscr{H}\Pi$ and $\mathrm{d}\langle\Pi\rangle\mathrm{d}t = 0$, where $\langle\Pi\rangle$ is the expectation value of Π in a certain state. Hence show that a state with even (odd) parity at one time continues to have even (odd) parity for all time. Generalize this argument to give a proof of the fundamental theorem of this section (see problem 17.3).

30.3 Show that Maxwell's equations have linearly independent solutions represented by the vector potentials

$$\mathbf{A}^E_{LM} = C_L k^{-1} \operatorname{curl} \mathbf{I}_{\mathrm{orb}} u_{LM},$$
$$\mathbf{A}^M_{LM} = iC_L \mathbf{I}_{\mathrm{orb}} u_{LM},$$

where $u_{LM} = j_L(kr)Y_{LM}(\theta, \phi) \exp(ikct)$, and

$$\mathbf{I}_{\text{orb}} = -i\mathbf{r} \wedge \text{grad}$$

is the vector operator (8.25). Show that

$$\mathbf{I}^2\mathbf{A}^\sigma_{LM} = (I_x{}^2 + I_y{}^2 + I_z{}^2)\mathbf{A}^\sigma_{LM} = L(L+1)\mathbf{A}^\sigma_{LM}, \quad (30.30)$$

and that the $\mathbf{A}^\sigma_{LM}$ transform according to (8.18). Here **I** is the total infinitestimal rotation operator.

$$\mathbf{I} = \mathbf{I}_{\text{orb}} + \mathbf{I}_{\text{comp}}.$$

It includes the change $\mathbf{I}_{\text{comp}}$ in the co-ordinate axes to which the components of the vector **A** are referred, as well as the change $\mathbf{I}_{\text{orb}}$ in the orbital variables. (See § 32 for a somewhat more detailed description of the difference between $\mathbf{I}_{\text{orb}}$ and $\mathbf{I}_{\text{comp}}$.) Also show

$$\mathbf{I}^2_{\text{comp}}\mathbf{A}^\sigma_{LM} = S(S+1)\mathbf{A}^\sigma_{LM} = 2\mathbf{A}^\sigma_{LM}. \quad (30.31)$$

The fields $\mathbf{A}^E_{LM}$, $\mathbf{A}^M_{LM}$ are called pure electric and magnetic multipole fields, and basic photon states can be defined in terms of them. (30.30) shows that the total angular momentum is $\sqrt{[L(L+1)]}\hbar$, and (30.31) that in some sense there is a spin angular momentum of one unit. However, the separation of the total angular momentum into orbital and spin is not as definite as in the case of an electron, for the $\mathbf{A}^\sigma_{LM}$ are not eigenfunctions of $I_{z\ \text{comp}}$, $I_{z\ \text{orb}}$, nor I^2_{orb} (Siegbahn 1955, p. 376).

30.4 Neglecting the mass difference between protons and neutrons, show that the probability of a nucleus making an electric dipole γ-ray transition involves the operator

$$\sum_k (\tau_{zk} + \tfrac{1}{2})\mathbf{r}_k = \sum \tau_{zk}\mathbf{r}_k,$$

where $\mathbf{r}_k$ is the kth nucleon co-ordinate relative to the centre of mass. Hence, show that the 5·11 MeV excited state of B^{10} with $J = 2$, $w = -1$, $T = 0$ cannot make an electric dipole transition to the ground state $J = 3$, $w = +1$, $T = 0$.

30.5 What would the ratio of the cross sections be for the reactions (30.25) if only the $T = \frac{1}{2}$ state contributed to the scattering? What is the ratio of the cross sections for $p + d \to \pi^+ + H^3$ and $p + d \to \pi^0 + He^3$ assuming charge-independent forces?

30.6* Discuss the assignment of quantum numbers to the 5·11 and 5·16 MeV excited states of B^{10}, particularly with reference to the reaction of $Li^6(\alpha\gamma)B^{10}$ (Jones and Wilkinson 1953, Ajzenberg and Lauritson 1955, Radicati 1953).

30.7* Set up the selection rules governing the emission of a γ-ray (Siegbahn 1955, p. 373; Gell-Mann and Telegdi 1953).

30.8* Find and discuss some examples in Ajzenberg and Lauritson (1955) of selection rules being used to determine the quantum numbers of excited states of nuclei.

30.9* A nucleus in an excited state undergoes two successive γ-decays. Outline the detailed theory of the angular correlation between the γ-rays, discussing in particular the importance of symmetry properties (Siegbahn 1955, p. 531).

30.10 A_+, A_0 and A_- are the amplitudes for the reactions

$$\pi^+ + p \to \Sigma^+ + K^+,$$
$$\pi^- + p \to \Sigma^0 + K^0,$$
$$\pi^- + p \to \Sigma^- + K^+.$$

Using the isotopic spin assignments of Table 35 (p. 332) and the assumption of charge independent forces, show

$$\sqrt{2}A_0 = A_+ - A_-$$

and hence that the differential cross-sections satisfy

$$[2\sigma_0]^{\frac{1}{2}} \leqslant [\sigma_+]^{\frac{1}{2}} + [\sigma_-]^{\frac{1}{2}}$$

at all angles of scattering.

Chapter VIII

RELATIVISTIC QUANTUM MECHANICS

31. The Representations of the Lorentz Group

Proper Lorentz transformations

If a flash of light is emitted at the origin of co-ordinate at time $t = 0$, the light wave will travel radially outwards in all directions with a uniform velocity c. The equation of the wave front is

$$x^2 + y^2 + z^2 - c^2t^2 = 0. \tag{31.1}$$

Let us now consider an observer moving with uniform velocity $\mathbf{v}$ relative to the co-ordinate system Ox, Oy, Oz, and let the origin of his new co-ordinate system x', y', z' be chosen to coincide with O at time $t' = t = 0$. Now according to the principle of the special theory of relativity, the velocity of light is independent of the motion of the observer, so that an observer at rest in the second system of co-ordinates also sees a spherical wave front

$$x'^2 + y'^2 + z'^2 - c^2t'^2 = 0. \tag{31.2}$$

Thus from (31.1) and (31.2) a transformation of co-ordinates from x, y, z, t to x', y', z', t' moving with relative velocity $\mathbf{v}$, leaves the quantity $x^2 + y^2 + z^2 - c^2t^2$ invariant. Such a transformation is called a Lorentz transformation, in particular a proper Lorentz transformation if it is achievable physically and does not involve an inversion of the space or time co-ordinates. If we put $T = ict$, the invariant quantity becomes $x^2 + y^2 + z^2 + T^2$. Now the characteristic feature of a rotation in three-dimensional space is that it leaves $x^2 + y^2 + z^2$ invariant (cf. equation (3.4); problem 3.7). Thus by analogy *we can regard the proper Lorentz transformations as rotations in the four-dimensional* x, y, z, T *space.*

For instance the transformation $L(v,z)$ to a system x', y', z', T' moving with velocity v along the Oz axis is†

$$z = \frac{(z' + vt')}{\sqrt{(1 - v^2/c^2)}}, \qquad t = \frac{t' - vz'/c^{2'}}{\sqrt{(1 - v^2/c^2)}},$$

† This can be proved by a more detailed consideration of (31.1) and (31.2). See any text discussing the special theory of relativity, e.g. Goldstein 1950, p. 188 ; Stratton 1941, p. 76.

and this can be written in the form of a rotation in the zT-plane by an angle θ.

$$\begin{aligned} z &= \frac{z'}{\sqrt{(1-\beta^2)}} - \frac{i\beta T'}{\sqrt{(1-\beta^2)}} = z' \cos\theta - T' \sin\theta \\ T &= \frac{i\beta z'}{\sqrt{(1-\beta^2)}} + \frac{T'}{\sqrt{(1-\beta^2)}} = z' \sin\theta + T' \cos\theta \end{aligned} \tag{31.3}$$

$$\tan\theta = i\beta \quad \text{or} \quad \tanh(i\theta) = -\beta \quad \text{where } \beta = v/c. \tag{31.4}$$

Thus by analogy with (8.3) we can write

$$L(v, z) = \exp(i\theta I_{zT}) \tag{31.5}$$

in terms of an infinitesimal rotation operator I_{zT} in the zT-plane. Here θ is restricted to be an imaginary number as shown by (31.4) so that the transformation is completely real when expressed in terms of x, y, z and t. Clearly the ordinary rotations in x, y, z space (which we shall call pure rotations) can be included as a special kind of Lorentz transformation, and the infinitesimal rotation operators I_x, I_y, I_z of § 8 become I_{yz}, I_{zx}, I_{xy} in the notation of (31.5), since a rotation about the x-axis is a rotation in the yz-plane. If the two equations (31.3) are written in reverse order, they can be interpreted as a transformation $\exp(-i\theta I_{Tz})$, so that $I_{Tz} = -I_{zT}$ and in general

$$I_{\mu\nu} = -I_{\nu\mu}. \tag{31.6}$$

Thus symbols such as I_{xx} which have not been defined so far must be considered as zero. We therefore obtain *six* independent infinitesimal Lorentz transformations I_{xy}, I_{yz}, I_{zx}, I_{xT}, I_{yT}, I_{zT}.

It is not difficult to prove that the product of two Lorentz transformations is also a Lorentz transformation. In this way by combining the simple transformation we have mentioned, we can build up a whole group of transformations. This is called the *proper Lorentz group* I and contains as a subgroup all proper rotations. However it does not contain the time inversion $t = -t'$ nor the space inversion Π since these cannot be represented purely in terms of rotations about the x, y, z, T axes.

Commutation relations

All the commutation relations between the above six infinitesimal rotation operators about the x, y, z, T axes can be deduced directly from the commutation relations between the ordinary infinitesimal rotation operators of § 8. For consider the rotation $R(\alpha,\ x) = \exp(i\alpha I_{yz})$,

$$\begin{aligned} x &= x' \\ y &= y' \cos\alpha - z' \sin\alpha \\ z &= y' \sin\alpha + z' \cos\alpha \end{aligned} \tag{31.7}$$

and two other rotations $R(\beta,\ y) = \exp(i\beta I_{zx})$ and $R(\gamma,\ z) = \exp(i\gamma I_{xy})$. The relationship between these three rotations is completely determined by the three sets of equations analogous to (31.7). In particular in § 8 we derived the commutation relation

$$I_{xy}I_{yz} - I_{yz}I_{xy} = iI_{zx} \tag{31.8}$$

from these equations by a procedure which effectively consists of writing down the relation between $R(\alpha,\ x)$, $R(\beta,\ y)$, $R(\gamma,\ z)$ when α, β, γ have some special simple values. We can now replace z by T in (31.7) and throughout the argument and obtain from (31.8)

$$I_{xy}I_{yT} - I_{yT}I_{xy} = iI_{Tx} = -iI_{xT}.$$

in general

$$\boxed{I_{\nu\mu}I_{\nu\rho} - I_{\nu\rho}I_{\mu\nu} = -iI_{\mu\rho} \qquad \text{(suffices not summed).}} \tag{31.9a}$$

Transformations such as $\exp(i\alpha I_{xy})$ and $\exp(i\theta I_{zT})$ commute since they alter only the x, y and z, T co-ordinates respectively, whence from (31.5) and (8.3)

$$\boxed{I_{\mu\nu}I_{\rho\sigma} - I_{\rho\sigma}I_{\mu\nu} = 0 \qquad (\mu,\ \nu,\ \rho,\ \sigma \text{ all different}).} \tag{31.9b}$$

Any desired commutation relation can now be derived from (31.9a), (31.9b) or a trivial relation such as

$$I_{\mu\nu}I_{\mu\nu} - I_{\mu\nu}I_{\mu\nu} = 0 \qquad \text{(not summed).} \tag{31.9c}$$

Irreducible representations

Instead of using the above infinitesimal Lorentz transformations, it is more convenient to define the linear combinations

$$\begin{gathered} A'_x = \tfrac{1}{2}(I_{yz} + I_{xT}), \quad A'_y = \tfrac{1}{2}(I_{zx} + I_{yT}), \quad A'_z = \tfrac{1}{2}(I_{xy} + I_{zT}), \\ A'_+ = A'_x + iA'_y, \quad A'_- = A'_x - iA'_y \\ A_+ = -A'_-, \quad A_- = -A'_+, \quad A_z = -A'_z \\ B_x = \tfrac{1}{2}(I_{yz} - I_{xT}), \quad B_y = \tfrac{1}{2}(I_{zx} - I_{yT}), \quad B_z = \tfrac{1}{2}(I_{xy} - I_{zT}), \\ B_+ = B_x + iB_y, \quad B_- = B_x - iB_y. \end{gathered} \tag{31.10}$$

Now it can easily be verified from (31.9) that each A and A' operator commutes with each B operator, e.g.

$$\begin{aligned} 4(A'_xB_y - A'_yB_x) &= (I_{yz}I_{zx} - I_{zx}I_{yz}) + (I_{xT}I_{zx} - I_{zx}I_{xT}) \\ &\qquad - (I_{yz}I_{yT} - I_{yT}I_{yz}) - (I_{xT}I_{yT} - I_{yT}I_{xT}) \\ &= -iI_{yx} - iI_{Tz} - iI_{zT} - iI_{xy} = 0. \end{aligned}$$

Also each set A_+, A_-, A_z; A'_+, A'_-, A'_z; B_+, B_-, B_z satisfies the same commutation relations (8.10) as the infinitesimal rotation operators I_+, I_-, I_z

$$\begin{aligned} I_zI_+ - I_+I_z &= I_+, \\ I_zI_- - I_-I_z &= -I_-, \\ I_+I_- - I_-I_+ &= 2I_z. \end{aligned} \tag{31.11}$$

We can now use the operators A_+, A_-, A_z and B_+, B_-, B_z satisfying the same relations (31.11) to deduce the irreducible representations of the Lorentz group, in the same way as the infinitesimal rotation operators were used in § 8 with respect to the rotation group. Consider a given vector space which is invariant under the Lorentz group. It is then also invariant under the infinitesimal operators (31.10). We first reduce the space with respect to the commuting operators A_z and B_z, and from among the eigenvectors belonging to the highest eigenvalue j of A_z we pick out the one with the highest eigenvalue j' of B_z. As in § 8, we can construct from this vector a total of $(2j + 1)(2j' + 1)$ standard base vectors $u_{mm'}$ such that (cf. (8.18))

$$\boxed{\begin{aligned} A_+u_{mm'} &= \sqrt{[j(j+1) - m(m+1)]}\, u_{m+1,m'} \\ A_-u_{mm'} &= \sqrt{[j(j+1) - m(m-1)]}\, u_{m-1,m'} \\ A_zu_{mm'} &= m\, u_{mm'} \\ B_+u_{mm'} &= \sqrt{[j'(j'+1) - m'(m'+1)]}\, u_{m,m'+1} \\ B_-u_{mm'} &= \sqrt{[j'(j'+1) - m'(m'-1)]}\, u_{m,m'-1} \\ B_zu_{mm'} &= m'\, u_{mm'} \end{aligned}} \tag{31.12}$$

As in § 8, these vectors form the basis of an irreducible representation of the proper Lorentz group which we shall denote by $D^{(jj')}$, where j, j' are integers or half odd integers. Any finite representation is reducible into a sum of these.

Examples

A little manipulation shows that *the quantities x, y, z, t transform according to the representation* $D^{(\frac{1}{2}\frac{1}{2})}$. In fact the standard base vectors are

$$u_{++} = z + ct = z - iT, \qquad u_{+-} = x - iy,$$
$$u_{-+} = x + iy, \qquad u_{--} = -z + ct = -z - iT, \tag{31.13}$$

where for simplicity we have written $u_{\pm\pm}$ for $u_{\pm 1/2\ \pm 1/2}$. From (8.11) the infinitesimal Lorentz transformations can be expressed as differential operators

$$I_{xy} = -i\left(x\frac{\partial}{\partial y} - y\frac{\partial}{\partial x}\right), \qquad I_{zT} = -i\left(z\frac{\partial}{\partial T} - T\frac{\partial}{\partial z}\right), \text{ etc.} \tag{31.14}$$

whence

$$A_z u_{++} = \tfrac{1}{2}(-I_{xy} - I_{zT})(z - iT) = \tfrac{1}{2}(z - iT) = \tfrac{1}{2}u_{++}$$

Similarly it may be verified that each of the vectors (31.13) satisfies (31.12).

If $u_{\frac{1}{2}0}$, $u_{-\frac{1}{2}0}$ (abbreviated to u_{+0}, u_{-0}) transform as standard base vectors (31.12) according to $D^{(\frac{1}{2}0)}$, and u_{0+}, u_{0-} according to $D^{(0\frac{1}{2})}$, we can construct the set of base vectors

$$u_{mm'} = \frac{u_{+0}^{j+m}u_{-0}^{j-m}}{[(j+m)!(j-m)!]^{1/2}}\,\frac{u_{0+}^{j'+m'}u_{0-}^{j'-m'}}{[(j'+m')!(j'-m')!]^{1/2}}. \tag{31.15}$$

These transform as standard base vectors according to $D^{(jj')}$. This can easily be verified, for the infinitesimal operators being defined in (8.1) in the same way as a differential operator like d/dx, operate on products in a similar way (cf. problems 8.10, 8.11), e.g.

$$A_+ u_{+0}^n = n\,u_{+0}^{n-1} A_+ u_{+0} = 0, \qquad A_- u_{+0}^n = n\,u_{+0}^{n-1}\,u_{-0}, \qquad \text{etc.}$$

Quantities transforming like u_{+0}, u_{-0} or u_{0+}, u_{0-} are called spinors, though this term is often also applied more generally to quantities of the type (31.15).

Also the matrices representing the infinitesimal Lorentz transformations can be written down from (31.12). Hence the matrix representing any desired Lorentz transformation can be calculated using the method applied to $D^{(1/2)}$ in § 8.

The complete Lorentz group

The proper Lorentz group $\mathfrak{l}$ considered so far can be augmented by combining each proper Lorentz transformation with the space inversion Π, the time inversion† τ, and the space-time inversion $\Pi\tau = \tau\Pi$

$$\begin{array}{|lllll|} \Pi: & x = -x', & y = -y', & z = -z', & T = T', \\ \tau: & x = x', & y = y', & z = z', & T = -T', \\ \Pi\tau: & x = -x', & y = -y', & z = -z', & T = -T', \end{array} \tag{31.16}$$

Such transformations are called improper Lorentz transformations, and together with the proper ones they form the *complete Lorentz group* $\mathfrak{L}$.

Let L be any proper Lorentz transformation. Then L can be made up as a product of simple transformations $\exp(i\theta I_{\mu\nu})$ which are pure rotations or simple Lorentz transformations along one of the co-ordinate axes. In § 8 (equation (8.22)) this was shown for any rotation. In the present case we can rotate the system such that the relative velocity is along the z-axis, apply $L(v, z)$, and then rotate again to any desired orientation. Now if we reduce each of the angles θ in these transformations to zero, L changes continuously into the identity transformation E. This cannot be done for any of the transformations (31.16) since these are inversions which in some sense turn our co-ordinate system "inside-out". However, ΠL or $L\Pi$ can be reduced continuously to Π. In this way the complete Lorentz group is seen to contain four branches $\mathfrak{l}$, $\Pi\mathfrak{l}$, $\tau\mathfrak{l}$, $\Pi\tau\mathfrak{l}$ consisting of all transformations that can be reduced continuously to E, Π, τ, $\Pi\tau$ respectively. Any transformation from $\tau\mathfrak{l}$ combined with any transformation from $\Pi\mathfrak{l}$ gives one belonging to $\Pi\tau\mathfrak{l}$, etc. Thus, since $\Pi^2 = \tau^2 = (\Pi\tau)^2 = E$, the four branches $\mathfrak{l} + \Pi\mathfrak{l} + \tau\mathfrak{l} + \Pi\tau\mathfrak{l}$ together form a group, the complete Lorentz group $\mathfrak{L}$, as stated above. Similarly $\mathfrak{l}$ and $\Pi\mathfrak{l}$ form the space-inversion Lorentz group, and $\mathfrak{l} + \tau\mathfrak{l}$ and $\mathfrak{l} + \Pi\tau\mathfrak{l}$ also form groups. However, $\mathfrak{l} + \Pi\mathfrak{l} + \Pi\tau\mathfrak{l}$ does not form a group because it contains the elements Π and τ but not their product $\Pi\tau$. In the following

† It should be noted that τ is the simple time-*inversion* operator and not the usual time-*reversal* operator of quantum mechanics defined in § 19.

we shall only concern ourselves with the complete Lorentz group $\mathfrak{L}$, the restriction to the various subgroups not being difficult.

Representations derived from $D^{(\frac{1}{2}\frac{1}{2})}$

Consider first the group E, Π, τ, $\Pi\tau$, where $\Pi^2 = \tau^2 = (\Pi\tau)^2 = E$, $\Pi(\Pi\tau) = \tau$, etc. This group is Abelian and isomorphic with the point group 222. Its only single-valued irreducible representations are the four one-dimensional ones shown in Table 36, as can easily be shown from the relations of § 14 or by the methods of § 7 (problem 7.4).

TABLE 36

Single-valued Irreducible Representations of E, Π, τ, $\Pi\tau$

	$\chi(E)$	$\chi(\Pi)$	$\chi(\tau)$	$\chi(\Pi\tau)$
χ_0	1	1	1	1
χ_1	1	1	-1	-1
χ_2	1	-1	1	-1
χ_3	1	-1	-1	1

We have already shown that x, y, z, T or, in standard form, the linear combinations (31.13) transform under the proper Lorentz group according to the irreducible representation $D^{(\frac{1}{2}\frac{1}{2})}$. Now these base vectors are also transformed into one another by the operations Π, τ, $\Pi\tau$ (31.16), and therefore form the basis of a representation of the complete Lorentz group which we shall call $D^{(\frac{1}{2}\frac{1}{2},0)}$. For instance the matrix representing Π with respect to the standard base vectors (31.13) is

$$\begin{bmatrix} \cdot & \cdot & \cdot & 1 \\ \cdot & -1 & \cdot & \cdot \\ \cdot & \cdot & -1 & \cdot \\ 1 & \cdot & \cdot & \cdot \end{bmatrix}.$$

Now from this representation we can obtain three other representations $D^{(\frac{1}{2}\frac{1}{2};r)}$, $r = 1, 2, 3$ as follows. We simply multiply the matrices of all transformations T belonging to the branches $\mathfrak{l}$, $\Pi\mathfrak{l}$, $\tau\mathfrak{l}$, $\Pi\tau\mathfrak{l}$ by $\chi_r(E) = 1$, $\chi_r(\Pi)$, $\chi_r(\tau)$, $\chi_r(\Pi\tau)$ respectively, where the χ_r are a row of characters from Table 36. Thus

$$D^{(\frac{1}{2}\frac{1}{2};r)}(T) = \chi_r(\alpha) D^{(\frac{1}{2}\frac{1}{2};0)}(T) \tag{31.17}$$

where $\alpha = E, \Pi, \tau, \Pi\tau$ as T belongs to $\mathfrak{l}, \Pi\mathfrak{l}, \tau\mathfrak{l}, \Pi\tau\mathfrak{l}$. This really does give representations. For suppose T_1, T_2 belong to the branches $\Pi\mathfrak{l}$, $\tau\mathfrak{l}$: then T_1T_2 belongs to $\Pi\tau\mathfrak{l}$, and from Table 36

$$\chi_r(\Pi)D^{(\frac{1}{2}\frac{1}{2};0)}(T_1)\chi_r(\tau)D^{(\frac{1}{2}\frac{1}{2};0)}(T_2) = \chi_r(\Pi\tau)D^{(\frac{1}{2}\frac{1}{2};0)}(T_1T_2)$$

as required. It is shown in appendix H that the four representations $D^{(\frac{1}{2}\frac{1}{2};r)}$ $r = 0, 1, 2, 3$ are inequivalent and the only ones which reduce to $D^{(\frac{1}{2}\frac{1}{2})}$ for proper Lorentz transformations, so that we have exhausted all the possibilities in this direction. Components transforming like x, y, z, T according to the representation $D^{(\frac{1}{2}\frac{1}{2};0)}$ are said to form a *regular* four-vector; and if according to $D^{(\frac{1}{2}\frac{1}{2};r)}$, $r = 1, 2, 3$, then a *pseudo*-vector of type r (Watanabe 1951).

The representation $D^{(\frac{1}{2}0+0\frac{1}{2})}$

Let u_{+0}, u_{-0} and u_{0+}, u_{0-} transform respectively according to $D^{(\frac{1}{2}0)}$ and $D^{(0\frac{1}{2})}$ under the proper Lorentz group. We shall now obtain from them a representation of the complete Lorentz group. From (31.15) the vectors

$$u_{+0}u_{0+},\ u_0u_{0+},\ u_{+0}u_{0-},\ u_{-0}u_{0-} \tag{31.18}$$

transform under the proper Lorentz group according to $D^{(\frac{1}{2}\frac{1}{2})}$. Let us also suppose that u_{+0}, u_{-0}, u_{0+}, u_{0-} are so related that the vectors (31.18) are also transformed into one another by Π and τ. They therefore transform under the complete Lorentz group according to one of the representations $D^{(\frac{1}{2}\frac{1}{2};r)}$, and let us suppose for the present that its is $D^{(\frac{1}{2}\frac{1}{2};0)}$. Then as far as their transformation properties are concerned, the vectors (31.18) may be identified with the vectors (31.13). From this identification we get

$$\begin{aligned} \tau\, u_{+0}u_{0+} &= -u_{-0}u_{0-}, & \tau\, u_{+0}u_{0-} &= u_{+0}u_{0-}, \\ \tau\, u_{-0}u_{0+} &= u_{-0}u_{0+}, & \tau\, u_{-0}u_{0-} &= -u_{+0}u_{0+}. \end{aligned} \tag{31.19}$$

From these relations it can be seen that if τ induces a linear transformation among the u_{+0}, u_{-0}, u_{0+}, u_{0-}, then we must have

$$\begin{aligned} \tau u_{+0} &= a u_{0-}, & \tau u_{0+} &= -(1/a)u_{-0}, \\ \tau u_{-0} &= -a u_{0+}, & \tau u_{0-} &= (1/a)u_{+0}, \end{aligned} \tag{31.20a}$$

where a is a numerical constant. In order to obtain a particular representation we shall put $a = 1$ and thus

$$\begin{aligned} \tau u_{+0} &= u_{0-}, & \tau u_{0+} &= -u_{-0}, \\ \tau u_{-0} &= u_{0+}, & \tau u_{0-} &= u_{+0}. \end{aligned} \tag{31.20b}$$

Similarly we can consider the effect of Π, and from (31.13) and (31.18) we obtain

$$\begin{aligned} \Pi u_{+0}u_{0+} &= u_{-0}u_{0-}, & \Pi u_{+0}u_{0-} &= -u_{+0}u_{0-}, \\ \Pi u_{-0}u_{0+} &= -u_{-0}u_{0+}, & \Pi u_{-0}u_{0-} &= u_{+0}u_{0+}, \end{aligned} \tag{31.21}$$

and analogously to† (31.20b)

$$\begin{aligned} \Pi u_{+0} &= -iu_{0-}, & \Pi u_{0+} &= iu_{-0}, \\ \Pi u_{-0} &= iu_{0+}, & \Pi u_{0-} &= -iu_{+0}. \end{aligned} \tag{31.22}$$

Thus *the vector space* $(u_{+0}, u_{-0}, u_{0+}, u_{0-})$ *is invariant under the complete Lorentz group, and forms the basis of a representation which we shall denote by* $D^{(\frac{1}{2}0+0\frac{1}{2};4)}$.

Furthermore the representation $D^{(\frac{1}{2}0+0\frac{1}{2};4)}$ is irreducible. For suppose it were reducible: then considering the proper Lorentz transformations it could only be reducible into $D^{(\frac{1}{2}0)} + D^{(0\frac{1}{2})}$ but, from (31.20), (31.22), the subspaces (u_{+0}, u_{-0}) and (u_{0+}, u_{0-}) are not invariant under Π and τ. It can also be shown now that the different representations obtained by choosing different values of a in (31.19) are all equivalent. They can all be obtained from our particular one ($a = 1$) by a continuous change of the parameter a. By continuity the representations cannot suddenly change from being equivalent to being inequivalent, so that they are all equivalent, (unless there were an infinite number of inequivalent representations of dimension 4).

Above we deduced a total of four representations $D^{(\frac{1}{2}\frac{1}{2};r)}$, $r = 0, 1, 2, 3$ from the representation $D^{(\frac{1}{2}\frac{1}{2};0)}$ by introducing the factors χ_r (Table 36) into the matrices of the various branches (equation (31.17)). We can now do the same thing to $D^{(\frac{1}{2}0+0\frac{1}{2};4)}$. However the four representations turn out to be all the same and just $D^{(\frac{1}{2}0+0\frac{1}{2};4)}$! For the pairs u_{+0}, u_{-0} and u_{0+}, u_{0-} both transform under pure rotations according to the representation $D^{(1/2)}$ of the rotation group (problem 31.3), so that like $D^{(1/2)}$ the representations $D^{(\frac{1}{2}0)}$ and $D^{(0\frac{1}{2})}$ of $\mathfrak{l}$ are double valued. Each element T of the group $\mathfrak{L}$ is *already* represented in $D^{(\frac{1}{2}0+0\frac{1}{2};4)}$ by two matrices $+D(T)$ and $-D(T)$, so that introducing the various changes of sign of Table 36 does not give new representations. However, because of the double-valuedness we can obtain some other representations by using the double-valued χ_r of Table 37 instead of the single-valued ones of Table 36. Thus we get the four representations $D^{(\frac{1}{2}0+0\frac{1}{2};r)}$, $r = 4, 5, 6, 7$.

† The constants in (31.20) and (31.22) are not completely independent because $\Pi\tau$ has to be represented by a multiple of the unit matrix (appendix H).

TABLE 37

Double-valued Irreducible Representations of E, Π, τ, $\Pi\tau$

	$\chi(E)$	$\chi(\Pi)$	$\chi(\tau)$	$\chi(\Pi\tau)$
χ_4	± 1	± 1	± 1	± 1
χ_5	± 1	± 1	$\pm i$	$\pm i$
χ_6	± 1	$\pm i$	± 1	$\pm i$
χ_7	± 1	$\pm i$	$\pm i$	± 1

This brings up a point which has some important physical consequences. The four representations $D^{(\frac{1}{2}\frac{1}{2};r)}$, $r = 0, 1, 2, 3$ are all single-valued and differ by changes in sign which certainly are often important. However the $D^{(\frac{1}{2}0+0\frac{1}{2};r)}$, $r = 4, 5, 6, 7$ are double-valued with no distinction between transformations differing in sign. This suggests that for physical applications at least, factors of $\pm i$ may also be unimportant. I.e. there may be no point in distinguishing between the various $D^{(\frac{1}{2}0+0\frac{1}{2};r)}$ since these differ only by factors of i in some of the branches of $\mathfrak{L}$. This is definitely known to be the case when dealing with a single spinor field like in the theory of electrons, positrons and electromagnetic radiation: it makes no difference to any of the physically meaningful results which representation $D^{(\frac{1}{2}0+0\frac{1}{2};r)}$ one uses to describe the electron-positron field† (Jauch and Rohrlich 1955; see also discussion below equation (11.23a)). From now on therefore we shall adopt the representation with $r = 6$ as our standard one because it gives the various factors i in the most convenient places, but we shall usually refer to it simply as $D^{(\frac{1}{2}0+0\frac{1}{2})}$ so as not to imply any significant distinction from the other $D^{(\frac{1}{2}0+0\frac{1}{2};r)}$.

For the usual quantum mechanical applications it is convenient to introduce in the vector space $(u_{+0}, u_{-0}, u_{0+}, u_{0-})$ a new set of base vectors

$$\begin{aligned} u_+ &= \frac{1}{\sqrt{2}}(u_{0+} + u_{-0}), & v_+ &= \frac{1}{\sqrt{2}}(u_{0+} - u_{-0}), \\ u_- &= \frac{1}{\sqrt{2}}(u_{0-} - u_{+0}), & v_- &= \frac{1}{\sqrt{2}}(u_{0-} + u_{+0}). \end{aligned} \tag{31.23}$$

† However it may be physically meaningful to distinguish between the $D^{(\frac{1}{2}0+0\frac{1}{2};r)}$, $r = 4, 5, 6, 7$ in a theory involving more than one type of spinor field, e.g. in a theory of electrons, nucleons, neutrinos, μ mesons and their antiparticles all in interaction with one another (Yang and Tiomno 1950). We shall not consider such a situation.

Adopting $D^{(\frac{1}{2}0+0\frac{1}{2};6)}$ now instead of $D^{(\frac{1}{2}0+0\frac{1}{2};4)}$ in accordance with our convention, we introduce a factor i in (31.22) and get

$$\boxed{\begin{array}{llll} \Pi u_+ = u_+, & \Pi u_- = u_-, & \Pi v_+ = -v_+, & \Pi v_- = -v_-, \\ \tau u_+ = -u_+, & \tau u_- = -u_-, & \tau v_+ = v_+, & \tau v_- = v_-. \end{array}} \quad \begin{array}{r} (31.24) \\ (31.25) \end{array}$$

It can easily be shown that *the pairs* u_+, u_- *and* v_+, v_- *each transform under pure rotations as standard base vectors according to the representation* $D^{(1/2)}$ *of the rotation group.* For instance I_z of § 8 is I_{xy} in the present notation, and we have

$$\begin{aligned} I_z u_+ = I_{xy} u_+ &= (B_z - A_z)(\sqrt{\tfrac{1}{2}})(u_{0+} + u_{-0}) && \text{from (31.10), (31.23),} \\ &= \tfrac{1}{2} u_+. && \text{from (31.12), (31.23).} \end{aligned}$$

In this way the relations (8.18) with $j = \frac{1}{2}$, m $= \pm\frac{1}{2}$ may be verified for each of the pairs u_+, u_- and v_+, v_-. For the sake of future reference, we also note that the matrices representing I_{xy}, I_{yz} and I_{zT} with respect to u_+, u_-, v_+, v_- are $\frac{1}{2}M$, $\frac{1}{2}N$ and $\frac{1}{2}K$ respectively, where

$$M = \begin{bmatrix} 1 & . & . & . \\ . & -1 & . & . \\ . & . & 1 & . \\ . & . & . & -1 \end{bmatrix}, \qquad N = \begin{bmatrix} . & 1 & . & . \\ 1 & . & . & . \\ . & . & . & 1 \\ . & . & 1 & . \end{bmatrix},$$

$$K = \begin{bmatrix} . & . & -1 & . \\ . & . & . & 1 \\ -1 & . & . & . \\ . & 1 & . & . \end{bmatrix}. \qquad (31.26)$$

Other representations

The following is a brief description of all the finite irreducible representations of the complete Lorentz group $\mathfrak{L}$. They are derived and described in detail in appendix H. Consider the base vectors $u_{mm'}$ transforming according to $D^{(jj')}$ under $\mathfrak{l}$. Then the vectors $\tau u_{mm'}$ transform according to $D^{(j'j)}$. Thus when $j \neq j'$, to get a representation of $\mathfrak{L}$ we have to take the $u_{mm'}$ and $\tau u_{mm'}$ together to form a representation $D^{(jj'+j'j;r)}$ of dimension $2(2j+1)(2j'+1)$. When j is an integer and j' half an odd integer or vice versa, the representation is double-valued. In this case r takes the values 4, 5, 6 or 7, giving representations related to one another in the same way as the $D^{(\frac{1}{2}0+0\frac{1}{2};r)}$ by (31.17) and Table 37. When j, j' are both integers or both half odd integers, the representations $D^{(jj'+j'j;r)}$ are single-valued and r takes the values 0 and 1. The representations with $r = 2, 3$ can also be constructed from (31.17) but are equivalent to those with $r = 1, 0$ respectively. When $j = j'$, we obtain four representations of $\mathfrak{L}$, i.e. $D^{(jj;r)}$, $r = 0, 1, 2, 3$. These have dimension $(2j + 1)^2$ and are directly analogous to the $D^{(\frac{1}{2}\frac{1}{2};r)}$. In all cases products of regular vectors are taken to transform according to representations with $r = 0$.

References

For further details about the proper and complete Lorentz groups the reader is referred to appendix H and to the following: Van der Waerden 1932; Boerner 1955; Watanabe 1951; Jauch and Rohrlich 1955, pp. 82 ff. (time-reversal); Murnaghan 1938; Harish-Chandra 1947; Wigner 1939; Heine 1957a.

Summary

Lorentz transformations can be regarded as rotations in the four-dimensional space $x, y, z, T = ict$. This leads to the irreducible representations $D^{(jj')}$ of the proper Lorentz group in the same way as the representations of the rotation group were determined in § 8. $D^{(jj')}$ has dimension $(2j + 1)(2j' + 1)$ and a standard set of base vectors $u_{mm'}$ transforms according to (31.12). In particular x, y, z, t transform according to $D^{(\frac{1}{2}\frac{1}{2})}$ under proper Lorentz transformations, the linear combinations (31.13) being standard base vectors. The complete Lorentz group $\mathfrak{L}$ includes the space and time inversions. There are four types of regular and pseudo-vectors corresponding to the representations $D^{(\frac{1}{2}\frac{1}{2};r)}$, $r = 0, 1, 2, 3$ of $\mathfrak{L}$. Thus x, y, z, t (or ict) form a regular vector transforming according to $D^{(\frac{1}{2}\frac{1}{2};0)}$. The spinors u_+, u_-, v_+, v_- used in an electron-positron wave function may be taken to transform according to any one of the representations $D^{(\frac{1}{2}0+0\frac{1}{2};r)}$, $r = 4, 5, 6, 7$ so that we shall write just $D^{(\frac{1}{2}0+0\frac{1}{2})}$. All the other irreducible representations have also been described.

PROBLEMS

31.1 Show that the combination of two Lorentz transformations is itself a Lorentz transformation.

31.2 Show that $\exp(i\theta_1 I_{zT}) \times \exp(i\theta_2 I_{zT}) = \exp[i(\theta_1 + \theta_2)I_{zT}]$, and hence deduce the relativistic law of composition of velocities $v = (v_1 + v_2)(1 + v_1v_2/c^2)$.

31.3 Show that $I_{yz} = \frac{1}{2}(-A_+ - A_- + B_+ + B_-)$ and hence calculate in the manner of § 8 the matrix representing $R(\theta, x)$ in the representation $D^{(\frac{1}{2}0)}$ in standard form.

31.4 Using the method of problem 8.16, show that if $u_{mm'}$ transforms as a standard base vector according to $D^{(jj')}$, then $u^*_{mm'}$ transforms like $u_{m'm}$ according to $D^{(j'j)}$. Hence show from (31.23) that u_+^*, u_-^*, v_+^*, v_-^* transform in the same way as v_-, $-v_+$, $-u_-$, u_+.

31.5 In the manner of § 9, show that

$$D^{(\frac{1}{2}\frac{1}{2})} \times D^{(\frac{1}{2}\frac{1}{2})} = D^{(11)} + D^{(10)} + D^{(01)} + D^{(00)}.$$

According to what representation do the six components of the angular momentum transform?

31.6 Show that the representation $D^{(jj')}$ is single-valued if j and j' are either both integers or both half odd integers. Hint: use the reduction of product representations such as in the previous problem.

31.7 Write down some physical examples of the various regular and pseudo scalars, vectors and tensors.

31.8 Show directly that u_{+0}, u_{-0} transforming according to $D^{(\frac{1}{2}0)}$ under proper Lorentz transformations, cannot possibly form the basis for a representation of the complete Lorentz group. Hint: u_{+0}, u_{-0} transform according to $D^{(\frac{1}{2})}$ under pure rotations: Hence from Schur's lemma, Π must be represented by a multiple of the unit matrix. Also $\Pi L(v) = L(-v)\Pi$, but this leads to a contradiction since $L(v)$ and $L(-v)$ are not represented by the same matrix.

31.9 Show that a set of vectors transforming according to $D^{(jj')}$ under the proper Lorentz group transform under pure rotations according to the representation $D^{(j)} \times D^{(j')}$ of the rotation group. Hence show that $D^{(jj')}$ is double-valued if j is an integer and j' half an odd integer.

32. The Dirac Equation

Relativistic wave equations

To set up a relativistic form of quantum mechanics to describe electrons or any other kinds of particles, we proceed as follows in complete analogy to the usual non-relativistic form. In relativity, the momenta p_x, p_y, p_z are the first three components of a four-vector p_μ whose fourth component is $p_T = iE_K/c$ where E_K is the kinetic energy.† The square of the length of this four-vector is a

† In accordance with the previous section, we take x, y, z, $ict = T$ as the four co-ordinates. We shall use μ etc. to denote x, y, z or T, e.g. p_μ and $\partial/\partial\mu$. The suffix j will be used to denote x, y or z (not T), and bold type an ordinary three-vector: thus $\mathbf{p}^2 = p_j p_j$.

constant $-m^2c^2$ (Goldstein 1950, p. 203).

$$p_\mu p_\mu = -m^2c^2 \tag{32.1a}$$

or

$$E_K{}^2 = \mathbf{p}^2c^2 + m^2c^4. \tag{32.1b}$$

In quantum mechanics we have to interpret the p_μ as operators, and corresponding to (32.1a) we have the relativistic wave equation

$$(p_\mu p_\mu + m^2c^2)\psi = 0. \tag{32.2a}$$

The quantity $p_\mu p_\mu$ is a world scalar, so that (32.2) is invariant under all proper Lorentz transformations. As usual, we adopt the Schrödinger representation and put

$$p_\mu = -i\hbar\frac{\partial}{\partial\mu}, \tag{32.3}$$

so that (32.2) becomes the differential equation

$$-\hbar^2\left(\frac{\partial^2}{\partial x^2}+\frac{\partial^2}{\partial y^2}+\frac{\partial^2}{\partial z^2}-\frac{1}{c^2}\frac{\partial^2}{\partial t^2}\right)\psi + m^2c^2\psi = 0, \tag{32.2b}$$

to replace the usual non-relativistic Schrödinger equation. If now we put $\psi = \psi(\mathbf{r}, t)$ and regard ψ as a wave function describing the state of one particle, we would appear to have a satisfactory, relativistically invariant theory.

For one particle this is more or less true, but a dilemma immediately appears when we try to discuss more than one particle this way. For two particles, for example, we would need six spatial co-ordinates $\mathbf{r}_1$ and $\mathbf{r}_2$, and to preserve the fundamental relativistic symmetry between space and time, this would automatically demand the introduction of two time co-ordinates t_1 and t_2. Now physically in any one frame of reference, time is something which is the same for all particles, and no physically acceptable and useful way has yet been discovered of developing a theory with more than one independent time co-ordinate. However, a way out of the difficulty has been found by developing the quantum theory of fields, and it is easy to see in outline how, by considering the electromagnetic interaction energy between two moving charges e_1 and e_2. The Coulomb energy is e_1e_2/r_{12}, and the magnetic energy can also be expressed in terms of $\mathbf{r}_1$, $\mathbf{r}_2$ and the momenta $\mathbf{p}_1$, $\mathbf{p}_2$. This gives the above type of description which depends explicitly on the number of particles, and which cannot be made truly relativistic because of the single time co-ordinate. However, if we describe the system in terms of the electric and magnetic fields $\mathscr{E}$ and $\mathbf{H}$, the total energy becomes $\frac{1}{2}\int(\mathscr{E}^2 + \mathbf{H}^2)\,dv$, quite independent of the number of particles. Moreover, $\mathscr{E}$ and $\mathbf{H}$ depend only on the one set of variables $\mathbf{r}$, t. We shall not follow this field approach any further here because it would rapidly lead us outside the scope of this book, but merely mention two facts that arise out of it. Firstly, a system with any number of particles can be described in terms of a quantum field operator $\Psi(\mathbf{r}, t)$ which, for freely moving particles, satisfies the equation (32.2) already derived, in this case called the field equation. Secondly, if the system contains only one particle, its motion can be described to a good approximation by (32.2) considered now as a wave equation for a one-particle wave function $\psi(\mathbf{r}, t)$. We shall therefore discuss (32.2) as a wave equation, remembering that this is only an approximation as a one-particle wave equation, but that at the same time it has a much wider applicability as a field equation. It makes no difference to the symmetry properties, of course, which interpretation we have in mind.

In (32.2) the wave function ψ (or field Ψ) may be a multi-component function with components $\psi_\alpha(\mathbf{r}, t)$. These can be the spin components of a spinor ψ as in § 11, or the components of a relativistic vector or tensor function (like the components $\mathscr{E}_x$, $\mathscr{E}_y$, $\mathscr{E}_z$ of the electric field $\mathscr{E}$ in three dimensions). With such multi-component functions, a Lorentz transformation has a two-fold effect. It first transforms the variables $\mathbf{r}$, t to new ones $\mathbf{r}'$, t', and at the same time it transforms the components ψ_α among one another. Here $\mathbf{r}'$, t' are just the co-ordinates in the new reference frame, of the same point P whose co-ordinates are $\mathbf{r}$, t in the old frame. This part of the transformation is not very interesting, so that we shall ignore it in what follows and just write $\psi_\alpha(P)$ or ψ_α without specifying which co-ordinates $\mathbf{r}$, t or $\mathbf{r}'$, t' are being used for P. However, the transformation of the different components ψ_α is more interesting. We note that it has to belong to one of the representations of the Lorentz group, because every transformation must preserve the *type* of function ψ that we are dealing with. In § 11 the use of representation $D^{(\frac{1}{2})}$ of the rotation group brought with it the whole concept of spin angular momentum. So here, we obtain different theories describing particles with different properties according to which irreducible representation the ψ_α belong to. The representations $D^{(00;r)}$ of the complete Lorentz group (§ 31) give scalar and pseudo-scalar "mesons", the π-mesons for example being such pseudo-scalar particles. A theory associated with $D^{(\frac{1}{2}\frac{1}{2};r)}$ is said to describe generally vector or pseudo-vector "mesons". In particular photons belong to the representation $D^{(\frac{1}{2}\frac{1}{2};1)}$ since this is the representation appropriate to the relativistic electromagnetic vector potential, which is the field that gets quantized and actually satisfies (32.2) with $m = 0$. The double-valued representations $D^{(\frac{1}{2}0+0\frac{1}{2};r)}$ are used in theories to describe all the spin $\frac{1}{2}$ particles such as electrons and positrons, protons, neutrons, anti-nucleons and μ-mesons. As mentioned in §31, all physical consequences are independent of r in $D^{(\frac{1}{2}0+0\frac{1}{2};r)}$, so that we just write $D^{(\frac{1}{2}0+0\frac{1}{2})}$.

In this section we shall develop the theory appropriate to electrons. From § 11, we know that an electron has an intrinsic degree of freedom described by the spin functions u_+, u_-, which transform according to $D^{(\frac{1}{2})}$ under purely spatial rotations. Under the proper Lorentz group this becomes the representation $D^{(\frac{1}{2}0)}$ (see problem 31.8). However, all electromagnetic forces are known to be invariant under space inversion, so that we should really consider the complete Lorentz group. The representation $D^{(\frac{1}{2}0)}$ has to be enlarged then to $D^{(\frac{1}{2}0+0\frac{1}{2})}$, and we shall now proceed to derive the theory based on this representation. Instead of using the pure component form ψ_α for the wave function (or field), we shall switch at this stage to the alternative form of writing ψ (see § 11) in terms of some set of base spinors,

$$\psi = \psi_\alpha u_\alpha = \psi_1 u_+ + \psi_2 u_- + \psi_3 v_+ + \psi_4 v_-, \tag{32.4}$$

where the ψ_α are functions of x, y, z, t only, and where for convenience we shall take the u_α to be the functions u_+, u_-, v_+, v_- transforming according to $D^{(\frac{1}{2}0+0\frac{1}{2};6)}$ as defined by (31.23) to (31.26).† As discussed already in § 8, the spinors u_α cannot be regarded as functions of x, y, z, t because of their peculiar transformation properties, e.g. the double-valuedness of the representation. It may be helpful to think of them as functions of some hidden degrees of freedom which will turn out to be the spin direction and whether the particle is an electron or a positron.

The Dirac equation

We return now to the wave equation (32.2). As it stands, it contains one unsatisfactory feature in that it involves the *second* differential coefficient $\partial^2\psi/\partial t^2$. This implies that to determine $\psi(\mathbf{r}, t)$ completely, it is necessary to specify both $\psi(\mathbf{r}, t = 0)$ and $\partial\psi(\mathbf{r}, t = 0)/\partial t$ as initial conditions, whereas physically just the state ψ of a system as one time should be sufficient to determine

† Where it is useful to employ the summation convention, we shall write the spinors u_+, u_-, v_+, v_- as u_α, $\alpha = 1$ to 4, always with α or β etc. for the spinor suffices. Thus the $\Gamma_{\mu,\alpha\beta}$ below transform under the complete Lorentz group according to $D^{(\frac{1}{2}\frac{1}{2};0)}$ as regards the μ suffix and according to $D^{(\frac{1}{2}0+0\frac{1}{2})}$ as regards α and β.

the state of all later times.† We therefore make (32.2) linear in $\partial\psi/\partial t$ by factorizing it. If we write

$$(\gamma_\mu p_\mu + imc)(\gamma_\mu p_\mu - imc)\psi = 0, \tag{32.5}$$

then this is equivalent to (32.2a) provided we have

condition I: $\quad \gamma_\mu{}^2 = 1,\ \gamma_\mu\gamma_\nu = -\gamma_\nu\gamma_\mu$ *for* $\mu \neq \nu$. $\quad$ (32.6)

It is also necessary that the γ_μ commute with the p_ν, i.e. they cannot depend on $\mathbf{r}$, t. Nor can the γ_μ depend on the p_ν since this would destroy the required linearity in the p_ν. We conclude that the γ_μ operate only on the extra spin variables of the spinors u_ν, and that they just transform the u_α into one another in order to preserve the form (32.4) of the wave function. I.e. we have

condition II: the γ_μ *operate only on the* u_α *and leave the vector space* (u_+, u_-, v_+, v_-) *invariant.*

We now suppose that, because of the factorization (32.5), ψ satisfies

$$\boxed{(\gamma_\mu p_\mu - imc)\psi = 0.} \tag{32.7}$$

This is the Dirac equation. Clearly every solution of (32.7) is also a solution of (32.5) and (32.2). However, the converse is not necessarily true for all mathematical solutions of (32.2), but since (32.7) has the required linearity in $p_T = -(\hbar/c)\partial/\partial t$ we presume that (32.7) gives all the physically relevant solutions of (32.2). To keep the operator in (32.7) still relativistically invariant, we further require

condition III: the γ_μ *transform according to* $D^{(\frac{1}{2}\frac{1}{2};0)}$ *under the complete Lorentz group* $\mathfrak{L}$.

Incidentally, it is interesting to note that in the case of the electromagnetic field we can use (32.2) direct without linearizing, since the time derivatives are determined by the fields through Maxwell's equations, e.g. $\partial\mathbf{B}/\partial t = -\text{curl}\ \mathscr{E}$.

The γ_μ *operators*

We now proceed to write down explicitly a set of operators, γ_μ, verify that these do in fact satisfy the conditions I to III, and then

† Note that in quantum mechanics the *state* of a system is not just a specification of the positions of particles, but includes as complete a description of all other variables such as the momenta, etc., as the uncertainty principle allows.

prove that the conditions I to III define the γ_μ uniquely so that the operators we have written down are not *some* set of γ_μ operators but the *only* set. Consider the operators γ_μ defined by the relations

$$\boxed{\gamma_\mu u_\beta = \Gamma_{\mu,\alpha\beta} u_\alpha,} \tag{32.8}$$

$$\boxed{\begin{aligned}
\Gamma_x &= \begin{bmatrix} \cdot & \cdot & \cdot & i \\ \cdot & \cdot & i & \cdot \\ \cdot & -i & \cdot & \cdot \\ -i & \cdot & \cdot & \cdot \end{bmatrix}, \quad
\Gamma_y = \begin{bmatrix} \cdot & \cdot & \cdot & 1 \\ \cdot & \cdot & -1 & \cdot \\ \cdot & -1 & \cdot & \cdot \\ 1 & \cdot & \cdot & \cdot \end{bmatrix}, \\
\Gamma_z &= \begin{bmatrix} \cdot & \cdot & i & \cdot \\ \cdot & \cdot & \cdot & -i \\ -i & \cdot & \cdot & \cdot \\ \cdot & i & \cdot & \cdot \end{bmatrix}, \quad
\Gamma_T = \begin{bmatrix} 1 & \cdot & \cdot & \cdot \\ \cdot & 1 & \cdot & \cdot \\ \cdot & \cdot & -1 & \cdot \\ \cdot & \cdot & \cdot & -1 \end{bmatrix}.
\end{aligned}} \tag{32.9a}$$

In an abbreviated notation these matrices can be written

$$\Gamma_j = \begin{bmatrix} \cdot & i\sigma_j \\ +i\sigma_j & \cdot \end{bmatrix}, \quad \Gamma_T = \begin{bmatrix} \mathbf{1} & \cdot \\ \cdot & -\mathbf{1} \end{bmatrix}, \tag{32.9b}$$

where σ_j are the Pauli matrices and $\mathbf{1}$ the unit matrix

$$\sigma_x = \begin{bmatrix} \cdot & 1 \\ 1 & \cdot \end{bmatrix}, \quad \sigma_y = \begin{bmatrix} \cdot & -i \\ i & \cdot \end{bmatrix},$$
$$\sigma_z = \begin{bmatrix} 1 & \cdot \\ \cdot & -1 \end{bmatrix}, \quad \mathbf{1} = \begin{bmatrix} 1 & \cdot \\ \cdot & 1 \end{bmatrix}. \tag{32.10}$$

Since the operators (32.8) are defined in terms of their effect on u_+, u_-, v_+, v_-, condition II is automatically satisfied. It is also easy to verify using the relation

$$\sigma_x\sigma_y = -\sigma_y\sigma_x = i\sigma_z, \text{ etc.} \tag{32.11}$$

that the matrices (32.9) and hence the operators (32.8) satisfy condition I.

We now establish the transformation properties of the γ_μ under $\mathfrak{L}$, and hence verify condition III. Since the γ_μ are not expressible in terms of x, y, z, T, one cannot transform them by a simple substitution of variables as in § 3. However, one can still define indirectly what is meant by transforming such an operator. Suppose ψ and ϕ are two functions related by $\phi = \gamma\psi$, and denote transformed quantities by primes. Transforming this equation in the sense of §§ 3 and 5 means obtaining from it $\phi' = \gamma'\psi'$, and if we know the transformation properties of ϕ and ψ we can determine those of γ. Expressed in a slightly different notation, if T is any transformation,

$$T(\gamma\psi) = \gamma'(T\psi).$$

The operator satisfying this equation is

$$\gamma' = T\gamma T^{-1}, \tag{32.12}$$

which gives the transformation of an operator γ (cf. equation (5.4†)). In our case if γ is represented by a matrix Γ (32.8) and T a proper or improper Lorentz transformation, then γ' is represented by the matrix

$$[D^{(\frac{1}{2}0+0\frac{1}{2})}(T)]\,\Gamma\,[D^{(\frac{1}{2}0+0\frac{1}{2})}(T)]^{-1}.$$

We confine ourselves from now on to the base vectors u_+, u_-, v_+, v_- and use the same symbol T to denote the Lorentz transformation and the matrix representing it with respect to these base vectors.

The transformation of the γ_μ under Π and τ can be derived from (31.24) and (31.25).

$$\Pi\Gamma_T\Pi^{-1} = \Gamma_T, \qquad \Pi\Gamma_j\Pi^{-1} = \Gamma_j,$$
$$\tau\,\Gamma_T\,\tau^{-1} = -\Gamma_T, \qquad \tau\,\Gamma_j\,\tau^{-1} = \Gamma_j. \tag{32.13}$$

Thus under Π, τ, $\Pi\tau$ the γ_μ transform in the same way as x, y, z, T, i.e. as a regular vector. Turning now to the proper Lorentz transformations, the matrices L representing them can be calculated from (31.26). Consider the rotation $R(\theta, z)$ represented by the matrix $\exp(\frac{1}{2}i\theta M)$ where M is given by (31.26). From (32.9) $M = i\Gamma_y\Gamma_x$ so that from (32.6) M commutes with Γ_z and Γ_T. Thus

$$R\Gamma_z R^{-1} = RR^{-1}\Gamma_z, \qquad R\Gamma_T R^{-1} = \Gamma_T.$$

Also from (32.6) M anticommutes with Γ_x and Γ_y,

$$M\Gamma_x = i\Gamma_y\Gamma_x\Gamma_x = -i\Gamma_x\Gamma_y\Gamma_x = -\Gamma_x M, \qquad M\Gamma_y = -\Gamma_y M.$$

Also $M^2 = E$. By expanding the exponentials we obtain

$$\Gamma_x \exp(-\tfrac{1}{2}i\theta M) = \exp(\tfrac{1}{2}i\theta M)\Gamma_x$$

and
$$\begin{aligned} e^{\frac{1}{2}i\theta M}\Gamma_x\, e^{-\frac{1}{2}i\theta M} &= e^{\frac{1}{2}i\theta M} e^{\frac{1}{2}i\theta M}\Gamma_x \\ &= (E\cos\theta + iM\sin\theta)\Gamma_x \\ &= \Gamma_x\cos\theta - \Gamma_y\sin\theta. \end{aligned} \tag{32.14}$$

Thus the Γ_μ transform according to $D^{(\frac{1}{2}\frac{1}{2})}$ under the spatial Lorentz transformation $R(\theta, z)$. The same applies to the transformations

$R(\theta, x)$ and $L(v, z)$ which are represented by $\exp(\frac{1}{2}i\theta N)$ and $\exp(\frac{1}{2}i\theta K)$ (31.26). This follows immediately from the above analysis by noting that $N = i\Gamma_z\Gamma_y$, $K = i\Gamma_T\Gamma_z$ and by changing the appropriate suffices. Now any proper Lorentz transformation can be expressed as a product of $R(\theta, z)$, $R(\theta, x)$ and $L(v, z)$. It therefore follows from all the above that the γ_μ transform under the complete Lorentz group according to $D^{(\frac{1}{2}\frac{1}{2};0)}$ in the same way as x, y, z, T and the p_μ. This verifies condition III.

Uniqueness of the γ_μ

It remains to prove that the operators γ_μ defined by (32.8), (32.9) are the only ones satisfying all of the conditions I, II and III. Let $\bar{\gamma}_\mu$ be another set of operators satisfying the three conditions. From condition II we can write

$$\bar{\gamma}_\mu u_\beta = \bar{\Gamma}_{\mu,\alpha\beta} u_\alpha, \tag{32.15}$$

where from condition I the matrices $\bar{\Gamma}_\mu$ satisfy the relations (32.6). We now use condition III to prove $\bar{\Gamma}_\mu = \pm\Gamma_\mu$.

The first step is to show that the matrices Γ_μ are linearly independent, in fact that all the 16 matrices

$$E,\ \Gamma_\mu,\ \Gamma_\mu\Gamma_\nu,\ \Gamma_\mu\Gamma_\nu\Gamma_\rho,\ \Gamma_x\Gamma_y\Gamma_z\Gamma_T \tag{32.16}$$

are linearly independent. Using (32.6) we note that each of these matrices except E can be written as a product of two anticommuting matrices $AB = -BA$. E.g.

$$\Gamma_x = (\Gamma_x\Gamma_y)\Gamma_y = -\Gamma_y(\Gamma_x\Gamma_y),$$
$$\Gamma_x\Gamma_y\Gamma_z = (\Gamma_x\Gamma_y\Gamma_z\Gamma_T)\Gamma_T = -\Gamma_T(\Gamma_x\Gamma_y\Gamma_z\Gamma_T).$$

The trace (sum of the diagonal elements) of any such matrix is zero, since

$$A_{\alpha\beta}B_{\beta\alpha} = \text{Trace}(AB) = \text{Trace}(-BA) = -B_{\beta\alpha}A_{\alpha\beta}.$$

Let us denote the 16 matrices (32.16) by Γ_r, $r = 1$ to 16. Then from (32.6) any product $\Gamma_q\Gamma_r$ is another Γ_s. In particular $\Gamma_r\Gamma_r = \pm E$. Consider now the matrix

$$Q = \sum_1^{16} a_s\Gamma_s.$$

We have that

$$\text{Trace}\ (\Gamma_r Q) = a_r\ \text{Trace}\ (\pm E)$$

which is non-zero unless $a_r = 0$. Hence $Q \neq 0$ unless all $a_r = 0$, $r = 1$ to 16. I.e. the matrices Γ_r are linearly independent.

The conclusion of the last paragraph does not depend on the number of vectors u_α in (32.15) and applies to square matrices of any order n by n satisfying (32.6). The maximum number of linearly independent matrices of order n by n is n^2, and since there are 16 matrices $\bar{\Gamma}_r$ they must be at least of order 4×4. Thus there is no representation of the operators γ_μ (32.6) of dimension less than four, and the matrices Γ_μ defined by (32.9) in terms of the four vectors $u_\alpha = u_+, u_-, v_+, v_-$ are irreducible. Equation (14.14) then shows that there is, apart from equivalence, only this one irreducible representation. Thus Γ_μ and $\bar{\Gamma}_\mu$ defined by (32.9), (32.15) are equivalent, and from (5.15) there exists a matrix P such that

$$\bar{\Gamma}_r = P\Gamma_r P^{-1}. \tag{32.17}$$

The last step of the proof is to determine P. Let us apply any proper or improper Lorentz transformation T. In the notation of (32.12) let

$$\Gamma'_s = T\Gamma_s T^{-1} = D_{rs}\Gamma_r,$$

where from condition III D_{rs} is determined by the representation $D^{(\frac{1}{2}\frac{1}{2};0)}$. Similarly

$$\bar{\Gamma}'_s = T\bar{\Gamma}_s T^{-1} = TP\Gamma_s P^{-1}T^{-1},$$

and

$$\bar{\Gamma}'_s = D_{rs}\bar{\Gamma}_r = PD_{rs}\Gamma_r P^{-1} = PT\Gamma_s T^{-1}P^{-1}.$$

Comparing these two equations, we obtain

$$Q\Gamma_s = \Gamma_s Q \text{ where } Q = T^{-1}P^{-1}TP. \tag{32.18}$$

Since this is true for any Γ_s, by Schur's lemma (appendix D) we have $Q = \lambda E$. Now $\det|Q| = 1 = \lambda^4$, whence $\lambda = \pm 1$ or $\pm i$. When $T = E$, then $\lambda = 1$ so that, by continuity, $\lambda = 1$ for all proper Lorentz transformations L, and from (32.18)

$$PL = LP. \tag{32.19}$$

The conditions of Schur's lemma are not completely fulfilled by (32.19) since the u_α and hence the matrices L are not irreducible under *proper* Lorentz transformations. Thus Schur's lemma only gives

$$P = \begin{bmatrix} \eta_1 & \cdot \\ \cdot & \eta_2 \end{bmatrix}$$

where P is now expressed in abbreviated form with respect to the base vectors u_{+0}, u_{-0} and u_{0+}, u_{0-}. In this representation Π becomes $\begin{bmatrix} \cdot & 1 \\ 1 & \cdot \end{bmatrix}$, and from $\Pi^{-1}P^{-1}\Pi P = \lambda E$ we easily obtain, by trying the four possibilities $\lambda = \pm 1$ or $\pm i$, that $\eta_1 = \eta_2$ or $-\eta_2$. In terms of u_+, u_-, v_+, v_- these two possibilities become

$$P = \eta_2 \begin{bmatrix} 1 & \cdot & \cdot & \cdot \\ \cdot & 1 & \cdot & \cdot \\ \cdot & \cdot & 1 & \cdot \\ \cdot & \cdot & \cdot & 1 \end{bmatrix} \quad \text{or} \quad \eta_2 \begin{bmatrix} \cdot & \cdot & 1 & \cdot \\ \cdot & \cdot & \cdot & 1 \\ 1 & \cdot & \cdot & \cdot \\ \cdot & 1 & \cdot & \cdot \end{bmatrix},$$

whence from (32.9) and (32.17) we obtain $\Gamma_\mu = \Gamma_\mu$ or $-\Gamma_\mu$ respectively. The minus sign in the latter case just corresponds to using the reversed order of factors in (32.5), and can be shown to lead to no new consequences of physical significance.

Thus apart from the trivial alternative of sign, the γ_μ operators are uniquely defined by the conditions I, II and III. We are therefore completely justified in using the specific form (32.9) for the γ_μ without any loss of generality, although some authors prefer to derive all results using the defining properties I, II and III directly without ever writing down the actual matrices (32.9) explicitly (Jauch and Rohrlich 1955).

Plane wave solutions

Since (32.1) applies to a free particle without any interactions, we expect the Dirac equation (32.7) to have plane wave solutions

$$(a_1 u_+ + a_2 u_- + a_3 v_+ + a_4 v_-)e^{i\mathbf{k}\cdot\mathbf{r}}e^{-iEt/\hbar} \tag{32.20}$$

where the a_α are constants giving the amplitudes of the four components ψ_α (32.4). Substituting (32.8), (32.9), (32.20) into (32.7), we obtain

$$\begin{aligned} (E - mc^2)a_1 + c\hbar k_z a_3 + c\hbar(k_x - ik_y)a_4 &= 0 \\ (E - mc^2)a_2 + c\hbar(k_x + ik_y)a_3 - c\hbar k_z a_4 &= 0 \\ -c\hbar k_z a_1 - c\hbar(k_x - ik_y)a_2 - (E + mc^2)a_3 &= 0 \\ -c\hbar(k_x + ik_y)a_1 + c\hbar k_z a_2 - (E + mc^2)a_4 &= 0. \end{aligned}$$

For these equations to be consistent the determinant of the coefficients of the a_α must vanish, which gives

$$(E^2 - m^2c^4 - c^2\hbar^2k^2)^2 = 0$$

i.e.
$$\begin{aligned} E = E_+ &\equiv (m^2c^4 + c^2\hbar^2k^2)^{1/2} \approx mc^2 + \frac{1}{2m}\hbar^2k^2 \\ &\geqslant mc^2 \end{aligned}$$

or
$$\begin{aligned} E = E_- &\equiv -(m^2c^4 + c^2\hbar^2k^2)^{1/2} \approx -mc^2 - \frac{1}{2m}\hbar^2k^2 \\ &\leqslant -mc^2 \end{aligned}$$

in accordance with (32.1). Corresponding to E_+ we obtain two linearly independent solutions for the a_α,

$$a_1 = 1 \quad a_2 = 0 \quad a_3 = \frac{-c\hbar k_z}{E_+ + mc^2} \quad a_4 = \frac{-c\hbar(k_x - ik_y)}{E_+ + mc^2}$$

$$a_1 = 0 \quad a_2 = 1 \quad a_3 = \frac{-c\hbar(k_x + ik_y)}{E_+ + mc^2} \quad a_4 = \frac{c\hbar k_z}{E_+ + mc^2} \tag{32.21a}$$

and two solutions corresponding to E_-

$$a_1 = \frac{c\hbar k_z}{-E_- + mc^2} \quad a_2 = \frac{c\hbar(k_x + ik_y)}{-E_- + mc^2} \quad a_3 = 1 \quad a_4 = 0$$

$$a_1 = \frac{c\hbar(k_x - ik_y)}{-E_- + mc^2} \quad a_2 = \frac{-c\hbar k_z}{-E_- + mc^2} \quad a_3 = 0 \quad a_4 = 1 \tag{32.21b}$$

There are therefore two types of solution. The one type has a positive energy greater than mc^2, and the components of the wave function involving u_+ and u_- are large while those involving v_+, v_- are small of order $p/2mc$. The reverse situation applies to the other type of solution with negative energy.

Motion in an electromagnetic field

The properties of the particles to which the Dirac equation applies become much more apparent when we consider their interaction with an electromagnetic field. An arbitrary field can be described by the four-vector (A_x, A_y, A_z, $A_T = i\phi$) where $\mathbf{A}$ is the magnetic vector potential and ϕ the scalar electric potential (Stratton 1941, p. 73). In a field we still have the relation (32.1) and hence (32.7) provided E_K and $\mathbf{p}$ are the purely kinetic energy and kinetic momentum $m\mathbf{v}$. However the canonical momenta are given for an electron by

$$p_\mu \text{ (canonical)} = p_\mu \text{ (kinetic)} - eA_\mu/c$$

where e is the charge on a proton, and it is these canonical momenta that have to be replaced by the operators (32.3). Hence in (32.7) we put

$$p_\mu = -i\hbar\frac{\partial}{\partial\mu} + \frac{e}{c}A_\mu. \tag{32.22}$$

The equation still retains all its invariance properties under the proper Lorentz group. However, it is no longer exactly equivalent to (32.2) because the p_μ now do not commute, but we shall assume that the correct description of our particles in an electromagnetic field is obtained from (32.7) rather than (32.2). This is justified firstly by the fact that it correctly describes observed physical phenomena, and secondly because the corresponding extra terms in (32.2) are just the ones required to keep it Lorentz invariant and have no classical analogue.

If we put

$$\psi = (\psi_1 u_+ + \psi_2 u_- + \psi_3 v_+ + \psi_4 v_-)e^{-iEt/\hbar}$$

we can from (32.9b) write (32.7) as two equations for two functions ψ_l and ψ_s each having two components (ψ_1, ψ_2) and (ψ_3, ψ_4) respectively.

$$c\sigma_j\left(-i\hbar\frac{\partial}{\partial j} + \frac{e}{c}A_j\right)\psi_s + (E + e\phi - mc^2)\psi_l = 0 \quad (32.23a)$$

$$c\sigma_j\left(-i\hbar\frac{\partial}{\partial j} + \frac{e}{c}A_j\right)\psi_l + (E + e\phi + mc^2)\psi_s = 0 \quad (32.23b)$$

This shows that ψ_l is large and ψ_s is small for positive energy states and vice versa for negative energy, in accordance with (32.21). We can separate the large and small components more clearly. Suppose E is positive and equal to $E' + mc^2$: multiply (32.23a) by $(E + e\phi - mc^2)$, (32.23b) by $c\sigma_j(-i\hbar\partial/\partial_j + eA_j/c)$, subtract and divide by $-2mc^2$. Then we obtain

$$\left(-\frac{\hbar^2}{2m}\nabla^2 - e\phi - \frac{ei\hbar}{mc}\mathbf{A}\cdot\mathbf{grad} - \frac{ei\hbar}{2mc}\operatorname{div}\mathbf{A} + \frac{e^2A^2}{2mc^2}\right)\psi_l + \\ + \frac{e}{mc}\mathbf{s}\cdot\mathbf{H}\,\psi_l + \frac{ie}{mc}\mathbf{s}\cdot\mathscr{E}\psi_s - \frac{(E' + e\phi)}{2mc^2}\psi_s = E'\psi_l \quad (32.24)$$

Here $\mathbf{s} = \frac{1}{2}\hbar\boldsymbol{\sigma}$ is the spin angular momentum operator used in §11 (see particularly problem 11.1), and $\mathscr{E}$, $\mathbf{H}$ are the electric and magnetic fields. (32.24) shows that ψ_l almost satisfies an orthodox Schrödinger equation. The terms in brackets describe the interaction between the charged particle and the field in strict analogy with classical mechanics (Schiff 1955, pp. 138, 292). In particular, for a uniform magnetic field,

$$\frac{-ei\hbar}{mc}\mathbf{A}\cdot\mathbf{grad} = \frac{e}{2mc}\mathbf{l}\cdot\mathbf{H} \quad (32.25)$$

where $\mathbf{l}$ is the angular momentum (8.27). The term $(e/mc)\mathbf{s}\cdot\mathbf{H}$ gives an additional interaction with an intrinsic magnetic moment $(-e/mc)\mathbf{s}$. This is proportional to the spin angular momentum but twice as large as one would expect by analogy with the orbital angular momentum term (32.25). The next term gives the spin-orbit coupling as can be seen after some further manipulation. From (32.23b) neglecting $\mathbf{A}$ and small quantities of order $(e\phi + E')/mc^2$, we obtain

$$\psi_s = -\frac{\sigma_j}{2mc}\left(-i\hbar\frac{\partial}{\partial j}\right)\psi_l.$$

Putting also

$$\phi = \phi(|r|), \qquad \mathscr{E} = -\frac{1}{r}\frac{d\phi}{dr}\mathbf{r}, \qquad \mathbf{s} = \frac{1}{2}\hbar\boldsymbol{\sigma},$$

we have from (32.11)

$$(\mathbf{s}\cdot\mathscr{E})\,(\boldsymbol{\sigma}\cdot\ -i\hbar\,\mathbf{grad}) = i\mathscr{E}\wedge(-i\hbar\,\mathbf{grad})\cdot\mathbf{s} = \frac{-i}{r}\frac{d\phi}{dr}\mathbf{l}\cdot\mathbf{s},$$

$$\frac{ie}{mc}\mathbf{s}\cdot\mathscr{E}\,\psi_s = -\frac{e}{2m^2c^2}\frac{1}{v}\frac{d\phi}{dr}\mathbf{l}\cdot\mathbf{s}\,\psi_l, \tag{32.26}$$

which is the usual spin-orbit coupling term for an electron in a central field (11.21). The last term on the left of (32.24) is another small relativistic correction which shifts all the energy levels slightly but introduces no new effects. Thus (32.24) has the following three features.

(i) The large components $\psi_l = (\psi_1, \psi_2)$ occur in the wave function multiplied by the spinors u_+, u_-. Those transform according to $D^{(1/2)}$ under pure rotations as shown in the previous section. This point is most important because it means that

(ii) we can identify the operators $s_x = \frac{1}{2}\hbar\sigma_x$, etc., with the components of the intrinsic (spin) angular momentum operator. We recapitulate briefly the argument of § 11 (see particularly problem 11.1). In quantum mechanics the angular momentum in the absence of a classical analogue is defined through the infinitesimal rotation operators (appendix F). The infinitesimal rotations operating on u_+, u_- are represented by the matrices $\frac{1}{2}\sigma_x$, $\frac{1}{2}\sigma_y$, $\frac{1}{2}\sigma_z$. Hence, if we write the wave function in component form (ψ_1, ψ_2), the spin angular momentum operators are $s_x = \frac{1}{2}\hbar\sigma_x$, etc.

(iii) On neglecting the last term on the left of (32.24) and the approximations inherent in (32.26), ψ_l satisfies an orthodox two component Schrödinger equation. The Hamiltonian besides the usual classical terms includes the interaction of an intrinsic magnetic moment $(-e/mc)\mathbf{s}$ parallel to the spin of the particle, and also a spin-orbit coupling term (32.26).

This is precisely the description of an electron used in § 11 which is known to give agreement with experiment (apart from relativistic and radiative corrections). We therefore conclude that particles described by the Dirac equation really are electrons as expected.

Positrons and charge conjugation

In a field theoretical discussion, we find that a theory based on the Dirac equation not only describes states with any number of electrons but also states containing two types of particles, electrons and positrons, which have identical properties except for opposite charge. This property is connected with the fact that the field equation is invariant under an additional symmetry transformation called charge conjugation. When this operates on a state vector (wave function) describing the system, it turns it into a new state in which electrons are replaced by positrons and vice versa. All states have positive energy relative to the vacuum state (defined as the state with no particles).

Now our present one-electron wave equation type of approach is really rather inadequate for discussing these features of positrons and charge conjugation, but we can exhibit something of their properties by considering the negative energy solutions of the Dirac equation and going through certain mental gymnastics. To start with we note that there are an unlimited number of negative energy states (32.21b) with energy going to $-\infty$, so that at first sight an electron initially in any state would continuously radiate energy as it jumps to states of lower and lower negative energy without limit. However, electrons do not behave in this fashion experimentally, and we must suppose that all the negative energy states are ordinarily filled and that electrons satisfy the exclusion principle (obey Fermi statistics) which prevents them from jumping into the filled negative energy states.† The infinite charge density corresponding to all these filled states is somehow supposed to be unobservable, but if a negative energy state happens to be temporarily empty this will be observed as a net positive charge, namely as a positron. The energy is also positive, corresponding to the energy required to lift the electron out of the negative energy state.

† Incidentally this explains in part why particles with a spin of half an odd integer have to obey Fermi Dirac statistics, while particles of integer spin obey Bose–Einstein statistics (Pauli 1955).

To derive the behaviour of a positron, we have to study the dynamical properties of a negative energy state. Equation (32.23) shows that ψ_s is now large and ψ_l small in accordance with our experience with the free particle solutions (32.21b). If we put $E = -mc^2 - E''$ we can separate the large and small components as before, and obtain an equation (A) which we shall not write down but which is similar to (32.24). The similarity in fact becomes striking if we take the complex conjugate of (A) and put $\phi_l = (\psi_4^*, -\psi_3^*)$, $\phi_s = (-\psi_2^*, \psi_1^*)$. Then (A) becomes

$$\left(\frac{-\hbar^2}{2m}\nabla^2 + e\phi + \frac{ei\hbar}{mc}\mathbf{A}\cdot\mathbf{grad} + \frac{ei\hbar}{2mc}\operatorname{div}\mathbf{A} + \frac{e^2A^2}{2mc^2}\right)\phi_l -$$

$$- \cdot \frac{e}{mc}\mathbf{s}\cdot\mathbf{H}\phi_l - \frac{ie}{mc}\mathbf{s}\cdot\mathscr{E}\phi_s - \frac{(E''-e\phi)}{2mc^2}\phi_s = E''\phi_l \tag{32.27}$$

This is identical with (32.24) except that the sign of the charge has been reversed! The orbital motion of an electron is described by the orbital functions ψ_α or ϕ_l, ϕ_s, and hence (32.27) shows that an electron in a negative energy state moves in an electromagnetic fields as if it had a positive charge but otherwise the ordinary mass and spin of an electron. Returning to our picture with all negative energy states except ψ filled, all the electrons in the filled state, move as if they had a positive charge, so that the hole corresponding to the unfilled state also moves as if it had a positive charge. In short, a positron behaves in every way as an electron with a positive charge e.

The relationship between electrons and positrons can also be demonstrated using the charge conjugation operator Γ where †

$$\Gamma\psi_\alpha = \psi_\alpha^*$$

$$\Gamma u_\beta = C_{\alpha\beta}u_\alpha \tag{32.28}$$

where
$$C = \begin{bmatrix} \cdot & \cdot & \cdot & 1 \\ \cdot & \cdot & -1 & \cdot \\ \cdot & -1 & \cdot & \cdot \\ 1 & \cdot & \cdot & \cdot \end{bmatrix}.$$

This transformation turns a negative energy state ψ into a positive energy one as is seen in two ways. Firstly from (32.20), taking the complex conjugate changes the sign of the energy, and secondly

† The operator Γ should not be confused with the matrices Γ_μ of (32.9).

the operator C turns a wave function with large u_+, u_- components into one with large v_+, v_- ones. Now the probability density

$$\psi_\alpha^*(\mathbf{r}, t)\, \psi_\alpha(\mathbf{r}, t) = (\Gamma\psi_\alpha)^*(\Gamma\psi_\alpha) \qquad (\alpha \text{ summed}) \tag{32.29}$$

is the same for both ψ and $\Gamma\psi$, so that they have the same dynamical properties which we shall determine by deriving the equation satisfied by the positive energy wave function $\Gamma\psi$. The negative energy state ψ satisfies the Dirac equation (32.7) for an electron, which in component form becomes

$$(\Gamma_{\mu,\alpha\beta} p_\mu - imc)\psi_\beta u_\alpha = 0.$$

Taking the complex conjugate we obtain†

$$(\Gamma^*_{\mu,\alpha\beta} p_\mu^* + imc)\psi_\alpha^* u_\beta^* = 0 \tag{32.30}$$

It follows from problem 31.4 that the spinors u_+^*, u_-^*, v_+^*, v_-^* transform under all Lorentz transformations like v_-, $-v_+$, $-u_-$, u_+, i.e. u_α^* transforms like Cu_α. We must regard u_α^* as still a function of the same intrinsic variables as u_α and hence as a function in the vector space (u_+, u_-, v_+, v_-). Then by Schur's lemma u_α^* is proportional to Cu_α (problem D.2). It can also be verified from (32.3), (32.9), (32.28) that $\Gamma^*_{\mu,\alpha\beta} p_\mu^* = -(C\Gamma_\mu C^{-1})_{\alpha\beta} p_\mu$. Hence from (32.30) we obtain

$$[(C\Gamma_\mu C^{-1})_{\alpha\beta} p_\mu - imc]\psi_\alpha^* C u_\beta = 0.$$

Here by definition (32.14), the operator represented by the matrix $C\Gamma_\mu C^{-1}$ with respect to the vectors Cu_β is just γ_μ. We therefore have

$$\left[\gamma_\mu\left(-i\hbar\frac{\partial}{\partial\mu}\right) - imc\right]\Gamma\psi = 0. \tag{32.31}$$

Thus $\Gamma\psi$ satisfies the same equation as ψ does. However, if we include the interaction with an electromagnetic field, (32.7) and (32.31) become respectively

$$\left[\gamma_\mu\left(-i\hbar\frac{\partial}{\partial\mu}\right) - imc + \frac{e}{c}\gamma_\mu A_\mu\right]\psi = 0, \tag{32.32}$$

$$\left[\gamma_\mu\left(-i\hbar\frac{\partial}{\partial\mu}\right) - imc - \frac{e}{c}\gamma_\mu A_\mu\right]\Gamma\psi = 0 \tag{32.33a}$$

† Since the u_α transform with complex coefficient matrices under real rotations and Lorentz transformations, they must be regarded as complex quantities.

A comparison of these equations now shows that $\Gamma\psi$ describes the motion of a particle which behaves *dynamically* in exactly the same way as an electron with positive energy but having opposite charge, and hence from (32.29) so does the negative energy electron state ψ. As before, we suppose ψ is unoccupied whereas all the other negative energy states are filled with electrons, which is observed as a net positive charge. We have now shown therefore that this system can be described from the point of view of the total *energy*, the net *charge* and the *dynamical properties* by saying that we have a positron with positive energy.

We can now see vaguely how it is possible in field theory to obtain the more satisfactory and tidy kind of description of positrons already outlined at the beginning of this subsection. We first note that by using the pair of equations (32.32) and (32.33) instead of just (32.32) alone, we can discard all negative energy solutions as physically meaningless, because the negative energy solutions of each equation have been redescribed as the positive energy solutions of the other. Furthermore if we regard charge conjugation as also changing the sign of the electromagnetic field or the charge

$$\boxed{\Gamma\mathbf{A} = -A \text{ or } \Gamma e = -e \text{ but not both,}}$$

then (32.32) and (32.33) become formally the same equation and we have Γ as a symmetry property of the equation. Since $\Gamma(\Gamma\psi) = \psi$, the theory is then completely symmetrical between electrons and positrons.

Transformation properties of physical quantities

All quantities of physical interest such as matrix elements or the charge density, involve expressions of the form†

$$\psi^*\delta\psi = \psi_\alpha{}^*\Delta_{\alpha\beta}\psi_\beta. \tag{32.34}$$

Here δ is some operator and Δ the matrix representing it with respect to the u_α. In general δ may also involve multiplication by a function of μ or differentiation, but this generalization can easily be made at the end.

We shall now discuss the transformation properties of (32.34) and start by examining what precisely is meant by transforming it. Let P be the point x, y, z, T in our system of co-ordinates and x', y', z', T' for another observer whom we shall refer to as "him".

† In writing ψ^*, we are here interpreting u^* in the sense of (19.9), and not in the sense of (19.10) as was done when discussing charge conjugation.

Let L denote a proper Lorentz transformation and also the corresponding matrix. Then from the fundamental definition of transforming a function, $Lu_\beta = L_{\alpha\beta}u_\alpha$ means

$$(\text{our } u_\beta) = L_{\alpha\beta} (\text{his } u_\alpha),$$

and
$$(\text{our } \psi_\alpha) = L_{\alpha\beta}{}^{-1} (\text{his } \psi_\beta).$$

Hence we have

$$(\text{our } \psi_\alpha{}^*)\, \Delta_{\alpha\beta} (\text{our } \psi_\beta) = (\text{his } \psi_\alpha{}^*)(\tilde{L}^{*-1}\Delta L^{-1})_{\alpha\beta} (\text{his } \psi_\beta). \tag{32.35}$$

Now $\tilde{L}^{*-1}\Delta L^{-1}$ does not appear to be equal to anything special, certainly not to the transformed matrix $L\Delta L^{-1}$ (32.12) because Lorentz transformations are in general not unitary. However, we can simplify (32.35) if we write δ in the form

$$\delta = \gamma_T\, \gamma_r \tag{32.36}$$

where γ_r is any combination of the γ operators, and use the fact that γ_T happens to have the special property

$$\tilde{L}^{*-1}\gamma_T = \gamma_T L \tag{32.37}$$

to be verified presently. Then we have

$$\tilde{L}^{*-1}\Delta L^{-1} = \Gamma_T L \Gamma_r L^{-1}.$$

Suppose now that γ_r transform according to some representation D;

$$(\text{his } \gamma_s) = L(\text{our } \gamma_s)L^{-1} = D_{rs} (\text{our } \gamma_r).$$

Using the relation

$$(\text{his } \gamma_r) (\text{his } u_\beta) = \Gamma_{r,\alpha\beta} (\text{his } u_\alpha),$$

we finally obtain from (32.35)

$$\begin{aligned} &(\text{our } \psi^*) (\text{our } \gamma_T\gamma_r) (\text{our } \psi) \\ &= D_{rs} (\text{his } \psi_\alpha{}^*) (\Gamma_T\Gamma_r)_{\alpha\beta} (\text{his } \psi_\beta) \\ &= D_{rs} (\text{his } \psi^*) (\text{his } \gamma_T\gamma_r) (\text{his } \psi), \end{aligned}$$

that is,

$\psi^*\gamma_T\gamma_r\psi$ transforms under proper Lorentz transformations in the same way as γ_r.	(32.38)

For example the charge density $\rho = -e\psi^*\psi$ is not invariant under Lorentz transformations as we might naïvely expect. (Lemma

C.2, appendix C, is inapplicable because a Lorentz transformation is not unitary.) However, if we write ρ as

$$\rho = -e\psi^*\psi = -e\psi^*\gamma_T\gamma_T\psi,$$

we see that it transforms like the fourth component of a four-vector, as a charge density in relativity should (Stratton 1941, p. 70). However the total charge

$$-e \int \psi^*\psi \, dv = -e \int \psi^*\gamma_T\gamma_T\psi \, dx \, dy \, dz$$

is invariant; because from (32.38) this transforms like $\gamma_T \, dx \, dy \, dz$, which transforms in the same way as $dx \, dy \, dz \, dT$, which is invariant as can easily be shown in the manner of problem 32.6. Incidentally because of the importance of property (32.38), $\psi^*\gamma_T$ is abbreviated by many authors to $\bar{\psi}$.

It now remains to verify (32.37). In the notation of (31.26), consider $L = R(\theta, z) = \exp(\frac{1}{2}i\theta M)$, where $M = i\Gamma_y\Gamma_x = \tilde{M} = M^*$. Hence $\tilde{L}^* = \exp(-\frac{1}{2}i\theta M) = L^{-1}$, and $\Gamma_T M = M\Gamma_T$. Hence

$$\tilde{L}^*\Gamma_T L = L^{-1}\Gamma_T L = L^{-1}L\Gamma_T = \Gamma_T.$$

Thus (32.37) is satisfied for $R(\theta, z)$, and similarly for $R(\theta, x)$. Consider now $L(v, z) = \exp(\frac{1}{2}i\theta K)$ (31.26), where $K = i\Gamma_T\Gamma_z = \tilde{K} = K^*$, $K\Gamma_T = -\Gamma_T K$, and $i\theta$ is real. We have

$$\tilde{L}^*\Gamma_T L = L\Gamma_T L = LL^{-1}\Gamma_T = \Gamma_T,$$

and (32.37) is again satisfied. For a product $L = L_1L_2$ of two such Lorentz transformations we have

$$(\tilde{L}_1\tilde{L}_2)^*\Gamma_T(L_1L_2) = \tilde{L}_2^*\tilde{L}_1^*\Gamma_T L_1L_2 = \Gamma_T.$$

Since any proper Lorentz transformation can be written in the form $R(\theta_1, z)R(\theta_2, x)L(v, z)$, it follows that (32.37) is satisfied for all proper Lorentz transformations.

Summary

We have sought relativistic wave functions transforming according to $D^{(\frac{1}{2}0+0\frac{1}{2})}$ under the complete Lorentz group, and have arrived at the Dirac equation. This equation is invariant under the complete Lorentz group (with the proviso of problem 32.9). The invariance property is sufficient to define the γ_α operators without arbitrariness. The particles described by the Dirac equation have an intrinsic spin angular momentum and a magnetic moment. The equation exhibits two types of solution which are related by the charge conjugation transformation and which correspond to electrons and positrons; more precisely the absence of an electron from a negative energy state behaves in every way like a positively charged particle which we identify as the positron. The quantities like $\psi^*\gamma_T\gamma_r\psi$ occurring in physical expectation values transform in the same way as the γ_r.

References

The relativistic invariance of the Dirac equation is shown nicely by Van der Waerden (1932). A comprehensive field theoretical treatment is contained in Jauch and Rohrlich (1955).

Problems

32.1 Derive equations (32.33) in detail.

32.2 Calculate the matrix L representing $L(v, z)$ in the representation $D^{(\frac{1}{2}0+0\frac{1}{2})}$. Is L a unitary matrix? Why does lemma 2 of appendix C not apply? In the manner of problem A.9, prove $\tilde{L}L = E$ and $\det|L| = 1$. (Dirac 1947, p. 259.)

32.3 Using (32.22) find the extra terms required to make (32.2a) equivalent to (32.7) and Lorentz invariant.

32.4 A quantum mechanical operator such as the Hamiltonian which corresponds to a physically observable quantity, in this case the energy, is Hermitian. This means that if it is represented by a matrix A, then $\tilde{A}^* = A$ in the notation of appendix A (see Schiff 1955, p. 129). Show from (32.9) that the γ_μ are Hermitian. Show also that $\gamma_T\gamma_x$ is Hermitian, and note the relevance of this to problem 32.5.

32.5 Derive the electric current density in the usual way (Schiff 1955, p. 23) from (32.7) and show that it has components

$$J_x = -iec(\psi^*\gamma_T\gamma_x\psi), \text{ etc.}$$

Note that the fourth component of this vector is correctly related to the charge density ρ by

$$\rho = J_T/ic = -e\psi^*\gamma_T\gamma_T\psi = -e\psi^*\psi.$$

32.6 If $L\gamma_e L^{-1} = D_{\nu\mu}\gamma_\nu$ where L is a proper Lorentz transformation, and if $\gamma_5 = \gamma_x\gamma_y\gamma_z\gamma_T$, show that $L\gamma_5L^{-1} = \det|D_{\nu\mu}|\gamma_5 = \gamma_5$; i.e. γ_5 remains invariant under proper Lorentz transformations.

Hence show that the quantities 1, γ_μ, $\gamma_{\mu\nu} = -\frac{1}{2}i(\gamma_\mu\gamma_\nu - \gamma_\nu\gamma_\mu)$, $\gamma_5\gamma_\mu$, γ_5 transform respectively as a scalar, vector, skew-symmetric second rank tensor, pseudo-vector and pseudo-scalar under proper Lorentz transformations and space-inversion.

32.7 Show, using problem 32.6, that the transformation properties of

$$\tfrac{1}{2}i\hbar\psi^*\gamma_T\gamma_5\gamma_j\psi \qquad (j = x, y, z)$$

under proper and improper Lorentz transformations are just those of an angular momentum vector. This quantity is in fact the spin angular momentum density. Verify this for the plane wave states (32.20), and prove it generally from the definition of angular momentum (appendix F).

32.8 Show that the current density and spin density of a positively charged particle in the state $\Gamma\psi$, is exactly the same as that given by the absence of an electron from the negative energy state ψ.

32.9* Prove that the electromagnetic vector potential A_μ transforms as a pseudo-vector of type 1 (Table 36, § 31), whereas $\partial/\partial\mu$ transforms as an ordinary vector. Hence, show that the Dirac equation with an electromagnetic field is still invariant under space-inversion Π, but no longer under the simple time-*inversion* ($\mathbf{r}$, $t \to \mathbf{r}$, $-t$) of equation (32.13). However, following Jauch and Rohrlich (1955, p. 88), show that there exists a more complicated transformation called time-*reversal* which leaves the Dirac equation invariant, and establish its analogy to the non-relativistic transformation of § 19.

32.10 Derive the irreducible representations of the 16 operators γ_r corresponding to the matrices (32.16). Hint: show that the 32 operators γ_r, $-\gamma_r$ form a group: derive its irreducible representations D, and strike out those not satisfying $D(-\gamma_r) = -D(\gamma_r)$.

32.11* Let γ_r be the 16 operators corresponding to the matrices (32.16). Write down their multiplication table and show that they do *not* form a group because of the occurrence of minus signs. They form what is called a ring. Discuss to what extent the theory of representations developed in § 6 and appendices C and D applies specifically to groups and to what extent also to rings. In this way justify our use of (14.14) and Schur's lemma in proving the uniqueness of the γ_μ in the text. References: Van der Waerden 1932, p. 54; Jauch and Rohrlich 1955, appendix A; Boerner 1955, Chapters I and II.

32.12 Discuss the invariance of Maxwell's equations under rotations, space-inversion and Lorentz transformations. Consider only the transformation of the components $\mathscr{E}_x$, $\mathscr{E}_y$, $\mathscr{E}_z$, etc., among one another and not their dependence on **r**, t. (See problem 11.9 and Stratton 1941, p. 70.)

32.13 In the field theoretical formalism, the Dirac equation is derived as a variational equation from the Lagrangian

$$\mathscr{L} = -\psi^*\gamma_T\left(\gamma_\mu\frac{\partial}{\partial\mu} + \frac{mc}{\hbar}\right)\psi - \frac{ie}{\hbar c}\psi^*\gamma_T\gamma_\mu A_\mu\psi.$$

Verify that $\mathscr{L}$ is invariant under proper Lorentz transformations, space-inversion, and charge conjugation, but not under the simple time-inversion τ (problem 32.9). Note: in the case of charge conjugation, use $(\partial\psi_\alpha^*/\partial\mu)\psi_\beta + \psi_\beta(\partial\psi_\alpha^*/\partial\mu) = 0$ (see Schiff 1955, equation (47.8)), and $\psi^*\gamma_T\gamma_\mu(\partial\psi/\partial\mu) + (\partial\psi^*/\partial\mu)\gamma_T\gamma_\mu\psi = 0$ (derivable from the Dirac equation).

33. Beta Decay

The electron-nuclear interaction

Let us consider a nucleus, and suppose that the energy levels and eigenstates have been determined in principle from the forces between the nucleons as in Chapter VII. Now in practice it is observed that nuclei do not stay in such states indefinitely, as we would expect if they were true eigenstates for the system. For instance it may happen that the 1s atomic electron, which has a finite probability of being at the nucleus, gets absorbed (like a photon) while the nucleus makes a transition to a new state (Bethe and Morrison 1956, p. 230). Thus there must be some interaction term in our equations which couples the nuclear field with the electron field. In the previous section, it was shown how one can only obtain an adequate description of states containing several particles or a varying number of particles by using a field theory. These remarks apply particularly to the process considered above: we start off with a state containing Z electrons which then changes continuously into a state containing $Z - 1$ electrons. However, we shall not go into the intricacies of such a field theoretical calculation. For present purposes it is sufficient to note that the interaction between the electron and nucleon fields can be represented in our formalism by an interaction term $\mathscr{H}_{\text{int}}$ in the nuclear Hamiltonian, and that this interaction term gives rise in the usual manner of second order perturbation theory to a transition probability per unit time of (Schiff 1955, p. 199)

$$\frac{2\pi}{\hbar}\rho(E)\left|\int \psi^*_{\text{final}}\mathscr{H}_{\text{int}}\psi_{\text{initial}}\,dv\right|^2. \tag{33.1}$$

Here the ψ's are the initial and final wave functions of the whole nuclear-electron system and $\rho(E)$ the density of final states per unit energy.

In this section we shall not consider the electron absorption process mentioned above, but only the usual beta decay process in which an electron is emitted by the nucleus. We therefore write $\psi_{\text{final}} = \Psi_{\text{f}}\psi_{\text{e}}$ where Ψ_{f} is the final wave function for the nucleus and ψ_{e} a plane-wave wave function for the electron going off in some direction. It is known experimentally that to conserve spin, energy and momentum, the process must involve another particle, the neutrino, with zero mass, zero charge and spin $\frac{1}{2}$ which is difficult to observe directly (Bethe and Morrison 1956, p. 25). One can either consider a neutrino to be emitted in the process or an antineutrino to be absorbed. We shall adopt the latter description,

$$\text{neutron} + \text{antineutrino} \rightarrow \text{proton} + \text{electron}, \tag{33.2}$$

and therefore write the initial wave function as $\Psi_{\text{i}}\psi_\nu$, where Ψ_{i} is the initial nuclear wave function and ψ_ν is a plane-wave antineutrino wave function.

Now the form of $\mathscr{H}_{\text{int}}$ is not really known. For the sake of simplicity it is usually assumed that it does not depend on the total charge (like an electromagnetic force), or on the momentum and angular momentum of the nucleons (like a spin-orbit coupling). If, further, $\mathscr{H}_{\text{int}}$ is to be relativistically invariant, it can be shown that $\mathscr{H}_{\text{int}}$ must have the following form (problem 33.2).

$$\int \Psi_{\text{f}}^*\psi_{\text{e}}^*\mathscr{H}_{\text{int}}\psi_\nu\Psi_{\text{i}}\,dv = \int \Psi_{\text{f}}^*\,[\sum_n \mathscr{H}(\mathbf{r}_n)\tau_{+n}]\Psi_{\text{i}}\,dv \tag{33.3}$$

$$\begin{aligned}\mathscr{H}(\mathbf{r}) = {} & \gamma_T(C_{\text{S}}\psi_{\text{e}}^*\gamma_T\psi_\nu + C'_{\text{S}}\psi_{\text{e}}^*\gamma_T\gamma_5\psi_\nu) + \\ & + \gamma_T\gamma_\mu(C_{\text{V}}\psi_{\text{e}}^*\gamma_T\gamma_\mu\psi_\nu + C'_{\text{V}}\psi_{\text{e}}^*\gamma_T\gamma_\mu\gamma_5\psi_\nu) + \\ & + \tfrac{1}{2}\gamma_T\gamma_{\lambda\mu}(C_{\text{T}}\psi_{\text{e}}^*\gamma_T\gamma_{\lambda\mu}\psi_\nu + C'_{\text{T}}\psi_{\text{e}}^*\gamma_T\gamma_{\lambda\mu}\gamma_5\psi_\nu) + \\ & + \gamma_T\gamma_\mu\gamma_5(C_{\text{A}}\psi_{\text{e}}^*\gamma_T\gamma_\mu\gamma_5\psi_\nu + C'_{\text{A}}\psi_{\text{e}}^*\gamma_T\gamma_\mu\psi_\nu) + \\ & + \gamma_T\gamma_5(C_{\text{P}}\psi_{\text{e}}^*\gamma_T\gamma_5\psi_\nu + C'_{\text{P}}\psi_{\text{e}}^*\gamma_T\psi_\nu). \\ & \qquad (\lambda,\ \mu \text{ summed})\end{aligned} \tag{33.4}$$

Here the γ_μ are the operators (32.8), also $\gamma_5 = \gamma_x\gamma_y\gamma_z\gamma_T$, $\gamma_{\lambda\mu} = -\frac{1}{2}i(\gamma_\lambda\gamma_\mu - \gamma_\mu\gamma_\lambda)$. The operators outside the brackets operate on the nth nucleon wave function, and the C_{S}, etc., are arbitrary constants. The operators inside the brackets operate on the electron and neutrino wave functions, i.e. in component form with the summation convention

$$\psi_e{}^*\gamma\psi_\nu = \psi_{e,\alpha}{}^*(r)\Gamma_{\alpha\beta}\psi_{\nu,\beta}(r).$$

Also in (33.3) τ_{+n} is the isotopic spin operator of chapter VII. If the nth nucleon in the initial state is already a proton, it cannot beta decay and τ_{+n} gives zero. If it is a neutron, it can decay into a proton, and the τ_{+n} turns Ψ_i into a wave function with the same total charge and M_T as Ψ_f. If we wanted to describe positron emission and electron capture we would have to include in (33.3) also $\gamma_T C_S{}^*\psi_\nu{}^*\gamma_T\psi_e\tau_- + \ldots$.

Relativistic invariance

The relativistic invariance of the interaction (33.3) follows immediately from the work of the preceding section. Equation (32.38) shows that $\psi_1{}^*\gamma_T\gamma_r\psi_2$ transforms in the same way as γ_r under Lorentz transformations, where γ_r is any combination of the γ_μ operators, e.g. $\gamma_\lambda\gamma_\mu\gamma_5$. Thus, considering in (33.3) the term in C_V, $\psi_e{}^*\gamma_T\gamma_\mu\psi_\nu$ transforms like a four-vector and $\Psi_f{}^*\gamma_T\gamma_\mu\tau_+\Psi_i$ likewise transforms like a four-vector, so that the summation over μ gives us the invariant scalar product. Similarly, all the other terms in (33.3) are invariant under proper Lorentz transformations, each being the scalar product of two factors which transform in the same way. The suffices S, V, T, A, P stand for scalar, vector, tensor, axial vector (pseudo-vector), and pseudo-scalar, respectively, because of the transformation properties of the respective factors (problem 32.6). However, from (32.13), γ_5 changes sign under space-inversion since γ_x, γ_y, γ_z each do but γ_T does not, so that the two terms in each bracket of (33.3) have opposite parities, and the whole interaction is not invariant under space-inversion. The consequences of this are discussed below.

Allowed transitions: selection rules

As a first approximation the size of a nucleus is small compared with the wavelengths of the electron and neutrino, so that we may put $\psi_e(\mathbf{r}) = \text{const} = \psi_e(0)$, $\psi_\nu(\mathbf{r}) = \psi_\nu(0)$ over the nucleus in (33.4). This gives the probability for an *allowed* transition. If it is zero, the actual transition probability may still be non-zero though small through the variation of $\psi_e(\mathbf{r})$ over the nucleus, such transitions being termed *forbidden*. We shall return to these later and now discuss the nuclear part of (33.3) for an allowed transition. Consider the axial vector term. We have for the operators operating on the nuclear spin functions

$$\gamma_T\gamma_j\gamma_5 = i\begin{bmatrix}\sigma_j & 0\\ 0 & \sigma_j\end{bmatrix}, \ (j = x, y, z), \ \gamma_T\gamma_T\gamma_5 = \begin{bmatrix}0 & 1\\ 1 & 0\end{bmatrix} \tag{33.5}$$

in the abbreviated notation of (32.9b). Now protons and neutrons are spin $\frac{1}{2}$ particles, and from the previous section we would expect the one-particle wave functions to satisfy a Dirac equation. While this cannot be precisely right because the particles do not have the magnetic moment deduced from the equation, we shall assume that for our purposes it is sufficient to assume that the Dirac equation applies. Then the first two components of the wave function are large and the last two smaller by a factor $v(\text{nucleon})/c$. The operators $\gamma_T\gamma_j\gamma_5$ connect large with large components, and if they give a non-zero matrix element we have an *ordinary transition*. On the other hand $\gamma_T\gamma_T\gamma_5$ connects large with small components, and if it is the only term giving a non-zero probability, we have a *relativistic transition* with probability reduced by about $(v/c)^2 \approx 10^{-3}$. Similarly all transitions can be classified as ordinary or relativistic.

Let us now return to the operators $\gamma_T\gamma_j\gamma_5$ (33.5). For calculating the transition probability we may neglect the small components of the nuclear wave functions completely and write the operators simply as $i\sigma_j = (2i/\hbar)s_j$. They transform according to $D^{(1)}$ under rotations, whence in the manner of (13.12) we have the selection rule for the nuclear angular momentum quantum number:

$$\boxed{J \to J, J \pm 1 \text{ except } 0 \to 0, \ w_i = w_f.} \tag{33.6}$$

The selection rule for the parity w follows from the fact that $\gamma_T\gamma_j\gamma_5$ is invariant under the space inversion Π (γ_T is invariant, γ_j and γ_5 change sign). Similarly, the tensor interaction of (33.4) leads to ordinary transitions with the same selection rule (33.6), which is called the *Gamow–Teller* rule. However the operator for the scalar interaction has the form

$$\gamma_T = \begin{bmatrix}1 & 0\\ 0 & -1\end{bmatrix}$$

which is invariant under rotations, and hence has the *Fermi selection rule*

$$\boxed{J \to J, \quad w_i = w_f.} \tag{33.7}$$

The vector interaction has the same selection rule. The selection rules for all the interactions in allowed transitions are summarized in Table 38, the rules for *relativistic transitions* being

$$J \to J, J \pm 1 \text{ except } 0 \to 0, w_i = -w_f, \tag{33.8}$$

or

$$J \to J, \quad w_i = -w_f. \tag{33.9}$$

Note the difference between the parity rules for ordinary and relativistic transitions.

TABLE 38

Selection Rules for Allowed Transitions

Interaction	Selection rules Ordinary transition	 Relativistic transition
Scalar	equ. (33·7)	none
Vector	(33.7)	equ. (33·8)
Tensor	(33.6)	(33.8)
Axial vector	(33.6)	(33·9)
Pseudoscalar	none	(33.9)

Favoured and unfavoured transitions

So far we have classified transitions as allowed or "forbidden", and as ordinary or relativistic. We shall now subdivide each category further into favoured and unfavoured. Consider the allowed ordinary transitions, and let us suppose the Ψ_i, Ψ_f are expressed in terms of the shell model as linear combinations of determinants of one-nucleon wave functions. Now τ_{+n} operating on such a determinant gives identically zero if the nth nucleon belongs to a doubly-filled shell, i.e. a shell filled for both protons and neutrons, because there is no vacant proton state to go into. Thus doubly-filled shells do not contribute to beta decay, and for this reason transition probabilities do not increase rapidly with atomic number.

Using determinant wave functions, we may further reduce (33.3) to a sum of simple proton-neutron matrix elements

$$\int \psi_P{}^* \mathscr{H} \tau_+ \psi_N \, dv. \tag{33.10}$$

If we neglect the small components of the nuclear wave functions, $\mathscr{H}$ becomes a linear combination of 1, σ_x, σ_y and σ_z (problem 11.1). Thus $\mathscr{H}$ changes at most the spin of the particle, and (33.10) is zero unless ψ_P and ψ_N have the same n and l quantum numbers. For instance a $2p_{3/2}$ state can decay into a $2p_{1/2}$ or $2p_{3/2}$ state but not into $2s_{1/2}$. Thus Ψ_i and Ψ_f must belong to the same set of n, l quantum numbers for a transition to be *favoured*. Otherwise (33.3) is non-zero only because of configurational mixing, which reduces the transition probability by a factor of about 0.02 to 0.05 (Blatt and Weisskopf 1952, p. 721). An example of a favoured transition is

$$\text{He}^6\ (J = 0) \rightarrow \text{Li}^6\ (J = 1)$$

the configuration in both nuclei being $(1s_{1/2})^4(2p_{3/2})^2$. A useful measure of the size of the nuclear matrix element squared is the quantity $1/(ft)$ where t is the half-life and f a correction factor that takes into account the energy of the electron-neutrino pair. Favoured transitions have $\log_{10}ft \approx 3{\cdot}1 \pm 0{\cdot}3$ (Table 39). Now most nuclei

TABLE 39

Some Beta Decays

Initial nucleus	J_i	Final nucleus	J_f	$\log_{10}ft$
neutron	$\frac{1}{2}$	H^1	$\frac{1}{2}$	3·2
H^3	$\frac{1}{2}$	He^3	$\frac{1}{2}$	3·1
He^6	0	Li^6	1	2·8
C^{10}	0	B^{10**} (positron emission)	0	3·8
about 25% of all known decays				5·0 ± 0·5
Be^{10}	0	B^{10}	3	13·7

contain rather more neutrons than protons, so that when one neutron decays into a proton, the latter for energetic reasons would be expected to lie in a lower subshell, i.e. Ψ_i and Ψ_f belong to different configurations. We may conclude therefore that, except in light nuclei, most transitions will be unfavoured. Indeed the most common value of $\log_{10}ft$ is around 5, corresponding to allowed ordinary but unfavoured transitions.

In the case of the scalar and vector interactions, $\mathscr{H}$ reduces to a constant for allowed ordinary transitions, and the conditions for a favoured transition become even more stringent. (33.3) becomes

$$\int \Psi_f^* \sum \tau_{+n}\Psi_i(\alpha, T, M_T)\, dv = \int \Psi_f^* T_+ \Psi_i(\alpha, T, M_T)\, dv$$
$$= [T(T+1) - M_T(M_T+1)]^{1/2}\int \Psi_f^*\Psi_i(\alpha, T, M_T+1)\, dv,$$

where α stands for all other quantum numbers required to designate the initial state completely. This matrix element is zero, therefore unless Ψ_f belongs to the same T-multiplet as Ψ_i (§ 29). This situation is rare for the following reason. Neglecting the Coulomb forces and the proton-neutron mass difference, the initial and final states have the same energies if the nuclear forces are charge-independent. But when we include the effect of the Coulomb repulsion and the mass difference, the initial nucleus with lower Z has the lower energy, so that we would not ordinarily expect beta decay to be energetically possible. An exception is $H^3 \rightarrow He^3$ in a $T = \frac{1}{2}$, $J = \frac{1}{2}$ T-multiplet (Table 39). Instead the Coulomb forces favour positron emission, and an example being the decay $C^{10} \rightarrow B^{10**}$ (second excited state) in the T-multiplet $T = 1$, $J = 0$ shown in Fig. 44. Incidentally this $J = 0 \rightarrow J = 0$ transition shows that the Hamiltonian (33.4) must contain some scalar or vector term from (33.7), whereas the $\Delta J = 1$ transition $He^3 \rightarrow Li^3$ (Table 39) show that it must also contain a tensor or axial-vector term.

The present subdivision into favoured and unfavoured transitions can also be applied to relativistic and forbidden transitions, though in the latter case it becomes much more complicated and loses its usefulness.

Parity non-conservation

So far in the selection rules (33.6) to (33.9) we have only considered parity as far as it affects the nuclear matrix element. We shall now consider the effect of the space inversion Π on the whole interaction (33.4). Since the terms with unprimed coefficients contain each γ operator twice, once operating on the nuclear components and once on the electron-neutrino ones, they all remain invariant under Π. The terms with primed coefficients contain an extra factor γ_5, so that they change sign. Thus *the transition probability would be unaffected by Π if either all primed or all unprimed coefficients were zero*, because the probability depends on the square of the matrix element. However, the transition probability is not invariant under Π if (33.4) contains both non-zero primed and unprimed terms (Lee and Yang 1956). We shall now illustrate this asymmetry by a particular example.

A nucleus undergoes a decay with spin change $J \rightarrow J - 1$. By the use of magnetic fields and low temperatures, the nuclei have all been oriented with their magnetic moments along the z-axis, i.e. they are in the $M_J = J$ state, so that the transition must involve the change $M_J = J \rightarrow M_J = J - 1$. From the selection rules (33.6), (33.7), only the tensor and axial vector interactions can take part in this transition, and for the sake of simplicity we shall assume all coefficients in (33.4) to be zero except C_A, C'_A.

The axial vector interaction can be rewritten in terms of $\gamma = \gamma_x \pm i\gamma_y$ instead of γ_x, γ_y, and because of the change in M_J the only terms giving non-zero contributions to the matrix element (33.3) are

$$\gamma_T\gamma_-\gamma_5(C_A\psi_e{}^*\gamma_T\gamma_+\gamma_5\psi_\nu + C'_A\psi_e{}^*\gamma_T\gamma_+\gamma_\nu\psi_\nu). \tag{33.11}$$

We shall only calculate some probabilities for the cases when both electron and neutrino are travelling parallel to one another in the positive or negative z-direction (upper and lower signs below). We only consider allowed transitions, and from (32.21) the components of $\psi_e(\mathbf{r})$, $\psi_\nu(\mathbf{r})$ for electron spin parallel and neutrino spin antiparallel to the z-axis are

$$\psi_e(0)\text{: } (1, 0, \mp B, 0) \qquad \psi_\nu(0)\text{: } (0, \mp 1, 0, 1) \tag{33.12}$$

where $B = cp_e/(E_e + mc^2)$.

The neutrino function follows from (32.21) by putting $m = 0$ and noting that, in our formalism, ψ_ν denotes an antineutrino wave function (negative energy solution; see (33.2)). Substituting (33.12) into (33.11) we obtain for the bracketed quantity in (33.11) the values

$$-iC_A(1 + B) + iC'_A(1 + B), \qquad (\uparrow\uparrow;\uparrow\downarrow), \tag{33.13a}$$

$$iC_A(1 + B) + iC'_A(1 + B), \qquad (\downarrow\uparrow;\downarrow\downarrow). \tag{33.13b}$$

The set of arrows denote the directions of the electron and neutrino momenta and spins with respect to the z-axis as follows ($\mathbf{p}_e$, $\mathbf{s}_e$; $\mathbf{p}_\nu$, $\mathbf{s}_\nu$). Since the nucleus loses one unit of spin angular momentum in the process, (33.2) shows that all other arrangements of electron and neutrino spin give zero probability: in particular we have the value

$$\text{zero} \qquad \text{for } (\uparrow\uparrow;\downarrow\uparrow). \tag{33.13c}$$

In all these transitions the value of the nuclear matrix element is the same. Thus from (33.13) we obtain transition probabilities proportional to

$$(1 + B)^2(|C_A|^2 + |C'_A|^2 - 2|C_A||C'_A|), \qquad (\uparrow\uparrow;\uparrow\downarrow),$$

$$(1 + B)^2(|C_A|^2 + |C'_A| \ + 2|C_A||C'_A|), \qquad (\downarrow\uparrow;\downarrow\downarrow).$$

I.e. if neither C_A nor C'_A is zero, parity non-conversion in the interaction leads to different transition probabilities for the electron and neutrino travelling together along the positive and negative z-axes. Note that in (33.13) we are comparing states in which the spin directions are the same, because spin is invariant under Π (appendix F). However, even with a parity conserving interaction we would not expect equality between (33.13a) and (33.13c), for here we have reversed all momentum and spin directions. This corresponds to a 180° rotation about the x-axis, and since we are observing the decay of an oriented nucleus, the whole system does not have this rotational symmetry. Experimentally one usually measures only the rate at which electrons are emitted without measuring their spin, and this too is not equal in the positive and negative z-directions, as can easily be shown by integrating the transition probability over all orientations for the neutrino and summing over all spin combinations. We have therefore shown by direct calculation that the non-conservation of parity in the interaction (33.4) leads to an asymmetry in the observed angular distribution of the electron, an asymmetry that is absent for a parity conserving interaction.

Tests of parity conservation

In practice, it has only recently been discovered that parity is not conserved in the interaction, because previously no experiment had been done in which parity non-conservation could produce an asymmetry (Lee and Yang 1956). For example, in the above calculation, we did assume that we had oriented nuclei initially, and this is not easy to achieve experimentally. It is therefore important to be able to decide easily which experiments will show up an asymmetry due to parity non-conservation, and also which will give a crucial test of whether other symmetry properties like time-reversal or charge-conjugation apply.

Let us consider the decay of a single nucleus. The quantities that can readily be measured are the initial direction of the angular momentum $\mathbf{J}$, the momentum $\mathbf{p}_e$ and spin (polarization) $\mathbf{s}$ of the electron, and the momentum $\mathbf{p}_\nu$ of the neutrino. The last is obtained from the recoil momentum $\mathbf{p}_N$ of the nucleus by the conservation of momentum $\mathbf{p}_\nu = -\mathbf{p}_N - \mathbf{p}_e$. The parities are

$$\boxed{\mathbf{J} \text{ and } \mathbf{s} - \text{even}; \qquad \mathbf{p}_e \text{ and } \mathbf{p}_\nu - \text{odd}} \tag{33.14}$$

It may also be possible to obtain information about the final angular momentum of the nucleus from angular correlation measurements on a γ-ray emitted after the β-decay, but we shall not consider that here. Let us denote the quantum mechanical operators corresponding to these and any other variables by P_i, P_f with $i, f = 1, 2, 3 \ldots$, and denote their numerical values in a particular experiment by p_i, p_f. Here suffices i and f refer to measurements made before and after the decay. Furthermore let the initial state ψ_{initial}

be changed into ψ_{final} by the interaction $\mathscr{H}_{\text{int}}$. Now by making the measurements p_i on the system, we pick out from ψ_{initial} the component

$$\delta(P_i - p_i)\psi_{\text{initial}}$$

which is an eigenfunction of P_i with eigenvalue p_i. Here δ is the Dirac delta function, or rather a product of delta functions, one for each measured quantity. This state then decays into

$$\mathscr{H}_{\text{int}}\delta(P_i - p_i)\psi_{\text{initial}},$$

and by making the measurements p_f we pick out the component

$$\delta(P_f - p_f)\mathscr{H}_{\text{int}}\delta(P_i - P_i)\psi_{\text{initial}},$$

which has the amplitude

$$M = \int \psi^*{}_{\text{final}}\delta(P_f - p_f)\mathscr{H}_{\text{int}}\delta(P_i - p_i)\psi_{\text{initial}}\,dv \tag{33.15a}$$

in the final wave function. The probability of obtaining the set of measurements p_i, p_f is therefore proportional to $|M|^2$;

$$\text{Prob}(p_1, p_2, p_3, \ldots) \propto |M|^2. \tag{33.15b}$$

Let us further apply the inversion transformation Π under the integral sign in (33.15a). This does not change the value of the total integral (lemma 2, appendix C).

$$M = \int \Pi\psi^*_{\text{final}}\delta(\Pi P_f - p_f)(\Pi\mathscr{H}_{\text{int}})\delta(\Pi P_i - p_i)\Pi\psi_{\text{initial}}\,dv. \tag{33.16}$$

Note that here p_i, p_f are just constant parameters in the integral which remain untransformed. For the delta functions we have $\delta(P) = \delta(-P)$ and hence

$$\delta(\Pi P_r - p_r) = \delta(P_r - \Pi p_r)$$

whether $\Pi P_r = P_r$ or $-P_r$. Although the p_r are just numerical parameters, they do refer to physical quantities P_r and we have written Πp_r to denote $\pm p_r$ according to whether $\Pi P_r = \pm P_r$.

If we now suppose that $\mathscr{H}_{\text{int}}$ is a parity conserving interaction, $\Pi\mathscr{H}_{\text{int}} = \pm\mathscr{H}_{\text{int}}$, then ψ_{initial} and ψ_{final} have definite parities and (33.16) becomes

$$M = \int \psi^*_{\text{final}}\,\delta(P_f - p_f)\mathscr{H}_{\text{int}}\,\delta(P_i - p_i)\psi_{\text{initial}}\,dv.$$

Comparison with (33.15) gives for the probabilities

$$\boxed{\text{Prob}(p_1, p_2, p_3, \ldots) = \text{Prob}(\Pi p_1, \Pi p_2, \Pi p_3, \ldots).} \tag{33.17}$$

Physically this just means that we get the same results from two apparatuses which are inversion images of one another, which is just what one would expect from a parity conserving interaction. Returning now to the notation of (33.14), we can expand the probability (33.17) in terms of $\mathbf{J}$, $\mathbf{p}_e$, $\mathbf{s}$, $\mathbf{p}_\nu$ and this gives powers and combinations of the products

$$\mathbf{p}_e \cdot \mathbf{p}_\nu, \qquad \mathbf{J} \cdot \mathbf{s}, \qquad \mathbf{J} \cdot \mathbf{p}_e \wedge \mathbf{p}_\nu, \qquad \text{etc.} \tag{33.18a}$$

$$\mathbf{J} \cdot \mathbf{p}_e, \qquad \mathbf{J} \cdot \mathbf{p}_\nu, \qquad \mathbf{p}_e \cdot \mathbf{s}, \qquad \mathbf{J} \cdot \mathbf{p}_e \wedge \mathbf{s}, \qquad \text{etc.} \tag{33.18b}$$

From (33.14) and (33.17), if $\mathscr{H}_{\text{int}}$ is parity conserving, then only the even parity combinations (33.18a) can occur in the transition probability. Thus to demonstrate parity non-conservation, it is necessary to measure a combination such as $\mathbf{J}$ and $\mathbf{p}_e$ occurring in (33.18b), and to show that the probability distribution contains an odd power of $\cos\theta = \mathbf{J} \cdot \mathbf{p}_e/|J|\,|p_e|$, i.e. that $P(\theta) \neq P(\pi - \theta)$. On the other hand measurements of $\mathbf{p}_e$ and $\mathbf{p}_\nu$ alone can never show up an asymmetry due to parity non-conservation, since we cannot make any odd-parity combination out of these two vectors.

Forbidden transitions

If in (33.4) we put $\psi_e(\mathbf{r}) = \psi_e(0)$, $\psi_\nu(\mathbf{r}) = \psi_\nu(0)$, we obtain zero probability for many transitions. These are said to be *forbidden*, but because of the crudeness of our approximation they really have a small non-zero probability in general. Taking again plane-wave wave functions, we have for the electron-neutrino parts of (33.4)

$$\begin{aligned} \psi_e^*(\mathbf{r})\gamma\psi_\nu(\mathbf{r}) &= \psi_e^*(0)\gamma\psi_\nu(0)\ \exp -i(\mathbf{k}_e - \mathbf{k}_\nu) \cdot \mathbf{r} \\ &= [\psi_e^*(0)\gamma\psi_\nu(0)] \sum_{l=0}^{\infty} \sum_{m=-l}^{l} \frac{i^{-l}}{(2l+1)!!} Y_{lm}(\Theta, \Phi)[(r/\lambda\!\!\!^-)^l Y_{lm}^*(\theta, \phi)] \end{aligned} \tag{33.19}$$

where $1/\lambda\!\!\!^- = |\mathbf{k}_e - \mathbf{k}_\nu|$, and Θ, Φ give the direction of $\mathbf{k}_e - \mathbf{k}_\nu$. The term in the last square bracket enters into the nuclear matrix element. The $Y_{lm}^*(\theta, \phi)$ transform according to $D^{(l)}$ under spatial rotations and have parity $(-1)^l$. Thus the selection rules become

$$|J_i - J_f| \leqslant l \leqslant J_i + J_f, \text{ (Fermi)}$$

or $$|l - 1| \leqslant j \leqslant l + 1 \text{ with } |J_i - J_f| \leqslant j \leqslant J_i + J_f, \quad \text{(Gamow–Teller)} \qquad (33.20)$$

and $$w_i = (-1)^l \, w_f \text{ (ordinary transitions)}$$

$$w_i = (-1)^{l+1} \, w_f \text{ (relativistic transitions).}$$

From (33.19) introducing the factor $(r/\lambda\!\!\!^{-})^l$ into the nuclear matrix element reduces it considerably because $R/\lambda\!\!\!^{-} \ll 1$ where R is the radius of the nucleus. Consider for instance the decay

$$Be^{10}(J = 0, T = 1, w = +1) \rightarrow B^{10}(J = 3, T = 0, w = +1)$$

among the levels of Fig. 44. From (33.20) the matrix element with lowest l is for an ordinary transition of the Gamov–Teller type with $l = 2$ given by the tensor and axial vector interactions. Since the initial and final configurations are $(1s_{1/2})^4(2p_{3/2})^6$, the terms

$$\int \psi_P^* \sigma_j \tau_+ r^2 Y_{2m}(\theta, \phi) \psi_N \, dv$$

in the nuclear matrix element are non-zero without invoking configurational mixing, so that the transition can be regarded as favoured. Thus from (33.19) the matrix element is reduced by $(R/\lambda\!\!\!^{-})^l/(2l + 1)!!$ compared with an ordinary favoured allowed $(l = 0)$ transition. With $R/\lambda\!\!\!^{-} \approx 1.0 \times 10^{-2}$ and $l = 2$, this gives

$$\log_{10} ft \approx 3{\cdot}1 + 2 \log_{10}(15 \times 10^4) = 13{\cdot}5.$$

The closeness of the agreement with the experimental value 13·7 (Table 39) must be regarded as somewhat fortuitous.

References

For further details about beta decay the reader is referred to any text on nuclear physics, e.g. Bethe and Morrison (1956), Sachs (1953), Blatt and Weisskopf (1952). Siegbahn (1955) is an exhaustive handbook covering the situation prior to the discovery of parity non-conservation. The consequences of parity non-conservation are discussed by Lee and Yang (1956) and many papers since then.

Summary

Beta decays have been classified into allowed or forbidden, ordinary or relativistic, and favoured or unfavoured. Selection rules have been calculated and transition probabilities estimated accordingly. The asymmetry in the motion of the emitted electron due to parity non-conservation of the interaction has been illustrated.

Problems

33.1 Recapitulate problem 32.6 of the previous section.

33.2 Verify in detail that the β-decay interaction (33.4) is a scalar under *proper* Lorentz transformations. Also prove that any Lorentz invariant scalar interaction not involving derivatives must have that general form (33.4) if we always couple protons with neutrons, electrons with neutrinos. Hint: any 4×4 matrix can be written as a linear combination of the 16 products of Γ_μ matrices (§ 32) which form a reducible representation of the Lorentz group.

33.3 Prove in detail the selection rules for ordinary relativistic transitions (33.8), (33.9) and for forbidden transitions (33.20).

33.4 Classify the following transitions as allowed or "forbidden" of order l, ordinary or relativistic, favoured or unfavoured: $Be^7 \to Li^7$, $Si^{31} \to P^{31}$, $N^{17} \to O^{17}$, $Mg^{27} \to Al^{27}$. Use the shell model (§ 29) to determine the initial and final J and parity. Note that in an odd A nucleus, the nucleons almost always pair themselves so that the total J and parity is that of the last unpaired nucleon. It may be assumed that this rule applies to all the above nuclei.

33.5* A given atom can decay either by the nucleus capturing a 1s electron or by emitting a positron. Outline how you would calculate the ratio of the two probabilities, and discuss whether it will depend on the relative magnitude of the C's in (33.4) (Blatt and Weisskopf 1952, p. 684; Bethe and Morrison 1956, p. 230).

33.6 With the notation of § 31,

$$\psi = \psi_1 u_{+0} + \psi_2 u_{-0} + \psi_3 u_{0+} + \psi_4 u_{0-}$$

is a spin $\frac{1}{2}$ wave function. Show that there is nothing inconsistent in having $\psi_3 = c\psi_1^*$, $\psi_4 = c\psi_2^*$, $c = \pm 1$ as regards transformation properties under proper Lorentz transformations and under space inversion. Express such a ψ in terms of u_+, u_-, v_+, v_- (31.23), and verify from (32.23), with charge put equal to zero, that the Dirac equation can have solutions of this type. Wave functions of this kind can be used to describe neutrinos in a "two-component" theory (cf. Case 1957).

33.7 Complete the list of (33.18) up to all triple products and write down which ones are not invariant under time-reversal. Hence discuss possible experiments to test whether or not the beta-decay interaction is invariant under time-reversal symmetry (Jackson *et al.* 1957).

34. Positronium

A positron and an electron can form a bound system similar to a hydrogen atom, known as positronium. In this section we shall discuss the symmetries describing its different energy levels, and whether the system in a given state annihilates into two photons or into three. A thorough analysis of the interaction between electrons, positrons and electromagnetic radiation belongs to the realm of field theory (Jauch and Rohrlich 1955, p. 274), but it is possible to point out all the (known) symmetries in the non-relativistic limit without such an amount of elaborate apparatus. Only an occasional reference to the work of §§ 31 and 32 is necessary.

Symmetry transformations

Let us consider two charges e_1 and e_2 each of which for the present may be an electron or a positron. The Hamiltonian for their interaction is

$$\mathscr{H} = -\frac{\hbar^2}{2m}\nabla_1{}^2 - \frac{\hbar^2}{2m}\nabla_2{}^2 + \frac{e_1 e_2}{r_{12}} + \mathscr{H}_{\text{spin}}, \tag{34.1}$$

where the spin-dependent interaction is

$$\begin{aligned}\mathscr{H}_{\text{spin}} = &-\frac{e_1 e_2}{mc_2}\left[\frac{(\mathbf{r}_1-\mathbf{r}_2)\wedge(\mathbf{v}_1-\mathbf{v}_2)}{r_{12}{}^3} - \frac{(\mathbf{r}_1-\mathbf{r}_2)\wedge\mathbf{v}_1}{2r_{12}{}^3}\right]\cdot\mathbf{s}_1 - \\ &-\frac{e_1 e_2}{mc^2}\left[\frac{(\mathbf{r}_2-\mathbf{r}_1)\wedge(\mathbf{v}_2-\mathbf{v}_1)}{r_{12}{}^3} - \frac{(\mathbf{r}_2-\mathbf{r}_1)\wedge\mathbf{v}_2}{2r_{12}{}^3}\right]\cdot\mathbf{s}_2 + \\ &+\frac{e_1 e_2}{m^2c^2}\left[\frac{\mathbf{s}_1\cdot\mathbf{s}_2}{r_{12}{}^3} - \frac{3\mathbf{s}_1\cdot(\mathbf{r}_2-\mathbf{r}_1)\mathbf{s}_2\cdot(\mathbf{r}_2-\mathbf{r}_1)}{r_{12}{}^5}\right].\end{aligned} \tag{34.2}$$

This may be derived in the same way as the spin-orbit coupling in § 11 (Heisenberg 1926).

The Hamiltonian (34.1), and indeed any Hamiltonian describing only electrons, positrons and radiation, is invariant under space inversion, time-reversal, charge conjugation and permutations. Time-reversal will not interest us in this section. The *charge conjugation* operation Γ replaces each electron by a positron and vice versa, i.e. $\Gamma e_i = -e_i$, and this is seen to leave $\mathscr{H}$ invariant. Now the state of our system depends on whether each particle is an electron or a positron, so that we include e_1 and e_2 explicitly in the wave function and write†

$$\Gamma\psi(\mathbf{r}_1, \mathbf{r}_2, \sigma_1, \sigma_2, e_1, e_2) = \psi(r_1, r_2, \sigma_1, \sigma_2, -e_1, -e_2). \quad (34.3)$$

The Hamiltonian is also invariant under the *space inversion* Π (32.16). The effect of Π on the orbital variables is to replace $\mathbf{r}_i$ by $-\mathbf{r}_i$, but we must also consider its effect on the spin functions. We saw in § 32 that in relativistic quantum theory the wave function of an electron can be written

$$\psi_{l+}u_+ + \psi_{l-}u_- + \psi_{s+}v_+ + \psi_{s-}v_-, \quad (34.4)$$

where u_+, u_- and v_+, v_- transform according to $D^{(1/2)}$ under pure rotations. For an electron with a small energy the component ψ_{l+}, ψ_{l-} are large of order unity, and ψ_{s+}, ψ_{s-} are very small of order v/c. For a state describing what is called a positron the components in v_+, v_- are the large ones (see equations (32.23)). Hence as in Chapter II, we shall in our non-relativistic discussion neglect the small components, and write electron and positron wave functions purely in terms of u_+, u_- and v_+, v_-. In § 11 the convention was made that u_+, u_- are invariant under Π, and in § 31 it was shown that therefore v_+, v_- change sign under Π (cf. equation (31.24))‡

$$\Pi v_+ = -v_+, \qquad \Pi v_- = -v_-. \quad (34.5)$$

Thus if ψ describes a state with p positrons ($p = 0$, 1 or 2 in our case), it will contain products of p spin function $v_{\pm i}$, and from (34.5)

$$\Pi\psi(\mathbf{r}_1, \mathbf{r}_2, \sigma_1, \sigma_2, e_1, e_2) = (-1)^p\psi(-\mathbf{r}_1, -\mathbf{r}_2, \sigma_1, \sigma_2, e_1, e_2). \quad (34.6)$$

† This charge conjugation operator is the same as that defined by (32.28). It leaves the spin σ of each particle invariant (problem 32.8). There is no complex conjugation in (34.3) because there are two particles and the complex conjugation of (32.28) comes in twice.

‡ This can also be proved more directly as follows. Since Π commutes with the full rotation group, by Schur's lemma (appendix D) it is represented by $\lambda_1 E$ and $\lambda_2 E$ with respect to (u_+, u_-) and $v_+, v_-)$, where each $\lambda = \pm 1$ or $\pm i$ since $\Pi^2 = \pm E$. Hence with respect to (u_+, u_-, v_+, v_-) Π is represented by the diagonal matrices

$$(\pm 1 \text{ or } \pm i) \operatorname{diag}[1, 1, 1, 1] \quad \text{(A)}$$

or

$$(\pm 1 \text{ or } \pm i) \operatorname{diag}[1, 1, -1, -1]. \quad \text{(B)}$$

The other combinations of λ_1, λ_2 contradict $\Pi^2 = \pm E$. If $L(\mathbf{v})$ is a Lorentz transformation with velocity $\mathbf{v}$, then $\Pi L(\mathbf{v}) = L(-\mathbf{v})\Pi$. Writing $L(\pm\mathbf{v}) = \exp(\pm i\theta I)$ as in (31.5) shows that the matrices representing $L(\mathbf{v})$ and $L(-\mathbf{v})$ are not equal so that Π cannot have the form (A). In accordance with the convention of chapter II and of (31.24), we choose $+1$ in (B), which proves (34.5).

The Hamiltonian (34.1) is further invariant under the *permutation* operation P,

$$P\mathbf{r}_1 = \mathbf{r}_2,\ P\sigma_1 = \sigma_2,\ Pe_1 = e_2,\ P\mathbf{r}_2 = \mathbf{r}_1,\ P\sigma_2 = \sigma_1,\ Pe_2 = e_1.$$

There are only two particles so that the eigenfunctions of (34.1) can be sorted out to be symmetric or antisymmetric with respect to P (§ 7). If the particles are both electrons or both positrons, the wave function must be antisymmetric according to the usual statement of the exclusion principle (§12). In fact ψ must still be antisymmetric if it describes an electron, positron pair. This is not obvious in a purely non-relativistic discussion in which the electron and positron are being treated almost as unrelated particles. However it follows in a relativistic theory because the difference between an electron and a positron function is a quantitative one in the relative sizes of the four components in (34.4). Thus whatever e_1 and e_2 are

$$\begin{aligned} P\psi(\mathbf{r}_1, \mathbf{r}_2, \sigma_1, \sigma_2, e_1, e_2) &= \psi(\mathbf{r}_2, \mathbf{r}_1, \sigma_2, \sigma_1, e_2, e_1) \\ &= -\psi(\mathbf{r}_1, \mathbf{r}_2, \sigma_1, \sigma_2, e_1, e_2) \end{aligned} \tag{34.7}$$

Now (34.1) can be expressed in terms of the position $\frac{1}{2}(\mathbf{r}_1 + \mathbf{r}_2)$ and velocity $\frac{1}{2}(\mathbf{v}_1 + \mathbf{v}_2)$ of the centre of mass, and the relative position $(\mathbf{r}_1 - \mathbf{r}_2)$ and velocity $(\mathbf{v}_1 - \mathbf{v}_2)$ of the two particles. Let us assume from now on that the centre of mass is at rest. Everything then depends only on $\mathbf{r}_1 - \mathbf{r}_2$ as regards space variables and

$$\begin{aligned} \psi(\mathbf{r}_1, \mathbf{r}_2, \sigma_1, \sigma_2, e_1, e_2) &= f(\mathbf{r}_1 - \mathbf{r}_2) \\ &= \psi(-\mathbf{r}_2, -\mathbf{r}_1, \sigma_1, \sigma_2, e_1, e_2). \end{aligned} \tag{34.8}$$

Also the first two terms of (34.2) reduce to

$$-\frac{3}{4}\frac{e_1e_2}{mc^2}\frac{(\mathbf{r}_1 - \mathbf{r}_2) \wedge (\mathbf{v}_1 - \mathbf{v}_2)}{{r_{12}}^3}(\mathbf{s}_1 + \mathbf{s}_2),$$

and $\mathscr{H}$ (34.1) becomes invariant under the *spin interchanges* transformation Σ,

$$\Sigma\mathbf{r}_i = \mathbf{r}_i, \qquad \Sigma e_i = e_i, \qquad \Sigma\sigma_1 = \sigma_2, \qquad \Sigma\sigma_2 = \sigma_1,$$

i.e.

$$\Sigma\,\psi(\mathbf{r}_1, \mathbf{r}_2, \sigma_1, \sigma_2, e_1, e_2) = \psi(\mathbf{r}_1, \mathbf{r}_2, \sigma_2, \sigma_1, e_1, e_2). \tag{34.9}$$

Symmetry classification of the eigenstates

The transformations Γ, Π, P, Σ all commute with one another and with the full rotation group. Since we have an even number of particles, we shall only require the single-valued irreducible representations $D^{(J)}$ of the rotation group, and it follows from the general theory of § 15 that each $D^{(J)}$ gives rise to 2^4 irreducible representations of the complete symmetry group of (34.1) by associating $D^{(J)}$ with a set of characters

$$\chi(\Gamma) = \pm 1, \qquad \chi(\Pi) = \pm 1, \qquad \chi(P) = \pm 1, \qquad \chi(\Sigma) = \pm 1,$$

such that the matrix $\chi(\alpha)D^{(J)}(R)$ represents the transformation αR where R is a proper rotation and $\alpha = \Gamma, \Pi, P, \Sigma$ or any combination of these. Because of the exclusion principle we have already restricted ourselves to the antisymmetrical representations with $\chi(P) = -1$. The fact that we wish to describe positronium consisting of one electron and one positron gives the further restrictions

$$e_1 = -e_2, \tag{34.10a}$$

$$p = 1. \tag{34.10b}$$

It follows that

$$\begin{aligned} \Sigma\Gamma\Pi\psi(\mathbf{r}_1, \mathbf{r}_2, \sigma_1, \sigma_2, e_1, e_2) &= (-1)^p\psi(-\mathbf{r}_1, -\mathbf{r}_2, \sigma_2, \sigma_1, -e_1, -e_2) \\ &\qquad \text{from (34.3), (34.6), (34.9),} \\ &= \psi(-\mathbf{r}_2, -\mathbf{r}_1, \sigma_1, \sigma_2, -e_2, -e_1) \text{ from (34.7), (34.10b),} \\ &= \psi(\mathbf{r}_1, \mathbf{r}_2, \sigma_1, \sigma_2, e_1, e_2) \text{ from (34.8), (34.10a).} \end{aligned}$$

Hence

$$\chi(\Sigma)\chi(\Gamma)\chi(\Pi) = 1. \tag{34.11}$$

As in the theory of atomic spectra, $\mathscr{H}_{\text{spin}}$ in (34.1) is relatively small and can be neglected in a first approximation. $\mathscr{H}$ is then invariant under separate rotations of the orbital variable $\mathbf{r}_1 - \mathbf{r}_2$ and of the spin co-ordinates, giving a configuration with degenerate wave functions transforming according to

$$D^{(l)} \times D^{(S)}, \text{ where } S = 0 \text{ or } 1. \tag{34.12}$$

Any $S = 0$ wave function contains the spin function

$$u_+v_- - u_-v_+$$

as a factor so that it is antisymmetric under the spin interchange Σ, and similarly the $S = 1$ ones are symmetric.

$$\chi(\Sigma) = -1 \text{ for } S = 0, \qquad \chi(\Sigma) = 1 \text{ for } S = 1. \tag{34.13}$$

The orbital parts of the wave functions contain the spherical harmonics and are multiplied by $(-1)^l$ under inversion, so that in view of (34.6) and (34.10b) we have

$$\chi(\Pi) = (-1)^{l+1}. \tag{34.14}$$

Now from (34.11)

$$\chi(\Gamma) = (-1)^{l+S} \tag{34.15}$$

Thus the symmetry properties of one configuration are completely determined by l and S with the help of (34.13), (34.14), (34.15).

$\mathscr{H}_{\text{spin}}$ is now considered as a perturbation, and this splits every configuration into terms characterized by a J quantum number. The configurational set of wave functions transforming according to (34.12) decomposes into several term sets transforming according to $D^{(J)}$ where the values of J are given by the vector coupling

TABLE 40

Energy Levels of Positronium

Level	l	S	J	$\chi(\Sigma)$	$\chi(P)$	$\chi(\Pi)$	$\chi(\Gamma)$
^1S	0	0	0	-1	-1	-1	$+1$
$^1P_1$	1	0	1	-1	-1	$+1$	-1
$^1D_1$	2	0	2	-1	-1	-1	$+1$
$^3S_1$	0	1	1	1	-1	-1	-1
^3P	1	1	0	1	-1	$+1$	$+1$
$^3P_1$	1	1	1	1	-1	$+1$	$+1$
$^3P_2$	1	1	2	1	-1	$+1$	$+1$

rule (9.2) as usual. In this way we obtain energy levels (terms) such as are shown in the usual spectroscopic notation (Chapter II) in Table 40. The table also shows the characters of P, Σ, Π, Γ as determined by (34.7) and (34.13) to (34.15).

Transition selection rules

If an atom makes a transition from one state to another with the emission of a photon, this transition is accompanied by a change in the charge distribution in the atom. In a semi-classical discussion of radiative transitions (Schiff 1955, p. 260), the current associated with this shift in the charge is considered to radiate the electromagnetic wave corresponding to the photon. If now the current is reversed corresponding to charge conjugation, it follows from

Maxwell's equations that all the fields are also reversed. Thus we can write

$$\Gamma\phi(\mathbf{r}) = -\phi(\mathbf{r}),$$

where $\phi(\mathbf{r})$ is a state function describing the electromagnetic field of one photon. Two crossed beams of light are known not to interact appreciably with one another, so that a multi-photon state is just a product of single free photon states,

$$\Phi = \phi_1(\mathbf{r}_1)\,\phi_2(\mathbf{r}_2)\phi_3(\mathbf{r}_3)\,\ldots\,\phi_n(\mathbf{r}_n)$$

and $$\Gamma\Phi = (-1)^n\Phi, \qquad \chi(\Gamma) = (-1)^n, \tag{34.16}$$

where n is the number of photons.

The Hamiltonian for a free electromagnetic field is quadratic in the field strength

$$\mathscr{H} = \frac{1}{8\pi}\int (H^2 + \mathscr{E}^2)\,dv,$$

and the interaction term between electrons and the field has been discussed in § 32. We see therefore that the total Hamiltonian $\mathscr{H}_{\text{tot}}$ for electrons, positrons and radiation is invariant under Γ. If ψ_{tot} is a total wave function, we have

$$\Gamma(\mathscr{H}_{\text{tot}}\psi_{\text{tot}}) = \mathscr{H}_{\text{tot}}(\Gamma\psi_{\text{tot}}).$$

Therefore Γ, considered as a quantum mechanical operator, commutes with the Hamiltonian and is a constant of the motion. This means that its expectation value $\chi(\Gamma)$ remains constant during any interaction process (see equation (17.1) or the fundamental theorem of § 30). In particular if the electron and positron in positronium annihilate into n photons, we must have from (34.16)

$$(-1)^n = \chi_i(\Gamma)$$

where $\chi_i(\Gamma)$ refers to the initial positronium state. For instance from Table 40, *the ground state* $^1S_0$ *annihilates into an even number of photons*, usually two. Similarly *the* $^3S_1$ *state annihilates into an odd number of photons*, the minimum number being three since the total momentum has also to be conserved.

Problems

34.1 Derive the selection rules for dipole transitions between the levels of positronium (Jauch and Rohrlich 1955, p. 279).

34.2 Write down for a system of two electrons the relations which correspond to (34.10), (34.13), (34.14). Show that there is no relation directly analogous to (34.11), but instead $\chi(\Pi)\chi(\Sigma) = \chi(P) = -1$, and construct a table of levels similar to Table 40. Hence show that the exclusion principle in this case excludes some of the terms that are allowed in Table 40, and illustrate this by writing down the form of some of the wave functions or by considering Slater's scheme (§ 12).

34.3 Show that the quantities e_1, e_2 in (34.1) may be regarded as vectors in a two dimensional pseudo spin space, and hence that the energy levels may be described by an additional pseudo spin quantum number $T = 0$ or 1. Using this formalism discuss the difference between the symmetry properties of states of positronium and of two electrons as noted in problem 34.2. Also write down the annihilation selection rules in this general formalism.

34.4* Write down in the notation of field theory the relations that correspond to equations (34.3), (34.6), (34.7), (34.8), (34.9), (34.10), and hence derive (34.11), (34.13), (34.14) (Jauch and Rohrlich 1955, p. 274). Note in particular how the commutation relations in field theory replace the exclusion principle, i.e. antisymmetry of ψ in the wave function formalism.

Appendix A

Matrix Algebra

The following gives the amount of matrix algebra assumed in the text. On some occasions when a more advanced result or concept is required, references are given to Margenau and Murphy (1943, Chapter 10).

Definition of a matrix

Consider the linear transformation of co-ordinates from (x, y, z) to (X, Y, Z)

$$\begin{aligned} x &= A_{11}X + A_{12}Y + A_{13}Z, \\ y &= A_{21}X + A_{22}Y + A_{23}Z, \\ z &= A_{31}X + A_{32}Y + A_{33}Z. \end{aligned} \tag{A.1}$$

This system of equations is one linear transformation and it is also often convenient to treat the coefficients $A_{11}, A_{12}, \ldots A_{33}$ as a whole. The square array of coefficients

$$\begin{bmatrix} A_{11}A_{12}A_{13} \\ A_{21}A_{22}A_{23} \\ A_{31}A_{32}A_{33} \end{bmatrix} \quad \text{or} \quad \begin{pmatrix} A_{11}A_{12}A_{13} \\ A_{21}A_{22}A_{23} \\ A_{31}A_{32}A_{33} \end{pmatrix}$$

is the *square matrix* A, the brackets indicating that the matrix refers to the array as a whole and incidentally not to the determinant of the coefficients. More generally, it is convenient to consider *rectangular matrices* with m rows and n columns, these then being of order $m \times n$ (pronounced m by n). A_{ij} is the $(ij)^{\text{th}}$ *element* of A, where i is the number of the row and j the number of the column in which A_{ij} is situated. Sometimes A is referred to as "the matrix A_{ij}", meaning the matrix whose $(ij)^{\text{th}}$ element is A_{ij}. In general it is not necessary for matrices to arise only in connection with linear transformations in the way they have been introduced here, but they frequently do either directly or indirectly.

Multiplication

Consider another linear transformation

$$\begin{aligned} X &= B_{11}\xi + B_{12}\eta + B_{13}\zeta, \\ Y &= B_{21}\xi + B_{22}\eta + B_{23}\zeta, \\ Z &= B_{31}\xi + B_{32}\eta + B_{33}\zeta. \end{aligned} \tag{A.2}$$

Suppose we have a function of x, y, z and we wish to express it first in terms of X, Y, Z and then in terms of ξ, η, ζ. We would first substitute for x, y, z using (A.1) and then use (A.2). If, however, we are only interested in the final result, we can shorten the calculation by eliminating X, Y, Z between (A.1) and (A.2) and using directly the resulting transformation

$$\begin{aligned} x &= (A_{11}B_{11} + A_{12}B_{21} + B_{13}A_{31})\xi + (A_{11}B_{12} + A_{12}B_{22} + \\ &\quad + A_{13}B_{32})\eta + (A_{11}B_{13} + A_{12}B_{23} + A_{13}B_{33})\zeta, \\ y &= (A_{21}B_{11} + A_{22}B_{21} + A_{23}B_{31})\xi + (A_{21}B_{12} + A_{22}B_{22} + \\ &\quad + A_{23}B_{32})\eta + (A_{21}B_{13} + A_{22}B_{23} + A_{23}B_{33})\zeta, \\ z &= (A_{31}B_{11} + A_{32}B_{21} + A_{33}B_{31})\xi + (A_{31}B_{12} + A_{32}B_{22} + \\ &\quad + A_{33}B_{32})\eta + (A_{31}B_{13} + A_{32}B_{23} + A_{33}B_{33})\zeta, \end{aligned} \tag{A.3}$$

which is a combination of (A.1) and (A.2). Denoting the coefficients in (A.3) by

$$C_{ij} = \sum_{k=1}^{3} A_{ik}B_{kj}, \tag{A.4}$$

we would like to express the relationship of the matrix C_{ij} to the matrices A and B. This is accomplished by defining C to be the matrix product

$$C = AB \tag{A.5}$$

of A and B in this order, where (A.4) is the rule for finding the elements of C from those of A and B. More generally, if A is a matrix of order $m \times p$ and B is of order $p \times n$, the *product* $C = AB$ *is defined as the matrix* C_{ij} *of order* $m \times n$ where

$$\boxed{C_{ij} = \sum_{k=1}^{p} A_{ik}B_{kj}.} \tag{A.6}$$

The order of multiplication is important, and the product $D = BA$ is a different matrix from C.

$$D_{ij} = \sum_{k} B_{ik}A_{kj}. \tag{A.7}$$

In fact unless $m = n$ it is impossible to carry out the summation in (A.7), and even then we see from (A.6) and (A.7) that $D_{ij} \neq C_{ij}$. Also (A.6) and (A.7) are similar to the laws for multiplying determinants, so that if A and B are square matrices of the same order, then

$$|AB| = |BA| = |A||B|, \tag{A.8}$$

where $|AB|$ is the determinant of the matrix AB.

With this definition of matrix multiplication, the transformation (A.1) can also be written completely in matrix form. Let q be the matrix

$$q = \begin{bmatrix} x \\ y \\ z \end{bmatrix}. \tag{A.9}$$

Such a matrix with one column is called a column matrix (or column vector) and is often written $\{xyz\}$ to save space. Similarly a $1 \times n$ matrix is a row matrix. In this notation (A.1) becomes

$$q = AQ \tag{A.10}$$

where $Q = \{XYZ\}$. Similarly (A.2) can be written

$$Q = B\nu \tag{A.11}$$

where $\nu = \{\xi\eta\zeta\}$. Then (A.3) becomes

$$q = AB\nu, \tag{A.12}$$

and this result also follows directly by substituting (A.11) in (A.10).

The double suffix summation convention, also known as the dummy suffix notation.

Although it is often convenient to write matrix products in the compact form (A.5), it is on other occasions useful to work in terms of the elements as in (A.6). Then summations over some of the suffices appear constantly, and it is customary to omit the summation signs but to consider any product to be automatically summed over every suffix such as k in (A.6) that occurs twice in one product. With this convention (A.4) and (A.6) become

$$C_{ij} = A_{ik}B_{kj}. \tag{A.13}$$

Before (A.1) or (A.10) can be written in this form, it is necessary to alter the notation slightly and to put

$$q_1 = x, \qquad q_2 = y, \qquad q_3 = z, \qquad Q_1 = X, \qquad \text{etc.} \tag{A.14}$$

Then (A.1) and (A.10), (A.2) and (A.11) become

$$q_i = A_{ij}Q_j, \qquad Q_j = B_{jk}\nu_k$$

whence by substituting for Q_j we have immediately

$$q_i = A_{ij}B_{jk}\nu_k = C_{ik}\nu_k \text{ from (A.13)},$$

in agreement with (A.3) and (A.12). Incidentally (A.13) can equally well be written $C_{ij} = A_{il}B_{lj}$ or $C_{kt} = A_{kl}B_{lt}$ or in any other equivalent form which retains the relationship between the suffices. In particular a repeated suffix can be called anything at all, since it is being summed over and thus would not appear in the relation if it were written out in full. Such changes are frequently necessary in making substitutions, to ensure that the same letter is not being used for indices that are not necessarily equal and that should not be summed. The particular symbol

$$\boxed{\begin{aligned} \delta_{ij} &= 1 \text{ for } i = j, \\ &= 0 \text{ for } i \neq j, \end{aligned}} \tag{A.15}$$

is often used. For instance $\delta_{ij}A_{jk} = A_{ik}$. It has the effect of replacing the suffix j by i or vice versa.

Derived matrices and special matrices

Starting with a matrix A of order $m \times n$ we can construct another matrix $\tilde{A}$ of order $n \times m$ called the *transpose of* A, such that the $(ij)^{\text{th}}$ element of $\tilde{A}$ is given by

$$\boxed{\tilde{A}_{ij} = A_{ji}.} \tag{A.16}$$

If A is a square matrix of order $n \times n$ and the determinant $|A| \neq 0$, it is possible to construct its *reciprocal or inverse* A^{-1} also of order $n \times n$ with the property

$$\boxed{AA^{-1} = A^{-1}A = E} \tag{A.17}$$

where E is the *unit matrix* (which is always square) of order n with elements

$$\boxed{E_{ij} = \delta_{ij},} \tag{A.18}$$

i.e. E has the form

$$\begin{bmatrix} 1 & 0 & 0 & . & . & . & 0 \\ 0 & 1 & 0 & . & . & . & 0 \\ 0 & 0 & 1 & . & . & . & 0 \\ . & . & . & . & & & . \\ . & . & . & & . & & . \\ . & . & . & & & . & . \\ 0 & 0 & 0 & . & . & . & 1 \end{bmatrix}.$$

It is possible to express the elements of A^{-1} in terms of the A_{ij} but for the present purposes it is immaterial to know what they are. From (A.17) and (A.8)

$$|A^{-1}| = \frac{1}{|A|}, \tag{A.19}$$

which shows why it is not possible to construct A^{-1} if $|A| = 0$.

A *unitary* matrix U is a square matrix with the property

$$U^{-1} = \tilde{U}^* \tag{A.20}$$

where * denotes taking the complex conjugate of every element. Thus from (A.17)

$$\boxed{\tilde{U}^*U = U\tilde{U}^* = E} \tag{A.21}$$

and from (A.8)

$$|U||U|^* = 1. \tag{A.22}$$

The *zero matrix* 0 is a matrix of any order with all elements zero.

Problems

A.1 Write down some matrices with simple numbers as elements and find their products. With square matrices check your results using (A.8), and also verify that in general $AB \neq BA$.

A.2 Write down the $(ij)^{\text{th}}$ element of each of the matrices $D_1 = A(BC)$ and $D_2 = (AB)C$ using the double suffix summation convention, and hence show that D_1 and D_2 are the same. Also write equations (A.17), (A.20) and (A.21) in terms of the elements in this way.

A.3 Using the summation convention show that $(\widetilde{AB}) = \tilde{B}\tilde{A}$ and that $(\tilde{\tilde{A}}) = A$.

A.4 Show that if A is a matrix of order $m \times n$ and E_m, E_n are respectively unit matrices of order $m \times m$ and $n \times n$, then

$$AE_n = E_mA = A.$$

A.5 By matrix multiplication show that $(AB)^{-1} = B^{-1}A^{-1}$ and that $(A^{-1})^{-1} = A$.

A.6 Show using (A.21) that the transformation matrix R of the transformation (2.2) is unitary. What are $|R|$ and R^{-1}? With the notation of (A.10) if $q = RQ$, show by multiplying by R^{-1} that $Q = R^{-1}q$ and verify this by solving the three simultaneous equations (2.2) for X, Y and Z in terms of x, y, z.

A.7 If $B = \tilde{U}^*AU$ where U is unitary, show that $|B| = |A|$ and $B_{tt} = A_{tt}$ (summed!).

A.8 Show that the product of any number of unitary matrices is also unitary.

A.9 If $q = AQ$ in the notation of equation (A.10), show using the result of problem A.3 that

$$\tilde{Q}^*\tilde{A}^*AQ = \tilde{q}^*q \tag{A.23}$$

and that each side of this equation is a one by one matrix, i.e. a single number. Show further that if A is unitary, then (A.23) when multiplied out becomes

$$x^2 + y^2 + z^2 = X^2 + Y^2 + Z^2. \tag{A.24}$$

Conversely, show that if equation (A.24) is to hold, then $\tilde{A}^*A$ must be the unit matrix E, i.e. A is unitary. This is in a simplified form the most important property of unitary matrices (cf. appendix C). Since the Hamiltonian of almost every conceivable physical system can be written so as to involve r or r_{ij} (cf. § 3), it follows at once that the symmetry transformations of such a Hamiltonian are unitary transformations.

A.10 The functions ϕ_i, $i = 1$ to n, are normalized and orthogonal. The n functions $\psi_j = A_{ij}\phi_i$ (summed) are also normalized and orthogonal to one another. Prove that the matrix A is unitary.

Appendix B

Homomorphism and Isomorphism

Definition. *Two groups* $\mathfrak{G}_1$ *and* $\mathfrak{G}_2$ *are* **homomorphic**† *if some element of* $\mathfrak{G}_2$ *can be associated with every element of* $\mathfrak{G}_1$, *such that if* $P_1Q_1 = R_1$ *then* $P_2Q_2 = R_2$, *where* P_2, Q_2, R_2 *are the elements of* $\mathfrak{G}_2$ *that correspond to the elements* P_1, Q_1, R_1 *of* $\mathfrak{G}_1$.

Definition. *Two groups* $\mathfrak{G}_1$ *and* $\mathfrak{G}_2$ *are* **isomorphic** *if there exists a one-to-one correspondence between the elements* $A_1, B_1, C_1, \ldots$ *of* $\mathfrak{G}_1$ *and* $A_2, B_2, C_2, \ldots$ *of* $\mathfrak{G}_2$, *such that if* $P_1Q_1 = R_1$ *then* $P_2Q_2 = R_2$ *and vice versa.* A relationship between the elements of $\mathfrak{G}_1$ and $\mathfrak{G}_2$ is said to be a *one-to-one correspondence* if the elements of $\mathfrak{G}_1$ and $\mathfrak{G}_2$ can be paired off together, A_1 and A_2, B_1 and B_2, etc., in such a way that each element of each group is paired with one and only one element of the other group. If the groups are finite this implies that they have the same number of elements. From the definitions, an isomorphism is automatically a homomorphism, and is in fact a special symmetrical kind of homomorphism which can be described approximately by saying that two isomorphic groups have exactly the same structure as regards the relationships of the elements in each group to one another.

The rotations (4.1), the linear transformations (4.13), the matrices of the coefficients in these transformations (the law of combination in this case being matrix multiplication in the order $FS = C$, cf. equation (4.19)), and the permutations (4.22) form groups which are all isomorphic with one another. This follows from the discussion of § 4 which establishes a one-to-one correspondence between elements of any two of these groups. As a first example of a homomorphism, consider the point-group 32 of rotations E, A, B, K, L, M (equation (4.1)) as $\mathfrak{G}_1$, and the group $\mathfrak{g} = (e, a, b, k, l, m)$ as $\mathfrak{G}_2$ where $e = a = b = 1$ and $k = l = m = -1$. The fact that the six elements of $\mathfrak{g}$ are not all different from one another does not violate any of the group properties of § 4.‡ Now if we

† Some authors denote by *simple isomorphism* and by *isomorphism* (or *multiple isomorphism*) what we call isomorphism and homomorphism respectively.

‡ Some authors adopt a more restrictive definition of a group than that given in § 4, by including the further condition that all the elements be distinct (Ledermann 1953, pp. 2, 3). Then the matrices of a representation do not always form a group, which would be inconvenient for the purposes of this book.

have a relation such as $KA = L$ (4.5), then $ka = l$ also holds as can easily be verified from the multiplication table of $\mathfrak{G}_1$ (Table 1), so that the groups are homomorphic. However, since $l = m$, we also have $ka = m$ although $KA \neq M$. Thus the "vice versa" condition of the definition of isomorphism is not satisfied. Our second example of homomorphism is closely related to the first. Consider the group of rotations 32 again as $\mathfrak{G}_1$ and the group $\mathfrak{g}' = (1, -1)$ as $\mathfrak{G}_2$. If we associate the element 1 of $\mathfrak{g}' = \mathfrak{G}_2$ with each of the elements E, A, B of $\mathfrak{G}_1$ and -1 with the elements K, L, M, then as before $S_1F_1 = C_1$ always implies $S_2F_2 = C_2$ so that the groups are homomorphic. However, the "vice versa" is again not satisfied and in addition the number of group elements is unequal so that the groups are not isomorphic. As a third example consider the roles of the groups 32 and $\mathfrak{g}'$ reversed, i.e. $\mathfrak{g}'$ as $\mathfrak{G}_1$ and 32 as $\mathfrak{G}_2$. Then we can associate E with $+1$ and K with -1, in which case $S_1F_1 = C_1$ always implies $S_2F_2 = C_2$ *and vice versa*, so that the groups are again at least homomorphic. However, they are still not isomorphic because although the "vice versa" condition is satisfied, not all the elements of $\mathfrak{G}_2$ have been paired off with elements of $\mathfrak{G}_1$.

Problems

B.1 Construct some other examples of isomorphism and homomorphism.

B.2 The matrices A, B, . . . form a group $\mathfrak{G}$. Show that the determinants $|A|$, $|B|$, . . . form a group which is homomorphic but not necessarily isomorphic with $\mathfrak{G}$.

B.3 Try to express in your own words the *additional* condition that a homomorphism must satisfy for it to be an isomorphism, and test your wording on the examples given above.

B.4 Compare the definitions of homomorphism and isomorphism in some of the texts on group theory given in the bibliography, testing them on the examples given above.

Appendix C

Theorems on Vector Spaces and Group Representations

In § 5 the results of certain theorems are appealed to, and these will now be proved. It should be noted that the following (rather clumsy) proofs have been produced only to show that the argument of the main text can be justified step by step. They are therefore as elementary as possible, involving only the minimum number of intermediate concepts and results. For much more comprehensive and elegant developments of the theory of representations, the reader is referred to the standard texts mentioned in the bibliography, e.g. Van der Waerden 1932, Wigner 1931, Speiser 1937 Weyl 1931.

Theorem 1. *The dimension of a vector space is unique.* Let ϕ_j $j = 1$ to n, be n linearly independent base vectors spanning the space $\mathfrak{R}(\phi_1 \ldots \phi_n)$. Further let ϕ'_i, $i = 1$ to m, be another set of m linearly independent vectors spanning exactly the same space $\mathfrak{R}$ It is required to prove that $m = n$.

Suppose $m < n$. Since the ϕ'_i span the space $\mathfrak{R}$, we can express the ϕ_j in terms of the ϕ'_i and we have the n relations

$$\phi_j = P_{ij}\phi'_i \tag{C.1}$$

where the P_{ij} are some coefficients. Let P be the square matrix of order $m \times m$ formed from the first m rows of the coefficients P_{ij}, $i = 1$ to m, $j = 1$ to m. Since the ϕ_j are linearly independent it is impossible to find m coefficients α_j such that

$$\alpha_j P_{ij} = 0, \tag{C.2}$$

as otherwise we would have $\alpha_j\phi_j = \alpha_j P_{ij}\phi'_i = 0$. Since (C.2) has only the solution $\alpha_j = 0$, all j, we have $\det|P| \neq 0$, and hence can solve the first m equations of (C.1) and obtain

$$\phi'_j = (P^{-1})_{ij}\phi_i.$$

This result can be substituted into the $(m + 1)^{\text{th}}$ equation of the set (C.1), which then gives a linear relation among the ϕ_i contrary to the hypothesis that they are linearly independent. Therefore $m \nless n$, and similarly $n \nless m$. Therefore $m = n$, which proves the theorem.

Lemma 1. *If $\mathfrak{R}$ is a vector space and $\mathfrak{r}$ a subspace of $\mathfrak{R}$, then it is possible to construct a unique vector space $\mathfrak{s}$ such that $\mathfrak{s}$ is orthogonal* *to* $\mathfrak{r}$ *and* $\mathfrak{R} = \mathfrak{r} + \mathfrak{s}$. Let the orthogonal, linearly independent functions $\phi_1, \ldots \phi_r$ span the space $\mathfrak{r}$. By taking any functions of $\mathfrak{R}$, it is possible to define further base vectors in $\mathfrak{R}$, and by the process of § 5 to make them orthogonal to the $\phi_1, \ldots \phi_r$ and to one another. Let this give the further base vectors $\phi_{r+1}, \ldots \phi_n$, where n is the dimension of $\mathfrak{R}$. Now the vector space $\mathfrak{s} = (\phi_{r+1}, \ldots \phi_n)$ is orthogonal to $\mathfrak{r}$ (i.e. any function of $\mathfrak{s}$ is orthogonal to any function of $\mathfrak{r}$). Further any function ϕ of $\mathfrak{R}$ can be written $\phi = \sum_1^n c_i\phi_i$, i.e. in the form $\phi = \phi^{(\mathfrak{r})} + \phi^{(\mathfrak{s})}$ where $\phi^{(\mathfrak{r})} = \sum_1^r c_i\phi_i$ is some function belonging to $\mathfrak{r}$ and $\phi^{(\mathfrak{s})} = \sum_{r+1}^n c_i\phi_i$ some function belonging to $\mathfrak{s}$. Hence $\mathfrak{R} = \mathfrak{r} + \mathfrak{s}$. Moreover, the splitting up of ϕ into $\phi^{(\mathfrak{r})}$ and $\phi^{(\mathfrak{s})}$ is unique if $\phi^{(\mathfrak{s})}$ has to be orthogonal to $\phi^{(\mathfrak{r})}$, so that $\mathfrak{s}$ is unique. This proves the lemma.

Lemma 2. *All symmetry transformations (except Lorentz transformations) leave the scalar product of two functions invariant.* We shall first prove the lemma for a rotation R. Let f be a function of x, y, z, and let us for the present consider R as a change to a new co-ordinate system X, Y, Z in the sense of §§ 2 and 3. Then f can be expressed in terms of X, Y, Z, say $f(x, y, z) = F(X, Y, Z)$. The volume element $\mathrm{d}\tau$ can be written as either $\mathrm{d}x\,\mathrm{d}y\,\mathrm{d}z$ or $\mathrm{d}X\,\mathrm{d}Y\,\mathrm{d}Z$ (Margenau and Murphy 1943, p. 190), and thus if the integrations are carried out over all space,

$$\int f(x, y, z)\,\mathrm{d}x\,\mathrm{d}y\,\mathrm{d}z = \int F(X, Y, Z)\,\mathrm{d}X\,\mathrm{d}Y\,\mathrm{d}Z. \tag{C.3}$$

Physically the reason for this equality is that we have not changed the value of the integrand at any point P, but simply referred to the same point P by two different labels (x, y, z) and (X, Y, Z). If we put $f = \phi^*\psi$ where ϕ and ψ are two functions, and interpret (C.3) in terms of the notation of equation (5.3), we obtain

$$\boxed{\int \phi^*\psi\,\mathrm{d}x\,\mathrm{d}y\,\mathrm{d}z = \int (R\phi)^*(R\psi)\,\mathrm{d}x\,\mathrm{d}y\,\mathrm{d}z.}$$

Thus the scalar product of ϕ and ψ remains invariant under the transformation R.

The same proof also applies to any transformation for which the volume elements are equal, $\mathrm{d}q_1\,\mathrm{d}q_2 \ldots \mathrm{d}q_{3n} = \mathrm{d}Q_1\,\mathrm{d}Q_2 \ldots \mathrm{d}Q_{3n}$ in the notation of equation (4.15), and this is clearly true for the other types of symmetry transformation that arise namely permutations, the inversion and reflections, though not for a Lorentz transformation (problem C.1). This establishes the lemma. Incidentally the reader should have no difficulty now in writing a general proof of the lemma using the above type of argument and the hint contained in problem C.3. Transformations which leave the scalar product of two functions invariant are called *unitary* transformations (cf. problem C.2), and it will be implicitly assumed in this appendix that all transformations mentioned are unitary.

Corollary to lemma 2. If the functions ϕ and ψ are orthogonal, then so are $T\phi$ and $T\psi$ where T is a unitary transformation.

Lemma 3. *If the vector space* $\mathfrak{R} = \mathfrak{r} + \mathfrak{s}$, *and* $\mathfrak{R}$ *and* $\mathfrak{r}$ *are each invariant under a group* $\mathfrak{G}$ *of (unitary) transformations, then* $\mathfrak{s}$ *is also invariant under* $\mathfrak{G}$. Let $\phi^{(\mathfrak{r})}$ and $\phi^{(\mathfrak{s})}$ be any two functions belonging to $\mathfrak{r}$ and $\mathfrak{s}$, and let $\phi_1^{(\mathfrak{r})} = T^{-1}\phi^{(\mathfrak{r})}$ where T is a transformation of $\mathfrak{G}$. Since T^{-1} also belongs to $\mathfrak{G}$, $\phi_1^{(\mathfrak{r})}$ belongs to $\mathfrak{r}$. Therefore $\phi^{(\mathfrak{s})}$ is orthogonal to $\phi_1^{(\mathfrak{r})}$, and by the corollary to lemma 2, $T\phi^{(\mathfrak{s})}$ is orthogonal to $T\phi_1^{(\mathfrak{r})} = \phi^{(\mathfrak{r})}$. Since $\mathfrak{R}$ is invariant under $\mathfrak{G}$, $T\phi^{(\mathfrak{s})}$ belongs to $\mathfrak{R}$. We therefore have that $T\phi^{(\mathfrak{s})}$ belongs to $\mathfrak{R}$ but is orthogonal to any function $\phi^{(\mathfrak{r})}$ of $\mathfrak{r}$. Hence $T\phi^{(\mathfrak{s})}$ belongs to $\mathfrak{s}$, and $\mathfrak{s}$ is invariant under $\mathfrak{G}$, which proves the lemma.

Theorem 2. *If* $\mathfrak{R}$ *is a vector space invariant under a group* $\mathfrak{G}$ *of transformations, then* $\mathfrak{R}$ *is either irreducible and contains no invariant subspace, or* $\mathfrak{R}$ *is reducible into a sum of orthogonal invariant subspaces.* We shall first point out just what there is to be proved. In § 5 we defined and investigated the consequences of having an irreducible or reducible vector space, and showed that a reducible space can be split up into a sum of invariant subspaces. Proceeding in this way we were not able to prove that the representation can always be reduced in such a way that the invariant subspaces are orthogonal to one another. For even if $\mathfrak{R}$ had been referred to orthogonal base vectors ϕ_i initially, this is no guarantee that after the equivalence transformation (5.15) the base vectors ϕ'_i of the reduced space would be orthogonal. The present theorem shows that the invariant subspaces can be made orthogonal to one another. We were also unable to prove in § 5 that the following situation never in fact arises. Consider $\mathfrak{R}$ transforming according to the representation D. Suppose that D is irreducible, but by choosing new base vectors ϕ'_i in $\mathfrak{R}$ and applying the equivalence transformation (5.15) it is possible to bring the matrices of D halfway towards reduced form, i.e. into the form

$$\begin{bmatrix} D_{ij}^{(1)}(T) & Q_{ij}(T) \\ 0 & D_{ij}^{(2)}(T) \end{bmatrix},$$

where $Q_{ij}(T)$ is not zero and $D_{ij}^{(1)}(T)$ of order $n_1 \times n_1$. Then the first n_1 base vectors $\phi'_1 \ldots \phi'_{n_1}$ transform according to the representation $D_{ij}^{(1)}(T)$ and form an invariant subspace. However, the remaining ϕ'_i do not transform according to $D_{ij}^{(2)}(T)$ because $Q_{ij}(T)$ is not zero. Thus we would have an irreducible vector space containing an invariant subspace. This situation does not correspond to either of the two possibilities allowed by the theorem and cannot in fact arise.

From § 5, a reducible vector space always contains an invariant subspace. Hence a space that does not contain an invariant subspace is irreducible. Now consider on the other hand that $\mathfrak{R}$ does contain an invariant subspace $\mathfrak{r}$. Then by lemma 3 it also contains another invariant subspace $\mathfrak{s}$ orthogonal to $\mathfrak{r}$ and such that $\mathfrak{R} = \mathfrak{r} + \mathfrak{s}$. Let $\phi'_1, \ldots \phi'_r$ and $\phi'_{r+1}, \ldots \phi'_n$ be two sets of base vectors spanning $\mathfrak{r}$ and $\mathfrak{s}$. Then the transformation in $\mathfrak{R}$ to the base vectors $\phi'_1, \ldots \phi'_n$ is an equivalence transformation of type (5.15) which brings all the matrices of the representation D into reduced form. The process can then be continued until none of the subspaces can be reduced further. We have therefore shown that if a vector space $\mathfrak{R}$ does not contain an invariant subspace, then it is irreducible; also if it does contain an invariant subspace, then it is reducible into a series of orthogonal, invariant subspaces. Since $\mathfrak{R}$ must either contain or not contain an invariant subspace, this exhausts all the possibilities, which proves the theorem.

Schur's lemma fits into the logical sequence of steps at this stage. In view of the importance of this lemma, its proof has been placed in a separate appendix (appendix D).

Lemma 4. *If a reducible representation D of a group $\mathfrak{G}$ is reduced into irreducible components $D^{(i)}$ one way*

$$D = D^{(1)} + D^{(2)} + \ldots + D^{(r)},$$

and also in another way

$$D = D^{(\alpha)} + D^{(\beta)} + \ldots + D^{(\rho)},$$

then $D^{(\alpha)}$ is equivalent to one of the representations $D^{(1)}, \ldots D^{(r)}$. Let $\mathfrak{R}$ be a vector space transforming according to D. Let the base vectors $\phi_i^{(1)}, \phi_j^{(2)}, \ldots \phi_k^{(r)}$, $i = 1$ to n_1, $j = 1$ to $n_2, \ldots, k = 1$ to n_r in $\mathfrak{R}$ transform according to $D^{(1)}, D^{(2)}, \ldots, D^{(r)}$ and similarly let the vectors $\phi_i^{(\alpha)}, \phi_j^{(\beta)}, \ldots \phi_k^{(\rho)}$ in $\mathfrak{R}$ transform according to $D^{(\alpha)}, D^{(\beta)}, \ldots D^{(\rho)}$. Each of these two sets of vectors is linearly independent in accordance with an equivalence transformation of the form (5.15) with $\det|P| \neq 0$, so that none of the vectors are zero. The $\phi_i^{(\alpha)}$ can be expressed in terms of the base vectors $\phi_i^{(1)}, \ldots \phi_k^{(r)}$; thus

$$\phi_j^{(\alpha)} = P_{ij}^{(1)}\phi_i^{(1)} + P_{ij}^{(2)}\phi_i^{(2)} + \ldots + P_{ij}^{(r)}\phi_i^{(r)}.$$

If T is any transformation of $\mathfrak{G}$, $T\phi_j^{(\alpha)}$ can now be written in two ways,

$$T\phi_j^{(\alpha)} = P_{ij}^{(1)} D_{ki}^{(1)}(T)\phi_k^{(1)} + \dots$$

and

$$T\phi_j^{(\alpha)} = D_{ij}^{(\alpha)}(T)\phi_i^{(\alpha)}$$
$$= D_{ij}^{(\alpha)}(T) P_{ki}^{(1)}\phi_k^{(1)} + \dots$$

Hence each of the matrices $P^{(s)}$ satisfies the conditions for Schur's lemma (appendix D) and is therefore zero unless $D^{(\alpha)}$ is equivalent to $D^{(s)}$. Since the $\phi_i^{(\alpha)}$ are not zero, $D^{(\alpha)}$ is equivalent to at least one of the representations $D^{(1)}, \dots D^{(r)}$, which proves the lemma.

Lemma 5. *Functions transforming according to different irreducible representations of a group are orthogonal.* Let $\phi_i^{(1)}$ $(i = 1 \text{ to } n)$ be a set of normalized orthogonal functions and $\phi_j^{(\alpha)}$ another such set, transforming respectively according to the non-equivalent irreducible representations $D^{(1)}$ and $D^{(\alpha)}$. Take any functions and orthogonalize them to the $\phi_i^{(1)}$ and to one another by the process described in § 5 to give a set $\phi_{n+1}, \phi_{n+2}, \dots$ which together with the $\phi_i^{(1)}$ form a complete set. Then

$$\phi_j^{(\alpha)} = \sum_{1}^{n} P_{ij}\phi_i^{(1)} + \sum_{n+1}^{\infty} P_{ir}\phi_r.$$

As in lemma 4, $P_{ij} = 0$ since $D^{(1)}$ and $D^{(\alpha)}$ are not equivalent. But P_{ij} is just $\int \phi_i^{(1)*}\phi_j^{(\alpha)}\, dv$ integrated over all space, so that $\phi_i^{(1)}$ and $\phi_j^{(\alpha)}$ are orthogonal for all i and j.

Theorem 3. *The reduction of a reducible representation is unique apart from equivalence.* In the notation of lemma 4, it is required to prove that the representations $D^{(1)}, D^{(2)}, \dots D^{(r)}$ and $D^{(\alpha)}, D^{(\beta)}, \dots D^{(\rho)}$ are pairwise equivalent. From lemma 4, each $D^{(\alpha)}, \dots D^{(\rho)}$ is equivalent to one of the $D^{(1)}, \dots D^{(r)}$, and this proves the uniqueness if none of the $D^{(1)}, \dots D^{(r)}$ is equivalent to another. Consider now the following case in which this is not so. Let $D^{(\alpha)}$ be equivalent to $D^{(1)}$ and $D^{(2)}$ but not to any of the $D^{(3)}, \dots D^{(r)}$ and let us make a trivial change of base vectors such that these representations become identical. Then in the notation of lemma 4, from Schur's lemma (appendix D) $P^{(1)} = aE$, $P^{(2)} = bE$, $P^{(3)} = P^{(4)} = \dots = P^{(r)} = 0$, i.e. $\phi_i^{(\alpha)} = a\phi_i^{(1)} + b\phi_i^{(2)}$. Now construct $\psi_i^{(2)} = b\phi_i^{(1)} - a\phi_i^{(2)}$. Then the vector space

$$(\dots \psi_i^{(2)} \dots) + (\dots \phi_i^{(3)} \dots) + \dots + (\dots \phi_i^{(r)} \dots) \quad \text{(C.4)}$$

is orthogonal to $(\dots \phi_i^{(\alpha)} \dots)$ by lemma 5 and by construction. Similarly from the $\phi_i^{(\beta)}, \dots \phi_i^{(\rho)}$ we can construct functions

$\psi_i^{(\beta)}, \ldots \psi_i^{(\rho)}$ transforming according to the same representations $D^{(\beta)}, \ldots D^{(\rho)}$ but orthogonal to the $\phi_i^{(\alpha)}$. Hence the vector space

$$(\ldots \psi_i^{(\beta)} \ldots) + (\ldots \psi_i^{(\gamma)} \ldots) + \ldots + (\ldots \psi_i^{(\rho)} \ldots)$$

is also orthogonal to $(\ldots \phi_i^{(\alpha)} \ldots)$, and hence by the uniqueness part of lemma 1 equal to the space (C.4). Hence

$$D^{(2)} + D^{(3)} + \ldots + D^{(r)} = D^{(\beta)} + D^{(\gamma)} + \ldots + D^{(\rho)}.$$

The process can now be repeated until all the component representations have been paired off. The proof can also be extended easily to the case where any number of the irreducible components are equivalent, which proves the theorem.

Problems

C.1 Show that the inversion transformation (3.11) and the permutation of two co-ordinates (3.10) are unitary transformations. If two transformations S and T are each unitary, show that the combined transformations TS and ST are also unitary. Hence show that all symmetry transformations mentioned in §§ 3 and 4 are unitary.

C.2 A unitary transformation T operating on a set of normalized orthogonal functions ϕ_i induces the transformations $T\phi_j = D_{ij}(T)\phi_i$. Show that $D(T)$ is a unitary matrix (appendix A).

C.3 Give a general proof of lemma 2 using problem A.9 (appendix A) and the form of the volume element in general co-ordinates (Margenau and Murphy 1943, pp. 187, 190).

C.4 With the notation of problem C.2, show that $\sum_i \phi_i^*\phi_i$ is invariant under T.

Appendix D

Schur's Lemma

Given: $\Gamma_{ij}^{(1)}(T)$ and $\Gamma_{ij}^{(2)}(T)$ are the matrices of two irreducible representations $\Gamma^{(1)}$ and $\Gamma^{(2)}$ (of dimensions n_1 and n_2) of some group, and there exists a matrix P (of order $n_2 \times n_1$) such that

$$P\Gamma^{(1)}(T) = \Gamma^{(2)}(T)P \tag{D.1}$$

for all group elements T.

The lemma states that

(i) if $\Gamma^{(1)}$ and $\Gamma^{(2)}$ are not equivalent, then $P = 0$;

(ii) if $\Gamma^{(1)}$ and $\Gamma^{(2)}$ are equivalent, then $P = 0$ or $\det|P| \neq 0$;

(iii) if $\Gamma_{ij}^{(1)}(T) = \Gamma_{ij}^{(2)}(T)$, i.e. all the matrices of $\Gamma^{(1)}$ and $\Gamma^{(2)}$ are equal and not just equivalent, then $P = 0$ or $P = \lambda E$ where λ is a constant and E is the unit matrix.

Proof. *Step* 1. Choose any set of n_2 linearly independent vectors u_i transforming according to $\Gamma^{(2)}$, i.e.

$$Tu_j = \Gamma_{ij}^{(2)}(T)u_i, \tag{D.2}$$

and define a set of n_1 vectors

$$v_j = P_{ij}u_i. \tag{D.3}$$

Let us also assume that $n_1 \leqq n_2$. This involves no loss of generality since if $n_1 > n_2$ we can replace P, $\Gamma^{(1)}(T)$, $\Gamma^{(2)}T$ by the transposed matrices $\tilde{P}$, $\tilde{\Gamma}^{(2)}(T)$, $\tilde{\Gamma}^{(1)}(T)$ respectively throughout the argument.

Step 2. Now

$$\begin{aligned} Tv_j &= TP_{ij}u_i && \text{from (D.3)}, \\ &= \Gamma_{ki}^{(2)}(T)P_{ij}u_k && \text{from (D.2)}, \\ &= P_{ki}\Gamma_{ij}^{(1)}(T)u_k && \text{from (D.1)}, \\ &= \Gamma_{ij}^{(1)}(T)v_i && \text{from (D.3)}; \end{aligned} \tag{D.4}$$

i.e. the vectors v_i transform according to the irreducible representation $\Gamma^{(1)}$.

Step 3. Suppose that $P \neq 0$. From (D.3) the vector space $(\ldots v_i \ldots)$, $i = 1$ to n_1, has a dimension $n'_1 \leqq n_1$ which is less than or equal to the dimension n_2 of the space $(\ldots u_i \ldots)$, $i = 1$ to n_2. Hence from step 2, we have inside the space $(\ldots u_i \ldots)$ an invariant subspace $(\ldots v_i \ldots)$ transforming according to $\Gamma^{(1)}$. If n'_1 were less than n_2, this would make the space $(\ldots u_i \ldots)$ and hence the representation $\Gamma^{(2)}$ reducible contrary to hypothesis. Hence $n'_1 = n_2$.

Step 4. Hence the space $(\ldots u_i \ldots)$ and the space $(\ldots v_i \ldots)$ are the same (appendix C, Theorem 1). Since also $n'_1 \leqq n_1 \leqq n_2$, from step 3, $n_1 = n_2$, and the representations $\Gamma^{(1)}$ and $\Gamma^{(2)}$ are equivalent. This is still on the assumption that $P \neq 0$. Thus if $\Gamma^{(1)}$ and $\Gamma^{(2)}$ are not equivalent, we must have $P = 0$, which proves part (i) of the lemma.

Step 5. Now suppose that $\Gamma^{(1)}$ and $\Gamma^{(2)}$ are equivalent. Thus $n_1 = n_2$, and by step 3 if $P \neq 0$ the vectors v_i are linearly independent. I.e. it is impossible to find a set of numbers α_j (not all zero) such that $\alpha_j v_j = 0$.

Step 6. It is therefore also impossible to find numbers α_j (not all zero) such that

$$\alpha_j P_{ij} = 0 \qquad \text{for all } i, \tag{D.5}$$

since (D.5) would imply $\alpha_j v_j = \alpha_j P_{ij} u_i = 0$ contrary to the result of step 5. Since the equations (D.5) have no solution except $\alpha_j = 0$ for all j, we have $|P| \neq 0$ (Margenau and Murphy 1943, p. 299), which proves part (ii).

Step 7. Consider now $\Gamma_{ij}{}^{(1)}(T) = \Gamma_{ij}{}^{(2)}(T)$, and assume $P \neq 0$. From step 6, $|P| \neq 0$.

Step 8. Hence we can find $\lambda \neq 0$ such that $|P - \lambda E| = 0$, for the constant term in this equation in λ is just $|P| \neq 0$.

Step 9. Consider $Q = P - \lambda E$ and the vectors $w_j = Q_{ij} u_i = v_j - \lambda u_j$. From (D.2) and (D.4) we have $Tw_j = \Gamma_{ij}{}^{(1)}(T) w_i$.

Step 10. Now the argument of steps 5 and 6 applied to Q and the vectors w_j gives that either $|Q| \neq 0$ contrary to our definition of Q from step 8, or $Q = 0$, i.e. $P = \lambda E$ which proves part (iii) of the lemma.

Corollary. Part (iii) of Schur's lemma can also be stated thus:

a matrix which commutes with every matrix of an irreducible representation of some group is a multiple of the unit matrix.

Problems

D.1 Trace the similarities and differences between the above proof of Schur's lemma with those found in Van der Waerden 1932, Wigner 1931, and effectively in Speiser 1937.

D.2 Two sets of functions ϕ_i and ϕ'_i, $i = 1$ to n, span the same vector space. They also transform according to exactly the same irreducible representation $\Gamma_{ij}(T)$ under a group of transformations T. Prove that $\phi_i = \lambda \phi'_i$ for all i, where λ is a constant.

Appendix E

Irreducible Representations of Abelian Groups

Theorem. *All irreducible representations of an Abelian group are one-dimensional.* Consider any representation $D_{ij}(A)$, $D_{ij}(B)$, . . of the Abelian group $\mathfrak{G}(A, B, \ldots)$. In our case all the matrices are unitary so that any *one* of them can be reduced to diagonal form by an equivalence transformation of the type (5.15) (see appendix C lemma 2 and problem C.2; Van der Waerden 1932, p. 26; Margenau and Murphy 1943, p. 316).

Step 1. Let $P^{-1}D(A)P = D'(A)$ be the diagonal matrix

$$\begin{bmatrix} \lambda_1 & \cdot & \cdot & \cdots & & & & & & \\ \cdot & \lambda_1 & \cdot & \cdots & & & & & & \\ \vdots & \vdots & \ddots & & & & & & & \\ \vdots & \vdots & & \lambda_1 & & & & & & \\ \vdots & \vdots & & & \lambda_2 & & & & & \\ \vdots & \vdots & & & & \ddots & & & & \\ \vdots & \vdots & & & & & \lambda_2 & & & \\ \vdots & \vdots & & & & & & \lambda_3 & & \\ \vdots & \vdots & & & & & & & \ddots & \end{bmatrix} \qquad \text{(E.1}$$

and put $P^{-1}D(B)P = D'(B)$. Since $\mathfrak{G}$ is Abelian

$$D'(A)D'(B) = D'(B)D'(A). \qquad \text{(E.2}$$

Now the left side of (E.2) is just the matrix $D'(B)$ with each row multiplied by the corresponding λ_m of $D'(A)$, and the right side is $D'(B)$ with each column multiplied by the corresponding λ_m of $D'(A)$. Thus $D'_{ij}(B) = 0$ if the λ_m's in the i and j^{th} rows of $D'(A)$ are not equal, i.e. if $D'_{ii}(A) \neq D'_{jj}(A)$ (i, j not summed). Thus $D'(B)$ has the form (E.3) where the blocks correspond to equal diagonal elements in $D'(A)$.

Step 2. $D'(B)$ can now be reduced to diagonal form by a transformation $P'^{-1}D'(B)P'$ with P' also having the form (E.3). This does not alter the $D'(A)$ because we are only making transformations between rows having the same λ_m. Thus both $D(A)$ and $D(B)$ have been reduced simultaneously to diagonal form.

Similarly all the other matrices can be reduced. For consider $D(T)$ and suppose that the preceding transformations have transformed it to $D''(T)$. Then by step 1, $D_{ij}''(T) = 0$ if rows i and j contain different diagonal elements in *any* of the matrices that have already been reduced to the diagonal form (E.1). If they do not, then by step 2, $D_{ij}''(T)$ can be reduced to zero by a transformation

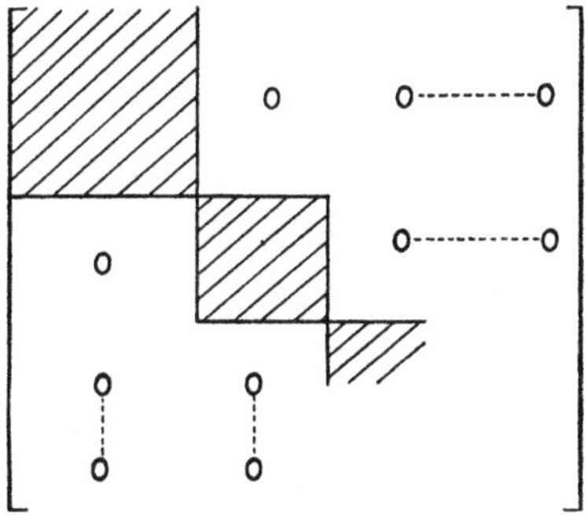

(E.3)

of the type (5.15), without altering the matrices that have already been diagonalized. Hence all matrices can be simultaneously diagonalized.

Thus any representation of an Abelian group can be reduced to a sum of one-dimensional representations, and these are therefore the only possible irreducible ones.

Appendix F

Momenta and Infinitesimal Transformations

Let $q_1, q_2, \ldots q_N$ be a complete set of co-ordinates describin[g] a system, and let T be the transformation to new co-ordinate[s] Q_n where

$$q_1 = Q_1 + \delta Q_1, \qquad q_n = Q_n, \qquad n = 2, 3, \ \ldots N,$$

and where in accordance with the convention of § 5 we replace th[e] Q_n by q_n after the transformation. Then analogously to (8.1) (8.2) we can define the infinitesimal transformation I_1 by

$$T = 1 + iI_1\delta q_1 + 0(\delta q_1{}^2). \tag{F.1}$$

We shall prove in this appendix that I_1 is related to the momentu[m] operator p_1 which is conjugate to the co-ordinate q_1 in the usua[l] sense (Schiff 1955, p. 133), in fact that we have (cf. equation 8.29)

$$\boxed{p_1 = \hbar I_1.} \tag{F.2}$$

Let S be any operator, ψ some function, and let the suffix t denot[e] transformed quantities after applying T. Then

$$T(S\psi) = S_t\psi_t = S_tT\psi.$$

Comparing left and right members of this equation, we must hav[e] (cf. footnote to equation (5.4))

$$\boxed{S_t = TST^{-1}.} \tag{F.3}$$

The variable q_1 is an operator in quantum mechanics as well as a co-ordinate occurring in a wave function. Hence putting q_1 fo[r] S and using (F.1), we obtain from (F.3)

$$q_1 + \delta q_1 = (1 + iI_1\delta q_1)q_1(1 - iI_1\delta q_1) + 0(\delta q_1{}^2),$$

i.e.

$$q_1I_1 - I_1q_1 = i. \tag{F.4a}$$

Similarly

$$q_nI_1 - I_1q_n = 0, \qquad n \neq 1;$$
$$p_nI_1 - I_1p_n = 0, \qquad \text{all } n. \tag{F.4b}$$

Now (F.4) shows that $\hbar I_1$ has exactly the same commutation relations as p_1 (Schiff 1955, p. 135). Thus the difference

$$p_1 - \hbar I_1 = f(p_n, q_n) \tag{F.5}$$

commutes with all p_n and all q_n. Since it commutes with p_n, f cannot depend on q_n which is the only fundamental variable not commuting with p_n. Similarly f does not depend on p_n and is therefore a constant. We now operate with (F.5) on the wave function ψ = constant. This represents a free particle at rest since it depends on no co-ordinates and hence cannot represent any motion or changing co-ordinates. Consequently $p_n\psi = 0$, and also $I_1\psi = 0$. Hence $f\psi = 0$, and since f is a constant, it must be zero. The result (F.2) now follows.

Another important result also follows from (F.3). The space inversion Π (3.11) commutes with every rotation $R(\alpha, \xi)$, so that from (8.3) it commutes with every infinitesimal rotation I_ξ. Thus

$$\boxed{I_{\xi t} = \Pi I_\xi \Pi^{-1} = \Pi \Pi^{-1} I_\xi = I_\xi,} \tag{F.6}$$

and I_ξ is invariant under Π. Thus from (F.2) all the angular momentum vectors are invariant under Π. As regards the orbital angular momentum **L** this fact is already obvious from (8.27), but it also applies to **J** (8.30) and the spin angular momentum operators introduced in § 11.

Appendix G

The Simple Harmonic Oscillator

The Hamiltonian

Consider the Hamiltonian

$$\boxed{\mathscr{H} = \frac{1}{2m}(p^2 + m^2\omega^2q^2),} \tag{G.1}$$

where m is the mass of a particle, q its position co-ordinate, and p the conjugate momentum. If this Hamiltonian is treated classically, it can easily be shown that the particle executes simple harmonic motion with angular frequency ω. Quantum mechanically, the operators p and q have to satisfy the commutation relation

$$qp - pq = i\hbar.$$

Instead of using p and q it is more convenient to introduce the operator

$$\boxed{a = (2m\hbar\omega)^{-1/2}(p + im\omega q),} \tag{G.2a}$$

and its Hermitian conjugate operator a^* defined by

$$\int (a^*\psi)^*\phi \, \mathrm{d}v = \int \psi^* a\phi \, \mathrm{d}v. \tag{G.3}$$

Since p and q are both self-conjugate, we have

$$a^* = (2m\hbar\omega)^{-1/2}(p - im\omega q). \tag{G.2b}$$

It is also convenient to define the operator

$$N = aa^*. \tag{G.4}$$

We then have

$$\begin{aligned} N = aa^* &= (2m\hbar\omega)^{-1}(p + im\omega q)(p - im\omega q) \\ &= (2m\hbar\omega)^{-1}[p^2 + m^2\omega^2q^2 + im\omega(qp - pq)] \\ &= (2m\hbar\omega)^{-1}[p^2 + m^2\omega^2q^2 - m\hbar\omega]. \end{aligned}$$

Thus

$$\boxed{\mathscr{H} = (N + \tfrac{1}{2})\hbar\omega = (aa^* + \tfrac{1}{2})\hbar\omega} \tag{G.5}$$

and in the same way it can be verified that

$$\boxed{\begin{aligned} a^*a - aa^* &= 1, \\ a^*a^n - a^na^* &= na^{n-1}. \end{aligned}} \qquad \begin{aligned} &\text{(G.6a)} \\ &\text{(G.6b)} \end{aligned}$$

Eigenfunctions and eigenvalues

Consider a function ψ_0 with the property

$$\boxed{a^*\psi_0 = 0.} \qquad \text{(G.7)}$$

From (G.4) we have $N\psi_0 = 0$, so that ψ_0 is an eigenfunction of N belonging to the eigenvalue $n = 0$. Consider now the function

$$\boxed{\psi_n = (n!)^{-1/2}a^n\psi_0,} \qquad \text{(G.8)}$$

where n is a positive integer. From (G.6) and (G.7) we have

$$\begin{aligned} N\psi_n &= (n!)^{-1/2}aa^*a^n\psi_0 \\ &= (n!)^{-1/2}a[na^{n-1} + a^na^*]\psi_0 \\ &= n\psi_n, \end{aligned}$$

so that ψ_n is an eigenfunction of N belonging to the eigenvalue n. It now follows from (G.5) that the ψ_n are also eigenfunctions of $\mathscr{H}$ belonging to the eigenvalues

$$E_n = (n + \tfrac{1}{2})\hbar\omega,$$

where n is a positive integer or zero. It can easily be verified that the ψ_n's are normalized and orthogonal (problems G.1 and G.2).

Physically we say that the energy level E_n contains n quanta of energy $\hbar\omega$, and (G.8) is described by saying that each factor of a "creates" an extra quantum out of the ground state ψ_0. It is this property (and see also (G.11)) which gives a and a^* their names of *creation* and *destruction operators*. N is the operator that counts the number of quanta.

Completeness of eigenfunctions

We shall now show, following Dirac (1958), that no further eigenfunctions of N and $\mathscr{H}$ exist. Let ψ_k be a normalized eigenfunction of N belonging to some eigenvalue k, which for the present is quite unrestricted. Then

$$\int \psi_k^* N\psi_k \, dv = k;$$

and from (G.3) and (G.4)

$$\int \psi_k^* N \psi_k \, dv = \int (a^* \psi_k)^* (a^* \psi_k) \, dv \geqq 0.$$

Comparing these, we conclude that

$$k \geqq 0, \tag{G.9}$$

$$\text{and if } a^* \psi_k = 0, \text{ then } k = 0. \tag{G.10}$$

We also have from (G.4)

$$N(a^* \psi_{k'}) = aa^* a^* \psi_{k'} = a^* N \psi_{k'} - a^* \psi_{k'} = (k' - 1) a^* \psi_{k'}. \tag{G.11}$$

Hence $a^* \psi_{k'}$ belongs to the eigenvalue $k' - 1$, and by repeated multiplication by a^* we can produce eigenfunctions $\psi_k = (a^*)^n \psi_{k'}$ belonging to the eigenvalues $k = k', k' - 1, k' - 2, k' - 3, \ldots$ If k' is not an integer, then k never has the value zero, and from (G.10) we never have $a^* \psi_k = 0$. Thus the series of eigenfunctions ψ_k goes on for ever, and sooner or later we reach a negative eigenvalue k, which is contrary to (G.9). We conclude therefore that k' can only have positive integer values, in which case our sequence of eigenfunctions terminates with the lowest one ψ_0 satisfying (G.10).

Analogously to (G.11), we have

$$N a \psi_{k'} = (k' + 1) a \psi_{k'},$$

so that $a\psi_{k'}$ is an eigenfunction belonging to the eigenvalue $k' + 1$. Thus starting from ψ_0, we can, by repeated multiplication by a, produce *one* eigenfunction belonging to each eigenvalue n. We have therefore derived the series of eigenfunctions ψ_n (G.8) instead of just conjuring them out of a hat. It now remains to show that each eigenvalue is non-degenerate, in which case it follows that the ψ_n are *all* the eigenfunctions. We show this by noting that N is invariant under the transformations

$$\begin{aligned} R(\theta)&:\!- \quad a \to e^{i\theta} a, \quad a^* \to e^{-i\theta} a; \\ \Pi&:\!- \quad a \to -a, \quad a^* \to -a^*; \end{aligned} \tag{G.12}$$

where θ is any (real) angle. These transformations commute, and there do not appear to be any others, so that the *complete* symmetry group of N is Abelian. All its irreducible representations are one-dimensional (appendix E), and we conclude that the eigenvalues of N are non-degenerate because there are no further symmetry properties that could produce any degeneracy.

Matrix elements

Any operator F can be expressed in terms of p and q, and hence as a power series in a and a^*. With the help of the commutation relation (G.6), any term in the expansion can be written $f_{rs}(a^*)^r a^s$. The matrix elements of this can be calculated easily from (G.8). We have

$$\begin{aligned}\langle m|(a^*)^r a^s|n\rangle \\ &= (m!n!)^{-1/2}\int (a^m\psi_0)^*(a^*)^r a^s a^n \psi_0 \, dv \\ &= (m!n!)^{-1/2}\int \psi_0^*(a^*)^{m+r}a^{s+n}\psi_0 \, dv. \qquad \text{(G.13)}\end{aligned}$$

This transforms under (G.12) according to

$$\exp i(s+n-m-r)\theta,$$

so that (G.13) is zero unless $m - n = s - r$ (from equation (13.8c) of the fundamental theorem of § 13). From (G.6b), (G.13) becomes for $m - n = s - r$

$$\langle m|(a^*)^r a^s|n\rangle = (m!n!)^{-1/2}(s+n)!\int \psi_0^*\psi_0 \, dv = \frac{(s+n)!}{(m!n!)^{1/2}}. \qquad \text{(G.14)}$$

Problems

G.1 Using (G.13), show that the ψ_m and ψ_n ($m \neq n$) are orthogonal. Also verify using (G.3) that the wave functions (G.8) are normalized if we assume ψ_0 is.

G.2 Calculate the matrix elements of q.

G.3 Calculate $\int \psi_n^* q^2 \psi_n \, dv$ and compare this with the mean value of q^2 for a classical oscillator with the same energy.

G.4 Show that $\int \psi_m^* q^r \psi_n \, dv = 0$ unless $m - n = r,\ r - 2,\ r - 4, \ldots, -r$.

Appendix H

The Irreducible Representations of the Complete Lorentz Group

In this appendix we shall construct systematically all the finite irreducible representations of the complete Lorentz group $\mathfrak{L}$, following the notation of § 31.

Step 1. Consider a vector space which is invariant under $\mathfrak{L}$. We can reduce it according to the proper Lorentz group $\mathfrak{l}$, and let $u_{mm'}$ be a set of standard base vectors in it transforming according to $D^{(jj')}$. If j, j' are both integers or both half odd integers, $D^{(jj')}$ is single-valued, and otherwise double-valued (problems 31.6 and 31.9).

Step 2. Now Π_T commutes with every proper and improper Lorentz transformation so that by Schur's lemma it is represented by a multiple aE of the unit matrix E. Also $(\Pi_T)^2$ is the identity transformation, so that $a = 1$ or -1 in single-valued representations. In double-valued representations we have $a = \pm 1$ as one possibility because of the double-valuedness, and $a = \pm i$ as another.

Step 3. Now we have that

$$\tau\, L(v,z)\tau = L(-v,z), \quad \text{i.e.} \quad \tau \exp(i\theta I_{zT}) = \exp(-i\theta I_{zT})\, \tau,$$

whence by expanding the exponential we obtain

$$I_{zT}\, \tau = -\tau I_{zT}.$$

Also τ commutes with a pure rotation and hence with I_{xy}. Therefore equations (31.10), (31.12) imply

$$\begin{aligned} A_z(\tau u_{mm'}) &= (-\tfrac{1}{2}I_{xy}\tau - \tfrac{1}{2}I_{zT}\tau)u_{mm'} \\ &= (-\tfrac{1}{2}I_{xy} + \tfrac{1}{2}\tau I_{zT})u_{mm'} \\ &= -\tau B_z u_{mm'} = -m'(\tau u_{mm'}), \\ B_z(\tau u_{mm'}) &= -m(\tau u_{mm'}). \end{aligned}$$

Thus the $\tau u_{mm'}$ behave as base vectors $U_{-m',-m}$. By operating also with A_+, A_-, B_+, B_- it is easily shown that the vectors $\delta\tau u_{mm'}$ in fact transform like (31.12) according to $D^{(j'j)}$ under $\mathfrak{l}$, where $\delta = (-1)^{j+j'-m-m'}$.

Step 4. We also have $\Pi L(v, z)\Pi = L(-v, z)$ and Π commutes with every pure rotation. Step 3 therefore applies also to $\Pi u_{mm'}$

Now $\Pi u_{mm'} = \tau(\Pi\tau)u_{mm'} = a\tau u_{mm'}$ where a is given by step 2, so that $\Pi u_{mm'}$ is proportional to $\tau u_{mm'}$. The vectors $u_{mm'}$ and $\tau u_{mm'}$ therefore span a vector space which is invariant under $\mathfrak{L}$. Also the matrix representing Π is always uniquely determined by those representing $\Pi\tau$ and τ.

Step 5. If $j = j'$, the $u_{mm'}$ and $\tau u_{mm'}$ both transform under $\mathfrak{l}$ according to $D^{(jj)}$ and may be related. In fact if we put

$$U_{mm'} = u_{mm'} + \delta\tau u_{-m',-m}, \qquad V_{mm'} = u_{mm'} - \delta\tau u_{-m',-m},$$

we have since $D^{(jj)}$ is single-valued and $\tau^2 u_{mm'} = u_{mm'}$,

$$\tau U_{mm'} = \delta U_{-m',-m}, \qquad \tau V_{mm'} = -\delta V_{-m',-m}.$$

Thus the $U_{mm'}$ span a vector space of dimension $(2j + 1)^2$ which is invariant under $\mathfrak{L}$, and so do the $V_{mm'}$. Coupled with the two possibilities of step 2, this gives four different representations of $\mathfrak{L}$. We denote them by $D^{(jj;r)}$, $r = 0, 1, 2, 3$, and they are related to one another in the same way (31.17) as the $D^{(\frac{1}{2}\frac{1}{2};r)}$ discussed there in detail.

Step 6. If $j \neq j'$ the $u_{mm'}$ and $\tau u_{mm'}$ transform under $\mathfrak{l}$ according to different irreducible representations and are therefore orthogonal. Hence to obtain a vector space invariant under $\mathfrak{L}$ we must take the $2(2j + 1)(2j' + 1)$ vectors $u_{mm'}$, $\tau u_{mm'}$ together. Since τ^2 is the identity transformation, τ is represented with respect to these base vectors by $D(\tau)$ where

$$D(\tau) = \begin{bmatrix} 0 & E \\ E & 0 \end{bmatrix}$$

for single-valued representations, and

$$D(\tau) = \pm \begin{bmatrix} 0 & E \\ E & 0 \end{bmatrix} \quad \text{(H.1a)} \qquad \text{or} \qquad \pm \begin{bmatrix} 0 & E \\ -E & 0 \end{bmatrix} \quad \text{(H.1b)}$$

for double-valued ones. Coupled with two alternatives of step 2, this gives two different representations $D^{(jj'+j'j;r)}$, $r = 0, 1$ when single-valued, and four representations $D^{(jj'+j'j;r)}$, $r = 4, 5, 6, 7$ when double-valued. In the case of (H.1b) it is convenient to use $i\tau u_{mm'}$ instead of $\tau u_{mm'}$ as base vectors, in which case $D(\tau)$ becomes

$$\pm i \begin{bmatrix} 0 & E \\ E & 0 \end{bmatrix}.$$

The representations $D^{(jj'+j'j;r)}$ are therefore related like the $D^{(\frac{1}{2}0+0\frac{1}{2};r)}$ by

$$D^{(jj'+j'j;r)}(T) = \chi_r(\alpha)D^{(jj'+j'j;4)}(T),$$

where $\alpha = E, \Pi, \tau, \Pi\tau$ as T belongs to the branches I, ΠI, τI, $\Pi\tau$ and the $\chi_r(\alpha)$ are given by Table 37. The single-valued ones ar related to one another like $D^{(\frac{1}{2}\frac{1}{2};r)}$, $r = 0, 1$ (equation (31.17)): th representations corresponding to $r = 2, 3$ are equivalent to thos with $r = 1, 0$ respectively when $j \neq j'$. In all cases the represen tation with $r = 0$ or 4 is defined to be the one obtained from (31.15 where u_{+0}, u_{-0}, u_{0+}, u_{0-} transform according to $D^{(\frac{1}{2}0+0\frac{1}{2};\,4)}$. Th single-valued ones can also be obtained by reducing

$$D^{(\frac{1}{2}\frac{1}{2};\,0)} \times D^{(\frac{1}{2}\frac{1}{2};\,0)} \times D^{(\frac{1}{2}\frac{1}{2};\,0)} \times \ldots$$

Step 7. We now show that all the above representations are in equivalent. Clearly this is so for those belonging to different pairs c values of j, j' because they are inequivalent under proper Lorent transformations. Suppose now for $j = j'$

$$D^{(jj;r)}(T) = PD^{(jj;s)}(T)P^{-1}, \qquad \text{(H.2}$$

i.e. that the two representations r and s (for short) are equivalent By construction r and s are identical for proper Lorentz transfor mations L, and from (H.2) P commutes with the irreducible repre sentation $D^{(jj;r)}(L) = D^{(jj;s)}(L)$ of the proper Lorentz group. Henc by Schur's lemma $P = cE$ where c is a constant. Then from (H.2 r and s are completely identical. Suppose similarly for $j \neq j'$

$$D^{(jj'+j'j;r)}(T) = PD^{(jj'+j'j;s)}(T)P^{-1}. \qquad \text{(H.3}$$

The representations are again equal for proper Lorentz transforma tions. However they become reducible

$$D^{(jj'+j'j;r)}(L) = \begin{bmatrix} L_1 & 0 \\ 0 & L_2 \end{bmatrix}$$

when we restrict ourselves to proper Lorentz transformations Hence if we write

$$P = \begin{bmatrix} P_1 & P_4 \\ P_3 & P_2 \end{bmatrix},$$

Schur's lemma gives $P_1 = c_1E$, $P_2 = c_2E$, $P_3 = P_4 = 0$. By usin this P and putting $T = \tau$ and $\Pi\tau$ in (H.3), it can easily be verifie that the constants c_1, c_2 cannot be chosen such that (H.3) is satisfie if $r \neq s$. Thus all the representations described are inequivalent.

Step 8. Given a finite vector space $\mathfrak{R}$ which is irreducible unde $\mathfrak{L}$, we can construct in this space one of the above representation by the procedure we have used. Using A_z, B_z we first pick out fron $\mathfrak{R}$ a vector $u_{jj'}$ with the highest eigenvalues $m = j$, $m' = j'$. Fron it using A_-, B_- (31.12) we construct all the $u_{mm'}$. Then from steps 4 and 5 we obtain in a unique way a definite set of base vectors. Now in steps 2, 5, 6 we listed *all* possible alternatives, so that our base vectors must transform according to one of the representations $D^{(jj;r)}$, $r = 0, 1, 2, 3$ or $D^{(jj'+j'j;r)}$, $r = 0, 1$ (single-valued) or $r = 4, 5, 6, 7$ (double-valued). Hence the irreducible space $\mathfrak{R}$ transforms according to a representation which is equivalent to one of these. It follows that we have derived all the finite irreducible representations of the complete Lorentz group.

Appendix I

Table of Wigner Coefficients $(jj'mm'|JM)$

Note: Reading down one column in the table gives the coefficients in the expansion

$$W_M^{(J)} = \sum_{m,m'} (jj'mm'|JM) U_m^{(j)} V_{m'}^{(j')}, \tag{20.1}$$

and reading across one row gives the coefficients in

$$U_m^{(j)} V_{m'}^{(j')} = \sum_{J,M} (jj'mm'|JM) W_M^{(J)}. \tag{20.5}$$

The tables are taken from Cohen (1949)†, who also gives a table for $D^{(3/2)} \times D^{(5/2)}$ not included here. Algebraic tables have been given by the following:
$D^{(l)} \times D^{(1/2)}$, $D^{(l)} \times D^{(1)}$, $D^{(l)} \times D^{(2)}$, Condon and Shortley (1951);
$D^{(l)} \times D^{(3/2)}$, Cohen (1949);
$D^{(l)} \times D^{(5/2)}$, Melvin and Swamy (1957);
$D^{(l)} \times D^{(3)}$, Falkoff *et al.* (1952).

Extensive numerical tables in decimal form have been published by Rose (1957) and Simon (1954).

$D_{\frac{1}{2}} \times D_{\frac{1}{2}}$

	W_1^1	W_0^1	W_0^0	W_{-1}^1
$U_{1/2}^{1/2}\ V_{1/2}^{1/2}$	1			
$U_{1/2}^{1/2}\ V_{-1/2}^{1/2}$		$\sqrt{\frac{1}{2}}$	$\sqrt{\frac{1}{2}}$	
$U_{-1/2}^{1/2}\ V_{1/2}^{1/2}$		$\sqrt{\frac{1}{2}}$	$-\sqrt{\frac{1}{2}}$	
$U_{-1/2}^{1/2}\ V_{-1/2}^{1/2}$				1

† The tables are reproduced here through the kind permission of Dr. Cohen.

$D_{\frac{1}{2}} \times D_1$

	$W^{3/2}_{3/2}$	$W^{3/2}_{1/2}$	$W^{1/2}_{1/2}$	$W^{3/2}_{-1/2}$	$W^{1/2}_{-1/2}$	$W^{3/2}_{-3/2}$
$U^1_1 \quad V^{1/2}_{1/2}$	1					
$U^1_1 \quad V^{1/2}_{-1/2}$		$\sqrt{\frac{1}{3}}$	$\sqrt{\frac{2}{3}}$			
$U^1_0 \quad V^{1/2}_{1/2}$		$\sqrt{\frac{2}{3}}$	$-\sqrt{\frac{1}{3}}$			
$U^1_0 \quad V^{1/2}_{-1/2}$				$\sqrt{\frac{2}{3}}$	$\sqrt{\frac{1}{3}}$	
$U^1_{-1} \quad V^{1/2}_{1/2}$				$\sqrt{\frac{1}{3}}$	$-\sqrt{\frac{2}{3}}$	
$U^1_{-1} \quad V^{1/2}_{-1/2}$						1

$D_{\frac{1}{2}} \times D_{\frac{3}{2}}$

	W^2_2	W^2_1	W^1_1	W^2_0	W^1_0	W^2_{-1}	W^1_{-1}	W^2_{-2}
$U^{3/2}_{3/2} \quad V^{1/2}_{1/2}$	1							
$U^{3/2}_{3/2} \quad V^{1/2}_{-1/2}$		$\sqrt{\frac{1}{4}}$	$\sqrt{\frac{3}{4}}$					
$U^{3/2}_{1/2} \quad V^{1/2}_{1/2}$		$\sqrt{\frac{3}{4}}$	$-\sqrt{\frac{1}{4}}$					
$U^{3/2}_{1/2} \quad V^{1/2}_{-1/2}$				$\sqrt{\frac{1}{2}}$	$\sqrt{\frac{1}{2}}$			
$U^{3/2}_{-1/2} \quad V^{1/2}_{1/2}$				$\sqrt{\frac{1}{2}}$	$-\sqrt{\frac{1}{2}}$			
$U^{3/2}_{-1/2} \quad V^{1/2}_{-1/2}$						$\sqrt{\frac{3}{4}}$	$\sqrt{\frac{1}{4}}$	
$U^{3/2}_{-3/2} \quad V^{1/2}_{1/2}$						$\sqrt{\frac{1}{4}}$	$-\sqrt{\frac{3}{4}}$	
$U^{3/2}_{-3/2} \quad V^{1/2}_{-1/2}$								1

$D_{\frac{1}{2}} \times D_2$

	$W^{5/2}_{5/2}$	$W^{3/2}_{3/2}$	$W^{3/2}_{3/2}$	$W^{5/2}_{1/2}$	$W^{3/2}_{1/2}$	$W^{5/2}_{-1/2}$	$W^{3/2}_{-1/2}$	$W^{5/2}_{-3/2}$	$W^{3/2}_{-3/2}$	$W^{5/2}_{-5/2}$
$U^2_2 \quad V^{1/2}_{1/2}$	1									
$U^2_2 \quad V^{1/2}_{-1/2}$		$\sqrt{\frac{1}{5}}$	$\sqrt{\frac{4}{5}}$							
$U^2_1 \quad V^{1/2}_{1/2}$		$\sqrt{\frac{4}{5}}$	$-\sqrt{\frac{1}{5}}$							
$U^2_1 \quad V^{1/2}_{-1/2}$				$\sqrt{\frac{2}{5}}$	$\sqrt{\frac{3}{5}}$					
$U^2_0 \quad V^{1/2}_{1/2}$				$\sqrt{\frac{3}{5}}$	$-\sqrt{\frac{2}{5}}$					
$U^2_0 \quad V^{1/2}_{-1/2}$						$\sqrt{\frac{3}{5}}$	$\sqrt{\frac{2}{5}}$			
$U^2_{-1} \quad V^{1/2}_{1/2}$						$\sqrt{\frac{2}{5}}$	$-\sqrt{\frac{3}{5}}$			
$U^2_{-1} \quad V^{1/2}_{-1/2}$								$\sqrt{\frac{4}{5}}$	$\sqrt{\frac{1}{5}}$	
$U^2_{-2} \quad V^{1/2}_{1/2}$								$\sqrt{\frac{1}{5}}$	$-\sqrt{\frac{4}{5}}$	
$U^2_{-2} \quad V^{1/2}_{-1/2}$										1

$D_{\frac{1}{2}} \times D_{\frac{5}{2}}$

	W^3_3	W^3_2	W^2_2	W^3_1	W^2_1	W^3_0	W^2_0	W^3_{-1}	W^2_{-1}	W^3_{-2}	W^2_{-2}	W^3_{-3}
$U^{5/2}_{5/2}\ V^{1/2}_{1/2}$	1											
$U^{5/2}_{5/2}\ V^{1/2}_{-1/2}$		$\sqrt{\frac{1}{6}}$	$\sqrt{\frac{5}{6}}$									
$U^{5/2}_{3/2}\ V^{1/2}_{1/2}$		$\sqrt{\frac{5}{6}}$	$-\sqrt{\frac{1}{6}}$									
$U^{5/2}_{3/2}\ V^{1/2}_{-1/2}$				$\sqrt{\frac{1}{3}}$	$\sqrt{\frac{2}{3}}$							
$U^{5/2}_{1/2}\ V^{1/2}_{1/2}$				$\sqrt{\frac{2}{3}}$	$-\sqrt{\frac{1}{3}}$							
$U^{5/2}_{1/2}\ V^{1/2}_{-1/2}$						$\sqrt{\frac{1}{2}}$	$\sqrt{\frac{1}{2}}$					
$U^{5/2}_{-1/2}\ V^{1/2}_{1/2}$						$\sqrt{\frac{1}{2}}$	$-\sqrt{\frac{1}{2}}$					
$U^{5/2}_{-1/2}\ V^{1/2}_{-1/2}$								$\sqrt{\frac{2}{3}}$	$\sqrt{\frac{1}{3}}$			
$U^{5/2}_{-3/2}\ V^{1/2}_{1/2}$								$\sqrt{\frac{1}{3}}$	$-\sqrt{\frac{2}{3}}$			
$U^{5/2}_{-3/2}\ V^{1/2}_{-1/2}$										$\sqrt{\frac{5}{6}}$	$\sqrt{\frac{1}{6}}$	
$U^{5/2}_{-5/2}\ V^{1/2}_{1/2}$										$\sqrt{\frac{1}{6}}$	$-\sqrt{\frac{5}{6}}$	
$U^{5/2}_{-5/2}\ V^{1/2}_{-1/2}$												1

$D_{\frac{1}{2}} \times D_3$

	$W^{7/2}_{7/2}$	$W^{7/2}_{5/2}$	$W^{5/2}_{5/2}$	$W^{7/2}_{3/2}$	$W^{5/2}_{3/2}$	$W^{7/2}_{1/2}$	$W^{5/2}_{1/2}$	$W^{7/2}_{-1/2}$	$W^{5/2}_{-1/2}$	$W^{7/2}_{-3/2}$	$W^{5/2}_{-3/2}$	$W^{7/2}_{-5/2}$	$W^{5/2}_{-5/2}$	$W^{7/2}_{-7/2}$
$U^3_3 \quad V^{1/2}_{1/2}$	1													
$U^3_3 \quad V^{1/2}_{-1/2}$		$\sqrt{\frac{1}{7}}$	$\sqrt{\frac{6}{7}}$											
$U^3_2 \quad V^{1/2}_{1/2}$		$\sqrt{\frac{6}{7}}$	$-\sqrt{\frac{1}{7}}$											
$U^3_2 \quad V^{1/2}_{-1/2}$				$\sqrt{\frac{2}{7}}$	$\sqrt{\frac{5}{7}}$									
$U^3_1 \quad V^{1/2}_{1/2}$				$\sqrt{\frac{5}{7}}$	$-\sqrt{\frac{2}{7}}$									
$U^3_1 \quad V^{1/2}_{-1/2}$						$\sqrt{\frac{3}{7}}$	$\sqrt{\frac{4}{7}}$							
$U^3_0 \quad V^{1/2}_{1/2}$						$\sqrt{\frac{4}{7}}$	$-\sqrt{\frac{3}{7}}$							
$U^3_0 \quad V^{1/2}_{-1/2}$								$\sqrt{\frac{4}{7}}$	$\sqrt{\frac{3}{7}}$					
$U^3_{-1} \quad V^{1/2}_{1/2}$								$\sqrt{\frac{3}{7}}$	$-\sqrt{\frac{4}{7}}$					
$U^3_{-1} \quad V^{1/2}_{-1/2}$										$\sqrt{\frac{5}{7}}$	$\sqrt{\frac{2}{7}}$			
$U^3_{-2} \quad V^{1/2}_{1/2}$										$\sqrt{\frac{2}{7}}$	$-\sqrt{\frac{5}{7}}$			
$U^3_{-2} \quad V^{1/2}_{-1/2}$												$\sqrt{\frac{6}{7}}$	$\sqrt{\frac{1}{7}}$	
$U^3_{-3} \quad V^{1/2}_{1/2}$												$\sqrt{\frac{1}{7}}$	$-\sqrt{\frac{6}{7}}$	
$U^3_{-3} \quad V^{1/2}_{-1/2}$														1

$D_{\frac{1}{2}} \times D_{\frac{7}{2}}$

	W^4_4	W^4_3 W^3_3	W^4_2 W^3_2	W^4_1 W^3_1	W^4_0 W^3_0	W^4_{-1} W^3_{-1}	W^4_{-2} W^3_{-2}	W^4_{-3} W^3_{-3}	W^4_{-4}
$U^{7/2}_{7/2}$ $V^{1/2}_{1/2}$	1								
$U^{7/2}_{7/2}$ $V^{1/2}_{-1/2}$ $U^{7/2}_{5/2}$ $V^{1/2}_{1/2}$		$\sqrt{\frac{1}{8}}$ $\sqrt{\frac{7}{8}}$ $\sqrt{\frac{7}{8}}$ $-\sqrt{\frac{1}{8}}$							
$U^{7/2}_{5/2}$ $V^{1/2}_{-1/2}$ $U^{7/2}_{3/2}$ $V^{1/2}_{1/2}$			$\sqrt{\frac{1}{4}}$ $\sqrt{\frac{3}{4}}$ $\sqrt{\frac{3}{4}}$ $-\sqrt{\frac{1}{4}}$						
$U^{7/2}_{3/2}$ $V^{1/2}_{-1/2}$ $U^{7/2}_{1/2}$ $V^{1/2}_{1/2}$				$\sqrt{\frac{3}{8}}$ $\sqrt{\frac{5}{8}}$ $\sqrt{\frac{5}{8}}$ $-\sqrt{\frac{3}{8}}$					
$U^{7/2}_{1/2}$ $V^{1/2}_{-1/2}$ $U^{7/2}_{-1/2}$ $V^{1/2}_{1/2}$					$\sqrt{\frac{1}{2}}$ $\sqrt{\frac{1}{2}}$ $\sqrt{\frac{1}{2}}$ $-\sqrt{\frac{1}{2}}$				
$U^{7/2}_{-1/2}$ $V^{1/2}_{-1/2}$ $U^{7/2}_{-3/2}$ $V^{1/2}_{1/2}$						$\sqrt{\frac{5}{8}}$ $\sqrt{\frac{3}{8}}$ $\sqrt{\frac{3}{8}}$ $-\sqrt{\frac{5}{8}}$			
$U^{7/2}_{-3/2}$ $V^{1/2}_{-1/2}$ $U^{7/2}_{-5/2}$ $V^{1/2}_{1/2}$							$\sqrt{\frac{3}{4}}$ $\sqrt{\frac{1}{4}}$ $\sqrt{\frac{1}{4}}$ $-\sqrt{\frac{3}{4}}$		
$U^{7/2}_{-5/2}$ $V^{1/2}_{-1/2}$ $U^{7/2}_{-7/2}$ $V^{1/2}_{1/2}$								$\sqrt{\frac{7}{8}}$ $\sqrt{\frac{1}{8}}$ $\sqrt{\frac{1}{8}}$ $-\sqrt{\frac{7}{8}}$	
$U^{7/2}_{-7/2}$ $V^{1/2}_{-1/2}$									1

$D_1 \times D_1$

	W^2_2	W^2_1	W^1_1	W^2_0	W^1_0	W^0_0	W^2_{-1}	W^1_{-1}	W^2_{-2}
$U^1_1 \; V^1_1$	1								
$U^1_1 \; V^1_0$		$\sqrt{\frac{1}{2}}$	$\sqrt{\frac{1}{2}}$						
$U^1_0 \; V^1_1$		$\sqrt{\frac{1}{2}}$	$-\sqrt{\frac{1}{2}}$						
$U^1_1 \; V^1_{-1}$				$\sqrt{\frac{1}{6}}$	$\sqrt{\frac{1}{2}}$	$\sqrt{\frac{1}{3}}$			
$U^1_0 \; V^1_0$				$\sqrt{\frac{2}{3}}$	0	$-\sqrt{\frac{1}{3}}$			
$U^1_{-1} \; V^1_1$				$\sqrt{\frac{1}{6}}$	$-\sqrt{\frac{1}{2}}$	$\sqrt{\frac{1}{3}}$			
$U^1_0 \; V^1_{-1}$							$\sqrt{\frac{1}{2}}$	$\sqrt{\frac{1}{2}}$	
$U^1_{-1} \; V^1_0$							$\sqrt{\frac{1}{2}}$	$-\sqrt{\frac{1}{2}}$	
$U^1_{-1} \; V^1_{-1}$									1

$D_1 \times D_{\frac{3}{2}}$

	$W^{5/2}_{5/2}$	$W^{5/2}_{3/2}$	$W^{3/2}_{3/2}$	$W^{5/2}_{1/2}$	$W^{3/2}_{1/2}$	$W^{1/2}_{1/2}$	$W^{5/2}_{-1/2}$	$W^{3/2}_{-1/2}$	$W^{1/2}_{-1/2}$	$W^{5/2}_{-3/2}$	$W^{3/2}_{-3/2}$	$W^{5/2}_{-5/2}$
$U^{3/2}_{3/2}\ V^1_1$	1											
$U^{3/2}_{3/2}\ V^1_0$		$\sqrt{\frac{2}{5}}$	$\sqrt{\frac{3}{5}}$									
$U^{3/2}_{1/2}\ V^1_1$		$\sqrt{\frac{3}{5}}$	$-\sqrt{\frac{2}{5}}$									
$U^{3/2}_{3/2}\ V^1_{-1}$				$\sqrt{\frac{1}{10}}$	$\sqrt{\frac{2}{5}}$	$\sqrt{\frac{1}{2}}$						
$U^{3/2}_{1/2}\ V^1_0$				$\sqrt{\frac{3}{5}}$	$\sqrt{\frac{1}{15}}$	$-\sqrt{\frac{1}{3}}$						
$U^{3/2}_{-1/2}\ V^1_1$				$\sqrt{\frac{3}{10}}$	$-\sqrt{\frac{8}{15}}$	$\sqrt{\frac{1}{6}}$						
$U^{3/2}_{1/2}\ V^1_{-1}$							$\sqrt{\frac{3}{10}}$	$\sqrt{\frac{8}{15}}$	$\sqrt{\frac{1}{6}}$			
$U^{3/2}_{-1/2}\ V^1_0$							$\sqrt{\frac{3}{5}}$	$-\sqrt{\frac{1}{15}}$	$-\sqrt{\frac{1}{3}}$			
$U^{3/2}_{-3/2}\ V^1_1$							$\sqrt{\frac{1}{10}}$	$-\sqrt{\frac{2}{5}}$	$\sqrt{\frac{1}{2}}$			
$U^{3/2}_{-1/2}\ V^1_{-1}$										$\sqrt{\frac{3}{5}}$	$\sqrt{\frac{2}{5}}$	
$U^{3/2}_{-3/2}\ V^1_0$										$\sqrt{\frac{2}{5}}$	$-\sqrt{\frac{3}{5}}$	
$U^{3/2}_{-3/2}\ V^1_{-1}$												1

$D_1 \times D_2$

	W^3_3	W^3_2	W^2_2	W^3_1	W^2_1	W^1_1	W^3_0	W^2_0	W^1_0	W^3_{-1}	W^2_{-1}	W^1_{-1}	W^3_{-2}	W^2_{-2}	W^3_{-3}
$U^2_2\ V^1_1$	1														
$U^2_2\ V^1_0$		$\sqrt{\frac{1}{3}}$	$\sqrt{\frac{2}{3}}$												
$U^2_1\ V^1_1$		$\sqrt{\frac{2}{3}}$	$-\sqrt{\frac{1}{3}}$												
$U^2_2\ V^1_{-1}$				$\sqrt{\frac{1}{15}}$	$\sqrt{\frac{1}{3}}$	$\sqrt{\frac{3}{5}}$									
$U^2_1\ V^1_0$				$\sqrt{\frac{8}{15}}$	$\sqrt{\frac{1}{6}}$	$-\sqrt{\frac{3}{10}}$									
$U^2_0\ V^1_1$				$\sqrt{\frac{6}{15}}$	$-\sqrt{\frac{1}{2}}$	$\sqrt{\frac{1}{10}}$									
$U^2_1\ V^1_{-1}$							$\sqrt{\frac{1}{5}}$	$\sqrt{\frac{1}{2}}$	$\sqrt{\frac{3}{10}}$						
$U^2_0\ V^1_0$							$\sqrt{\frac{3}{5}}$	0	$-\sqrt{\frac{2}{5}}$						
$U^2_{-1}\ V^1_1$							$\sqrt{\frac{1}{5}}$	$-\sqrt{\frac{1}{2}}$	$\sqrt{\frac{3}{10}}$						
$U^2_0\ V^1_{-1}$										$\sqrt{\frac{6}{15}}$	$\sqrt{\frac{1}{2}}$	$\sqrt{\frac{1}{10}}$			
$U^2_{-1}\ V^1_0$										$\sqrt{\frac{8}{15}}$	$-\sqrt{\frac{1}{6}}$	$-\sqrt{\frac{3}{10}}$			
$U^2_{-2}\ V^1_1$										$\sqrt{\frac{1}{15}}$	$-\sqrt{\frac{1}{3}}$	$\sqrt{\frac{3}{5}}$			
$U^2_{-1}\ V^1_{-1}$													$\sqrt{\frac{2}{3}}$	$\sqrt{\frac{1}{3}}$	
$U^2_{-2}\ V^1_0$													$\sqrt{\frac{1}{3}}$	$-\sqrt{\frac{2}{3}}$	
$U^2_{-2}\ V^1_{-1}$															1

$D_1 \times D_{\frac{5}{2}}$

	$W^{7/2}_{7/2}$	$W^{7/2}_{5/2}$	$W^{5/2}_{5/2}$	$W^{7/2}_{3/2}$	$W^{5/2}_{3/2}$	$W^{3/2}_{3/2}$	$W^{7/2}_{1/2}$	$W^{5/2}_{1/2}$	$W^{3/2}_{1/2}$	$W^{7/2}_{-1/2}$	$W^{5/2}_{-1/2}$	$W^{3/2}_{-1/2}$	$W^{7/2}_{-3/2}$	$W^{5/2}_{-3/2}$	$W^{3/2}_{-3/2}$	$W^{7/2}_{-5/2}$	$W^{5/2}_{-5/2}$	$W^{7/2}_{-7/2}$
$U^{5/2}_{5/2} V^1_1$	1																	
$U^{5/2}_{5/2} V^1_0$		$\sqrt{\frac{2}{7}}$	$\sqrt{\frac{5}{7}}$															
$U^{5/2}_{3/2} V^1_1$		$\sqrt{\frac{5}{7}}$	$\sqrt{\frac{2}{7}}$															
$U^{5/2}_{5/2} V^1_{-1}$				$\sqrt{\frac{1}{21}}$	$\sqrt{\frac{2}{7}}$	$\sqrt{\frac{2}{3}}$												
$U^{5/2}_{3/2} V^1_0$				$\sqrt{\frac{10}{21}}$	$\sqrt{\frac{9}{35}}$	$-\sqrt{\frac{4}{15}}$												
$U^{5/2}_{1/2} V^1_1$				$\sqrt{\frac{10}{21}}$	$-\sqrt{\frac{16}{35}}$	$\sqrt{\frac{1}{15}}$												
$U^{5/2}_{3/2} V^1_{-1}$							$\sqrt{\frac{1}{7}}$	$\sqrt{\frac{16}{35}}$	$\sqrt{\frac{2}{5}}$									
$U^{5/2}_{1/2} V^1_0$							$\sqrt{\frac{4}{7}}$	$\sqrt{\frac{1}{35}}$	$-\sqrt{\frac{2}{5}}$									
$U^{5/2}_{-1/2} V^1_1$							$\sqrt{\frac{2}{7}}$	$-\sqrt{\frac{18}{35}}$	$\sqrt{\frac{1}{5}}$									
$U^{5/2}_{1/2} V^1_{-1}$										$\sqrt{\frac{2}{7}}$	$\sqrt{\frac{18}{35}}$	$\sqrt{\frac{1}{5}}$						
$U^{5/2}_{-1/2} V^1_0$										$\sqrt{\frac{4}{7}}$	$-\sqrt{\frac{1}{35}}$	$-\sqrt{\frac{2}{5}}$						
$U^{5/2}_{-3/2} V^1_1$										$\sqrt{\frac{1}{7}}$	$-\sqrt{\frac{16}{35}}$	$\sqrt{\frac{2}{5}}$						
$U^{5/2}_{-1/2} V^1_{-1}$													$\sqrt{\frac{10}{21}}$	$\sqrt{\frac{16}{35}}$	$\sqrt{\frac{1}{15}}$			
$U^{5/2}_{-3/2} V^1_0$													$\sqrt{\frac{10}{21}}$	$-\sqrt{\frac{9}{35}}$	$-\sqrt{\frac{4}{15}}$			
$U^{5/2}_{-5/2} V^1_1$													$\sqrt{\frac{1}{21}}$	$-\sqrt{\frac{2}{7}}$	$\sqrt{\frac{2}{3}}$			
$U^{5/2}_{-3/2} V^1_{-1}$																$\sqrt{\frac{5}{7}}$	$\sqrt{\frac{2}{7}}$	
$U^{5/2}_{-5/2} V^1_0$																$\sqrt{\frac{2}{7}}$	$-\sqrt{\frac{5}{7}}$	
$U^{5/2}_{-5/2} V^1_{-1}$																		1

$D_1 \times D_3$

	W^4_4	W^4_3	W^3_3	W^4_2	W^3_2	W^2_2	W^4_1	W^3_1	W^2_1	W^4_0	W^3_0	W^2_0	W^4_{-1}	W^3_{-1}	W^2_{-1}	W^4_{-2}	W^3_{-2}	W^2_{-2}	W^4_{-3}	W^3_{-3}	W^4_{-4}
$U^3_3 V^1_1$	1																				
$U^3_3 V^1_0$		$\sqrt{\frac{1}{4}}$	$\sqrt{\frac{3}{4}}$																		
$U^3_2 V^1_1$		$\sqrt{\frac{3}{4}}$	$-\sqrt{\frac{1}{4}}$																		
$U^3_3 V^1_{-1}$				$\sqrt{\frac{1}{28}}$	$\sqrt{\frac{1}{4}}$	$\sqrt{\frac{5}{7}}$															
$U^3_2 V^1_0$				$\sqrt{\frac{3}{7}}$	$\sqrt{\frac{1}{3}}$	$-\sqrt{\frac{5}{21}}$															
$U^3_1 V^1_1$				$\sqrt{\frac{15}{28}}$	$-\sqrt{\frac{5}{12}}$	$\sqrt{\frac{1}{21}}$															
$U^3_2 V^1_{-1}$							$\sqrt{\frac{3}{28}}$	$\sqrt{\frac{5}{12}}$	$\sqrt{\frac{10}{21}}$												
$U^3_1 V^1_0$							$\sqrt{\frac{15}{28}}$	$\sqrt{\frac{1}{12}}$	$-\sqrt{\frac{8}{21}}$												
$U^3_0 V^1_1$							$\sqrt{\frac{10}{28}}$	$-\sqrt{\frac{6}{12}}$	$\sqrt{\frac{3}{21}}$												
$U^3_1 V^1_{-1}$										$\sqrt{\frac{3}{14}}$	$\sqrt{\frac{1}{2}}$	$\sqrt{\frac{2}{7}}$									
$U^3_0 V^1_0$										$\sqrt{\frac{8}{14}}$	0	$-\sqrt{\frac{3}{7}}$									
$U^3_{-1} V^1_1$										$\sqrt{\frac{3}{14}}$	$-\sqrt{\frac{1}{2}}$	$\sqrt{\frac{2}{7}}$									
$U^3_0 V^1_{-1}$													$\sqrt{\frac{10}{28}}$	$\sqrt{\frac{6}{12}}$	$\sqrt{\frac{3}{21}}$						
$U^3_{-1} V^1_0$													$\sqrt{\frac{15}{28}}$	$-\sqrt{\frac{1}{12}}$	$-\sqrt{\frac{8}{21}}$						
$U^3_{-2} V^1_1$													$\sqrt{\frac{3}{28}}$	$-\sqrt{\frac{5}{12}}$	$\sqrt{\frac{10}{21}}$						
$U^3_{-1} V^1_{-1}$																$\sqrt{\frac{15}{28}}$	$\sqrt{\frac{5}{12}}$	$\sqrt{\frac{1}{21}}$			
$U^3_{-2} V^1_0$																$\sqrt{\frac{3}{7}}$	$-\sqrt{\frac{1}{3}}$	$-\sqrt{\frac{5}{21}}$			
$U^3_{-3} V^1_1$																$\sqrt{\frac{1}{28}}$	$-\sqrt{\frac{1}{4}}$	$\sqrt{\frac{5}{7}}$			
$U^3_{-2} V^1_{-1}$																			$\sqrt{\frac{3}{4}}$	$\sqrt{\frac{1}{4}}$	
$U^3_{-3} V^1_0$																			$\sqrt{\frac{1}{4}}$	$-\sqrt{\frac{3}{4}}$	
$U^3_{-3} V^1_{-1}$																					1

$D_{\frac{3}{2}} \times D_{\frac{3}{2}}$

	W^3_3	W^3_2	W^2_2	W^3_1	W^2_1	W^1_1	W^3_0	W^2_0	W^1_0	W^0_0	W^3_{-1}	W^2_{-1}	W^1_{-1}	W^3_{-2}	W^2_{-2}	W^3_{-3}
$U^{3/2}_{3/2} V^{3/2}_{3/2}$	1															
$U^{3/2}_{3/2} V^{3/2}_{1/2}$		$\sqrt{\frac{1}{2}}$	$\sqrt{\frac{1}{2}}$													
$U^{3/2}_{1/2} V^{3/2}_{3/2}$		$\sqrt{\frac{1}{2}}$	$-\sqrt{\frac{1}{2}}$													
$U^{3/2}_{3/2} V^{3/2}_{-1/2}$				$\sqrt{\frac{1}{5}}$	$\sqrt{\frac{1}{2}}$	$\sqrt{\frac{3}{10}}$										
$U^{3/2}_{1/2} V^{3/2}_{1/2}$				$\sqrt{\frac{3}{5}}$	0	$-\sqrt{\frac{2}{5}}$										
$U^{3/2}_{-1/2} V^{3/2}_{3/2}$				$\sqrt{\frac{1}{5}}$	$-\sqrt{\frac{1}{2}}$	$\sqrt{\frac{3}{10}}$										
$U^{3/2}_{3/2} V^{3/2}_{-3/2}$							$\sqrt{\frac{1}{20}}$	$\sqrt{\frac{1}{4}}$	$\sqrt{\frac{9}{20}}$	$\sqrt{\frac{1}{4}}$						
$U^{3/2}_{1/2} V^{3/2}_{-1/2}$							$\sqrt{\frac{9}{20}}$	$\sqrt{\frac{1}{4}}$	$-\sqrt{\frac{1}{20}}$	$-\sqrt{\frac{1}{4}}$						
$U^{3/2}_{-1/2} V^{3/2}_{1/2}$							$\sqrt{\frac{9}{20}}$	$-\sqrt{\frac{1}{4}}$	$-\sqrt{\frac{1}{20}}$	$\sqrt{\frac{1}{4}}$						
$U^{3/2}_{-3/2} V^{3/2}_{3/2}$							$\sqrt{\frac{1}{20}}$	$-\sqrt{\frac{1}{4}}$	$\sqrt{\frac{9}{20}}$	$-\sqrt{\frac{1}{4}}$						
$U^{3/2}_{1/2} V^{3/2}_{-3/2}$											$\sqrt{\frac{1}{5}}$	$\sqrt{\frac{1}{2}}$	$\sqrt{\frac{3}{10}}$			
$U^{3/2}_{-1/2} V^{3/2}_{-1/2}$											$\sqrt{\frac{3}{5}}$	0	$-\sqrt{\frac{2}{5}}$			
$U^{3/2}_{-3/2} V^{3/2}_{1/2}$											$\sqrt{\frac{1}{5}}$	$-\sqrt{\frac{1}{2}}$	$\sqrt{\frac{3}{10}}$			
$U^{3/2}_{-1/2} V^{3/2}_{-3/2}$														$\sqrt{\frac{1}{2}}$	$\sqrt{\frac{1}{2}}$	
$U^{3/2}_{-3/2} V^{3/2}_{-1/2}$														$\sqrt{\frac{1}{2}}$	$-\sqrt{\frac{1}{2}}$	
$U^{3/2}_{-3/2} V^{3/2}_{-3/2}$																1

$D_{\frac{3}{2}} \times D_2$

	$W^{7/2}_{7/2}$	$W^{7/2}_{5/2}$	$W^{5/2}_{5/2}$	$W^{7/2}_{3/2}$	$W^{5/2}_{3/2}$	$W^{3/2}_{3/2}$	$W^{7/2}_{1/2}$	$W^{5/2}_{1/2}$	$W^{3/2}_{1/2}$	$W^{1/2}_{1/2}$	$W^{7/2}_{-1/2}$	$W^{5/2}_{-1/2}$	$W^{3/2}_{-1/2}$	$W^{1/2}_{-1/2}$	$W^{7/2}_{-3/2}$	$W^{5/2}_{-3/2}$	$W^{3/2}_{-3/2}$	$W^{7/2}_{-5/2}$	$W^{5/2}_{-5/2}$	$W^{7/2}_{-7/2}$
$U^2_2 V^{3/2}_{3/2}$	1																			
$U^2_2 V^{3/2}_{1/2}$		$\sqrt{\frac{3}{7}}$	$\sqrt{\frac{4}{7}}$																	
$U^2_1 V^{3/2}_{3/2}$		$\sqrt{\frac{4}{7}}$	$-\sqrt{\frac{3}{7}}$																	
$U^2_2 V^{3/2}_{-1/2}$				$\sqrt{\frac{1}{7}}$	$\sqrt{\frac{16}{35}}$	$\sqrt{\frac{2}{5}}$														
$U^2_1 V^{3/2}_{1/2}$				$\sqrt{\frac{4}{7}}$	$\sqrt{\frac{18}{35}}$	$-\sqrt{\frac{2}{5}}$														
$U^2_0 V^{3/2}_{3/2}$				$\sqrt{\frac{2}{7}}$	$-\sqrt{\frac{18}{35}}$	$\sqrt{\frac{1}{5}}$														
$U^2_2 V^{3/2}_{-3/2}$							$\sqrt{\frac{1}{35}}$	$\sqrt{\frac{6}{35}}$	$\sqrt{\frac{2}{5}}$	$\sqrt{\frac{2}{5}}$										
$U^2_1 V^{3/2}_{-1/2}$							$\sqrt{\frac{12}{35}}$	$\sqrt{\frac{5}{14}}$	0	$-\sqrt{\frac{3}{10}}$										
$U^2_0 V^{3/2}_{1/2}$							$\sqrt{\frac{18}{35}}$	$-\sqrt{\frac{3}{35}}$	$-\sqrt{\frac{1}{5}}$	$\sqrt{\frac{1}{5}}$										
$U^2_{-1} V^{3/2}_{3/2}$							$\sqrt{\frac{4}{35}}$	$-\sqrt{\frac{27}{70}}$	$\sqrt{\frac{2}{5}}$	$-\sqrt{\frac{1}{10}}$										
$U^2_1 V^{3/2}_{-3/2}$											$\sqrt{\frac{4}{35}}$	$\sqrt{\frac{27}{70}}$	$\sqrt{\frac{2}{5}}$	$\sqrt{\frac{1}{10}}$						
$U^2_0 V^{3/2}_{-1/2}$											$\sqrt{\frac{18}{35}}$	$\sqrt{\frac{3}{35}}$	$-\sqrt{\frac{1}{5}}$	$-\sqrt{\frac{1}{5}}$						
$U^2_{-1} V^{3/2}_{1/2}$											$\sqrt{\frac{12}{35}}$	$-\sqrt{\frac{5}{14}}$	0	$\sqrt{\frac{3}{10}}$						
$U^2_{-2} V^{3/2}_{3/2}$											$\sqrt{\frac{1}{35}}$	$-\sqrt{\frac{6}{35}}$	$\sqrt{\frac{2}{5}}$	$-\sqrt{\frac{2}{5}}$						
$U^2_0 V^{3/2}_{-3/2}$															$\sqrt{\frac{2}{7}}$	$\sqrt{\frac{18}{35}}$	$\sqrt{\frac{1}{5}}$			
$U^2_{-1} V^{3/2}_{-1/2}$															$\sqrt{\frac{4}{7}}$	$-\sqrt{\frac{1}{35}}$	$-\sqrt{\frac{2}{5}}$			
$U^2_{-2} V^{3/2}_{1/2}$															$\sqrt{\frac{1}{7}}$	$-\sqrt{\frac{16}{35}}$	$\sqrt{\frac{2}{5}}$			
$U^2_{-1} V^{3/2}_{-3/2}$																		$\sqrt{\frac{4}{7}}$	$\sqrt{\frac{3}{7}}$	
$U^2_{-2} V^{3/2}_{-1/2}$																		$\sqrt{\frac{3}{7}}$	$-\sqrt{\frac{4}{7}}$	
$U^2_{-2} V^{3/2}_{-3/2}$																				1

$D_2 \times D_2$

	W^4_4	W^4_3	W^3_3	W^4_2	W^3_2	W^2_2	W^4_1	W^3_1	W^2_1	W^1_1	W^4_0	W^3_0	W^2_0	W^1_0	W^0_0
$U^2_2 V^2_2$	1														
$U^2_2 V^2_1$		$\sqrt{\frac{1}{2}}$	$\sqrt{\frac{1}{2}}$												
$U^2_1 V^2_2$		$\sqrt{\frac{1}{2}}$	$-\sqrt{\frac{1}{2}}$												
$U^2_2 V^2_0$				$\sqrt{\frac{3}{14}}$	$\sqrt{\frac{1}{2}}$	$\sqrt{\frac{2}{7}}$									
$U^2_1 V^2_1$				$\sqrt{\frac{8}{14}}$	0	$-\sqrt{\frac{3}{7}}$									
$U^2_0 V^2_2$				$\sqrt{\frac{3}{14}}$	$-\sqrt{\frac{1}{2}}$	$\sqrt{\frac{2}{7}}$									
$U^2_2 V^2_{-1}$							$\sqrt{\frac{1}{14}}$	$\sqrt{\frac{3}{10}}$	$\sqrt{\frac{6}{14}}$	$\sqrt{\frac{2}{10}}$					
$U^2_1 V^2_0$							$\sqrt{\frac{6}{14}}$	$\sqrt{\frac{2}{10}}$	$-\sqrt{\frac{1}{14}}$	$-\sqrt{\frac{3}{10}}$					
$U^2_0 V^2_1$							$\sqrt{\frac{6}{14}}$	$-\sqrt{\frac{2}{10}}$	$-\sqrt{\frac{1}{14}}$	$\sqrt{\frac{3}{10}}$					
$U^2_{-1} V^2_2$							$\sqrt{\frac{1}{14}}$	$-\sqrt{\frac{3}{10}}$	$\sqrt{\frac{6}{14}}$	$-\sqrt{\frac{2}{10}}$					
$U^2_2 V^2_{-2}$											$\sqrt{\frac{1}{70}}$	$\sqrt{\frac{1}{10}}$	$\sqrt{\frac{4}{14}}$	$\sqrt{\frac{4}{10}}$	$\sqrt{\frac{1}{5}}$
$U^2_1 V^2_{-1}$											$\sqrt{\frac{16}{70}}$	$\sqrt{\frac{4}{10}}$	$\sqrt{\frac{1}{14}}$	$-\sqrt{\frac{1}{10}}$	$-\sqrt{\frac{1}{5}}$
$U^2_0 V^2_0$											$\sqrt{\frac{36}{70}}$	0	$-\sqrt{\frac{4}{14}}$	0	$\sqrt{\frac{1}{5}}$
$U^2_{-1} V^2_1$											$\sqrt{\frac{16}{70}}$	$-\sqrt{\frac{4}{10}}$	$\sqrt{\frac{1}{14}}$	$\sqrt{\frac{1}{10}}$	$-\sqrt{\frac{1}{5}}$
$U^2_{-2} V^2_2$											$\sqrt{\frac{1}{70}}$	$-\sqrt{\frac{1}{10}}$	$\sqrt{\frac{4}{14}}$	$-\sqrt{\frac{4}{10}}$	$\sqrt{\frac{1}{5}}$
$U^2_1 V^2_{-2}$															
$U^2_0 V^2_{-1}$															
$U^2_{-1} V^2_0$															
$U^2_{-2} V^2_1$															
$U^2_0 V^2_{-2}$															
$U^2_{-1} V^2_{-1}$															
$U^2_{-2} V^2_0$															
$U^2_{-1} V^2_{-2}$															
$U^2_{-2} V^2_{-1}$															
$U^2_{-2} V^2_{-2}$															

	W^4_{-1}	W^3_{-1}	W^2_{-1}	W^1_{-1}	W^4_{-2}	W^3_{-2}	W^2_{-2}	W^4_{-3}	W^3_{-3}	W^4_{-4}
$U^2_2 V^2_2$										
$U^2_2 V^2_1$										
$U^2_1 V^2_2$										
$U^2_2 V^2_0$										
$U^2_1 V^2_1$										
$U^2_0 V^2_2$										
$U^2_2 V^2_{-1}$										
$U^2_1 V^2_0$										
$U^2_0 V^2_1$										
$U^2_{-1} V^2_2$										
$U^2_2 V^2_{-2}$										
$U^2_1 V^2_{-1}$										
$U^2_0 V^2_0$										
$U^2_{-1} V^2_1$										
$U^2_{-2} V^2_2$										
$U^2_1 V^2_{-2}$	$\sqrt{\frac{1}{14}}$	$\sqrt{\frac{3}{10}}$	$\sqrt{\frac{6}{14}}$	$\sqrt{\frac{2}{10}}$						
$U^2_0 V^2_{-1}$	$\sqrt{\frac{6}{14}}$	$\sqrt{\frac{2}{10}}$	$-\sqrt{\frac{1}{14}}$	$-\sqrt{\frac{3}{10}}$						
$U^2_{-1} V^2_0$	$\sqrt{\frac{6}{14}}$	$-\sqrt{\frac{2}{10}}$	$-\sqrt{\frac{1}{14}}$	$\sqrt{\frac{3}{10}}$						
$U^2_{-2} V^2_1$	$\sqrt{\frac{1}{14}}$	$-\sqrt{\frac{3}{10}}$	$\sqrt{\frac{6}{14}}$	$-\sqrt{\frac{2}{10}}$						
$U^2_0 V^2_{-2}$					$\sqrt{\frac{3}{14}}$	$\sqrt{\frac{1}{2}}$	$\sqrt{\frac{2}{7}}$			
$U^2_{-1} V^2_{-1}$					$\sqrt{\frac{8}{14}}$	0	$-\sqrt{\frac{3}{7}}$			
$U^2_{-2} V^2_0$					$\sqrt{\frac{3}{14}}$	$-\sqrt{\frac{1}{2}}$	$\sqrt{\frac{2}{7}}$			
$U^2_{-1} V^2_{-2}$								$\sqrt{\frac{1}{2}}$	$\sqrt{\frac{1}{2}}$	
$U^2_{-2} V^2_{-1}$								$\sqrt{\frac{1}{2}}$	$-\sqrt{\frac{1}{2}}$	
$U^2_{-2} V^2_{-2}$										1

Appendix J

Notation for the Thirty-two Crystal Point-groups

International symbol	Full symmetry symbol	Schoenflies symbol
1	1	C_1
$\bar{1}$	$\bar{1}$	$S_2(C_i)$
2	2	C_2
m	m	$C_{1h}(C_s)$
2/m	$\frac{2}{m}$	C_{2h}
2mm	2mm	C_{2v}
222	222	$D_2(V)$
mmm	$\frac{2}{m}\frac{2}{m}\frac{2}{m}$	$D_{2h}(V_h)$
4	4	C_4
$\bar{4}$	$\bar{4}$	S_4
4/m	$\frac{4}{m}$	C_{4h}
4mm	4mm	C_{4v}
$\bar{4}$ 2m	$\bar{4}$ 2m	$D_{2d}(V_d)$
4 2 2	4 2 2	D_4
4/mmm	$\frac{4}{m}\frac{2}{m}\frac{2}{m}$	D_{4h}
3	3	C_3
$\bar{3}$	$\bar{3}$	$S_6(C_{3i})$
3m	3m	C_{3v}
$\bar{3}$m	$\bar{3}\frac{2}{m}$	D_{3d}
32	32	D_3
6	6	C_6
$\bar{6}$	$\bar{6}$	C_{3h}
6/m	$\frac{6}{m}$	C_{6h}

International symbol	Full symmetry symbol	Schoenflies symbol
6 mm	6 mm	C_{6v}
$\bar{6}$ m 2	$\bar{6}$ m2	D_{3h}
6 2 2	6 2 2	D_6
6/mmm	$\frac{6}{m}\frac{2}{m}\frac{2}{m}$	D_{6h}
2 3	2 3	T
m 3	$\frac{2}{m}\bar{3}$	T_h
$\bar{4}$ 3 m	$\bar{4}$ 3 m	T_d
4 3 2	4 3 2	O
m 3 m	$\frac{4}{m}\bar{3}\frac{2}{m}$	O_h

Appendix K

Character Tables for the Crystal Point-groups

Notation

The point-groups are designated by their *international symbols* which are explained in § 16. The corresponding *Schoenflies* symbols may be found in appendix J.

The *classes* of group elements are indicated by one typical element from each class. These elements are given by using an extension of the international notation for the groups. The symbol n now stands for a rotation by $360/n$ degrees and n^r for 360 r/n degrees. The rotation axis is given by a suffix, e.g. x, y, z, and d meaning an axis at 45° to Ox and Oy. The symbol m_z thus means a reflection in a plane perpendicular to the z-axis. $\bar{n}$ is an improper rotation by $360/n°$, i.e. a proper rotation by this angle followed by inversion through the origin. The inversion Π (3.11) itself is thus denoted by $\bar{1}$.

The *irreducible representations* are labelled systematically as follows. A and B always denote one-dimensional representations, B being used if a rotation by $360/n°$ about the principal axis (chosen as z-axis) has the character -1. E is used for a two-dimensional representation, and T for a three-dimensional one. A pair of complex conjugate one-dimensional representations are always bracketed together and regarded as a two-dimensional representation E, because for most purposes they behave as such due to time-reversal symmetry (§§ 19 and 23). If there are two representations in which the characters of m_z differ in sign, they are distinguished by $'$ and $''$. Subscripts g and u (German, gerade and ungerade) refer to positive and negative characters of the inversion Π (or $\bar{1}$ in the above notation). When this system allows several different labellings, the use of u and g takes preference over $'$ and $''$, which in turn takes preference over suffices 1, 2, etc. We have followed this scheme consistently in this appendix, though this has meant introducing minor differences from the notation of other authors.

Some point-groups are *direct products* $\mathfrak{g} \times \bar{1}$ (§ 15), where $\mathfrak{g}$ is a proper point-group and $\bar{1}$ the inversion group (E, Π). Their character tables are not given explicitly but may easily be constructed as follows. Each representation D of $\mathfrak{g}$ gives rise to two representations D_g and D_u of the group $\mathfrak{g} \times \bar{1}$. In these the proper rotations R

have the same characters as in D, but the improper rotations ΠR have the characters

$$\chi(\Pi R) = \chi(R) \text{ in } D_{\text{g}},$$
$$= -\chi(R) \text{ in } D_{\text{u}}.$$

The classes are as follows. Suppose $R_1, R_2, \ldots$ form a class of $\mathcal{g}$. Then the proper rotations $R_1, R_2, \ldots$ of the group $\mathcal{g} \times \bar{1}$ again form a class, and the corresponding improper rotations ΠR_1, $\Pi R_2, \ldots$ form another class. Thus $\mathcal{g} \times \bar{1}$ has twice as many classes as $\mathcal{g}$, which checks with the fact that it also has twice as many representations (cf. equation (14.16)). The above construction for the character table of $\mathcal{g} \times \bar{1}$ follows deductively from the theory of § 15, or can be justified *a posteriori* by noting that the character table satisfies all the orthogonality requirements, etc., of § 14.

On the right side of each character table are listed the quantities $x, y, z, x^2, y^2, z^2, xy, yz, zx, I_x, I_y, I_z$ in the row of the representation to which they belong. Here I_x, I_y, I_z are the infinitesimal rotation operators which transform as a pseudovector. The chief use of this is in connection with the *vibrational modes of molecules* as summarized in Table 23. In the case of groups $\mathcal{g} \times \bar{1}$ whose character tables are not given in full, the quantities x, y, z always transform according to the appropriate *ungerade* representations, D_{u}, and x^2, $y^2, z^2, xy, yz, zx, I_x, I_y, I_z$ according to gerade representations D_{g}.

Triclinic, monoclinic and orthorhombic point-groups

$\bar{1}$	E	$\bar{1}$	
A_{g}	1	1	x^2; y^2; z^2; xy; yz; zx; I_x; I_y; I_z.
A_{u}	1	-1	x; y; z.

2	E	2_z	
A	1	1	z; y^2; z^2; z^2; xy; I_z.
B	1	-1	x; y; xz; yz; I_x; I_y.

m	E	m_z	
A'	1	1	x; y; x^2; y^2; z^2; xy; I_z.
A''	1	-1	z; yz; xz; I_x; I_y.

222	E	2_x	2_y	2_z	
A	1	1	1	1	x^2; y^2; z^2.
B_1	1	−1	−1	1	z; xy; I_z
B_2	1	−1	1	−1	y; xz; I_y
B_3	1	1	−1	−1	x; yz; I_x

Here A, B_1, B_2, B_3 are used for 222 because any one of the two-fold axes can be considered the principal one.

$$\text{mmm} = 222 \times \bar{1}$$

2mm	E	2_z	m_y	m_x	
A_1	1	1	1	1	z; x^2; y^2; z^2.
A_2	1	1	−1	−1	xy; I_z.
B_1	1	−1	1	−1	x; xz; I_y.
B_2	1	−1	−1	1	y; yz; I_x.

2/m	E	2_z	m_z	$\bar{1}$	
A_g	1	1	1	1	x^2; y^2; z^2; xy; I_z.
B_g	1	−1	−1	1	yz; xz; I_x; I_y.
A_u	1	1	−1	−1	z;
B_u	1	−1	1	−1	x; y;

Tetragonal point-groups

4	E	2_z	4_z	4_z^3	
A	1	1	1	1	z; $x^2 + y^2$; z^2; I_z.
B	1	1	−1	−1	$x^2 - y^2$; xy.
E {	1	−1	i	$-i$	{ x, y; xz, yz;
	1	−1	$-i$	i	I_x, I_y.

$4/\mathrm{m} = 4 \times \bar{1}$

$\bar{4}$	E	2_z	$\bar{4}_z$	$\bar{4}_z^3$	
A	1	1	1	1	$x^2 + y^2$; z^2; I_z.
B	1	1	−1	−1	z; $x^2 - y^2$; xy.
E {	1	−1	i	$-i$	{ x, y; xz, yz;
	1	−1	$-i$	i	I_x, I_y.

422	E	2_z	4_z	2_x	2_d	
A_1	1	1	1	1	1	$x^2 + y^2$; z^2.
A_2	1	1	1	−1	−1	z; I_z.
B_1	1	1	−1	1	−1	$x^2 - y^2$.
B_2	1	1	−1	−1	1	xy.
E	2	−2	0	0	0	x, y; xz, yz; I_x, I_y.

$4/\mathrm{mmm} = 422 \times \bar{1}$

4mm	E	2_z	4_z	m_x	m_g	
A_1	1	1	1	1	1	z; $x^2 + y^2$; z^2.
A_2	1	1	1	−1	−1	I_z.
B_1	1	1	−1	1	−1	$x^2 - y^2$.
B_2	1	1	−1	−1	1	xy.
E	2	−2	0	0	0	x, y; xz, yz; I_x, I_y.

$\bar{4}$2m	E	2_z	$\bar{4}_z$	2_x	m_d	
A_1	1	1	1	1	1	$x^2 + y^2$; z^2.
A_2	1	1	1	−1	−1	I_z.
B_1	1	1	−1	1	−1	$x^2 - y^2$.
B_2	1	1	−1	−1	1	z; xy.
E	2	−2	0	0	0	x, y; xz, yz; I_x, I_y.

Trigonal and hexagonal point-groups

3	E	3_z	3_z^2	
A	1	1	1	$z;\ x^2+y^2;\ z^2;\ I_z.$
E	1	ω	ω^2	$x,\ y;\ x^2 - {}^2,\ xy;$
	1	ω^2	ω	$xz,\ yz;\ I_x,\ I_y.$

where $\omega = \exp(2\pi i/3)$.

$$\bar{3} = 3 \times \bar{1}$$

32	E	3_z	2_y	
A_1	1	1	1	$x^2+y^2;\ z^2.$
A_2	1	1	-1	$z;\ I_z.$
E	2	-1	0	$x,\ y;\ x^2-y^2,\ xy;\ xz,\ yz;\ I_x,\ I_y.$

The representations A_1, A_2, E of this group are the same as $\mathscr{I}$, $\mathscr{A}$, Γ of Tables 3 and 8.

$$\bar{3}\mathrm{m} = 32 \times \bar{1}$$

3m	E	3_z	m_x	
A_1	1	1	1	$z;\ x^2+y^2;\ z^2.$
A_2	1	1	-1	$I_z.$
E	2	-1	0	$x,\ y;\ x^2-y^2;\ xy;\ xz,\ yz;\ I_x,\ I_y.$

6	E	6_z	3_z	2_z	3_z^2	6_z^5	
A	1	1	1	1	1	1	$z;\ x^2+y^2;\ z^2;\ I_z.$
B	1	-1	1	-1	1	-1	
E_1	1	$-\omega^2$	ω	-1	ω^2	$-\omega$	$x,\ y;\ xz,\ yz;$
	1	$-\omega$	ω^2	-1	ω	$-\omega^2$	$I_x,\ I_y.$
E_2	1	ω	ω^2	1	ω	ω^2	$x^2-y^2,\ xy.$
	1	ω^2	ω	1	ω^2	ω	

where $\omega = \exp(2\pi i/3)$.

$6/m = 6 \times \bar{1}$

$\bar{6}$	E	3_z	3_z^2	m_z	$\bar{6}_z$	$\bar{6}_z^5$	
A'	1	1	1	1	1	1	$x^2 + y^2$; z^2; I_z.
A''	1	1	1	−1	−1	−1	z.
E'	1	ω	ω^2	1	ω	ω^2	x, y; $x^2 - y^2, xy$.
	1	ω^2	ω	1	ω^2	ω	
E''	1	ω	ω^2	−1	$-\omega$	$-\omega^2$	xz, yz; I_x, I_y.
	1	ω^2	ω	−1	$-\omega^2$	$-\omega$	

where $\omega = \exp(2\pi i/3)$.

622	E	2_z	3_z	6_z	2_y	2_x	
A_1	1	1	1	1	1	1	$x^2 + y^2$; z^2.
A_2	1	1	1	1	−1	−1	z; I_z.
B_1	1	−1	1	−1	1	−1	
B_2	1	−1	1	−1	−1	1	
E_1	2	−2	−1	1	0	0	x, y; xz, yz; I_x, I_y.
E_2	2	2	−1	−1	0	0	$x^2 - y^2, xy$.

$6/mmm = 622 \times \bar{1}$

6mm	E	2_z	3_z	6_z	m_y	m_x	
A_1	1	1	1	1	1	1	z; $x^2 + y^2$; z^2.
A_2	1	1	1	1	−1	−1	I_z.
B_1	1	−1	1	−1	−1	1	
B_2	1	−1	1	−1	1	−1	
E_1	2	−2	−1	1	0	0	x, y; xz, yz; I_x, I_y.
E_2	2	2	−1	−1	0	0	$x^2 - y^2, xy$.

$\bar{6}$m2	E	m_z	3_z	$\bar{6}_z$	2_y	m_x	
A'_1	1	1	1	1	1	1	$x^2 + y^2$; z^2.
A'_2	1	1	1	1	−1	−1	I_z.
A''_1	1	−1	1	−1	1	−1	
A''_2	1	−1	1	−1	−1	1	z.
E'	2	2	−1	−1	0	0	x, y; $x^2 - y^2, xy$.
E''	2	−2	−1	1	0	0	xz, yz; I_x, I_y.

Cubic point-groups

23	E	2_z	3	3^2	
A	1	1	1	1	$x^2+y^2+z^2=r^2$.
E	1	1	ω	ω^2	$x^2-y^2,\ 3z^2-r^2$.
	1	1	ω^2	ω	
T	3	−1	0	0	$x, y, z; xy, yz, zx; I_x, I_y, I_z$.

where $\omega = \exp(2\pi i/3)$.

$$\mathrm{m3} = 23 \times \bar{1}$$

432	E	3	2_z	2_d	4_z	
A_1	1	1	1	1	1	$x^2+y^2+z^2=r^2$.
A_2	1	1	1	−1	−1	
E	2	−1	2	0	0	$x^2-y^2,\ 3z^2-r^2$.
T_1	3	0	−1	−1	1	$x, y, z; I_x, I_y, I_z$.
T_2	3	0	−1	1	−1	xy, yz, zx.

$$\mathrm{m3m} = 432 \times \bar{1}$$

$\bar{4}$3m	E	3	2_z	m_g	$\bar{4}_z$	
A_1	1	1	1	1	1	$x^2+y^2+z^2=r^2$.
A_2	1	1	1	−1	−1	
E	2	−1	2	0	0	$x^2-y^2,\ 3z^2-r^2$.
T_1	3	0	−1	−1	1	I_x, I_y, I_z.
T_2	3	0	−1	1	−1	$x, y, z; xy, yz, zx$.

Appendix L

Character Tables for the Axial Rotation Group and Derived Groups

Notation

The notation follows that of appendix K as closely as possible. The groups ∞m and ∞/mm are denoted by $C_{\infty v}$ and $D_{\infty h}$ in the Schoenflies notation.

∞	E	$R(\phi, z)$	
A	1	1	z; z^2; $x^2 + y^2$; I_z.
E_1	1	$\exp(i\phi)$	x, y; xz, yz;
	1	$\exp(-i\phi)$	I_x, I_y.
E_2	1	$\exp(i2\phi)$	$x^2 - y^2, xy$.
	1	$\exp(-i2\phi)$	
....	..	....	
E_m	1	$\exp(im\phi)$	
	1	$\exp(-im\phi)$	

m	E	$R(\phi, z)$	m_x	
$A_1 = \Sigma^+$	1	1	1	z; z^2; $x^2 + y^2$.
$A_2 = \Sigma^-$	1	1	-1	I_z.
$E_1 = \Pi$	2	$2 \cos \phi$	0	x, y; xz, yz; I_x, I_y.
$E_2 = \Delta$	2	$2 \cos 2\phi$	0	$x^2 - y^2, xy$.
....	..	....	..	
E_m	2	$2 \cos m\phi$	0	

∞/mm	E	$R(\phi, z)$	2_x	$\bar{1}$	$\bar{1}R(\phi, z)$	m_x	
$A_{1g} = \Sigma_g^+$	1	1	1	1	1	1	z^2; $x^2 + y^2$.
$A_{1u} = \Sigma_u^-$	1	1	1	-1	-1	-1	
$A_{2g} = \Sigma_g^-$	1	1	-1	1	1	-1	I_z.
$A_{2u} = \Sigma_u^+$	1	1	-1	-1	-1	1	z.
$E_{1g} = \Pi_g$	2	$2\cos\phi$	0	2	$2\cos\phi$	0	$xz, yz, I_x; I_y$.
$E_{1u} = \Pi_u$	2	$2\cos\phi$	0	-2	$-2\cos\phi$	0	x, y.
$E_{2g} = \Delta_g$	2	$2\cos 2\phi$	0	2	$2\cos 2\phi$	0	$x^2 - y^2, xy$.
$E_{2u} = \Delta_u$	2	$2\cos 2\phi$	0	-2	$-2\cos 2\phi$	0	
......	...	...	...	...	...	...	
E_{mg}	2	$2\cos m\phi$	0	2	$2\cos m\phi$	0	
E_{mu}	2	$2\cos m\phi$	0	-2	$-2\cos m\phi$	0	

$$\infty/m = \infty \times \bar{1}$$

$\infty 2$ is isomorphic with ∞m.

GENERAL REFERENCES

Group theory applied to quantum mechanics

B. L. VAN DER WAERDEN; *Die Gruppentheoretische Methode in der Quantenmechanik*, 1932. After a summary of the relevant parts of quantum mechanics in Chapter I, Chapter II develops in 32 pages all the necessary general theory of vector spaces and group representation from first principles. The whole treatment is extremely concise, neat and elegant, yet without being more general and abstract than is required for the purpose at hand. The style is terse, occasionally to the point of making the argument difficult to follow on a first reading. The properties of group characters are developed but not used. In succeeding chapters the irreducible representations of the rotation group and the Lorentz group are derived, and then applied to the theory of atomic spectra. The final chapter deals with the applications to diatomic molecules. It appears that at least three different unpublished translations of this book are in existence for purpose of private study.

E. WIGNER; *Gruppentheorie und ihre Anwendung auf die Quantenmechanik der Atmospektren* (translated *Group Theory and Its Application to the Quantum Mechanics of Atomic Spectra*), 1931. Although the subject matter is very similar to that of the previous book, the treatment is somewhat different. The representation theory is developed from the point of view of the group characters, even for continuous groups. In the discussion of the rotation group this makes for rather more algebraic manipulation. The exclusion principle is taken into account via a detailed discussion of the representations of the permutation group (as outlined in § 28), because it had not been realized at the time that this could be avoided. Altogether the book is a mine of information on the detailed manipulation of the rotation and permutation groups, but this has the consequence that the wood is often lost for the trees and the present author finds the book rather unsuitable as a first introduction to the subject of Group Theory in Quantum Mechanics in general.†

H. WEYL; *Gruppentheorie und Quantenmechanik* (translated *The Theory of Groups and Quantum Mechanics*), 1931. This book is considerably more difficult reading than either of the previous two. However, after having mastered it, many people feel that they have a much deeper and more satisfying understanding of the whole subject than they had before, for in this book the actual structure of the quantum theory is developed from a group-theoretical point of view. As regards specific applications, the rotation, Lorentz and permutation groups are discussed in detail and applied to atomic spectra and to Dirac's relativistic theory of photons and electrons.

H. EYRING, J. WALTER and G. E. KIMBALL; *Quantum Chemistry*, 1944. From the present point of view, the importance of this book lies in the fact that it is the first and best known standard general text book which has dared to assume that elementary group theory is, or should be, part of a modern student's equipment just as much as being able to write down the

† Note added in proof; An English translation in a revised and enlarged edition has recently been published (E. P. Wigner, 1959. *Group Theory and Its Application to the Quantum Mechanics of Atomic Spectra*, Academic Press, New York and London). Three new chapters have been added: one on Racah Coefficients; one on Time Reversal Symmetry; and one on the Physical Interpretation and Classical Limits of Wigner and Racah Coefficients.

Schrödinger equation. The elements of group theory are given in very condensed form in one chapter near the middle of the book, very much from the point of view of getting some useful algebraic relations about group characters for use as a tool later on in the calculation of molecular energy levels and vibrations. The more general group-theoretical attitude towards degeneracy and selection rules is not stressed.

L. D. LANDAU and E. M. LIFSHITZ; *Quantum Mechanics, Non-Relativistic Theory*, 1958. This general quantum mechanics text also includes a chapter, which is very clearly written, on group-theoretical symmetry arguments.

C. ECKART; *The Application of Group Theory to the Quantum Dynamics of Monatomic Systems, Rev. Mod. Phys.* **2**, 304, 1930. This review article of 75 pages develops the theory from first principles, together with the simpler and more important applications to atomic spectra.

H. MARGENAU and G. M. MURPHY; *The Mathematics of Physics and Chemistry*, 1943. This volume includes a chapter on group theory, which contains some useful information about groups and their irreducible representations. The discussion of its application to quantum mechanics is so brief as to be of not much help to the reader.

S. BHAGAVANTAM and T. VENKATARAYUDU; *Theory of Groups and Its Application to Physical Problems*, 1948. The book includes an introduction to group theory and its use in physics, including quantum mechanics. The principal application is to vibrations in molecules and solids. The discussion of solids is incorrect in parts (at least it seems so to the present author).

Mathematical theory of group representations

A SPEISER; *Theorie der Gruppen von Endlicher Ordnung*, 1937. From the point of view of the student of quantum mechanics the importance of this book lies primarily in its discussion of group representations, which forms a very satisfactory alternative reference to Van der Waerden (1932) as an elementary but systematic and rigorous presentation of the basic theory. The emphasis is on finite groups, group characters and their orthogonality relations, so that the treatment appeals particularly to those who have in mind the applications to molecular problems. Chapter 6 with its beautiful illustrations of two-dimensional space-groups is also noteworthy.

G. BIRKHOFF and S. MACLANE; *A Survey of Modern Algebra*, 1941. Chapters 6 to 9 of this standard text book give an introductory account of the algebra of groups, vector spaces, matrices and linear transformations. Although not all the aspects required in quantum mechanical applications are included, the treatment is particularly noteworthy for its simple style and for the many examples which illustrate the meaning of the various terms as they are introduced.

W. LEDERMAN; *Introduction to the Theory of Finite Groups*, 1953. This gives a good introduction to groups, but does not mention group representations.

F. D. MURNAGHAN; *The Theory of Group Representations*, 1938. A rather advanced account of the theory of group representations is given, with particular reference to the permutation, rotation, Lorentz and crystallographic point-groups.

H. BOERNER; *Darstellungen von Gruppen mit Berücksichtigung der Bedürfnisse der Modernen Physik*, 1955. The mathematical theory is presented of the representations of the permutation group and of continuous groups of non-singular linear transformations in n-dimensions (including the unitary, rotation, Lorentz but not simplectic groups).

BIBLIOGRAPHY

A. Abragam, J. Horowitz and M. H. L. Pryce; *Proc. Roy. Soc. A*, **230**, 169, 1955.

A. Abragam and M. H. L. Pryce; *Proc. Roy. Soc. A*, **205**, 135, 1951a.

A. Abragam and M. H. L. Pryce; *Proc. Roy. Soc. A*, **206**, 173, 1951b.

F. Ajzenberg and T. Lauritson; *Rev. Mod. Phys.* **27**, 77, 1955.

G. A. Baker; *Phys. Rev.* **103**, 1119, 1956.

D. Bell; *Rev. Mod. Phys.* **26**, 311, 1954.

H. A. Bethe; *Ann. der Phys.* **3**, 133, 1929.

*H. A. Bethe and P. Morrison; *Elementary Nuclear Theory*, 2nd ed., Wiley & Sons, New York, 1956.

S. Bhagavantam and T. Venkatarayudu; *Theory of Groups and Its Application to Physical Problems.* Andhra University, Waltair, 1948.

L. C. Biedenharm, J. M. Blatt and E. M. Rose; *Rev. Mod. Phys.* **24**, 249, 1952.

G. Birkhoff and S. MacLane; *A Survey of Modern Algebra.* Macmillan, New York, 1941.

*J. M. Blatt and V. F. Weisskopf; *Theoretical Nuclear Physics.* Wiley & Sons, New York, 1952.

B. Bleaney and K. W. H. Stevens; *Rep. Prog. Phys.* **16**, 108, 1953.

L. J. Bodi and C. F. Curtiss; *J. Chem. Phys.* **25**, 1117, 1956.

H. Boerner; *Darstellungen von Gruppen mit Berücksichtigung der Bedürfnisse der Modernen Physik.* Springer Verlag, Berlin, 1955.

*M. Born; *Moderne Physik.* Translated *Atomic Physics.* Hafner Publishing Co., New York, 1933.

M. Born and E. Openheimer; *Ann. der Phys.* **84**, 457, 1927.

L. P. Bouckaert, R. Smoluchowski and E. P. Wigner; *Phys. Rev.* **50**, 58, 1936.

K. D. Bowers and J. Owen; *Rep. Prog. Phys.* **18**, 304, 1955.

S. F. Boys, G. B. Cook, C. M. Reeves and I. Shavitt; *Nature*, **178**, 1207, 1956.

H. C. Brinkman; *Applications of Spinor Invariants in Atomic Physics.* North Holland Publishing Co., Amsterdam, 1956.

M. J. Buerger; *X-ray Crystallography.* Wiley & Sons, New York, 1942.

M. J. Buerger; *Elementary Crystallography.* Wiley & Sons, New York, 1956.

W. G. Cady; *Piezoelectricity.* McGraw Hill, New York, 1947.

J. Callaway; *Phys. Rev.* **103**, 1219, 1957.

K. M. Case; *Phys. Rev.* **107**, 307, 1957.

H. B. G. Casimir; *Rotations of a Rigid Body in Quantum Mechanics.* J. B. Wolters, Groningen, 1931.

H. B. G. Casimir; *On the Interaction Between Atomic Nuclei and Electrons.* Teylers Tweede Genootschop, 1936.

E. R. Cohen; *Thesis*, submitted to California Institute of Technology, 1949. The relevant tables and text are to be published as *Tables of Clebsch-Gordan Coefficients* by Atomics International and the U.S. Atomic Energy Commission, 1958–59.

M. H. Cohen; (to be published). 1959.

E. U. CONDON and G. H. SHORTLEY; *The Theory of Atomic Spectra*, corrected ed. Cambridge University Press, 1951.

C. A. COULSON; *Valence*. Clarendon Press, Oxford, 1952.

C. A. COULSON and I. FISHER; *Phil. Mag.* **40**, 386, 1949.

C. A. COULSON and H. C. LONGUET-HIGGINS; *Proc. Roy. Soc. A*, **191**, 39; **192**, 16, 1947–48; and later papers in the same series.

P. A. M. DIRAC; *Proc. Roy. Soc. A*, **114**, 710, 1927.

P. A. M. DIRAC; *The Principles of Quantum Mechanics*, 4th ed. Oxford University Press, 1958.

G. DRESSELHAUS; *Phys. Rev.* **100**, 580, 1955.

C. V. DURELL and A. ROBSON; *Advanced Algebra*, Vol. 2. Bell & Sons, London, 1937.

C. ECKART; *Rev. Mod. Phys.* **2**, 304, 1930.

A. R. EDWARDS; *Angular Momentum in Quantum Mechanics*. Princeton University Press, 1957.

L. EISENBUD and E. P. WIGNER; *Proc. Nat. Acad. Sci. U.S.A.* **27**, 281, 1941.

R. J. ELLIOTT, *Phys. Rev.* **96**, 266 and 280, 1954.

R. J. ELLIOTT and K. W. H. STEVENS; *Proc. Roy. Soc. A*, **215**, 437, 1952.

H. EYRING, J. WALTER and G. E. KIMBALL; *Quantum Chemistry*. Wiley & Sons, New York, 1944.

D. L. FALKOFF, G. S. COLLADAY and R. E. SELLS; *Canad. J. Phys.* **30**, 253, 1952.

V. FOCK; *Zeit. f. Phys.* **98**, 145, 1935.

G. FROBENIUS and I. SCHUR; *Sitz. Berichte Preuss. Akad. Wiss.* p. 186, 1906.

F. G. FUMI; *Nuovo Cimento*, **9**, 739, 1952a.

F. G. FUMI; *Acta Cryst.* **5**, 44, 1952b.

F. G. FUMI; *Acta Cryst.* **5**, 691, 1952c.

M. GELL-MANN and V. L. TELEGDI; *Phys. Rev.* **91**, 169, 1953.

M. GELL-MANN and K. M. WATSON; *Annual Rev. Nucl. Sci.* **4**, 219, 1954.

H. GOLDSTEIN; *Classical Mechanics*. Addison-Wesley Press, Cambridge, Mass., 1951.

E. F. GROSS; *Nuovo Cimento Suppl.* **3**, 672, 1956.

G. G. HALL; *Phil. Mag.* **43**, 338, 1952.

HARISH-CHANDRA; *Proc. Roy. Soc. A*, **189**, 372, 1947.

D. R. HARTREE; *The Calculation of Atomic Structures*. Wiley & Sons, New York, 1957.

R. F. S. HEARMON; *Adv. in Phys.* **5**, 323, 1956.

V. HEINE; *Phys. Rev.* **107**, 620, 1957a.

V. HEINE; *Proc. Roy. Soc. A*, **240**, 361, 1957b.

V. HEINE; *Phys. Rev.* **107**, 1002, 1957c.

V. HEINE; *Nature*, **181**, 525, 1958.

W. HEISENBERG; *Zeit. f. Phys.* **39**, 499, 1926.

*W. HEISENBERG; *The Physical Principles of Quantum Theory*. Chicago University Press, 1930.

E. M. HENLEY, M. A. RUDERMAN and J. STEINBERGER; *Annual Rev. Nucl. Science*, **3**, 1, 1953.

F. HERMAN; *Rev. Mod. Phys.* **30**, 102, 1958.

C. HERRING; *Phys. Rev.* **52**, 361, 1937a.

C. HERRING; *Phys. Rev.* **52**, 365, 1937b.

C. HERRING; *Phys. Rev.* **57**, 1169, 1940.

C. HERRING; *J. Franklin Inst.* **233**, 525, 1942.

C. HERRING; *Phys. Rev.* **95**, 954, 1954.

G. HERZBERG; *Infrared and Raman Spectra of Polyatomic Molecules* van Nostrand, New York, 1945.
W. V. HOUSTON; *Rev. Mod. Phys.* **20**, 161, 1948.
F. HUND; *Zeit. f. Phys.* **99**, 119, 1936.
J. D. JACKSON, S. B. TREIMAN and H. W. WYLD; *Phys. Rev.* **106**, 517, 1957.
H. A. JAHN; *Proc. Roy. Soc. A*, **164**, 117, 1938.
H. A. JAHN and E. TELLER; *Proc. Roy. Soc. A*, **161**, 220, 1937.
J. M. JAUCH and F. ROHRLICH; *The Theory of Photons and Electrons.* Addison-Wesley Press, Cambridge, Mass., 1955.
D. F. JOHNSTON; *Proc. Roy. Soc. A*, **243**, 546, 1958.
G. A. JONES and D. H. WILKINSON; *Phys. Rev.* **90**, 722, 1953.
E. O. KANE; *J. Phys. Chem. Solids*, **1**, 82 and 249, 1956.
C. KITTEL; *Introduction to Solid State Physics*, 2nd ed. Wiley & Sons, New York, 1956.
M. J. KLEIN; *Am. J. Phys.* **20**, 65, 1952.
H. KOPFERMAN; *Kernmomente.* Akademische Verlagsgesellschaft, 1940.
G. F. KOSTER; *Solid State Physics*, **5**, 174, 1957.
H. A. KRAMERS; *Proc. Amsterdam Acad.* **33**, 959, 1930.
L. D. LANDAU and E. M. LIFSHITZ; *Quantum Mechanics, Non-Relativistic Theory.* Pergamon Press, London, 1958.
W. LEDERMAN; *Introduction to the Theory of Finite Groups*, 2nd ed. Oliver & Boyd, Edinburgh, 1953.
T. D. LEE and C. N. YANG; *Phys. Rev.* **104**, 256, 1956. See also C. S. WU, E. AMBLER *et al.*; *Phys. Rev.* **105**, 1413, 1957.
D. E. LITTLEWOOD; *The Theory of Group Characters.* Oxford University Press, 1940.
R. MALM and D. R. INGLIS; *Phys. Rev.* **95**, 993, 1954.
H. MARGENAU and G. M. MURPHY; *The Mathematics of Physics and Chemistry.* D. van Nostrand Co., New York, 1943.
M. A. MELVIN and N. V. V. J. SWAMY; *Phys. Rev.* **107**, 186, 1957.
A. C. MENZIES; *Rept. Prog. Phys.* **16**, 83, 1953.
E. A. MILNE; *Vectorial Mechanics.* Methuen & Co., London, 1948.
N. F. MOTT and H. JONES; *The Theory of the Properties of Metals and Alloys.* Clarendon Press, Oxford, 1936.
N. F. MOTT and K. W. H. STEVENS; *Phil. Mag.* **2**, 1364, 1957.
*F. D. MURNAGHAN; *The Theory of Group Representations.* Johns Hopkins Press, Baltimore, 1938.
W. A. NIERENBERG; *Annual Rev. Nucl. Sci.* **7**, 1957.
P. NOZIERES; *Phys. Rev.* **109**, 1510, 1958.
R. OLSON and S. RODRIGUES; *Phys. Rev.* **108**, 1212, 1957.
K. ONO, S. KOIDE, H. SEKIYAMA and H. ABE; *Phys. Rev.* **96**, 38, 1954.
W. OPECHOWSKI; *Physica*, **7**, 552, 1940.
A. W. OVERHAUSER; *Phys. Rev.* **101**, 1702, 1956.
W. PAULI; *Rev. Mod. Phys.* **13**, 221, 1941.
W. PAULI; *Niels Bohr and the Development of Physics*, p. 31. Pergamon Press, London, 1955.
L. PAULING; *The Nature of the Chemical Bond.* Cornell University Press, 1939.
L. PAULING and S. GOUDSMIT; *The Structure of Line Spectra.* McGraw-Hill Book Co., New York, 1930.
R. E. PEIERLS; *Quantum Theory of Solids.* Clarendon Press, Oxford, 1955.
F. C. PHILLIPS; *An Introduction to Crystallography.* Longmans, Green & Co., London, 1946.

G. Placzek; in *Marx Handbuch der Radiologie*, 2nd ed. vol. **6**, part 2, p. 209, Akademische Verlagsgesellschaft, Leipzig, 1934.
G. W. Pratt; *Phys. Rev.* **88**, 1217, 1952.
G. W. Pratt; *Phys. Rev.* **102**, 1303, 1956.
M. H. L. Pryce; *Proc. Phys. Soc. A*, **63**, 25, 1950.
G. Racah; *Nuovo Cimento*, **14**, 322, 1937.
G. Racah; *Phys. Rev.* **62**, 438, 1942.
L. A. Radicati; *Proc. Phys. Soc. A*, **66**, 139, 1953.
N. F. Ramsey; *Nuclear Moments*, or in *Experimental Nuclear Physics* (Segre ed.) vol. **1**. Wiley & Sons, New York, 1953.
C. C. J. Roothan; *Rev. Mod. Phys.* **23**, 69, 1951.
*M. E. Rose; *Elementary Theory of Angular Momentum.* Wiley & Sons, New York, 1957.
J. Rosenthal and G. M. Murphy; *Rev. Mod. Phys.* **8**, 317, 1936.
R. G. Sachs; *Nuclear Theory.* Cambridge, Mass., 1953.
L. I. Schiff; *Quantum Mechanics*, 2nd ed. McGraw-Hill Book Co., New York, 1955.
C. Schwartz; *Phys. Rev.* **97**, 380, 1955.
J. Schwinger; *On Angular Momentum*, NYO–3071. U.S. Atomic Energy Commission, Oak Ridge, 1952.
F. Seitz; *Zeit. f. Krist.* **88**, 433, 1934; **90**, 289, 1935; **91**, 336, 1935; **94**, 100, 1936.
F. Seitz; *Annals of Math.* **37**, 17, 1936.
F. Seitz; *The Modern Theory of Solids.* McGraw-Hill Book Co., New York, 1940.
K. Siegbahn; (editor), *Beta and Gamma Ray Spectroscopy.* North Holland Publishing Co., Amsterdam, 1955.
A. Simon; *Numerical Table of the Clebsch-Gordan Coefficients*, ORNL–1718. U.S. Atomic Energy Commission, Oak Ridge, 1954.
J. C. Slater; *Phys. Rev.* **34**, 1293, 1929.
J. C. Slater; *Introduction to Chemical Physics.* McGraw-Hill Book Co., New York, 1939.
J. C. Slater; *Phys. Rev.* **81**, 385, 1951.
A. Speiser; *Die Theorie der Gruppen von Endlicher Ordnung*, 3rd ed. Springer Verlag, Berlin, 1937.
H. Sponer and E. Teller; *Rev. Mod. Phys.* **13**, 76, 1941.
M. J. Stephen; *Proc. Phys. Soc.* **71**, 485, 1958.
K. W. H. Stevens; *Proc. Phys. Soc. A*, **65**, 209, 1952.
J. A. Stratton; *Electromagnetic Theory.* McGraw-Hill Book Co., New York, 1941.
L. H. Thomas; *Nature*, **117**, 514, 1926.
G. E. Uhlenbeck and S. Goudsmit; *Naturwiss.* **13**, 953, 1925; also *Nature*, **117**, 264, 1926.
B. L. van der Waerden; *Die Gruppentheoretische Methode in der Quantenmechanik.* Springer Verlag, Berlin, 1932.
F. C. Von der Lage and H. A. Bethe; *Phys. Rev.* **71**, 612, 1947.
S. Watanabe; *Phys. Rev.* **84**, 1008, 1951.
*H. Weyl; *Gruppentheorie und Quantenmechanik*, 2nd ed., 1931. (Translated *Theory of Groups and Quantum Mechanics.* Dover, New York.)
E. Wigner; *Gruppentheorie und ihre Anwendung auf die Quantenmechanik der Atomspektren.* Vieweg & Sohn, Braunschweig, 1931.† (Translated *Group Theory and Its Application to the Quantum Mechanics of Atomic Spectra.* Academic Press, New York.)

† See also note added in proof on p. 457.

E. P. WIGNER; *Nachr. Akad, Wiss. Göttingen, Math. Phys. Kl.* **31**, 546, 1932.
E. WIGNER; *Annals of Math.* **40**, 149, 1939.
E. P. WIGNER; *On the Matrices which Reduce the Kronecker Products of Representations of Simply Reducible Groups.* Microwave Research Institute, Polytechnic Institute of Brooklyn, New York, 1952.
*E. G. WILSON, J. C. DECIUS and P. C. CROSS; *Molecular Vibrations.* McGraw-Hill Book Co., New York, 1955.
C. N. YANG and J. TIOMNO; *Phys. Rev.* **79**, 495, 1950.
W. H. ZACHARIASEN; *The Theory of X-ray Diffraction in Crystals.* Wiley & Sons, New York, 1945.

INDEX

A

Abelian group, 14, **49**, 124, 420.
Accidental degeneracy, **44**, 222, 278, 301.
Ammonia, 241.
Angular distributions, 339.
Angular momentum, 62–4, 80: coupling of, 67–72, 176–89.
Anharmonicity, 256.
Antisymmetric representation, 51, 90.
Axial rotation group, **15**, 55, 67, 218: representations, **49**, 136, **455–6**.

B

Band structure, 271–303: sticking together of bands, 285–93.
Base vectors, 29: "standard," 58, 66, 354.
Basis, for representation, 28.
Benzene, 213, 227, 244.
Beta decay, 164, 384–396.
Bloch function (orbital), 216, **266**, 281.
Body centred cubic structure, 265, 268, 280.
Brillouin zone, 267: special points, 273: extended zone scheme 282: reduced zone scheme, 268.

C

Carbon, 210.
Carbon dioxide, 241, 263.
Central self-consistent field Hamiltonian, 75.
Cerium ethylsulphate, 149.
Characters, 114–124: for character tables of particular groups, *see under* names of groups, particularly Point-group.
Charge conjugation, 376, 384, 397.
Charge independence of nuclear forces 321–333.
Chromous sulphate, 155.
Classes, **114**, 122, 124–5.
Clebsch-Gordon, *see* Wigner coefficients.
Combination tone, 252.
Commutation, 14, 146, 419.
Commutation relations, **54**, **352–4**, 367, 368, 423.
Compatibility relations, 277.
Complex conjugate representations, 171, 235, 448.
Configurational interaction, 77, 206.
Configurations, atomic, **75**, 91, 92, 98: of the elements, 93: molecular, 221: nuclear, 327.
Constants of the motion, 144–7, 337, 348.
Continuous groups, 113, 125, 144.
Covalent bond, 208–21.
Crystal field splittings, 47, 112, **148–163**.
Cubic group, *see* point-group.
Cyclic group, 48.

D

$D^{(j)}$, 58.
Degeneracy, 2, 29, **42**: accidental, **44**, 222, 278, 301: splitting by perturbations, 44: due to time reversal, 171–3, 291, 448: in atoms, 75–7, 84–5, 90–3: in molecular vibrations, 235, **258–61**, 263: in solids, **278–9**, **285–8**, 291–303: due to isotopic spin, 325.
Determinantal wave functions, 91, 206.
Deuteron, **328–30**, 333, 337–8.
Dimension of vector space, 26, 30, 412.
Dirac equation, 366–383.
Dirac method, 143–7.
Direct inspection method, 310.
Double suffix notation, 406.
Double-valued (spin) representations, **59–60**, **137–41**, 226, 293, 359–61, 363.

E

Elastic constants, 305–12.
Electromagnetic field, 349: vector potential, 383.

Element of a group, 14.
Energy levels, crossing, 222: *see also* Degeneracy.
Equivalence, of representations, **31–3**, **115**, 119, 123: transformation, 32.
Equivalent orbitals, 94.
Eulerian angles, 60.
Exchange, 91.
Exclusion principle, electrons, 89–99, 376, 403: nucleons, 315–21: positronium, 399, 403.

F

Face centred cubic structure, 265, 268, 280, 281.
Favoured transitions, 388.
Finite groups, 113–25, 145: *see also* Point-groups.
Forbidden transitions, 386, 394–5.
Free electron model, 282.
Full rotation and reflection group, 15, 125–8.
Full rotation group, 15, 51–72: irreducible representations, 55–61: characters, 117: reduction of product representation, 67–72.
Fundamental theorems, of group theory, 412–19: on degeneracy of energy levels, 41–7: on matrix elements, 100–3, 120: on reactions, 335–7, 348: on tensor components, 309.

G

g-factor, **111**, 153, 160–2.
Gamma rays, 349.
Gamma matrices, 367–87: definition, 367–8: transformation properties, 369: uniqueness, 370–2.
General references, 457–8.
Germanium, 296.
Glide planes, **285**, 285–302.
Group, definition, 14: for specific groups and types of group, *see under* their respective name, e.g. Point-group.
Group of **k**, 275, 299.

H

Hamiltonian, of free atom, 73: symmetry transformations of, 6–11, 19.
Hartree's equations, 74.
Hartree-Fock equations, 91, 96.
Heitler-London wave function, 218.
Homomorphism, 410.
Hund's rule, 97.
Hydrogen molecule, 10, **206–8**, 216–8, 228.
Hyperfine structure, 189–205: s-electron effect, 198.

I

"Identical" representations, 33.
Identity representation, **26**, 103, 120.
Improper rotations, **9**, 129, 131, 238.
Indium antimonide, 296.
Infinitesimal rotation operators, 52–72, 106, 351–5: definition, 52: commutation relations, 54–5, 352–4: relation to angular momentum, 62–4: relation to spin, 80, 375.
Infinitesimal transformations, 422–3.
Infra-red spectra, 239–41, 245–64.
Inversion (in space), 9, 86, 356: group, 41: centre in point-groups, 448: centre in solids, 295: in polar coordinates, 10: *see also* Parity and Time inversion.
Irreducible representations, definition, 34: in vector space, 35: specification by characters, 115: number of, 122: test for, 124: of specific groups, *see* name of group or type of group, e.g. Point-group.
Isomorphism, 20–2, 410–11.
Isotopic spin, formalism, 313–21: and nuclear forces, 322–34: and reactions, 337–9, 346–9.

J

Jahn-Teller effect, 175.
jj-coupling, 86.

K

Kramers' theorem, 166, 169–70.

L

Landé splitting factor, 111.
Level, of a term, 85: *see also* Degeneracy and Energy level.

Lorentz group, **351–63**, 428: transformation under, 365–70, **379–81**, 385.

M

Magnetic field, Hamiltonian, 153, 374: Zeeman effect, 111: Zeeman splitting in crystals, 148–62: effect of time reversal, 170, 383.
Magnetic moment, of electron, 78, 375: of nucleus, 190, 192.
Matrix algebra, 404.
Matrix elements, 2, 99–112: Fundamental theorem on selection rules, 100–3, 120: with Wigner and Racah coefficients, 183–6.
Mesons, **331–2**, 334, 338, 346.
Molecular, wave functions, 206–228: vibrations, **229–44**, 245–64.
Multicomponent functions, 81–3, 88, 365–6.

N

Negative energy states, 373, 376.
Neutrino, 385, 391: two component, 396.
Normal coordinates, 232–3, 239.
Nuclear, forces, 321–33: spin, 190: moments, 192.

O

Octahedral complexes, 227.
Operator equivalents, **153**, 163.
Orbital angular momentum, 80: " quenched," 159.
Orbitals, atomic, 74: in molecules, 206–28.
Orthogonality relations, 117–8, 123.
Overtones, 252.
Oxygen molecule, 220, 229.
Ozone, 16, 233–40, 248–52, 259–60, **262**: *see also* " 32."

P

Paramagnetic ions, 148–163.
Parity, 86: of spin functions, 87, 361, 398: of spherical harmonics, 87: atomic, 87, 104: nuclear, 190, 328, 341–3: of fundamental particles, 331, 334, 360: non-conservation of, 386, **390**, 395. *See also* Inversion.
Paschen-Back effect, 111.
Periodic boundary conditions, 266.
Periodic table, 92.
Permutation (symmetric) group, 9, 15: of order two, 41: of order three, 21: of order n, representations, 50: *see also* Exclusion principle.
Perturbation, splitting of levels, 44–6.
Perturbation theory, degenerate, 106: spin-Hamiltonian form, 160, 223: for band structures, 283.
Piezoelectric constants, 312.
Point-groups, 15, 128–142: list of, **130**, 131, 135, **446**: international and Shoenflies notation, 129–31, 446: character tables, 136–7, **448–54**, **455–6**: double-valued representations, 137–41: point group three-two, 13–5, 27 (irred. reps.), 115 (characters).
Positron, 376: positronium, 397–403.
Product groups, 125–8.
Product representation, 67, 116: symmetrical, 260, 308.
Projection operators, 119.
Pseudo-vector, 89, 358, 383.

Q

Quadrupole moment, 201, 329.
Quantum numbers, 75–6, 84.
Quenching of orbital moment, 159.

R

Racah coefficients, 183–8, 202.
Raman spectra, 239–41, 248–64: in solids, 284.
Reactions, nuclear, 334–50: fundamental theorem on, 335–7, 348: selection rules, 337–8, 341–3: angular distribution of products 339–46: relative cross sections, 338–9, 346–8.
Reciprocal lattice, **267**, 280.
Reduced wave vector, 268.
Reducibility, 33–7, 416.
Reduction of a representation, 67, 115.
Reflection, 9.
Representation(s), 25–9: reducible, irreducible, 33–7, 416: reduction

of, 67, 115: equivalence, 31–3: "identical," 33: identity representation, **26**, 103, 120: complex conjugate, 171, 235: unfaithful, 26: double-valued (spin), **59**, **137**, 226, 293, 359, 363: regular, 121: of particular groups, *see* group or type of group, e.g. Point-groups.
Relativistic, wave equation, 363–6: transitions, 387, 395.
Rotations, 3–7, 11, **52–4**: notation, 4: composition of, 65: Eulerian angles, 60: *see also* Improper rotations, Infinitesimal rotation operators, Full rotation group, Axial rotation group, Point-groups.
Russell-Saunders coupling, 85, 97.

S

Schur's lemma, 101, 118, 415, **418**.
Screw axes, **285–7**, 296, 299–301.
Selection rules, Fundamental theorem for, 99, 102–3, 120: electric dipole transitions, 103: magnetic dipole, 110: molecular spectra, 240, 247, 249: beta decay, 386–8, 395: gamma emission, 349, 350: reactions, 337–9: positronium, 401.
Self-consistent field, 74, 91.
Shell model, 326.
Simple harmonic oscillator, 6, 233, **252–4**, 256–61, 264, **424–7**.
Space-groups, 128, **265**, **284**, 265–303: summary of irreducible representations, 273–6, 299–300: spin representations, 293.
Spectroscopic notation, 75, 218.
Spherical harmonics, 58–9: parity, 87: characters, 117: matrix elements, 149–53.
Spin, electron, 78–9, 365–6, 375: nuclear, 190, 330: other particles, 331, 365: functions, 82–3 (definition), 60, 83 (transformation), 87, 360, 361, 398 (parity), 167 (time-reversal): operators, 79, 88: matrices, 88.
Spin—, -Hamiltonian, 156–64, 223: -orbit coupling, **81**, **107**, 226, 293, **375**, 397 (electrons), 329, 334 (nuclei): -spin coupling, 81: -representations (double-valued), **59–60**, **137–41**, 226, 293, 359–61, 363.
Spinor, 66, **83**, 177, 355: fields, 360, 365: invariants, 177.
Splitting of levels, 1, 44–7: *see also* Degeneracy and Perturbation theory.
Square lattice, 268–283.
Star of **k**, 274, 299.
Stereogram, 129–30.
Sub-group, 20.
Sum rule, 109.
Symmetric group, *see* Permutation group.
Symmetrical product, 260, 308.
Symmetry, consequences of, 1–3: transformations, 3–6: of Hamiltonian, 6–12, 16–20.

T

Tensors, 70, 72, 184, 304–12.
Terms (energy levels), atomic, 83–6, 93–7, 107–110: molecular, 221, 225, 229: positronium, 401.
32 (point-group "three-two"), 13–5: irreducible representations, 27: characters, 115: *see also* Point-groups.
Time-inversion, **164**, 356, 428. *See also* Time-reversal.
Time-reversal, 164–75: and complex representations, 235, 448: in band structures, 290, 295: symmetry of tensors, 312: Dirac equation, 383: beta decay, 396. *See also* Time-inversion.
Transformations, 3–6, 12, 24, 30: of Hamiltonian, 6–12, 16–9: of wave functions, 8, 24, 41: of operators, 25 footnote: of multi-component functions, 81–3, 88, 365–6: equivalence, 32: unitary, 409, 414: relativistic, 379.
Transitions, *see* Selection rules.
Translation group, 265.

U

Unfavoured transitions, 388.
Unitary, matrix, 408: transformations, 409, **414**.

V

Valence, bond orbital, 208, 213.
Vector coupling, *see* Wigner coefficient.
Vector potential, 383.
Vector space, definition, 29: reducibility, 35: mathematical theory, 412–7: application to quantum mechanics, 41.
Vectors in ordinary space, 59, 70, 358: pseudo-, 89, 358, 383.
Vibrations, of molecules, **229–44**, 245–64: in solids, 284.

W

Wave functions, transformation of, 8, 24, 41: multicomponent, 81–3, 88, 365–6.
Wave vector, 266: reduced, 268, 282.
Wigner coefficients, 70–2, 176–89: tables of, 432–45.

X Y Z

Zeeman effect, 111: in crystals, 148–62.

A CATALOG OF SELECTED

DOVER BOOKS

IN SCIENCE AND MATHEMATICS

Astronomy

BURNHAM'S CELESTIAL HANDBOOK, Robert Burnham, Jr. Thorough guide to the stars beyond our solar system. Exhaustive treatment. Alphabetical by constellation: Andromeda to Cetus in Vol. 1; Chamaeleon to Orion in Vol. 2; and Pavo to Vulpecula in Vol. 3. Hundreds of illustrations. Index in Vol. 3. 2,000pp. 6⅛ x 9¼.
Vol. I: 0-486-23567-X
Vol. II: 0-486-23568-8
Vol. III: 0-486-23673-0

EXPLORING THE MOON THROUGH BINOCULARS AND SMALL TELESCOPES, Ernest H. Cherrington, Jr. Informative, profusely illustrated guide to locating and identifying craters, rills, seas, mountains, other lunar features. Newly revised and updated with special section of new photos. Over 100 photos and diagrams. 240pp. 8¼ x 11. 0-486-24491-1

THE EXTRATERRESTRIAL LIFE DEBATE, 1750–1900, Michael J. Crowe. First detailed, scholarly study in English of the many ideas that developed from 1750 to 1900 regarding the existence of intelligent extraterrestrial life. Examines ideas of Kant, Herschel, Voltaire, Percival Lowell, many other scientists and thinkers. 16 illustrations. 704pp. 5⅜ x 8½. 0-486-40675-X

THEORIES OF THE WORLD FROM ANTIQUITY TO THE COPERNICAN REVOLUTION, Michael J. Crowe. Newly revised edition of an accessible, enlightening book recreates the change from an earth-centered to a sun-centered conception of the solar system. 242pp. 5⅜ x 8½. 0-486-41444-2

A HISTORY OF ASTRONOMY, A. Pannekoek. Well-balanced, carefully reasoned study covers such topics as Ptolemaic theory, work of Copernicus, Kepler, Newton, Eddington's work on stars, much more. Illustrated. References. 521pp. 5⅜ x 8½.
0-486-65994-1

A COMPLETE MANUAL OF AMATEUR ASTRONOMY: TOOLS AND TECHNIQUES FOR ASTRONOMICAL OBSERVATIONS, P. Clay Sherrod with Thomas L. Koed. Concise, highly readable book discusses: selecting, setting up and maintaining a telescope; amateur studies of the sun; lunar topography and occultations; observations of Mars, Jupiter, Saturn, the minor planets and the stars; an introduction to photoelectric photometry; more. 1981 ed. 124 figures. 25 halftones. 37 tables. 335pp. 6½ x 9¼. 0-486-40675-X

AMATEUR ASTRONOMER'S HANDBOOK, J. B. Sidgwick. Timeless, comprehensive coverage of telescopes, mirrors, lenses, mountings, telescope drives, micrometers, spectroscopes, more. 189 illustrations. 576pp. 5⅜ x 8¼. (Available in U.S. only.)
0-486-24034-7

STARS AND RELATIVITY, Ya. B. Zel'dovich and I. D. Novikov. Vol. 1 of *Relativistic Astrophysics* by famed Russian scientists. General relativity, properties of matter under astrophysical conditions, stars, and stellar systems. Deep physical insights, clear presentation. 1971 edition. References. 544pp. 5⅜ x 8¼. 0-486-69424-0

Chemistry

THE SCEPTICAL CHYMIST: THE CLASSIC 1661 TEXT, Robert Boyle. Boyle defines the term "element," asserting that all natural phenomena can be explained by the motion and organization of primary particles. 1911 ed. viii+232pp. 5⅜ x 8½. 0-486-42825-7

RADIOACTIVE SUBSTANCES, Marie Curie. Here is the celebrated scientist's doctoral thesis, the prelude to her receipt of the 1903 Nobel Prize. Curie discusses establishing atomic character of radioactivity found in compounds of uranium and thorium; extraction from pitchblende of polonium and radium; isolation of pure radium chloride; determination of atomic weight of radium; plus electric, photographic, luminous, heat, color effects of radioactivity. ii+94pp. 5⅜ x 8½. 0-486-42550-9

CHEMICAL MAGIC, Leonard A. Ford. Second Edition, Revised by E. Winston Grundmeier. Over 100 unusual stunts demonstrating cold fire, dust explosions, much more. Text explains scientific principles and stresses safety precautions. 128pp. 5⅜ x 8½. 0-486-67628-5

THE DEVELOPMENT OF MODERN CHEMISTRY, Aaron J. Ihde. Authoritative history of chemistry from ancient Greek theory to 20th-century innovation. Covers major chemists and their discoveries. 209 illustrations. 14 tables. Bibliographies. Indices. Appendices. 851pp. 5⅜ x 8½. 0-486-64235-6

CATALYSIS IN CHEMISTRY AND ENZYMOLOGY, William P. Jencks. Exceptionally clear coverage of mechanisms for catalysis, forces in aqueous solution, carbonyl- and acyl-group reactions, practical kinetics, more. 864pp. 5⅜ x 8½. 0-486-65460-5

ELEMENTS OF CHEMISTRY, Antoine Lavoisier. Monumental classic by founder of modern chemistry in remarkable reprint of rare 1790 Kerr translation. A must for every student of chemistry or the history of science. 539pp. 5⅜ x 8½. 0-486-64624-6

THE HISTORICAL BACKGROUND OF CHEMISTRY, Henry M. Leicester. Evolution of ideas, not individual biography. Concentrates on formulation of a coherent set of chemical laws. 260pp. 5⅜ x 8½. 0-486-61053-5

A SHORT HISTORY OF CHEMISTRY, J. R. Partington. Classic exposition explores origins of chemistry, alchemy, early medical chemistry, nature of atmosphere, theory of valency, laws and structure of atomic theory, much more. 428pp. 5⅜ x 8½. (Available in U.S. only.) 0-486-65977-1

GENERAL CHEMISTRY, Linus Pauling. Revised 3rd edition of classic first-year text by Nobel laureate. Atomic and molecular structure, quantum mechanics, statistical mechanics, thermodynamics correlated with descriptive chemistry. Problems. 992pp. 5⅜ x 8½. 0-486-65622-5

FROM ALCHEMY TO CHEMISTRY, John Read. Broad, humanistic treatment focuses on great figures of chemistry and ideas that revolutionized the science. 50 illustrations. 240pp. 5⅜ x 8½. 0-486-28690-8

Engineering

DE RE METALLICA, Georgius Agricola. The famous Hoover translation of greatest treatise on technological chemistry, engineering, geology, mining of early modern times (1556). All 289 original woodcuts. 638pp. 6¾ x 11. 0-486-60006-8

FUNDAMENTALS OF ASTRODYNAMICS, Roger Bate et al. Modern approach developed by U.S. Air Force Academy. Designed as a first course. Problems, exercises. Numerous illustrations. 455pp. 5⅜ x 8½. 0-486-60061-0

DYNAMICS OF FLUIDS IN POROUS MEDIA, Jacob Bear. For advanced students of ground water hydrology, soil mechanics and physics, drainage and irrigation engineering and more. 335 illustrations. Exercises, with answers. 784pp. 6⅛ x 9¼. 0-486-65675-6

THEORY OF VISCOELASTICITY (Second Edition), Richard M. Christensen. Complete consistent description of the linear theory of the viscoelastic behavior of materials. Problem-solving techniques discussed. 1982 edition. 29 figures. xiv+364pp. 6⅛ x 9¼. 0-486-42880-X

MECHANICS, J. P. Den Hartog. A classic introductory text or refresher. Hundreds of applications and design problems illuminate fundamentals of trusses, loaded beams and cables, etc. 334 answered problems. 462pp. 5⅜ x 8½. 0-486-60754-2

MECHANICAL VIBRATIONS, J. P. Den Hartog. Classic textbook offers lucid explanations and illustrative models, applying theories of vibrations to a variety of practical industrial engineering problems. Numerous figures. 233 problems, solutions. Appendix. Index. Preface. 436pp. 5⅜ x 8½. 0-486-64785-4

STRENGTH OF MATERIALS, J. P. Den Hartog. Full, clear treatment of basic material (tension, torsion, bending, etc.) plus advanced material on engineering methods, applications. 350 answered problems. 323pp. 5⅜ x 8½. 0-486-60755-0

A HISTORY OF MECHANICS, René Dugas. Monumental study of mechanical principles from antiquity to quantum mechanics. Contributions of ancient Greeks, Galileo, Leonardo, Kepler, Lagrange, many others. 671pp. 5⅜ x 8½. 0-486-65632-2

STABILITY THEORY AND ITS APPLICATIONS TO STRUCTURAL MECHANICS, Clive L. Dym. Self-contained text focuses on Koiter postbuckling analyses, with mathematical notions of stability of motion. Basing minimum energy principles for static stability upon dynamic concepts of stability of motion, it develops asymptotic buckling and postbuckling analyses from potential energy considerations, with applications to columns, plates, and arches. 1974 ed. 208pp. 5⅜ x 8½. 0-486-42541-X

METAL FATIGUE, N. E. Frost, K. J. Marsh, and L. P. Pook. Definitive, clearly written, and well-illustrated volume addresses all aspects of the subject, from the historical development of understanding metal fatigue to vital concepts of the cyclic stress that causes a crack to grow. Includes 7 appendixes. 544pp. 5⅜ x 8½. 0-486-40927-9

ROCKETS, Robert Goddard. Two of the most significant publications in the history of rocketry and jet propulsion: "A Method of Reaching Extreme Altitudes" (1919) and "Liquid Propellant Rocket Development" (1936). 128pp. 5⅜ x 8½. 0-486-42537-1

STATISTICAL MECHANICS: PRINCIPLES AND APPLICATIONS, Terrell L. Hill. Standard text covers fundamentals of statistical mechanics, applications to fluctuation theory, imperfect gases, distribution functions, more. 448pp. 5⅜ x 8½. 0-486-65390-0

ENGINEERING AND TECHNOLOGY 1650–1750: ILLUSTRATIONS AND TEXTS FROM ORIGINAL SOURCES, Martin Jensen. Highly readable text with more than 200 contemporary drawings and detailed engravings of engineering projects dealing with surveying, leveling, materials, hand tools, lifting equipment, transport and erection, piling, bailing, water supply, hydraulic engineering, and more. Among the specific projects outlined-transporting a 50-ton stone to the Louvre, erecting an obelisk, building timber locks, and dredging canals. 207pp. 8⅜ x 11¼. 0-486-42232-1

THE VARIATIONAL PRINCIPLES OF MECHANICS, Cornelius Lanczos. Graduate level coverage of calculus of variations, equations of motion, relativistic mechanics, more. First inexpensive paperbound edition of classic treatise. Index. Bibliography. 418pp. 5⅜ x 8½. 0-486-65067-7

PROTECTION OF ELECTRONIC CIRCUITS FROM OVERVOLTAGES, Ronald B. Standler. Five-part treatment presents practical rules and strategies for circuits designed to protect electronic systems from damage by transient overvoltages. 1989 ed. xxiv+434pp. 6⅛ x 9¼. 0-486-42552-5

ROTARY WING AERODYNAMICS, W. Z. Stepniewski. Clear, concise text covers aerodynamic phenomena of the rotor and offers guidelines for helicopter performance evaluation. Originally prepared for NASA. 537 figures. 640pp. 6⅛ x 9¼. 0-486-64647-5

INTRODUCTION TO SPACE DYNAMICS, William Tyrrell Thomson. Comprehensive, classic introduction to space-flight engineering for advanced undergraduate and graduate students. Includes vector algebra, kinematics, transformation of coordinates. Bibliography. Index. 352pp. 5⅜ x 8½. 0-486-65113-4

HISTORY OF STRENGTH OF MATERIALS, Stephen P. Timoshenko. Excellent historical survey of the strength of materials with many references to the theories of elasticity and structure. 245 figures. 452pp. 5⅜ x 8½. 0-486-61187-6

ANALYTICAL FRACTURE MECHANICS, David J. Unger. Self-contained text supplements standard fracture mechanics texts by focusing on analytical methods for determining crack-tip stress and strain fields. 336pp. 6⅛ x 9¼. 0-486-41737-9

STATISTICAL MECHANICS OF ELASTICITY, J. H. Weiner. Advanced, self-contained treatment illustrates general principles and elastic behavior of solids. Part 1, based on classical mechanics, studies thermoelastic behavior of crystalline and polymeric solids. Part 2, based on quantum mechanics, focuses on interatomic force laws, behavior of solids, and thermally activated processes. For students of physics and chemistry and for polymer physicists. 1983 ed. 96 figures. 496pp. 5⅜ x 8½. 0-486-42260-7

Mathematics

FUNCTIONAL ANALYSIS (Second Corrected Edition), George Bachman and Lawrence Narici. Excellent treatment of subject geared toward students with background in linear algebra, advanced calculus, physics and engineering. Text covers introduction to inner-product spaces, normed, metric spaces, and topological spaces; complete orthonormal sets, the Hahn-Banach Theorem and its consequences, and many other related subjects. 1966 ed. 544pp. 6⅛ x 9¼. 0-486-40251-7

ASYMPTOTIC EXPANSIONS OF INTEGRALS, Norman Bleistein & Richard A. Handelsman. Best introduction to important field with applications in a variety of scientific disciplines. New preface. Problems. Diagrams. Tables. Bibliography. Index. 448pp. 5⅜ x 8½. 0-486-65082-0

VECTOR AND TENSOR ANALYSIS WITH APPLICATIONS, A. I. Borisenko and I. E. Tarapov. Concise introduction. Worked-out problems, solutions, exercises. 257pp. 5⅜ x 8¼. 0-486-63833-2

AN INTRODUCTION TO ORDINARY DIFFERENTIAL EQUATIONS, Earl A. Coddington. A thorough and systematic first course in elementary differential equations for undergraduates in mathematics and science, with many exercises and problems (with answers). Index. 304pp. 5⅜ x 8½. 0-486-65942-9

FOURIER SERIES AND ORTHOGONAL FUNCTIONS, Harry F. Davis. An incisive text combining theory and practical example to introduce Fourier series, orthogonal functions and applications of the Fourier method to boundary-value problems. 570 exercises. Answers and notes. 416pp. 5⅜ x 8½. 0-486-65973-9

COMPUTABILITY AND UNSOLVABILITY, Martin Davis. Classic graduate-level introduction to theory of computability, usually referred to as theory of recurrent functions. New preface and appendix. 288pp. 5⅜ x 8½. 0-486-61471-9

ASYMPTOTIC METHODS IN ANALYSIS, N. G. de Bruijn. An inexpensive, comprehensive guide to asymptotic methods–the pioneering work that teaches by explaining worked examples in detail. Index. 224pp. 5⅜ x 8½ 0-486-64221-6

APPLIED COMPLEX VARIABLES, John W. Dettman. Step-by-step coverage of fundamentals of analytic function theory–plus lucid exposition of five important applications: Potential Theory; Ordinary Differential Equations; Fourier Transforms; Laplace Transforms; Asymptotic Expansions. 66 figures. Exercises at chapter ends. 512pp. 5⅜ x 8½. 0-486-64670-X

INTRODUCTION TO LINEAR ALGEBRA AND DIFFERENTIAL EQUATIONS, John W. Dettman. Excellent text covers complex numbers, determinants, orthonormal bases, Laplace transforms, much more. Exercises with solutions. Undergraduate level. 416pp. 5⅜ x 8½. 0-486-65191-6

RIEMANN'S ZETA FUNCTION, H. M. Edwards. Superb, high-level study of landmark 1859 publication entitled "On the Number of Primes Less Than a Given Magnitude" traces developments in mathematical theory that it inspired. xiv+315pp. 5⅜ x 8½. 0-486-41740-9

CALCULUS OF VARIATIONS WITH APPLICATIONS, George M. Ewing. Applications-oriented introduction to variational theory develops insight and promotes understanding of specialized books, research papers. Suitable for advanced undergraduate/graduate students as primary, supplementary text. 352pp. 5⅜ x 8½. 0-486-64856-7

COMPLEX VARIABLES, Francis J. Flanigan. Unusual approach, delaying complex algebra till harmonic functions have been analyzed from real variable viewpoint. Includes problems with answers. 364pp. 5⅜ x 8½. 0-486-61388-7

AN INTRODUCTION TO THE CALCULUS OF VARIATIONS, Charles Fox. Graduate-level text covers variations of an integral, isoperimetrical problems, least action, special relativity, approximations, more. References. 279pp. 5⅜ x 8½. 0-486-65499-0

COUNTEREXAMPLES IN ANALYSIS, Bernard R. Gelbaum and John M. H. Olmsted. These counterexamples deal mostly with the part of analysis known as "real variables." The first half covers the real number system, and the second half encompasses higher dimensions. 1962 edition. xxiv+198pp. 5⅜ x 8½. 0-486-42875-3

CATASTROPHE THEORY FOR SCIENTISTS AND ENGINEERS, Robert Gilmore. Advanced-level treatment describes mathematics of theory grounded in the work of Poincaré, R. Thom, other mathematicians. Also important applications to problems in mathematics, physics, chemistry and engineering. 1981 edition. References. 28 tables. 397 black-and-white illustrations. xvii + 666pp. 6⅛ x 9¼. 0-486-67539-4

INTRODUCTION TO DIFFERENCE EQUATIONS, Samuel Goldberg. Exceptionally clear exposition of important discipline with applications to sociology, psychology, economics. Many illustrative examples; over 250 problems. 260pp. 5⅜ x 8½. 0-486-65084-7

NUMERICAL METHODS FOR SCIENTISTS AND ENGINEERS, Richard Hamming. Classic text stresses frequency approach in coverage of algorithms, polynomial approximation, Fourier approximation, exponential approximation, other topics. Revised and enlarged 2nd edition. 721pp. 5⅜ x 8½. 0-486-65241-6

INTRODUCTION TO NUMERICAL ANALYSIS (2nd Edition), F. B. Hildebrand. Classic, fundamental treatment covers computation, approximation, interpolation, numerical differentiation and integration, other topics. 150 new problems. 669pp. 5⅜ x 8½. 0-486-65363-3

THREE PEARLS OF NUMBER THEORY, A. Y. Khinchin. Three compelling puzzles require proof of a basic law governing the world of numbers. Challenges concern van der Waerden's theorem, the Landau-Schnirelmann hypothesis and Mann's theorem, and a solution to Waring's problem. Solutions included. 64pp. 5⅜ x 8½. 0-486-40026-3

THE PHILOSOPHY OF MATHEMATICS: AN INTRODUCTORY ESSAY, Stephan Körner. Surveys the views of Plato, Aristotle, Leibniz & Kant concerning propositions and theories of applied and pure mathematics. Introduction. Two appendices. Index. 198pp. 5⅜ x 8½. 0-486-25048-2

TENSOR CALCULUS, J.L. Synge and A. Schild. Widely used introductory text covers spaces and tensors, basic operations in Riemannian space, non-Riemannian spaces, etc. 324pp. 5⅝ x 8¼. 0-486-63612-7

ORDINARY DIFFERENTIAL EQUATIONS, Morris Tenenbaum and Harry Pollard. Exhaustive survey of ordinary differential equations for undergraduates in mathematics, engineering, science. Thorough analysis of theorems. Diagrams. Bibliography. Index. 818pp. 5⅜ x 8½. 0-486-64940-7

INTEGRAL EQUATIONS, F. G. Tricomi. Authoritative, well-written treatment of extremely useful mathematical tool with wide applications. Volterra Equations, Fredholm Equations, much more. Advanced undergraduate to graduate level. Exercises. Bibliography. 238pp. 5⅜ x 8½. 0-486-64828-1

FOURIER SERIES, Georgi P. Tolstov. Translated by Richard A. Silverman. A valuable addition to the literature on the subject, moving clearly from subject to subject and theorem to theorem. 107 problems, answers. 336pp. 5⅜ x 8½. 0-486-63317-9

INTRODUCTION TO MATHEMATICAL THINKING, Friedrich Waismann. Examinations of arithmetic, geometry, and theory of integers; rational and natural numbers; complete induction; limit and point of accumulation; remarkable curves; complex and hypercomplex numbers, more. 1959 ed. 27 figures. xii+260pp. 5⅜ x 8½. 0-486-63317-9

POPULAR LECTURES ON MATHEMATICAL LOGIC, Hao Wang. Noted logician's lucid treatment of historical developments, set theory, model theory, recursion theory and constructivism, proof theory, more. 3 appendixes. Bibliography. 1981 edition. ix + 283pp. 5⅜ x 8½. 0-486-67632-3

CALCULUS OF VARIATIONS, Robert Weinstock. Basic introduction covering isoperimetric problems, theory of elasticity, quantum mechanics, electrostatics, etc. Exercises throughout. 326pp. 5⅜ x 8½. 0-486-63069-2

THE CONTINUUM: A CRITICAL EXAMINATION OF THE FOUNDATION OF ANALYSIS, Hermann Weyl. Classic of 20th-century foundational research deals with the conceptual problem posed by the continuum. 156pp. 5⅜ x 8½. 0-486-67982-9

CHALLENGING MATHEMATICAL PROBLEMS WITH ELEMENTARY SOLUTIONS, A. M. Yaglom and I. M. Yaglom. Over 170 challenging problems on probability theory, combinatorial analysis, points and lines, topology, convex polygons, many other topics. Solutions. Total of 445pp. 5⅜ x 8½. Two-vol. set. Vol. I: 0-486-65536-9 Vol. II: 0-486-65537-7

INTRODUCTION TO PARTIAL DIFFERENTIAL EQUATIONS WITH APPLICATIONS, E. C. Zachmanoglou and Dale W. Thoe. Essentials of partial differential equations applied to common problems in engineering and the physical sciences. Problems and answers. 416pp. 5⅜ x 8½. 0-486-65251-3

THE THEORY OF GROUPS, Hans J. Zassenhaus. Well-written graduate-level text acquaints reader with group-theoretic methods and demonstrates their usefulness in mathematics. Axioms, the calculus of complexes, homomorphic mapping, *p*-group theory, more. 276pp. 5⅜ x 8½. 0-486-40922-8

Math–Decision Theory, Statistics, Probability

ELEMENTARY DECISION THEORY, Herman Chernoff and Lincoln E. Moses. Clear introduction to statistics and statistical theory covers data processing, probability and random variables, testing hypotheses, much more. Exercises. 364pp. 5⅜ x 8½. 0-486-65218-1

STATISTICS MANUAL, Edwin L. Crow et al. Comprehensive, practical collection of classical and modern methods prepared by U.S. Naval Ordnance Test Station. Stress on use. Basics of statistics assumed. 288pp. 5⅜ x 8½. 0-486-60599-X

SOME THEORY OF SAMPLING, William Edwards Deming. Analysis of the problems, theory and design of sampling techniques for social scientists, industrial managers and others who find statistics important at work. 61 tables. 90 figures. xvii +602pp. 5⅜ x 8½. 0-486-64684-X

LINEAR PROGRAMMING AND ECONOMIC ANALYSIS, Robert Dorfman, Paul A. Samuelson and Robert M. Solow. First comprehensive treatment of linear programming in standard economic analysis. Game theory, modern welfare economics, Leontief input-output, more. 525pp. 5⅜ x 8½. 0-486-65491-5

PROBABILITY: AN INTRODUCTION, Samuel Goldberg. Excellent basic text covers set theory, probability theory for finite sample spaces, binomial theorem, much more. 360 problems. Bibliographies. 322pp. 5⅜ x 8½. 0-486-65252-1

GAMES AND DECISIONS: INTRODUCTION AND CRITICAL SURVEY, R. Duncan Luce and Howard Raiffa. Superb nontechnical introduction to game theory, primarily applied to social sciences. Utility theory, zero-sum games, n-person games, decision-making, much more. Bibliography. 509pp. 5⅜ x 8½. 0-486-65943-7

INTRODUCTION TO THE THEORY OF GAMES, J. C. C. McKinsey. This comprehensive overview of the mathematical theory of games illustrates applications to situations involving conflicts of interest, including economic, social, political, and military contexts. Appropriate for advanced undergraduate and graduate courses; advanced calculus a prerequisite. 1952 ed. x+372pp. 5⅜ x 8½. 0-486-42811-7

FIFTY CHALLENGING PROBLEMS IN PROBABILITY WITH SOLUTIONS, Frederick Mosteller. Remarkable puzzlers, graded in difficulty, illustrate elementary and advanced aspects of probability. Detailed solutions. 88pp. 5⅜ x 8½. 65355-2

PROBABILITY THEORY: A CONCISE COURSE, Y. A. Rozanov. Highly readable, self-contained introduction covers combination of events, dependent events, Bernoulli trials, etc. 148pp. 5⅜ x 8¼. 0-486-63544-9

STATISTICAL METHOD FROM THE VIEWPOINT OF QUALITY CONTROL, Walter A. Shewhart. Important text explains regulation of variables, uses of statistical control to achieve quality control in industry, agriculture, other areas. 192pp. 5⅜ x 8½. 0-486-65232-7

Math–Geometry and Topology

ELEMENTARY CONCEPTS OF TOPOLOGY, Paul Alexandroff. Elegant, intuitive approach to topology from set-theoretic topology to Betti groups; how concepts of topology are useful in math and physics. 25 figures. 57pp. 5⅜ x 8½. 0-486-60747-X

COMBINATORIAL TOPOLOGY, P. S. Alexandrov. Clearly written, well-organized, three-part text begins by dealing with certain classic problems without using the formal techniques of homology theory and advances to the central concept, the Betti groups. Numerous detailed examples. 654pp. 5⅜ x 8½. 0-486-40179-0

EXPERIMENTS IN TOPOLOGY, Stephen Barr. Classic, lively explanation of one of the byways of mathematics. Klein bottles, Moebius strips, projective planes, map coloring, problem of the Koenigsberg bridges, much more, described with clarity and wit. 43 figures. 210pp. 5⅜ x 8½. 0-486-25933-1

THE GEOMETRY OF RENÉ DESCARTES, René Descartes. The great work founded analytical geometry. Original French text, Descartes's own diagrams, together with definitive Smith-Latham translation. 244pp. 5⅜ x 8½. 0-486-60068-8

EUCLIDEAN GEOMETRY AND TRANSFORMATIONS, Clayton W. Dodge. This introduction to Euclidean geometry emphasizes transformations, particularly isometries and similarities. Suitable for undergraduate courses, it includes numerous examples, many with detailed answers. 1972 ed. viii+296pp. 6⅛ x 9¼. 0-486-43476-1

PRACTICAL CONIC SECTIONS: THE GEOMETRIC PROPERTIES OF ELLIPSES, PARABOLAS AND HYPERBOLAS, J. W. Downs. This text shows how to create ellipses, parabolas, and hyperbolas. It also presents historical background on their ancient origins and describes the reflective properties and roles of curves in design applications. 1993 ed. 98 figures. xii+100pp. 6½ x 9¼. 0-486-42876-1

THE THIRTEEN BOOKS OF EUCLID'S ELEMENTS, translated with introduction and commentary by Sir Thomas L. Heath. Definitive edition. Textual and linguistic notes, mathematical analysis. 2,500 years of critical commentary. Unabridged. 1,414pp. 5⅜ x 8½. Three-vol. set.

Vol. I: 0-486-60088-2 Vol. II: 0-486-60089-0 Vol. III: 0-486-60090-4

SPACE AND GEOMETRY: IN THE LIGHT OF PHYSIOLOGICAL, PSYCHOLOGICAL AND PHYSICAL INQUIRY, Ernst Mach. Three essays by an eminent philosopher and scientist explore the nature, origin, and development of our concepts of space, with a distinctness and precision suitable for undergraduate students and other readers. 1906 ed. vi+148pp. 5⅜ x 8½. 0-486-43909-7

GEOMETRY OF COMPLEX NUMBERS, Hans Schwerdtfeger. Illuminating, widely praised book on analytic geometry of circles, the Moebius transformation, and two-dimensional non-Euclidean geometries. 200pp. 5⅝ x 8¼. 0-486-63830-8

DIFFERENTIAL GEOMETRY, Heinrich W. Guggenheimer. Local differential geometry as an application of advanced calculus and linear algebra. Curvature, transformation groups, surfaces, more. Exercises. 62 figures. 378pp. 5⅜ x 8½. 0-486-63433-7

History of Math

THE WORKS OF ARCHIMEDES, Archimedes (T. L. Heath, ed.). Topics include the famous problems of the ratio of the areas of a cylinder and an inscribed sphere; the measurement of a circle; the properties of conoids, spheroids, and spirals; and the quadrature of the parabola. Informative introduction. clxxxvi+326pp. 5⅜ x 8½. 0-486-42084-1

A SHORT ACCOUNT OF THE HISTORY OF MATHEMATICS, W. W. Rouse Ball. One of clearest, most authoritative surveys from the Egyptians and Phoenicians through 19th-century figures such as Grassman, Galois, Riemann. Fourth edition. 522pp. 5⅜ x 8½. 0-486-20630-0

THE HISTORY OF THE CALCULUS AND ITS CONCEPTUAL DEVELOPMENT, Carl B. Boyer. Origins in antiquity, medieval contributions, work of Newton, Leibniz, rigorous formulation. Treatment is verbal. 346pp. 5⅜ x 8½. 0-486-60509-4

THE HISTORICAL ROOTS OF ELEMENTARY MATHEMATICS, Lucas N. H. Bunt, Phillip S. Jones, and Jack D. Bedient. Fundamental underpinnings of modern arithmetic, algebra, geometry and number systems derived from ancient civilizations. 320pp. 5⅜ x 8½. 0-486-25563-8

A HISTORY OF MATHEMATICAL NOTATIONS, Florian Cajori. This classic study notes the first appearance of a mathematical symbol and its origin, the competition it encountered, its spread among writers in different countries, its rise to popularity, its eventual decline or ultimate survival. Original 1929 two-volume edition presented here in one volume. xxviii+820pp. 5⅜ x 8½. 0-486-67766-4

GAMES, GODS & GAMBLING: A HISTORY OF PROBABILITY AND STATISTICAL IDEAS, F. N. David. Episodes from the lives of Galileo, Fermat, Pascal, and others illustrate this fascinating account of the roots of mathematics. Features thought-provoking references to classics, archaeology, biography, poetry. 1962 edition. 304pp. 5⅜ x 8½. (Available in U.S. only.) 0-486-40023-9

OF MEN AND NUMBERS: THE STORY OF THE GREAT MATHEMATICIANS, Jane Muir. Fascinating accounts of the lives and accomplishments of history's greatest mathematical minds–Pythagoras, Descartes, Euler, Pascal, Cantor, many more. Anecdotal, illuminating. 30 diagrams. Bibliography. 256pp. 5⅜ x 8½. 0-486-28973-7

HISTORY OF MATHEMATICS, David E. Smith. Nontechnical survey from ancient Greece and Orient to late 19th century; evolution of arithmetic, geometry, trigonometry, calculating devices, algebra, the calculus. 362 illustrations. 1,355pp. 5⅜ x 8½. Two-vol. set. Vol. I: 0-486-20429-4 Vol. II: 0-486-20430-8

A CONCISE HISTORY OF MATHEMATICS, Dirk J. Struik. The best brief history of mathematics. Stresses origins and covers every major figure from ancient Near East to 19th century. 41 illustrations. 195pp. 5⅜ x 8½. 0-486-60255-9

Physics

OPTICAL RESONANCE AND TWO-LEVEL ATOMS, L. Allen and J. H. Eberly. Clear, comprehensive introduction to basic principles behind all quantum optical resonance phenomena. 53 illustrations. Preface. Index. 256pp. 5⅜ x 8½. 0-486-65533-4

QUANTUM THEORY, David Bohm. This advanced undergraduate-level text presents the quantum theory in terms of qualitative and imaginative concepts, followed by specific applications worked out in mathematical detail. Preface. Index. 655pp. 5⅜ x 8½. 0-486-65969-0

ATOMIC PHYSICS (8th EDITION), Max Born. Nobel laureate's lucid treatment of kinetic theory of gases, elementary particles, nuclear atom, wave-corpuscles, atomic structure and spectral lines, much more. Over 40 appendices, bibliography. 495pp. 5⅜ x 8½. 0-486-65984-4

A SOPHISTICATE'S PRIMER OF RELATIVITY, P. W. Bridgman. Geared toward readers already acquainted with special relativity, this book transcends the view of theory as a working tool to answer natural questions: What is a frame of reference? What is a "law of nature"? What is the role of the "observer"? Extensive treatment, written in terms accessible to those without a scientific background. 1983 ed. xlviii+172pp. 5⅜ x 8½. 0-486-42549-5

AN INTRODUCTION TO HAMILTONIAN OPTICS, H. A. Buchdahl. Detailed account of the Hamiltonian treatment of aberration theory in geometrical optics. Many classes of optical systems defined in terms of the symmetries they possess. Problems with detailed solutions. 1970 edition. xv + 360pp. 5⅜ x 8½. 0-486-67597-1

PRIMER OF QUANTUM MECHANICS, Marvin Chester. Introductory text examines the classical quantum bead on a track: its state and representations; operator eigenvalues; harmonic oscillator and bound bead in a symmetric force field; and bead in a spherical shell. Other topics include spin, matrices, and the structure of quantum mechanics; the simplest atom; indistinguishable particles; and stationary-state perturbation theory. 1992 ed. xiv+314pp. 6⅛ x 9¼. 0-486-42878-8

LECTURES ON QUANTUM MECHANICS, Paul A. M. Dirac. Four concise, brilliant lectures on mathematical methods in quantum mechanics from Nobel Prize-winning quantum pioneer build on idea of visualizing quantum theory through the use of classical mechanics. 96pp. 5⅜ x 8½. 0-486-41713-1

THIRTY YEARS THAT SHOOK PHYSICS: THE STORY OF QUANTUM THEORY, George Gamow. Lucid, accessible introduction to influential theory of energy and matter. Careful explanations of Dirac's anti-particles, Bohr's model of the atom, much more. 12 plates. Numerous drawings. 240pp. 5⅜ x 8½. 0-486-24895-X

ELECTRONIC STRUCTURE AND THE PROPERTIES OF SOLIDS: THE PHYSICS OF THE CHEMICAL BOND, Walter A. Harrison. Innovative text offers basic understanding of the electronic structure of covalent and ionic solids, simple metals, transition metals and their compounds. Problems. 1980 edition. 582pp. 6⅛ x 9¼. 0-486-66021-4

HYDRODYNAMIC AND HYDROMAGNETIC STABILITY, S. Chandrasekhar Lucid examination of the Rayleigh-Benard problem; clear coverage of the theory o instabilities causing convection. 704pp. 5⅜ x 8¼. 0-486-64071-X

INVESTIGATIONS ON THE THEORY OF THE BROWNIAN MOVEMENT Albert Einstein. Five papers (1905–8) investigating dynamics of Brownian motior and evolving elementary theory. Notes by R. Fürth. 122pp. 5⅜ x 8½. 0-486-60304-

THE PHYSICS OF WAVES, William C. Elmore and Mark A. Heald. Uniqu overview of classical wave theory. Acoustics, optics, electromagnetic radiation, more Ideal as classroom text or for self-study. Problems. 477pp. 5⅜ x 8½. 0-486-64926-

GRAVITY, George Gamow. Distinguished physicist and teacher takes reader friendly look at three scientists whose work unlocked many of the mysteries behin the laws of physics: Galileo, Newton, and Einstein. Most of the book focuses o Newton's ideas, with a concluding chapter on post-Einsteinian speculations concern ing the relationship between gravity and other physical phenomena. 160pp. 5⅜ x 8½ 0-486-42563-

PHYSICAL PRINCIPLES OF THE QUANTUM THEORY, Werner Heisenberg Nobel Laureate discusses quantum theory, uncertainty, wave mechanics, work c Dirac, Schroedinger, Compton, Wilson, Einstein, etc. 184pp. 5⅜ x 8½. 0-486-60113-

ATOMIC SPECTRA AND ATOMIC STRUCTURE, Gerhard Herzberg. One c best introductions; especially for specialist in other fields. Treatment is physica rather than mathematical. 80 illustrations. 257pp. 5⅜ x 8½. 0-486-60115-

AN INTRODUCTION TO STATISTICAL THERMODYNAMICS, Terrell I Hill. Excellent basic text offers wide-ranging coverage of quantum statistical mechar ics, systems of interacting molecules, quantum statistics, more. 523pp. 5⅜ x 8½. 0-486-65242-

THEORETICAL PHYSICS, Georg Joos, with Ira M. Freeman. Classic overvie covers essential math, mechanics, electromagnetic theory, thermodynamics, quar tum mechanics, nuclear physics, other topics. First paperback edition. xxiii + 885p 5⅜ x 8½. 0-486-65227-

PROBLEMS AND SOLUTIONS IN QUANTUM CHEMISTRY AN PHYSICS, Charles S. Johnson, Jr. and Lee G. Pedersen. Unusually varied problem detailed solutions in coverage of quantum mechanics, wave mechanics, angul momentum, molecular spectroscopy, more. 280 problems plus 139 supplementar exercises. 430pp. 6½ x 9¼. 0-486-65236-

THEORETICAL SOLID STATE PHYSICS, Vol. 1: Perfect Lattices in Equilibriun Vol. II: Non-Equilibrium and Disorder, William Jones and Norman H. Marc Monumental reference work covers fundamental theory of equilibrium properties perfect crystalline solids, non-equilibrium properties, defects and disordered system Appendices. Problems. Preface. Diagrams. Index. Bibliography. Total of 1,301pp. 5 x 8½. Two volumes. Vol. I: 0-486-65015-4 Vol. II: 0-486-65016

WHAT IS RELATIVITY? L. D. Landau and G. B. Rumer. Written by a Nobel Pri physicist and his distinguished colleague, this compelling book explains the speci theory of relativity to readers with no scientific background, using such famili objects as trains, rulers, and clocks. 1960 ed. vi+72pp. 5⅜ x 8½. 0-486-42806